Inverse Problems in Vibration

AF181723

SOLID MECHANICS AND ITS APPLICATIONS
Volume 119

Series Editor: G.M.L. GLADWELL
Department of Civil Engineering
University of Waterloo
Waterloo, Ontario, Canada N2L 3GI

Aims and Scope of the Series

The fundamental questions arising in mechanics are: *Why?*, *How?*, and *How n*
The aim of this series is to provide lucid accounts written by authoritative re
giving vision and insight in answering these questions on the subject of mecha
relates to solids.

The scope of the series covers the entire spectrum of solid mechanics. Thus it
the foundation of mechanics; variational formulations; computational m
statics, kinematics and dynamics of rigid and elastic bodies: vibrations of s
structures; dynamical systems and chaos; the theories of elasticity, plasti
viscoelasticity; composite materials; rods, beams, shells and membranes; s
control and stability; soils, rocks and geomechanics; fracture; tribology; exp
mechanics; biomechanics and machine design.

The median level of presentation is the first year graduate student. Some texts a
graphs defining the current state of the field; others are accessible to final ye
graduates; but essentially the emphasis is on readability and clarity.

For a list of related mechanics titles, see final pages.

Inverse Problems in Vibration

Second Edition

by

Graham M.L. Gladwell

University of Waterloo,
Department of Civil Engineering,
Waterloo, Ontario, Canada

KLUWER ACADEMIC PUBLISHERS

DORDRECHT / BOSTON / LONDON

A C.I.P. Catalogue record for this book is available from the Library of Congress.

ISBN 978-90-481-6701-2 (PB)
ISBN 978-1-4020-2721-5 (e-book)

Published by Kluwer Academic Publishers,
P.O. Box 17, 3300 AA Dordrecht, The Netherlands.

Sold and distributed in North, Central and South America
by Kluwer Academic Publishers,
101 Philip Drive, Norwell, MA 02061, U.S.A.

In all other countries, sold and distributed
by Kluwer Academic Publishers,
P.O. Box 322, 3300 AH Dordrecht, The Netherlands.

Printed on acid-free paper

All Rights Reserved
© 2010 Kluwer Academic Publishers
No part of this work may be reproduced, stored in a retrieval system, or transmitted
in any form or by any means, electronic, mechanical, photocopying, microfilming, record
or otherwise, without written permission from the Publisher, with the exception
of any material supplied specifically for the purpose of being entered
and executed on a computer system, for exclusive use by the purchaser of the work.

Printed in the Netherlands.

All appearance indicates neither a total exclusion nor a manifest presence of divinity, but the presence of a God who hides himself. Everything bears this character.

Pascal's *Pensées*, 555.

Contents

Preface

The last thing one settles in writing a book is what one should put in first.
Pascal's *Pensées*, 19

In 1902 Jacques Hadamard introduced the term *well-posed problem.* His definition, an abstraction from the known properties of the classical problems of mathematical physics, had three elements:

Existence: the problem has a solution

Uniqueness: the problem has only one solution

Continuity: the solution is a continuous function of the data.

Much of the research into theoretical physics and engineering before and after 1902 has concentrated on formulating problems, with properly chosen initial and/or boundary conditions, so that their solutions do have these characteristics: the problems are well posed.

Over the years it began to be recognized that there were important and apparently sensible questions that could be asked that did not fall into the category of well-posed problems. They were eventually called ill-posed problems. Many of these problems looked like a classical problem except that the roles of known and unknown quantitites had been reversed: the data, the known, were related to the outcome, the solution of a classical problem; while the unknowns were related to the data for the classical problem: they were thus called *inverse problems*, in contrast to the *direct* classical problems. (Later reflection suggested that the choice of which to be called direct and which to be called inverse was partly a historical accident.) For completeness, one should add that not all such inverse problems are ill-posed, and not all ill-posed problems are inverse problems! This book is about inverse problems in vibration, and many of these problems are ill-posed because they fail to satisfy one or more of Hadamard's criteria: they may not have a solution at all, unless the data are properly chosen; they may have many solutions; the solution may not be a continuous function of the data, in particular, as the data are varied by small amounts, it can leave the feasible region in which there is one or more solutions, and enter the region where there is no solution.

Classical vibration theory is concerned, in large part, with the infinitesimal undamped free vibration of various discrete or continuous bodies. This book is concerned only with such classical vibration theory. One of the basic problems in this theory is the determination of the natural frequencies (eigenfrequencies or simply eigenvalues) and normal modes of the vibrating body. A body that is modelled as a discrete system of rigid masses, rigid rods, massless springs, or as a finite element model (FEM) will be governed by an ordinary matrix differential equation in time t with constant coefficients. It will have a finite number of eigenvalues, and the normal modes will appear as vectors, called eigenvectors. A body that is modelled as a continuous system will be governed by a set of partial differential equations in time and one or more spatial variables. It will have an infinity of eigenvalues, and the normal modes will be functions, eigenfunctions, of the space variables.

In the context of classical theory, inverse problems are concerned with the construction of a model of a given type, i.e., a mass-spring system, a string, etc., that has given eigenvalues and/or eigenvectors or eigenfunctions, i.e., given *spectral data*. In general, if some such spectral data are given, there can be no system, a unique system, or many systems, having these properties. In the original, 1986, edition of this book, we were concerned exclusively with a stricter class of inverse problems, the so-called *reconstruction problems*. Here the data are such that there *is* only *one* vibrating system of the specified type which has the given spectral properties. In this new edition we have widened the scope of our study to include inverse problems that do not fall under this strict classification.

Before describing what the book is, we first say what it is not: it is not a book about computation. In Engineering, the almost universal approach to inverse problems is through least squares: find a system which minimizes the distance between the predicted and desired behaviours. While the early studies were examples of brute force, there is now an established and rigorous discipline governing such approaches, based on the work of Tikhonov, Morozov etc. See for example Kirsch (1996). We do not refer to any of this work in this book. Rather, we are concerned with basic analysis, qualitative properties, whether a problem has one or more solutions, etc. There are occasions when one method that we describe, that should theoretically lead to the construction of a solution, is found in practice to be ill-conditioned, and this has led to another, better behaved, procedure; in such a case we have presented both methods and discussed why one fails while the other succeeds; see for example Section 4.3. Because we are concerned with fundamental analysis, the range of physical systems that we can consider is relatively narrow; essentially it is confined to the basic elements of structures, rods, beams and membranes, and excludes structures composed of combinations of these elements. This restriction in scope is understandable; indeed, until the introduction of the finite element method and high-speed large-memory computing, the only direct vibration problems that could be solved were those involving those same structural elements in isolation. The study of inverse problems is at an earlier stage of evolution than that of direct problems.

The book falls into two parts: Chapters 1-9 are concerned with discrete systems, Chapter 10-14 with continuous systems.

Matrix analysis is the language of discrete systems, and it is developed, as needed, in Chapters 1 and 3. Thus, Chapter 1 provides the basic definitions and introduces quadratic forms, minimax theorems, eigenvalues, etc. Chapter 2 provides the basic physics of the vibrating systems that are analysed. Chapter 3 lays out the classical analysis of *Jacobi* matrices, the matrices that appear in the simplest kinds of vibrating systems, in-line sequences of masses connected by springs. Chapter 4 concerns inverse problems for Jacobi matrices. Chapter 5 provides an introduction to more general discrete systems, and the language of graph theory that is needed to analyse them.

Inverse problems in vibration are concerned with constructing a vibrating system of a particular type, e.g., a string, a beam, a membrane, that has specified (behavioural) properties. The system so constructed must be *realistic*: its defining parameters, masses, lengths, stiffnesses, etc., must be *positive.* Signs, positive and negative, lie at the heart of any deep discussion of inverse problems. Chapter 6, on Positivity, introduces the mathematics relating to different kinds of matrices: positive, totally positive, oscillatory, etc. This mathematics, due to Fekete, Perron, Gantmacher, Krein and others, was first applied to vibrating systems by Gantmacher and Krein in their classic *Oscillation Matrices and Kernels and Small Oscillations of Mechanical Systems* (1950), that has just recently (2002) been reprinted by the American Mathematical Society.

Sometimes the data that are supplied are insufficient to identify a unique vibrating system; there is then a family of systems having the specified properties - an isospectral family. Chapter 7 describes how one can form such *isospectral families*, and be sure that each member of the family has the necessary positivity properties. There are essentially two ways of forming families: algebraic, and differential. The former uses a carefully chosen rotation to go from one member to another. The latter uses the idea of *isospectral flow*; a matrix can flow, under so-called *Toda flow* along a path so that it retains the same eigenvalues and at the same time retains a particular structure and particular positivity properties.

Chapter 8 is concerned with one particular type of vibrating system: a beam vibrating in flexure. This problem had been a severe stumbling block in the early history of inverse problems.

Chapter 9 completes the first part of the blook with a study of modes, i.e., normal modes, and nodes. This analysis depends heavily on the positivity study of Chapter 6.

The second part of the book, Chapters 10-14, is concerned with continuous systems. The problems appear in two related forms, differential equations and integral equations. The integral equations, which use the Green's function for the system, are the easier to analyse, for it is the Green's function, Gantmacher and Krein's *kernel*, that has the all-important positivity properties. Moreover, the Green's function operator appearing in the integral equation is a concrete example of a positive compact self-adjoint operator in a Hilbert space, so that we may immediately make use of the well-developed theory of such operators, as described in Chapter 10.

Chapter 11 uses this theory, and the fundamental Gel'fand-Levitan transformation operator, to provide solutions to some inverse problems for the Sturm-Liouville equation. This equation, which appears in three related forms, is the governing equation for the vibrating string and rod. The Chapter describes the classical approach, as well as some recent techniques that are more readily adaptable to computation.

Chapter 12 discusses families of isospectral continuous systems. Chapter 13 applies the Gel'fand-Levitan transformation to the inverse problem for the continuous Euler-Bernoulli beam.

Chapter 14 is a short (too short) study of inverse nodal problems. While it is difficult in practice to measure a vibration mode, it is comparatively easy to locate the nodes of a particular mode. There is now a considerable body of research, due primarily to McLaughlin and Hald, that focuses on what nodal data is sufficient to identify, say, the mass distribution on a vibrating string, rod, or membrane, and how one can construct such a vibrating system from a knowledge of some nodes of some modes. Section 14.4 briefly reports on this research.

The book concludes with another short chapter on damage identification.

The history of mathematics and the physical sciences leads to an important far-reaching conclusion: the study of one topic can throw light on many other topics, even on some which at first seem have no connection with the original topic. The study of inverse problems in vibration provides a clear example of this connectedness. On the one hand, there are topics in inverse problems that are illumined by knowledge in other fields, notably linear algebra and operator theory; on the other hand the study of inverse vibration problems throws light on the classical direct problems by highlighting the fundamental qualitative properties of solutions.

A remark on the quotations from Pascal's Pensées is in order. I used the translation by W.F. Trotter that appeared in Everyman's Library, published by J.M. Dent & Sons in 1956. My copy is dated 26th April 1957 and contains an 8d (old pence) ticket for the London Transport bus No. 73 from Euston Road to Stoke Newington, reminding me that the Pensées were my daily bus reading to and from my 'digs' when I was Assistant Lecturer in Mathematics at University College London. I chose the Pensées for the chapter captions because it is clear from his writings that Pascal considered the search for God to be an inverse problem. His comments on the place of reason, heart and will in seeking a solution of the problem, though sometimes enigmatic, are as deep and relevant in 2004 as they were in 1654. I hope that these excerpts from the Pensées will whet readers' appetites for Pascal's writings.

The caption for Chapter 11 reminds me that many people have contributed to this book. Some were acknowledged in the Preface to the first edition. This new edition contains material taken from papers written with graduate students Brad Willms, Mohamed Movahheddy, Hongmei Zhu and with colleagues Brian

Davies, Josef Leydold, Peter Stadler and Antonino Morrassi. In addition to these, I have freely taken from papers by numerous colleagues worldwide, as referenced in the bibliography.

Parts of the book were read at the proof stage by Antonino Morrassi, Maeve McCarthy, Oscar Rojo and Michele Dilena. I thank them for pointing out many errors and shortcomings, some of which I have managed to correct.

The book was typed by Tracy Taves. Thank you for your stamina and your attention to detail. Colin Campbell helped us out with his understanding of the idiosyncracies of LaTeX.

Finally, I acknowledge the patience and understanding of my wife, Joyce, who saw me immersed in books in my study for years on end.

George Carrier once remarked that the aim of mathematics is insight, not numbers. It is the author's wish that this book will provide insight into the many interconnected topics in mathematics, physics and engineering that appear in the study of inverse problems in vibration.

G.M.L. Gladwell
Waterloo, Ontario
March, 2004

Chapter 1

Matrix Analysis

It is a bad sign when, on seeing a person, you remember his book. [1]
Pascal's Pensées

1.1 Introduction

The book relies heavily on matrix analysis. In this Chapter we shall present the basic definitions and properties of matrices, and provide proofs of some important theorems that will be used later. Since matrix analysis now has an established position in Engineering and Science, it will be assumed that the reader has had some exposure to it; the presentation in the early stages will therefore be brief. The reader may supplement the treatment here with standard texts.

1.2 Basic definitions and notation

We use the word *iff* to mean 'if and only if'. A *matrix* is a rectangular array of real or complex numbers together with a set of rules that specify how the numbers are to be manipulated.

A matrix A is said to have *order* $m \times n$ if it has m rows and n columns. The set of all *real* matrices, i.e., matrices with real entries, of order $m \times n$, is sometimes denoted by $\mathbb{R}^{m \times n}$. Following Horn and Johnson (1985) [183], we use the simpler notation $M_{m,n}$, and say $\mathbf{A} \in M_{m,n}$. We write

[1] Blaise Pascal (1623-1662) lived among the French intelligentia, and in that context it *was* a bad sign; one should be known for more than just a book one had written. When the first edition of this book was being translated into Chinese, the translator objected, for in 20th century China, it would be a *good* sign. If you met someone you knew who had written a book, you would mention it immediately!

$$\mathbf{A} = \begin{bmatrix} a_{11} & a_{12} & \cdots & a_{1n} \\ a_{21} & a_{22} & \cdots & a_{2n} \\ \cdot & \cdot & & \cdot \\ a_{m1} & a_{m2} & \cdots & a_{mn} \end{bmatrix}.$$

The entry in row i and column j is a_{ij}, and $\mathbf{A}$ is often written simply as

$$\mathbf{A} = (a_{ij}).$$

Two matrices $\mathbf{A}, \mathbf{B}$ are said to be *equal* if they have the same order $m \times n$, and if

$$a_{ij} = b_{ij}, \quad (i = 1, 2, \ldots, m;\ j = 1, 2, \ldots, n);$$

Then we write

$$\mathbf{A} = \mathbf{B}.$$

The *transpose* of the matrix $\mathbf{A}$ is the $n \times m$ matrix $\mathbf{A}^T$, whose rows are the columns of $\mathbf{A}$. We note that the transpose of $\mathbf{A}^T$ is $\mathbf{A}$; we say that $\mathbf{A}$ and $\mathbf{A}^T$ are *transposes* (of each other), and write this

$$(\mathbf{A}^T)^T = \mathbf{A}.$$

For example

$$\mathbf{A} = \begin{bmatrix} 1 & 2 & -4 \\ 2 & 6 & 7 \end{bmatrix}, \quad \mathbf{A}^T = \begin{bmatrix} 1 & 2 \\ 2 & 6 \\ -4 & 7 \end{bmatrix}$$

are transposes.

If $m = n$ then the $m \times n$ matrix $\mathbf{A}$ is said to be a *square* matrix of order n: $\mathbf{A} \in M_{n,n}$; we abbreviate $M_{n,n}$ to M_n; thus $\mathbf{A} \in M_n$. A square matrix that is equal to its transpose is said to be *symmetric*; in this case

$$\mathbf{A} = \mathbf{A}^T,$$

or alternatively

$$a_{ij} = a_{ji}, \quad (i, j = 1, 2, \ldots, n).$$

The set of real symmetric matrices of order n is denoted by S_n. The matrix

$$\mathbf{A} = \begin{bmatrix} 1 & 2 & 9 \\ 2 & 4 & 6 \\ 9 & 6 & 3 \end{bmatrix}$$

is symmetric. The square matrix $\mathbf{A}$ is said to be *diagonal* if it has non-zero entries only on the *principal diagonal* running from top left to bottom right. We write

$$\mathbf{A} = \begin{bmatrix} a_{11} & 0 & 0 & \cdots & 0 \\ 0 & a_{22} & 0 & \cdots & 0 \\ 0 & 0 & a_{33} & \cdots & 0 \\ 0 & 0 & 0 & \cdots & a_{nn} \end{bmatrix} = diag(a_{11}, a_{22}, \ldots, a_{nn}).$$

The *unit* matrix of order n is

$$\mathbf{I} = \mathbf{I}_n = \; diag(1, 1, \ldots, 1).$$

The elements of this matrix are denoted by the *Kronecker delta*

$$\delta_{ij} = \begin{cases} 1 & i = j \\ 0 & i \neq j \end{cases} . \tag{1.2.1}$$

The *zero* matrix of order $m \times n$ is the matrix with all its $m \times n$ entries zero.

A matrix with 1 column and n rows is called a *column vector* of order n, and is written

$$\mathbf{x} = \begin{bmatrix} x_1 \\ x_2 \\ \vdots \\ x_n \end{bmatrix} = \{x_1, x_2, \ldots, x_n\}.$$

The set of all such real vectors constitutes a linear vector space that we denote by V_n.

The transpose of a column vector is a *row vector*, written

$$\mathbf{x}^T = [x_1, x_2, \ldots, x_n].$$

Two matrices $\mathbf{A}, \mathbf{B}$ may be added or subtracted iff they have the same order $m \times n$. Their sum and difference are matrices $\mathbf{C}$ and $\mathbf{D}$ respectively of the same order $m \times n$, the elements of which are

$$c_{ij} = a_{ij} + b_{ij} \quad , \quad d_{ij} = a_{ij} - b_{ij}.$$

We write,

$$\mathbf{C} = \mathbf{A} + \mathbf{B} \quad , \quad \mathbf{D} = \mathbf{A} - \mathbf{B}.$$

The product of a matrix $\mathbf{A}$ by a *number* (or *scalar*) k is the matrix $k\mathbf{A}$ with elements ka_{ij}.

Two matrices $\mathbf{A}$ and $\mathbf{B}$ can be *multiplied* in the sense $\mathbf{AB}$ only if the number of columns of $\mathbf{A}$ is equal to the number of rows of $\mathbf{B}$. Thus if $\mathbf{A}$ has order $m \times n$, $\mathbf{B}$ has order $n \times p$ then

$$\mathbf{AB} = \mathbf{C},$$

where $\mathbf{C}$ has order $m \times p$. We write

$$\mathbf{A}(m \times n) \times \; \mathbf{B}(n \times p) = \mathbf{C}(m \times p). \tag{1.2.2}$$

The element in row i and column j of $\mathbf{C}$ is c_{ij}, and is equal to the sum of the elements of row i of $\mathbf{A}$ multiplied by the corresponding elements of column j of $\mathbf{B}$. Thus

$$c_{ij} = a_{i1}b_{1j} + a_{i2}b_{2j} + \ldots + a_{in}b_{nj} = \sum_{k=1}^{n} a_{ik}b_{kj}, \tag{1.2.3}$$

and for example

$$\begin{bmatrix} 2 & 3 & 1 \\ 1 & 6 & 7 \end{bmatrix} \begin{bmatrix} 1 & 2 & 1 & 0 \\ 0 & 1 & -1 & 0 \\ -1 & 0 & 2 & 1 \end{bmatrix} = \begin{bmatrix} 1 & 7 & 1 & 1 \\ -6 & 8 & 9 & 7 \end{bmatrix}.$$

The most important consequence of this definition is that matrix multiplication is (in general) *non-commutative*, i.e.,

$$\mathbf{AB} \neq \mathbf{BA}.$$

Indeed, if $\mathbf{A}$ is $(m \times n)$ and $\mathbf{B}$ is $(n \times p)$ then $\mathbf{BA}$ cannot be formed at all unless $m = p$. Even when $m = p$, the two matrices are not necessarily equal, as is shown by the example

$$\begin{aligned} \mathbf{A} &= \begin{bmatrix} 1 & 1 \\ 0 & -1 \end{bmatrix}, & \mathbf{B} &= \begin{bmatrix} 1 & 0 \\ -1 & 1 \end{bmatrix}, \\ \mathbf{AB} &= \begin{bmatrix} 0 & 1 \\ 1 & -1 \end{bmatrix}, & \mathbf{BA} &= \begin{bmatrix} 1 & 1 \\ -1 & -2 \end{bmatrix}. \end{aligned} \tag{1.2.4}$$

In addition, this definition implies that there are *divisors of zero*; i.e., there can be non-zero matrices $\mathbf{A}, \mathbf{B}$ such that

$$\mathbf{AB} = \mathbf{0}.$$

An example is provided by

$$\begin{bmatrix} 1 & 1 \\ 2 & 2 \end{bmatrix} \begin{bmatrix} 1 & 1 \\ -1 & -1 \end{bmatrix} = \begin{bmatrix} 0 & 0 \\ 0 & 0 \end{bmatrix}.$$

The product of $\mathbf{A}(m \times n)$ and a column vector $\mathbf{x}(n \times 1)$ is a column vector $\mathbf{y}(m \times 1)$ with elements

$$y_i = a_{i1}x_1 + a_{i2}x_2 + \ldots + a_{in}x_n \ , \quad (i = 1, 2, \ldots, m). \tag{1.2.5}$$

This means that the set of m equations

$$\begin{aligned} a_{11}x_1 + a_{12}x_2 + \ldots + a_{1n}x_n &= y_1, \\ a_{21}x_1 + a_{22}x_2 + \ldots + a_{2n}x_n &= y_2, \\ \cdots\cdots\cdots\cdots\cdots\cdots\cdots\cdots & \\ a_{m1}x_1 + a_{m2}x_2 + \ldots + a_{mn}x_n &= y_m, \end{aligned} \tag{1.2.6}$$

may be written as the single matrix equation

$$\begin{bmatrix} a_{11} & a_{12} & \cdots & a_{1n} \\ a_{21} & a_{22} & \cdots & a_{2n} \\ \cdot & \cdot & \cdots & \cdot \\ a_{m1} & a_{m2} & \cdots & a_{mn} \end{bmatrix} \begin{bmatrix} x_1 \\ x_2 \\ \cdot \\ x_n \end{bmatrix} = \begin{bmatrix} y_1 \\ y_2 \\ \cdot \\ y_m \end{bmatrix}, \tag{1.2.7}$$

or

$$\mathbf{Ax} = \mathbf{y}. \tag{1.2.8}$$

The product of an $(n \times 1)$ column vector $\mathbf{x}$ and its transpose $\mathbf{x}^T(1 \times n)$ is an $n \times n$ symmetric matrix

$$\mathbf{x}\mathbf{x}^T = \begin{bmatrix} x_1^2 & x_1x_2 & \cdots & x_1x_n \\ x_2x_1 & x_2^2 & \cdots & x_2x_n \\ \cdot & \cdot & \cdots & \cdot \\ x_nx_1 & x_nx_2 & \cdots & x_n^2 \end{bmatrix}. \tag{1.2.9}$$

On the other hand, the product of $\mathbf{x}^T(1 \times n)$ and $\mathbf{x}(n \times 1)$ is a (1×1) matrix, i.e., a scalar

$$\mathbf{x}^T\mathbf{x} = x_1^2 + x_2^2 + \ldots + x_n^2. \tag{1.2.10}$$

This quantity, which is positive iff the x_i (assumed to be real) are not all zero, is called the square of the L_2 *norm* of $\mathbf{x}$, i.e.,

$$||\mathbf{x}||^2 = \mathbf{x}^T\mathbf{x}\ , \quad ||\mathbf{x}|| = (x_1^2 + x_2^2 + \ldots + x_n^2)^{\frac{1}{2}}. \tag{1.2.11}$$

The scalar (or dot) product of $\mathbf{x}$ and $\mathbf{y}$ is defined to be

$$\mathbf{x}^T\mathbf{y} = \mathbf{y}^T\mathbf{x} = x_1y_1 + x_2y_2 + \ldots + x_ny_n. \tag{1.2.12}$$

Two vectors are said to be *orthogonal* if

$$\mathbf{x}^T\mathbf{y} = 0. \tag{1.2.13}$$

It has been noted that matrix multiplication is *non-commutative*. This holds even if the matrices are square (see (1.2.4)) or symmetric, as illustrated by

$$\begin{bmatrix} 1 & 2 \\ 2 & 2 \end{bmatrix}\begin{bmatrix} 1 & -1 \\ -1 & 1 \end{bmatrix} = \begin{bmatrix} -1 & 1 \\ 0 & 0 \end{bmatrix}, \begin{bmatrix} 1 & -1 \\ -1 & 1 \end{bmatrix}\begin{bmatrix} 1 & 2 \\ 2 & 2 \end{bmatrix} = \begin{bmatrix} -1 & 0 \\ 1 & 0 \end{bmatrix}. \tag{1.2.14}$$

This example, which shows that the product of two symmetric matrices is not (necessarily) symmetric, hints also that there might be a relation between the products **AB** and **BA**. This result is sufficiently important to be called:

Theorem 1.2.1

$$(\mathbf{AB})^T = \mathbf{B}^T\mathbf{A}^T, \tag{1.2.15}$$

so that when **A**, **B**, *are symmetric, then*

$$(\mathbf{AB})^T = \mathbf{BA}. \tag{1.2.16}$$

Proof. Consider the element in row i, column j on each side of (1.2.15). Suppose **A** is $(m \times n)$, **B** is $(n \times p)$, then **AB** is $m \times p$ and $(\mathbf{AB})^T$ is $p \times m$. Then

$$((\mathbf{AB})^T)_{ij} = (\mathbf{AB})_{ji} = \sum_{k=1}^{n} a_{jk}b_{ki},$$

and

$$\begin{aligned} (\mathbf{B}^T\mathbf{A}^T)_{ij} &= (\text{ row } i \text{ of } \mathbf{B}^T) \times (\text{ column } j \text{ of } A^T) \\ &= (\text{ column } i \text{ of } \mathbf{B}) \times (\text{ row } j \text{ of } A) \\ &= \textstyle\sum_{k=1}^{n} b_{ki}a_{jk} \quad \blacksquare \end{aligned}$$

Exercises 1.2

1. If

$$\mathbf{A} = \begin{bmatrix} 1 & 2 & 3 \\ 2 & 3 & 5 \\ 3 & 5 & 8 \end{bmatrix}$$

find a square matrix $\mathbf{B}$ such that $\mathbf{AB} = 0$. Show that if a_{33} is changed then the only possible matrix $\mathbf{B}$ would be the zero matrix.

2. Show that, whatever the matrix $\mathbf{A}$, the two matrices $\mathbf{AA}^T$ and $\mathbf{A}^T\mathbf{A}$ are symmetric. Are these two matrices equal?

3. Show that if $\mathbf{A}$, $\mathbf{B}$ are square and of order n, and $\mathbf{A}$ is symmetric, then $\mathbf{BAB}^T$ and $\mathbf{B}^T\mathbf{AB}$ are symmetric.

4. Show that if $\mathbf{A}$, $\mathbf{B}$, $\mathbf{C}$ can be multiplied in the order $\mathbf{ABC}$, then $(\mathbf{ABC})^T = \mathbf{C}^T\mathbf{B}^T\mathbf{A}^T$.

5. If $\mathbf{x}$ is complex, then its L_2 norm is defined by

$$||\mathbf{x}||^2 = |x_1|^2 + |x_2|^2 + \ldots + |x_n|^2.$$

Show that

$$||\mathbf{x}||^2 = \mathbf{x}^*\mathbf{x}$$

where $\mathbf{x}^* = \bar{\mathbf{x}}^T$, the complex conjugate transpose of $\mathbf{x}$.

1.3 Matrix inversion and determinants

In this section we shall be concerned almost exclusively with *square* matrices. The *determinant* of a (square) matrix A, denoted by $\det(\mathbf{A})$ or $|\mathbf{A}|$, is defined to be

$$\det(\mathbf{A}) = |\mathbf{A}| = \sum \pm\, a_{1i_1}a_{2i_2}\cdots a_{ni_n}\ ; \tag{1.3.1}$$

where the suffices $i_1, i_1, \ldots, i_n$ are a permutation of the numbers $1, 2, 3, \ldots, n$; the sign is $+$ if the permutation is even, and $-$ if it is odd, and the summation is carried out over all $n!$ permutations of $1, 2, 3, \ldots, n$. We note that each product in the sum contains just one element from each row and just one element from each column of $\mathbf{A}$. Thus for 2×2 and 3×3 matrices respectively

$$\begin{vmatrix} a_{11} & a_{12} \\ a_{21} & a_{22} \end{vmatrix} = a_{11}a_{22} - a_{12}a_{21},$$

$$\begin{vmatrix} a_{11} & a_{12} & a_{13} \\ a_{21} & a_{22} & a_{23} \\ a_{31} & a_{32} & a_{33} \end{vmatrix} = \begin{matrix} a_{11}a_{22}a_{33} & + & a_{12}a_{23}a_{31} & + & a_{13}a_{21}a_{32} \\ -a_{11}a_{23}a_{32} & - & a_{12}a_{21}a_{33} & - & a_{13}a_{22}a_{31} \end{matrix}. \tag{1.3.2}$$

The permutation $i_1, i_2, \ldots, i_n$ is even or odd according to whether it may be obtained from $1, 2, \ldots, n$ by means of an even or an odd number of interchanges, respectively. Thus $1,3,2,4$ and $2,3,1,4$ are respectively odd and even permutations of $1,2,3,4$ because

$$\begin{array}{l}(1,2,3,4) \to (1,3,2,4),\\ (1,2,3,4) \to (2,1,3,4) \to (2,3,1,4).\end{array}$$

We now list some of the properties of determinants.

Lemma 1.3.1 *If two rows (or columns) of* $\mathbf{A}$ *are interchanged, the determinant retains its numerical value, but changes sign.*

If the new matrix is called $\mathbf{B}$ then

$$b_{1i} = a_{2i}\ ,\ b_{2i} = a_{1i}\ ,\ b_{ji} = a_{ji}\ ,\ (j = 3,4,\ldots,n)$$

and

$$\begin{aligned}\det(\mathbf{B}) &= \sum \pm\, b_{1i_1} b_{2i_2} b_{3i_3} \cdots b_{ni_n},\\ &= \sum \pm\, a_{2i_1} a_{1i_2} a_{3i_3} \cdots a_{ni_n},\\ &= \sum \pm\, a_{1i_2} a_{2i_1} a_{3i_3} \cdots a_{ni_n}.\end{aligned}$$

But if $i_1, i_2, i_3, \ldots, i_n$ is even (odd) then $i_2, i_1, i_3, \ldots, i_n$ is odd (even), so that each term in $\det(\mathbf{B})$ appears in $\det(\mathbf{A})$ (and vice versa) with the opposite sign, so that $\det(\mathbf{B}) = -\det(\mathbf{A})$.

Lemma 1.3.2 *If two rows (columns) of* $\mathbf{A}$ *are identical then* $\det(\mathbf{A}) = 0$.

If the two rows (columns) are interchanged, then, on the one hand, $\det(\mathbf{A})$ is unchanged, while on the other, Lemma 1.3.1, $\det(\mathbf{A})$ changes sign. Thus $\det(\mathbf{A}) = -\det(\mathbf{A})$ and hence $\det(\mathbf{A}) = 0$.

Lemma 1.3.3 *If one row (column) of* $\mathbf{A}$ *is multiplied by* k *then the determinant is multiplied by* k.

Each term in the expansion is multiplied by k.

Lemma 1.3.4 *If two rows (columns) of* $\mathbf{A}$ *are proportional, then* $\det(\mathbf{A}) = 0$.

This follows from Lemmas 1.3.1, 1.3.3.

Lemma 1.3.5 *If one row (column) of* $\mathbf{A}$ *is added to another row (column) then the determinant is unchanged.*

If the matrix $\mathbf{B}$ is obtained, say, by adding row 2 to row 1 then

$$b_{1i} = a_{1i} + a_{2i}\ ,\ b_{ji} = a_{ji}\ ,\ j = 2,3,\ldots,n.$$

Thus

$$\begin{aligned}\det(\mathbf{B}) &= \sum \pm\, b_{1i_1} b_{2i_2} b_{3i_3} \cdots b_{ni_n} =\\ &= \sum \pm\, (a_{1i_1} + a_{2i_1}) a_{2i_2} a_{3i_3} \cdots a_{ni_n},\\ &= \sum \pm\, a_{1i_1} a_{2i_2} a_{3i_3} \cdots a_{ni_n} \pm \sum a_{2i_1} a_{2i_2} a_{3i_3} \cdots a_{ni_n},\end{aligned}$$

and the first sum is $\det(\mathbf{A})$ while the second, having its first and second rows equal is zero.

Lemma 1.3.6 *If a linear combination of rows (columns) of* $\mathbf{A}$ *is added to another row (column) then the determinant is unchanged.*

This follows directly from Lemma 1.3.5. We may now prove

Theorem 1.3.1 *If the rows (columns) of* $\mathbf{A}$ *are linearly dependent then* $\det(\mathbf{A}) = 0$.

Proof. Denote the rows by $\mathbf{a}_1^T, \mathbf{a}_2^T, \ldots, \mathbf{a}_n^T$. By hypotheses, there are scalars $c_1, c_2, \ldots, c_n$ *not all zero*, such that

$$c_1\mathbf{a}_1^T + c_2\mathbf{a}_2^T + \cdots + c_n\mathbf{a}_n^T = \mathbf{0}.$$

There is a c_i not zero; let it be c_m. Then

$$-\mathbf{a}_m^T = \sum_{\substack{i=1\\ i\neq m}}^{n} (c_i/c_m)\mathbf{a}_i^T.$$

If the sum on the right is added to row m of $\mathbf{A}$, the new matrix has a zero row, so that its determinant, which by Lemma 1.3.6 is $\det(\mathbf{A})$, is zero ■

Before proving the converse of this theorem, we need some more notation. A *minor* of order p of a matrix $\mathbf{A}$ is the determinant of a (square) submatrix of $\mathbf{A}$ formed by taking elements from p rows $i_1, i_2, \ldots, i_p$ and p columns $j_1, j_2, \ldots, j_p$. We denote the minor by

$$A(i_1, i_2, \ldots, i_p;\ j_1, j_2, \ldots, j_p)$$

Thus if

$$\mathbf{A} = \begin{bmatrix} 2 & 1 & 3 \\ -1 & 2 & 4 \\ 1 & 0 & 7 \end{bmatrix} \tag{1.3.3}$$

then

$$A(1;1) = 2, \quad A(1,2;1,2) = \begin{vmatrix} 2 & 1 \\ -1 & 2 \end{vmatrix} = 5, \quad A(1,2;2,3) = -2.$$

There is an important special case. The minor of order $n-1$ obtained by deleting the ith row and jth column of $\mathbf{A}$ is denoted by $\hat{a}_{ij}$. Thus for the $\mathbf{A}$ in (1.3.3),

$$\hat{a}_{11} = \begin{vmatrix} 2 & 4 \\ 0 & 7 \end{vmatrix} = 14, \quad \hat{a}_{12} = \begin{vmatrix} -1 & 4 \\ 1 & 7 \end{vmatrix} = -11, \quad \hat{a}_{13} = \begin{vmatrix} -1 & 2 \\ 1 & 0 \end{vmatrix} = -2.$$

The minors $\hat{a}_{ij}$ occur in the expansion of a determinant: for the determinant in (1.3.2) we may write

$$\begin{aligned} \det(\mathbf{A}) &= a_{11}(a_{22}a_{33} - a_{23}a_{32}) - a_{12}(a_{21}a_{33} - a_{23}a_{31}) + a_{13}(a_{21}a_{32} - a_{22}a_{31}) \\ &= a_{11}\hat{a}_{11} - a_{12}\hat{a}_{12} + a_{13}\hat{a}_{13} \end{aligned} \tag{1.3.4}$$

This is called the expansion of $\det(\mathbf{A})$ *along the first row.* Thus for $\mathbf{A}$ in (1.3.3) we have

$$33 = 2 \times 14 - 1 \times (-11) + 3 \times (-2).$$

The coefficients $\hat{a}_{11}, -\hat{a}_{12}, \hat{a}_{13}$ in (1.3.4) are called the *cofactors* of a_{11}, a_{12}, a_{13} respectively, and are denoted by A_{11}, A_{12}, A_{13} respectively. Thus we write (1.3.4) as

$$\begin{aligned}\det(\mathbf{A}) &= a_{11}A_{11} + a_{12}A_{12} + a_{13}A_{13} \\ &= \begin{vmatrix} a_{11} & a_{12} & a_{13} \\ a_{21} & a_{22} & a_{23} \\ a_{31} & a_{32} & a_{33} \end{vmatrix}\end{aligned} \qquad (1.3.5)$$

If we take the cofactors of the *first* row and multiply them by the elements of *another* row, say the *second* row, then we get zero:

$$\begin{vmatrix} a_{21} & a_{22} & a_{23} \\ a_{21} & a_{22} & a_{23} \\ a_{31} & a_{32} & a_{33} \end{vmatrix} = a_{21}A_{11} + a_{22}A_{12} + a_{23}A_{13} = 0 \qquad (1.3.6)$$

The determinant on the left is zero because it has two rows equal. These two results, (1.3.5) and (1.3.6), are special cases of

Theorem 1.3.2

$$\sum_{k=1}^{n} a_{ik}A_{jk} = \det(\mathbf{A})\delta_{ij} \qquad (1.3.7)$$

$$\sum_{k=1}^{n} A_{ki}a_{kj} = \det(\mathbf{A})\delta_{ij} \qquad (1.3.8)$$

where δ_{ij} is defined in (1.2.1).

Proof. When $i = j$, so that $\delta_{ii} = 1$, these equations merely state the definition of a cofactor. When $i \neq j$ they state that the determinant of a matrix with two rows (or columns) equal, is zero ■

Now compare equation (1.3.7) with (1.2.3). If we define a matrix $\mathbf{B}$ such that

$$b_{kj} = A_{jk} \qquad (1.3.9)$$

then we can write (1.3.7) as

$$\sum_{k=1}^{n} a_{ik}b_{kj} = \det(\mathbf{A})\delta_{ij} \qquad (1.3.10)$$

which, in matrix terms, states that

$$\mathbf{AB} = \det(\mathbf{A})\mathbf{I} \qquad (1.3.11)$$

Likewise, (1.3.8) may be written

$$\mathbf{BA} = \det(\mathbf{A})\mathbf{I}. \qquad (1.3.12)$$

The matrix $\mathbf{B}$ is called the *adjoint* (or *adjugate*) of $\mathbf{A}$ and is denoted by $adj(\mathbf{A})$. Thus equation (1.3.11), (1.3.12) state that

$$\mathbf{A}\ adj(\mathbf{A}) = adj(\mathbf{A})\mathbf{A} = \det(\mathbf{A})\mathbf{I}. \tag{1.3.13}$$

We are now in a position to prove the converse of Theorem 1.3.1, namely

Theorem 1.3.3 *If* $\det(\mathbf{A}) = 0$, *then the rows (columns) of* $\mathbf{A}$ *are linearly dependent.*

Proof. We prove the result for the columns. That for the rows may be proved likewise. We will prove it by induction on n. It certainly holds, trivially, when $n = 1$, for then $\det(\mathbf{A}) = a_{11}$. Let $\mathbf{a}_1, \mathbf{a}_2, \ldots, \mathbf{a}_n$ be the columns of $\mathbf{A}$, and suppose $\det(\mathbf{A}) = 0$. *Either* each set of $n-1$ vectors selected from $\mathbf{a}_1, \mathbf{a}_2, \ldots \mathbf{a}_n$ is a linearly dependent set, in which case the complete set is linearly dependent as required, *or* there is a set of $n-1$ vectors, which without loss of generality we may take to be $\mathbf{a}_1, \mathbf{a}_2, \ldots \mathbf{a}_{n-1}$, which is linearly independent. Now imagine creating a set of vectors $\mathbf{b}_1, \mathbf{b}_2, \ldots, \mathbf{b}_{n-1}$ by deleting the ith row of each of the vectors $\mathbf{a}_1, \mathbf{a}_2, \ldots, \mathbf{a}_{n-1}$. For at least one value of i the set $\mathbf{b}_1, \mathbf{b}_2, \ldots, \mathbf{b}_{n-1}$ must be linearly independent. By the inductive hypothesis, the $(n-1) \times (n-1)$ determinant formed from these vectors must be non-zero; at least one of the terms b_{kj} in equation (1.3.10) is non-zero. If $\det(\mathbf{A}) = 0$, equation (1.3.10) states that

$$\sum_{k=1}^{n} a_{ik} b_{kj} = 0 \quad i, j = 1, 2, \ldots, n \tag{1.3.14}$$

Since $\mathbf{a}_k = \{a_{1k}, a_{2k}, \ldots, a_{nk}\}$, we may write the n equations (1.3.14) obtained by taking $j = 1, 2, \ldots, n$, as

$$\sum_{k=1}^{n} b_{kj} \mathbf{a}_k = 0 \quad j = 1, 2, \ldots, n. \tag{1.3.15}$$

For at least one value of j, not all the b_{kj} are zero; the columns $\mathbf{a}_1, \mathbf{a}_2, \ldots, \mathbf{a}_n$ are linearly dependent ■

Theorem 1.3.4 *The matrix equations*

$$\mathbf{A}\mathbf{x} = \mathbf{0}, \quad \mathbf{y}^T\mathbf{A} = \mathbf{0}$$

have non-trivial solutions $\mathbf{x}$ *and* $\mathbf{y}$ *respectively iff* $\det(\mathbf{A}) = 0$.

Proof. The theorem is a corollary of Theorem 1.3.3. If $\mathbf{a}_1, \mathbf{a}_2, \ldots, \mathbf{a}_n$ are the columns of $\mathbf{A}$, then

$$\begin{aligned}\mathbf{A}\mathbf{x} &= [\mathbf{a}_1, \mathbf{a}_2, \ldots, \mathbf{a}_n]\{x_1, x_2, \ldots, x_n\} \\ &= x_1\mathbf{a}_1 + x_2\mathbf{a}_2 + \cdots + x_n\mathbf{a}_n.\end{aligned}$$

We can find $x_1, \ldots, x_n$, not all zero, such that

$$x_1\mathbf{a}_1 + x_2\mathbf{a}_2 + \cdots + x_n\mathbf{a}_n = \mathbf{0}$$

iff $\mathbf{a}_1, \mathbf{a}_2, \ldots, \mathbf{a}_n$ are linearly dependent. By Theorem 1.3.3 this happens iff $\det(\mathbf{A}) = 0$. This happens in turn iff the rows of $\mathbf{A}$ are linearly dependent, i.e., $\mathbf{y}^T\mathbf{A} = \mathbf{0}$ has a non-trivial solution ■

Theorem 1.3.5 *If* $\mathbf{A}, \mathbf{B}$ *are square matrices of order* n *then*

$$\det(\mathbf{AB}) = \det(\mathbf{A}).\det(\mathbf{B})$$

The proof of this result is left to Ex. 1.3.3.

The square matrix $\mathbf{A}$ is said to be *singular* if $\det(\mathbf{A}) = 0$, *non-singular* or *invertible* if $\det(\mathbf{A}) \neq 0$. Theorem 1.3.4 shows that if $\mathbf{A}$ is non-singular, then the equation $\mathbf{Ax} = \mathbf{0}$ has only the trivial solution $\mathbf{x} = \mathbf{0}$. Ex. 1.3.5 extends this result: if $\mathbf{A}$ is non-singular, then the matrix equations $\mathbf{AS} = \mathbf{0}$, $\mathbf{TA} = \mathbf{0}$ have only the trivial solutions $\mathbf{S} = \mathbf{0}$, $\mathbf{T} = \mathbf{0}$; when $\mathbf{A}$ is non-singular there are no *divisors of zero*.

The matrix $\mathbf{R}$ is said to be an *inverse* of $\mathbf{A}$ if $\mathbf{AR} = \mathbf{I}$.

Theorem 1.3.6 *If* $\mathbf{A}$ *has an inverse, it is unique, and* $\mathbf{RA} = \mathbf{I}$.

Proof. Suppose $\mathbf{AR} = \mathbf{I}$. Theorem 1.3.5 shows that

$$\det(\mathbf{A}).\det(\mathbf{R}) = \det(\mathbf{I}) = 1 \tag{1.3.16}$$

so that $\det(\mathbf{A}) \neq 0$: $\mathbf{A}$ is non-singular. If $\mathbf{R}_1, \mathbf{R}_2$ were two inverses, then $\mathbf{AR}_1 = \mathbf{I} = \mathbf{AR}_2$, so that $\mathbf{A}(\mathbf{R}_1 - \mathbf{R}_2) = \mathbf{0}$. But $\mathbf{A}$ is non-singular, so that $\mathbf{R}_1 - \mathbf{R}_2 = \mathbf{0}$: $\mathbf{R}_2 = \mathbf{R}_1$. Now if $\mathbf{AR} = \mathbf{I}$ then $\mathbf{ARA} = \mathbf{A}$, i.e., $\mathbf{A}(\mathbf{RA} - \mathbf{I}) = \mathbf{0}$. But $\mathbf{A}$ is non-singular, so that $\mathbf{RA} - \mathbf{I} = 0$, i.e., $\mathbf{RA} = \mathbf{I}$ ■

Theorem 1.3.6 shows that *if* $\mathbf{A}$ has an inverse, then $\mathbf{A}$ is non-singular. The logical negative of this statement is that if $\mathbf{A}$ is *singular* it does *not* have an inverse. We now prove the converse.

Theorem 1.3.7 *If* $\mathbf{A}$ *is non-singular, then it has an inverse.*

Proof. If $\mathbf{A}$ is non-singular, then $\det(\mathbf{A}) \neq 0$, and equation (1.3.13) may be written

$$\mathbf{AR} = \mathbf{RA} = \mathbf{I}, \tag{1.3.17}$$

where $\mathbf{R} = adj(\mathbf{A})/\det(\mathbf{A})$ ■

If $\mathbf{A}$ is non-singular, its unique inverse is denoted by $\mathbf{A}^{-1}$. We have

$$\mathbf{AA}^{-1} = \mathbf{A}^{-1}\mathbf{A} = \mathbf{I}. \tag{1.3.18}$$

Theorem 1.3.8 *The equation*

$$\mathbf{Ax} = \mathbf{b} \tag{1.3.19}$$

either has a unique solution, if $\mathbf{A}$ *is non-singular; or if* $\mathbf{A}$ *is singular, it has a solution only for certain* $\mathbf{b}$.

Proof. If $\mathbf{A}$ is non-singular then

$$\mathbf{x} = \mathbf{A}^{-1}(\mathbf{A}\mathbf{x}) = \mathbf{A}^{-1}\mathbf{b}$$

is the unique solution. If $\mathbf{A}$ is singular, then there is one (or more) $\mathbf{y}$ such that

$$\mathbf{y}^T\mathbf{A} = \mathbf{0}.$$

Then

$$\mathbf{y}^T(\mathbf{A}\mathbf{x}) = \mathbf{y}^T\mathbf{b} = 0$$

so that (1.3.19) has a solution only if $\mathbf{b}$ is orthogonal to any $\mathbf{y}$ which satisfies $\mathbf{y}^T\mathbf{A} = \mathbf{0}$. If $\mathbf{A}$ is singular then $\mathbf{A}\mathbf{x} = \mathbf{0}$ has one or more solutions $\mathbf{x}_1, \mathbf{x}_2, \ldots, \mathbf{x}_m$, so that if $\mathbf{x}_0$ is one solution satisfying $\mathbf{A}\mathbf{x}_0 = \mathbf{b}$, then

$$\mathbf{x} = \mathbf{x}_0 + \sum_{i=1}^{m} c_i\mathbf{x}_i \tag{1.3.20}$$

is also a solution for arbitrary $c_1, c_2, \ldots, c_m$ ■

Note that trying to solve $\mathbf{A}\mathbf{x} = \mathbf{b}$ by actually finding the inverse of $\mathbf{A}$, is an extremely wasteful and clumsy procedure. Finding $\mathbf{A}^{-1}$ is equivalent to solving $\mathbf{A}\mathbf{x} = \mathbf{b}$ *for all possible* $\mathbf{b}$, not just for the given $\mathbf{b}$. Techniques for solving $\mathbf{A}\mathbf{x} = \mathbf{b}$ form the subject matter of *numerical linear algebra*, for which see Bishop, Gladwell and Michaelson (1965) [33] or Golub and Van Loan (1983) [135]. Note also that we have not in fact shown how to find one solution $\mathbf{x}_0$ if $\mathbf{B}$ is in fact orthogonal to all solutions of $\mathbf{y}^T\mathbf{A} = \mathbf{0}$; this too is covered in numerical linear algebra.

In numerical linear algebra the starting point of almost all the procedures for solving linear equations such as (1.3.19), whether $\mathbf{A}$ is square or not, or of finding determinants, is *Gaussian elimination*. This is a systematic reduction of an array (a_{ij}) to (usually) upper triangular form by subtracting multiples of one equation from another. Lemma 1.3.6 shows that the determinant of coefficients is unchanged under such an operation.

The application of Gaussian elimination to the equations

$$\begin{aligned} x_1 + 3x_2 + 2x_3 &= 6 \\ 2x_1 + 5x_2 + 6x_3 &= 13 \\ 3x_1 + 4x_2 + 7x_3 &= 14 \end{aligned}$$

would proceed as follows; only the coefficients need be retained:

$$\begin{array}{cccccc} 1 & 3 & 2 & : & 6 \\ 2 & 5 & 6 & : & 13 \\ 3 & 4 & 7 & : & 14 \end{array} \quad \rightarrow \quad \begin{array}{ccccc} 1 & 3 & 2 & : & 6 \\ 0 & -1 & 2 & : & 1 \\ 0 & -5 & 1 & : & -4 \end{array} \quad \rightarrow \quad \begin{array}{ccccc} 1 & 3 & 2 & : & 6 \\ 0 & -1 & 2 & : & 1 \\ 0 & 0 & -9 & : & -9 \end{array}$$

The determinant of $\mathbf{A}$ is $1 \times (-1) \times (-9) = 9$. The last of the new equations gives $-9x_3 = -9$, $x_3 = 1$; when substituted in the new second equation this gives $-x_2 = 1 - 2x_3 = -1$, $x_2 = 1$; then $x_1 + 3 + 2 = 6$ gives $x_1 = 1$.

Exercises 1.3

1. Show that if $\mathbf{A}$ is upper (lower) triangular, i.e., $a_{ij} = 0$ if $j > i (j < i)$, then
$$\det(\mathbf{A}) = a_{11}a_{22}\ldots a_{nn}.$$

2. If
$$\mathbf{A} = \begin{bmatrix} 1 & 3 & 2 \\ 2 & 5 & 6 \\ 3 & 4 & 7 \end{bmatrix}$$
find $\mathbf{A}^{-1}$. Verify that $\mathbf{A}\mathbf{A}^{-1} = \mathbf{A}^{-1}\mathbf{A} = \mathbf{I}$.

3. Prove that if $\mathbf{A}, \mathbf{B}$ are square matrices of order n, then
$$\det(\mathbf{AB}) = \det(\mathbf{A}).\det(\mathbf{B}).$$
Hint: consider the $2n \times 2n$ matrix
$$\mathbf{C} = \begin{bmatrix} \mathbf{A} & \mathbf{0} \\ \mathbf{I} & \mathbf{B} \end{bmatrix}$$
Show that $\det(\mathbf{C}) = \det(\mathbf{A}).\det(\mathbf{B})$. Now subtract multiples of rows $(n+1)$ to $2n$ from rows 1 to n to delete all elements in the top left quarter of $\mathbf{C}$. The elements in the top right quarter will be those of $-\mathbf{AB}$.

4. Use Gaussian elimination to solve the equations
$$\begin{aligned} x_1 + 2x_2 + 4x_3 + 8x_4 &= -9 \\ x_2 + 3x_3 + 2x_4 &= 1 \\ x_1 + 2x_2 + 5x_3 + 6x_4 &= -3 \\ -x_1 + 3x_2 + 4x_3 + 7x_4 &= -10 \end{aligned}$$

5. Show that if $\mathbf{A}$ is non-singular, then the matrix equations $\mathbf{AS} = \mathbf{0}$ and $\mathbf{TA} = \mathbf{0}$ have only the trivial solution $\mathbf{S} = \mathbf{0}$, $\mathbf{T} = \mathbf{0}$, respectively.

1.4 Eigenvalues and eigenvectors

If $\mathbf{A}$ and $\mathbf{C}$ are square matrices of order n then the equation
$$\mathbf{Cx} = \lambda\mathbf{Ax} \tag{1.4.1}$$
will have a *non-trivial* solution $\mathbf{x}$ (i.e., one for which $||\mathbf{x}|| \neq 0$) iff the matrix $\mathbf{C} - \lambda\mathbf{A}$ is singular, i.e., the scalar λ satisfies the determinantal or *characteristic* equation
$$\det(\mathbf{C} - \lambda\mathbf{A}) = 0. \tag{1.4.2}$$
The roots of this equation are called the *eigenvalues* of the matrix pair $(\mathbf{C}, \mathbf{A})$; they may be real or complex. If λ is an eigenvalue, a vector $\mathbf{x}$ satisfying (1.4.1) is called an *eigenvector* corresponding to λ.

In many mathematical texts, attention is focused almost exclusively on the case when $\mathbf{A} = \mathbf{I}$. In this case λ is said to be an *eigenvalue* of $\mathbf{C}$. The problem (1.4.1) is called the *generalised eigenvalue problem.* In Mechanics there are many problems in which two matrices, $\mathbf{C}, \mathbf{A}$ appear, and it will be convenient to develop the theory for this case.

The eigenvalue theory for general, i.e., not necessarily symmetric matrices $\mathbf{C}, \mathbf{A}$, is extremely complicated. (See Ex. 1.4.8). However, for all, or almost all, the problems encountered in this book, the matrices $\mathbf{C}, \mathbf{A}$ have special properties: they are real and symmetric, and one at least is *positive definite*, defined as follows.

Suppose $\mathbf{A}$ is real and symmetric, and $\mathbf{x}$ is a real $n \times 1$ column vector. The quantity $\mathbf{x}^T\mathbf{A}\mathbf{x}$ is a scalar. Written in full it is

$$\mathbf{x}^T\mathbf{A}\mathbf{x} = a_{11}x_1^2+2a_{12}x_1x_2+\cdots+2a_{1n}x_1x_n+a_{22}x_2^2+\cdots+2a_{2n}x_2x_n+\cdots+a_{nn}x_n^2. \tag{1.4.3}$$

This is called a *quadratic form.* In many physical applications the kinetic energy and the potential energy of a mechanical system may be expressed as quadratic forms in the generalised velocities or displacements, respectively. The kinetic energy of a system is always positive, unless all the generalised velocities are zero. This leads us to a definition. The matrix $\mathbf{A}$ is said to be *positive definite* if $||\mathbf{x}|| \neq 0$ implies $\mathbf{x}^T\mathbf{A}\mathbf{x} > 0$. (Clearly, if $||\mathbf{x}|| = 0$, so that $x_1 = 0 = x_2 = \ldots x_n$, then $\mathbf{x}^T\mathbf{A}\mathbf{x} \equiv 0$.) If $\mathbf{A}$ satisfies the weaker condition, that $||\mathbf{x}|| \neq 0$ implies $\mathbf{x}^T\mathbf{A}\mathbf{x} \geq 0$, i.e., there is a vector $\mathbf{x}$ such that $||\mathbf{x}|| \neq 0$ and $\mathbf{x}^T\mathbf{A}\mathbf{x} = 0$, then $\mathbf{A}$ is said to be *positive semi-definite.* We will find later that the matrix associated with the potential energy of an unanchored system is positive semi-definite; there is a vector $\mathbf{x}$ corresponding to a rigid body displacement of the system, for which the potential (or *strain*) energy is zero.

Theorem 1.4.1 *If* $\mathbf{C}, \mathbf{A}$ *are real and symmetric, and* $\mathbf{A}$ *is positive definite, then the eigenvalues and eigenvectors of (1.4.1) are real.*

Proof. Suppose $\lambda, \mathbf{x}$ possibly complex, and with $||\mathbf{x}|| \neq 0$, are an eigenpair of (1.4.1), multiply both sides by $\mathbf{x}^* = \bar{\mathbf{x}}^T$ to obtain

$$\mathbf{x}^*\mathbf{C}\mathbf{x} = \lambda\mathbf{x}^*\mathbf{A}\mathbf{x} \tag{1.4.4}$$

The quantities $\mathbf{x}^*\mathbf{A}\mathbf{x}$ and $\mathbf{x}^*\mathbf{C}\mathbf{x}$ are both *real.* This is so because $\mathbf{x}^*\mathbf{A}\mathbf{x}$, for instance, is a scalar, and therefore equal to its own transpose. Thus

$$a = \mathbf{x}^*\mathbf{A}\mathbf{x} = (\mathbf{x}^*\mathbf{A}\mathbf{x})^T = \mathbf{x}^T\mathbf{A}^T\bar{\mathbf{x}} = \mathbf{x}^T\mathbf{A}\bar{\mathbf{x}} = (\overline{\mathbf{x}^*\mathbf{A}\mathbf{x}}) = \bar{a}$$

but if $a = \bar{a}$, then a is real. Similarly, $\mathbf{x}^*\mathbf{C}\mathbf{x}$ is real. Moreover, if $||\mathbf{x}|| \neq 0$, i.e., at least one element in $\mathbf{x}$ is not zero, then a is strictly positive, i.e., $a > 0$. For let $\mathbf{x} = \mathbf{u} + i\mathbf{v}$ where $\mathbf{u}, \mathbf{v}$ are real, then

$$\mathbf{x}^*\mathbf{A}\mathbf{x} = (\mathbf{u}^T - i\mathbf{v}^T)\mathbf{A}(\mathbf{u} + i\mathbf{v}) = \mathbf{u}^T\mathbf{A}\mathbf{u} + i\{\mathbf{u}^T\mathbf{A}\mathbf{v} - \mathbf{v}^T\mathbf{A}\mathbf{u}\} + \mathbf{v}^T\mathbf{A}\mathbf{v}.$$

But since $\mathbf{x}^*\mathbf{A}\mathbf{x}$ is real, the imaginary term is zero, and thus

$$\mathbf{x}^*\mathbf{A}\mathbf{x} = \mathbf{u}^T\mathbf{A}\mathbf{u} + \mathbf{v}^T\mathbf{A}\mathbf{v} \geq 0.$$

The inequality is strict because *either* at least one element of $\mathbf{u}$ is non-zero, in which case $\mathbf{u}^T\mathbf{A}\mathbf{u} > 0$; *or* if $\mathbf{u} \equiv \mathbf{0}$, at least one element of $\mathbf{v}$ is non-zero, in which case $\mathbf{v}^T\mathbf{A}\mathbf{v} > 0$.

Now return to equation (1.4.4); $\mathbf{x}^*\mathbf{C}\mathbf{x}$ and $\mathbf{x}^*\mathbf{A}\mathbf{x}$ are both real and $\mathbf{x}^*\mathbf{A}\mathbf{x}$ is positive. Hence

$$\lambda = \mathbf{x}^*\mathbf{C}\mathbf{x}/\mathbf{x}^*\mathbf{A}\mathbf{x}$$

is real. Since λ is real, the vector $\mathbf{x}$, obtained by solving a set of simultaneous linear equations with real coefficients, is real. Therefore, $\mathbf{x}^* = \mathbf{x}^T$, and we can write

$$\lambda = \mathbf{x}^T\mathbf{C}\mathbf{x}/\mathbf{x}^T\mathbf{A}\mathbf{x}. \ \blacksquare$$

This ratio is often called, and we will call it, the *Rayleigh Quotient* corresponding to equation (1.4.1). (It was Lord Rayleigh (Rayleigh (1894) [290]) who, in his classical treatise *Theory of Sound* used this quotient to take the first steps towards the variational treatment of eigenvalues. We discuss this further in Chapter 2.) We write

$$\lambda_R = R(\mathbf{x}) = \mathbf{x}^T\mathbf{C}\mathbf{x}/\mathbf{x}^T\mathbf{A}\mathbf{x} \tag{1.4.5}$$

Ex. 1.4.7 shows the necessity of having one of the matrices $\mathbf{A}, \mathbf{C}$, positive definite.

The conditions which must be satisfied if a (symmetric) matrix $\mathbf{A}$ is to be positive definite or positive semi-definite may be expressed in terms of the *principal minors* of $\mathbf{A}$. A principal minor of order p of a matrix $\mathbf{A}$ (symmetric or not) is a determinant of a submatrix formed from p rows $i_1, i_2, \ldots, i_p$ and the *same* p columns $i_1, i_2, \ldots, i_p$. Thus for $\mathbf{A}$ in (1.3.3),

$$\begin{vmatrix} 2 & 1 \\ -1 & 2 \end{vmatrix}, \begin{vmatrix} 2 & 3 \\ 1 & 7 \end{vmatrix}, \begin{vmatrix} 2 & 4 \\ 0 & 7 \end{vmatrix}, \begin{vmatrix} 2 & 1 & 3 \\ -1 & 2 & 4 \\ 1 & 0 & 7 \end{vmatrix}$$

are all principal minors. In the notation of Section 1.3, a principal minor is $A(i_1, i_2, \ldots, i_p;\ i_1, i_2, \ldots, i_p)$.

There is a special notation for the *leading principal minors* of $\mathbf{A}$, these are as follows:

$$D_1 = a_{11}\ ,\ D_2 = \begin{vmatrix} a_{11} & a_{12} \\ a_{21} & a_{22} \end{vmatrix}, \ldots\ D_n = |\mathbf{A}| = \det(\mathbf{A}). \tag{1.4.6}$$

Now we may state

Theorem 1.4.2 *The symmetric matrix* $\mathbf{A}$ *is positive definite iff the leading principal minors* $(D_i)_1^n$ *are all positive.* $\mathbf{A}$ *will be positive semi-definite iff* $(D_i)_1^{n-1} \geq 0,\ D_n = 0.$

This will not be proved until Chapter 5. Note that since $D_n = \det(\mathbf{A})$, this states that a positive-definite matrix is non-singular, and a positive semi-definite matrix is singular.

We may now refine Theorem 1.4.1 to give

Theorem 1.4.3 *If* $\mathbf{C}, \mathbf{A}$ *are real and symmetric and* $\mathbf{A}$ *is positive definite then equation (1.4.1) will have* n *real eigenvalues, although they need not be distinct. If* $\mathbf{C}$ *is positive definite they will be positive, if* $\mathbf{C}$ *is positive semi-definite they will be non-negative.*

Proof. Equation (1.4.2) may be expanded in terms of the coefficients $c_{ij} - \lambda a_{ij}$; the result is an nth degree polynomial equation for λ, namely

$$\triangle(\lambda) = \det(\mathbf{C} - \lambda\mathbf{A}) \equiv \triangle_0 + \triangle_1\lambda + \triangle_2\lambda^2 + \cdots + \triangle_n\lambda^n = 0. \tag{1.4.7}$$

Most of the coefficients $\triangle_i$ are complicated functions of a_{ij} and c_{ij}, but the first and last may be easily identified, namely

$$\triangle_0 = \det(\mathbf{C}), \quad \triangle_n = (-1)^n \det(\mathbf{A}). \tag{1.4.8}$$

Since $\mathbf{A}$ is positive-definite, $\det(\mathbf{A}) > 0$ so that $\triangle_n \neq 0$. This means that equation (1.4.7) is a proper equation of degree n with n roots $(\lambda_i)_1^n$. This proves the first part of the Theorem. If $\mathbf{C}$ is positive-definite, then both numerator and denominator of the Rayleigh Quotient (1.4.5) will be positive, so that $(\lambda_i)_1^n > 0$. If $\mathbf{C}$ is only positive semi-definite, then the numerator of the Rayleigh Quotient is only positive or zero (i.e., non-negative), so that the λ_i are non-negative. Moreover, since $\lambda_1\lambda_2\ldots\lambda_n = (-)^n\triangle_0/\triangle_n = \det(\mathbf{C})/\det(\mathbf{A})$ equation (1.4.7) will have at least one zero root when $\det(\mathbf{C}) = 0$ ■

Under the conditions of Theorem 1.4.3 the eigenvalues $(\lambda_i)_1^n$ may be labelled in *increasing* order:

$$0 \leq \lambda_1 \leq \lambda_2 \leq \ldots \leq \lambda_n. \tag{1.4.9}$$

Theorem 1.4.4 *Eigenvectors* $\mathbf{u}_i, \mathbf{u}_j$ *corresponding to two different eigenvalues* $\lambda_i, \lambda_j (\lambda_i \neq \lambda_j)$ *of the symmetric matrix pair* $(\mathbf{C}, \mathbf{A})$ *are orthogonal w.r.t. both* $\mathbf{A}$ *and to* $\mathbf{C}$*, i.e.,*

$$\mathbf{u}_i^T\mathbf{A}\mathbf{u}_j = 0 = \mathbf{u}_i^T\mathbf{C}\mathbf{u}_j. \tag{1.4.10}$$

Proof. By definition

$$\mathbf{C}\mathbf{u}_i = \lambda_i\mathbf{A}\mathbf{u}_i, \quad \mathbf{C}\mathbf{u}_j = \lambda_j\mathbf{A}\mathbf{u}_j. \tag{1.4.11}$$

Transpose the first equation and multiply it on the right by $\mathbf{u}_j^T$; multiply the second equation on the left by $\mathbf{u}_i^T$, to obtain

$$\begin{aligned}(\mathbf{u}_i^T\mathbf{C})\mathbf{u}_j &= \lambda_i(\mathbf{u}_i^T\mathbf{A})\mathbf{u}_j \\ \mathbf{u}_i^T(\mathbf{C}\mathbf{u}_j) &= \lambda_j\mathbf{u}_i^T(\mathbf{A}\mathbf{u}_j)\end{aligned}$$

Subtract these two equations to yield

$$(\lambda_i - \lambda_j)\mathbf{u}_i^T \mathbf{A}\mathbf{u}_j = 0.$$

But $\lambda_i - \lambda_j \neq 0$, so that $\mathbf{u}_i^T \mathbf{A}\mathbf{u}_j = 0$, and hence $\mathbf{u}_i^T \mathbf{C}\mathbf{u}_j = 0$. ■

Premultiplying equation (1.4.11) by $\mathbf{u}_i^T$ we find

$$c_i \equiv \mathbf{u}_i^T \mathbf{C}\mathbf{u}_i = \lambda_i \mathbf{u}_i^T \mathbf{A}\mathbf{u}_i = \lambda_i a_i \tag{1.4.12}$$

Sometimes, we will *normalise* an eigenvector $\mathbf{u}_i$ w.r.t. $\mathbf{A}$; then $a_i = 1, c_i = \lambda_i$.

An important corollary of this result is

Theorem 1.4.5 *If the symmetric matrix pair* $(\mathbf{C}, \mathbf{A})$ *has distinct eigenvalues* $(\lambda_i)_1^n$, *and* $\mathbf{A}$ *is positive-definite, then the eigenvectors* $\mathbf{u}_i$ *are linearly independent, and therefore span* V_n, *the space of* n*-vectors.*

Proof. The eigenvectors $\mathbf{u}_i$ are linearly independent; for suppose

$$\alpha_1 \mathbf{u}_1 + \alpha_2 \mathbf{u}_2 + \cdots + \alpha_n \mathbf{u}_n = \mathbf{0};$$

multiplying by $\mathbf{u}_i^T \mathbf{A}$ we have

$$\alpha_1(\mathbf{u}_i^T \mathbf{A}\mathbf{u}_1) + \alpha_2(\mathbf{u}_i^T \mathbf{A}\mathbf{u}_2) + \cdots + \alpha_n(\mathbf{u}_i^T \mathbf{A}\mathbf{u}_n) = 0.$$

But $\mathbf{u}_i^T \mathbf{A}\mathbf{u}_j = 0$ if $i \neq j$, so that only the ith term in this equation is non-zero, and hence

$$\alpha_i(\mathbf{u}_i^T \mathbf{A}\mathbf{u}_i) = 0.$$

Since $\mathbf{A}$ is positive definite, $\mathbf{u}_i^T \mathbf{A}\mathbf{u}_i > 0$ and $\alpha_i = 0$; all the $(\alpha_i)_1^n$ are zero; the $\mathbf{u}_i$ are linearly independent. Any vector $\mathbf{u} \epsilon V_n$ may be written uniquely as

$$\mathbf{u} = \sum_{j=1}^{n} \alpha_j \mathbf{u}_j \tag{1.4.13}$$

where

$$\alpha_j = \mathbf{u}_j^T \mathbf{A}\mathbf{u} / \mathbf{u}_j^T \mathbf{A}\mathbf{u}_j. \quad ■ \tag{1.4.14}$$

In this book we are not concerned with methods for *computing* eigenvalues and eigenvectors. A simple treatment of the classical techniques may be found in Bishop, Gladwell and Michaelson (1965) [33]. A comprehensive account of modern techniques is given by Golub and Van Loan (1983) [135]. The classical treatise on the *symmetric* eigenvalue problem is Parlett (1980) [264]. We are concerned only with the qualitative properties of eigenvalues.

Exercises 1.4

1. If

$$\mathbf{A} = \begin{bmatrix} 1 & -1 & & \\ -1 & 2 & -1 & \\ & -1 & 2 & -1 \\ & & -1 & 1 \end{bmatrix}$$

show that $\mathbf{A}$ is positive semi-definite. For what $\mathbf{x}$ is $\mathbf{A}\mathbf{x} = \mathbf{0}$?

2. Show that $\mathbf{A}^{-1}$ is positive definite iff $\mathbf{A}$ is positive definite.

3. Verify the conditions given in Theorem 1.4.2 for $\mathbf{A}$ to be positive definite, when $n = 2$, by writing

$$\begin{aligned} \mathbf{x}^T\mathbf{A}\mathbf{x} &= a_{11}x_1^2 + 2a_{12}x_1x_2 + a_{22}x_2^2 \\ &= a_{11}\left\{\left(x_1 + \tfrac{a_{12}}{a_{11}}\right)^2 + \left(\tfrac{a_{11}a_{22}-a_{12}^2}{a_{11}^2}\right)x_2^2\right\}. \end{aligned}$$

Extend the analysis to $n = 3$.

4. Find the eigenvalues and eigenvectors of the pair

$$\mathbf{C} = \begin{bmatrix} 2 & -1 & 0 \\ -1 & 2 & -1 \\ 0 & -1 & 2 \end{bmatrix}, \qquad \mathbf{A} = \begin{bmatrix} 1 & & \\ & 1 & \\ & & 1 \end{bmatrix}$$

Hint: replace the eigen-equation by the equivalent *recurrence relation* $-x_{r-1}+(2-\lambda)x_r-x_{r+1} = 0$ with appropriate end conditions for $r = 1, r = 3$, and seek a solution of the form $x_r = A\cos r\theta + B\sin r\theta$. Generalise this result.

5. Show that if $n = 3$, $\mathbf{A}$ is symmetric, and D_1, D_2, D_3 of equation (1.4.6) are all **positive**, then *all* the principal minors of $\mathbf{A}$ are positive. Hint: write $a_{11}\det(\mathbf{A})$ as a 2×2 determinant with elements which are minors of $\mathbf{A}$ of order 2. This is a particular case of a general result, see e.g., Gantmacher (1959) [97].

6. Show that the real symmetric matrix $\mathbf{A}$ has positive eigenvalues iff it is positive-definite.

7. Take

$$\mathbf{C} = \begin{bmatrix} 1 & -1 \\ -1 & -1 \end{bmatrix}, \quad \mathbf{A} = \begin{bmatrix} -1 & -1 \\ -1 & 1 \end{bmatrix}.$$

The eigenvalues are not real. Where does the argument used in the proof of Theorem 1.4.1 break down?

8. Take

$$\mathbf{C} = \begin{bmatrix} 0 & 1 \\ 0 & 0 \end{bmatrix}, \quad \mathbf{A} = \begin{bmatrix} 1 & 0 \\ 0 & 1 \end{bmatrix}.$$

Show that equation (1.4.1) has only one eigenvalue and one eigenvector, so that the eigenvectors do not span the space V_2. This is the kind of difficulty attending the non-symmetric eigenvalue problem.

Chapter 2

Vibrations of Discrete Systems

Our nature consists in motion; complete rest is death.
Pascal's *Pensées*, 129

2.1 Introduction

The formulation and solution of the equations governing the motion of a discrete vibrating system, i.e., one which has a finite number of degrees of freedom, have been fully considered elsewhere. See for example, Bishop and Johnson (1960) [34], Bishop, Gladwell and Michaelson (1965) [33], Meirovich (1975) [234]. In this chapter we shall give a brief account of those parts of the theory that will be needed for the solution of inverse problems.

Throughout this book we shall be concerned with the *infinitesimal* vibration of a *conservative* system about some datum configuration, which will usually be an equilibrium position.

Before embarking on a general discussion we shall first formulate the equations of motion for some simple vibrating systems.

2.2 Vibration of some simple systems

Figure 2.2.1 shows a vibrating system consisting of n masses connected by linear springs of stiffnesses $(k_r)_1^n$. The whole lies in a straight line on a smooth horizontal table and is excited by forces $(F_r(t))_1^n$.

Newton's equations of motion for the system are

$$m_r \ddot{u}_r = F_r + \theta_{r+1} - \theta_r \ , \quad r = 1, 2, \ldots, n-1, \tag{2.2.1}$$

$$m_n \ddot{u}_n = F_n - \theta_n, \tag{2.2.2}$$

where $\cdot$ denotes differentiation with respect to time. Hooke's law states that the spring forces are given by

$$\theta_r = k_r(u_r - u_{r-1}) \ , \quad r = 1, 2, \ldots, n. \tag{2.2.3}$$

If the left hand end is pinned then

$$u_0 = 0. \tag{2.2.4}$$

Forced vibration analysis concerns the solution of these equations for given forcing functions $F_r(t)$. *Free* vibration analysis consists in finding solutions to the equations which require no external excitation, i.e., $F_r(t) \equiv 0, \ r = 1, 2, \ldots, n$, and which satisfy the stated end conditions.

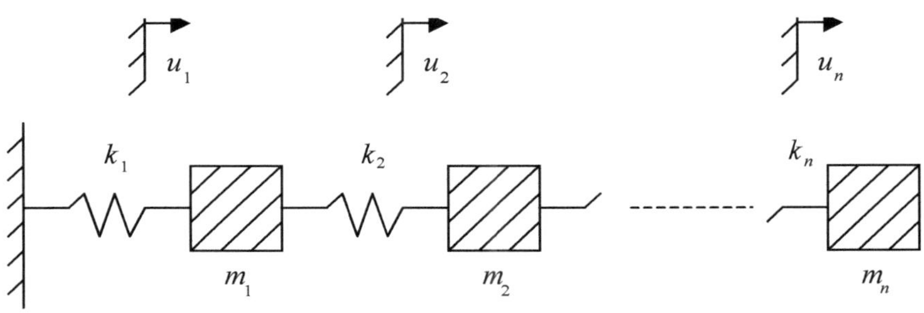

Figure 2.2.1 - n masses connected by springs

The system shown in Figure 2.2.1 has considerable engineering importance. It is the simplest possible discrete model for a rod vibrating in longitudinal motion. Here the masses and stiffnesses are obtained by *lumping* the continuously distributed mass and stiffness of the rod. Equations (2.2.1) - (2.2.4) also describe the torsional vibrations of the system shown in Figure 2.2.2., provided that the u_r, k_r, m_r are interpreted as torsional rotations, torsional stiffnesses and moments of inertia respectively. Such a discrete system provides a simple model for the torsional vibrations of a rod with a continuous distribution of inertia and stiffness.

There is a third system which is mathematically equivalent to equations (2.2.1) - (2.2.4). This is the transverse motion of the string shown in Figure 2.2.3 which is pulled taut by a tension T and which is loaded by masses $(m_r)_1^n$. (But note that the string shown in Figure 2.2.3 has its right hand end *fixed*, rather than free, as in Figures 2.2.1 and 2.2.2. In order to simulate a string with a free end, the last segment of the string must be attached to a massless ring that slides on a smooth vertical rod.) If in accordance with the assumption of infinitesimal vibration, the string departs very little from the straight line equilibrium position, then the equation governing the motion of mass m_r may be derived by considering Figure 2.2.4.

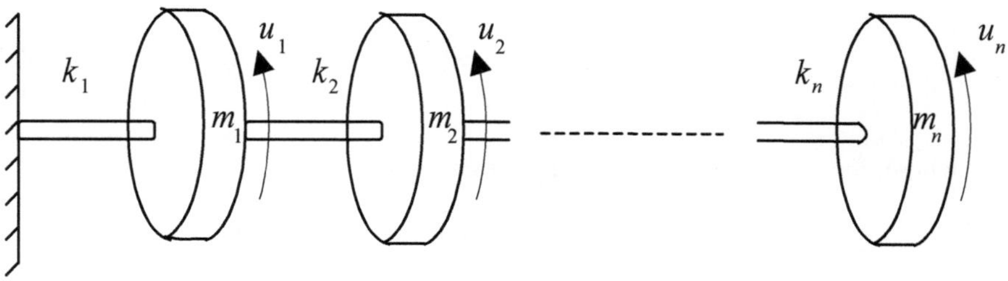

Figure 2.2.2 - A torsionally vibrating system

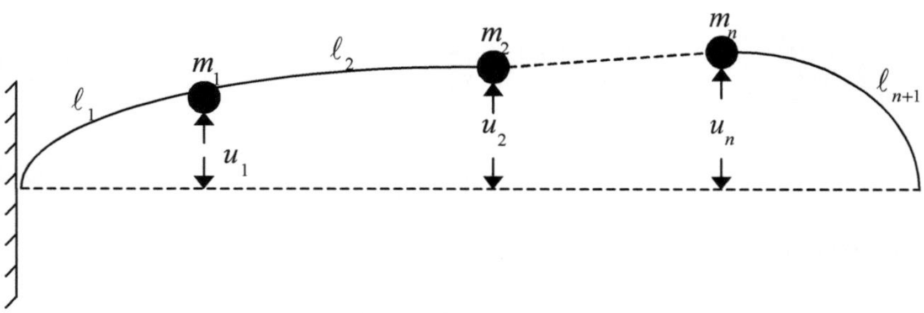

Figure 2.2.3 - n masses on a taut string

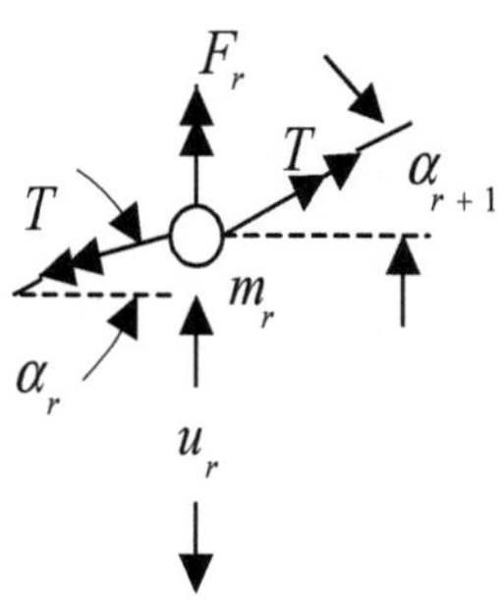

Figure 2.2.4 - The forces acting on the mass m_r

Newton's equation of motion yields

$$m_r\ddot{u}_r = F_r + T\sin\alpha_{r+1} - T\sin\alpha_r \ , \quad (2.2.5)$$
$$= F_r + \theta_{r+1} - \theta_r \ , \quad (2.2.6)$$

where, for small deflections, we may take $\sin\alpha_r = \alpha_r$,

$$\theta_r = T\alpha_r = k_r(u_r - u_{r-1}) \ , \quad k_r = T/\ell_r.$$

In order to express equations (2.2.1) - (2.2.3) in matrix form we use (2.2.3) to obtain

$$m_r\ddot{u}_r = F_r + k_{r+1}u_{r+1} - (k_{r+1}+k_r)u_r + k_r u_{r-1} \ , m_n\ddot{u}_n = F_n - k_n u_n + k_n u_{n-1} \ ,$$

which yields

$$\begin{bmatrix} m_1 & & & \\ & m_2 & & \\ & & \ddots & \\ & & & m_n \end{bmatrix}\begin{bmatrix} \ddot{u}_1 \\ \ddot{u}_2 \\ \cdot \\ \ddot{u}_n \end{bmatrix} + \begin{bmatrix} k_1+k_2 & -k_2 & 0 & \cdots & 0 & 0 \\ -k_2 & k_2+k_3 & -k_3 & \cdots & 0 & 0 \\ \cdot & \cdot & \cdot & \cdots & \cdot & \cdot \\ 0 & 0 & 0 & \cdots & -k_n & k_n \end{bmatrix}\begin{bmatrix} u_1 \\ u_2 \\ \cdot \\ u_n \end{bmatrix}$$
$$= \begin{bmatrix} F_1 \\ F_2 \\ \cdot \\ F_n \end{bmatrix}. \quad (2.2.7)$$

This equation may be written

$$\mathbf{M\ddot{u}} + \mathbf{Ku} = \mathbf{F} \quad (2.2.8)$$

where the matrices $\mathbf{M}, \mathbf{K}$ are called respectively the *inertia* (or *mass*) and the *stiffness* matrices of the system. Note that both $\mathbf{M}$ and $\mathbf{K}$ are *symmetric*; this is a property shared by the matrices corresponding to any conservative system. We note also that both $\mathbf{M}, \mathbf{K}$ are positive-definite. In this particular example the matrix $\mathbf{M}$ is *diagonal* while $\mathbf{K}$ is *tridiagonal*, i.e., its only non-zero elements are on the principal diagonal, and the two neighbouring diagonals, called the *codiagonals*.

Equation (2.2.3) can also be constructed by introducing $\boldsymbol{\theta} = \{\theta_1, \theta_2, \ldots, \theta_n\}$ and noting that

$$\begin{bmatrix} \theta_1 \\ \theta_2 \\ \cdot \\ \cdot \\ \theta_n \end{bmatrix} = \begin{bmatrix} k_1 & & & \\ & k_2 & & \\ & & \cdot & \\ & & & k_n \end{bmatrix}\begin{bmatrix} 1 & 0 & \ldots & 0 \\ -1 & 1 & \ldots & 0 \\ \cdot & \cdot & \ldots & \cdot \\ 0 & \cdot & \ldots & 0 \\ 0 & \cdot & \ldots & 1 \end{bmatrix}\begin{bmatrix} u_1 \\ u_2 \\ \cdot \\ \cdot \\ u_n \end{bmatrix}$$

which will be written

$$\boldsymbol{\theta} = \hat{\mathbf{K}}\mathbf{E}^T\mathbf{u} \ , \quad (2.2.9)$$

where

$$\mathbf{E} = \begin{bmatrix} 1 & -1 & 0 & \cdots & 0 \\ 0 & 1 & -1 & \cdots & 0 \\ \cdot & \cdot & \cdot & \cdot & \cdot \\ 0 & \cdots & & 1 & -1 \\ 0 & \cdots & & 0 & 1 \end{bmatrix}, \mathbf{E}^{-1} = \begin{bmatrix} 1 & 1 & 1 & \cdots & 1 \\ 0 & 1 & 1 & \cdots & 1 \\ \cdot & \cdot & \cdot & \cdot & \cdot \\ 0 & 0 & 0 & \cdots & 1 \\ 0 & 0 & 0 & \cdots & 1 \end{bmatrix} \tag{2.2.10}$$

and $\hat{\mathbf{K}} = diag(k_1, k_2, \ldots, k_n)$.

Using the matrix $\mathbf{E}$, we may write equation (2.2.1) - (2.2.2) in the form

$$\mathbf{M}\ddot{\mathbf{u}} = -\mathbf{E}\boldsymbol{\theta} + \mathbf{F} \ ,$$

so that on using (2.2.9) we find

$$\mathbf{M}\ddot{\mathbf{u}} + \mathbf{E}\hat{\mathbf{K}}\mathbf{E}^T\mathbf{u} = \mathbf{F} \ , \tag{2.2.11}$$

and

$$\mathbf{K} = \mathbf{E}\hat{\mathbf{K}}\mathbf{E}^T . \tag{2.2.12}$$

For free vibration analysis there are two important end conditions. The right hand end may be *free*, in which case there is *no* restriction on the $(u_i)_1^n$, or it may be *fixed*, in which case $u_n = 0$.

Exercises 2.2

1. Verify that the stiffness matrix in equation (2.2.7) satisfies the conditions of Theorem 1.4.2. Obtain a proof that applies to principal minors of any order i, such that $1 \le i \le n$.

2. Consider the multiple pendulum of Figure 2.2.5.

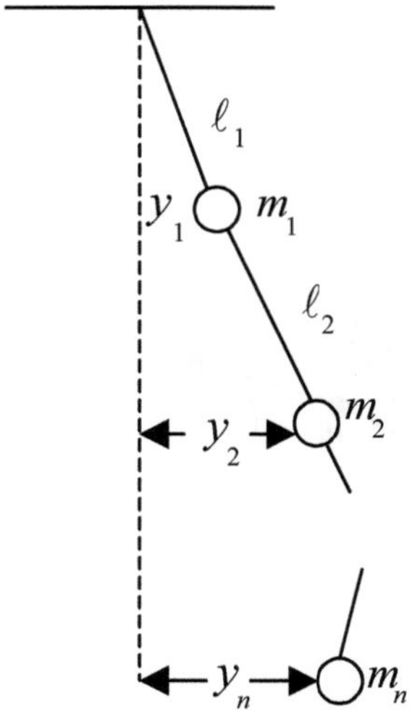

Figure 2.2.5 - A compound pendulum made up of n inextensible strings

Show that the kinetic and potential energies of the system for small oscillations are given by

$$\begin{aligned} 2T &= m_1\dot{y}_1^2 + m_2\dot{y}_2^2 + \ldots + m_n\dot{y}_n^2, \\ 2V &= \sigma_1\frac{y_1^2}{\ell_1} + \sigma_2\frac{(y_2-y_1)^2}{\ell_2} + \ldots + \sigma_n\frac{(y_n-y_{n-1})^2}{\ell_n} \end{aligned}$$

where $\sigma_r = g\sum_{s=r}^{n} m_s$.

2.3 Transverse vibration of a beam

Figure 2.3.1 shows a simple discrete model for the transverse vibration of a beam; it consists of $n+2$ masses $(m_r)_{-1}^{n}$ linked by massless rigid rods of lengths $(\ell_r)_0^n$ which are themselves connected by n rotational springs of stiffnesses $(k_r)_1^n$. The mass and stiffness of the beam, which are actually distributed along the length, have been lumped at $n+2$ points.

The discrete system is governed by a set of four first-order difference equations, which may be deduced from Figure 2.3.2.

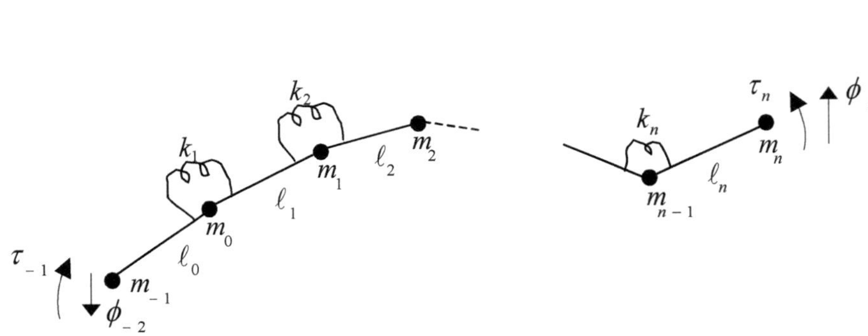

Figure 2.3.1 - A discrete model of a vibrating beam

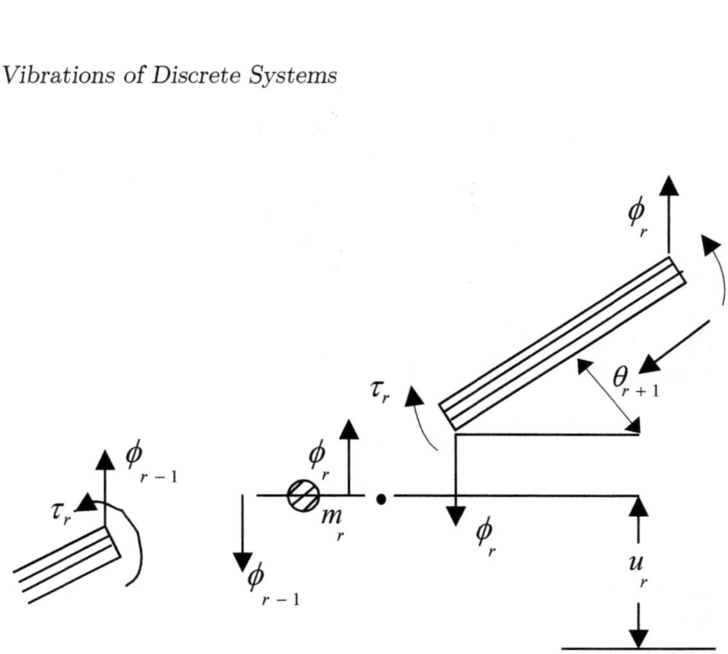

Figure 2.3.2 - The configuration around m_r

For small displacements, the rotations are

$$\theta_r = (u_r - u_{r-1})/\ell_r \ , \quad r = 0, 1, \ldots, n.$$

If the rth spring has rotational stiffness k_r, then the moment τ_r needed to produce a relative rotation $\theta_{r+1} - \theta_r$ of the two rigid rods on either side of m_r is

$$\tau_r = k_{r+1}(\theta_{r+1} - \theta_r) \ , \quad r = 0, 1, \ldots, n-1.$$

Equilibrium of the rod linking m_r and m_{r+1} yields the shearing forces

$$\phi_r = (\tau_r - \tau_{r+1})/\ell_{r+1} \ , \quad r = -1, 0, \ldots, n-1,$$

while Newton's equation of motion for mass m_r is

$$m_r \ddot{u}_r = \phi_r - \phi_{r-1} \ , \quad r = -1, 0, \ldots, n.$$

Here ϕ_{-2}, ϕ_n and τ_{-1}, τ_n denote external shearing forces and bending moments, respectively, applied to the ends.

Suppose that the left hand end is *clamped* so that

$$u_{-1} = 0 = u_0,$$

then only the masses $(m_r)_1^n$ move, and the governing equations may be written

$$\boldsymbol{\theta} = \mathbf{L}^{-1}\mathbf{E}^T\mathbf{u}, \tag{2.3.2}$$

$$\boldsymbol{\tau} = \hat{\mathbf{K}}\mathbf{E}^T\boldsymbol{\theta}, \tag{2.3.3}$$

$$\boldsymbol{\phi} = \mathbf{L}^{-1}\mathbf{E}\boldsymbol{\tau} - \ell_n^{-1}\tau_n\mathbf{e}_n, \tag{2.3.4}$$

$$\mathbf{M}\ddot{\mathbf{u}} = -\mathbf{E}\boldsymbol{\phi} + \phi_n\mathbf{e}_n, \tag{2.3.5}$$

where $\mathbf{u} = \{u_1, u_2, \ldots, u_n\}$, $\boldsymbol{\theta} = \{\theta_1, \theta_2, \ldots, \theta_n\}$, $\boldsymbol{\tau} = \{\tau_0, \tau_1, \ldots, \tau_{n-1}\}$, $\boldsymbol{\phi} = \{\phi_0, \phi_1, \ldots, \phi_{n-1}\}$, $\hat{\mathbf{K}} = diag(k_r)$, $\mathbf{L} = diag(\ell_r)$, $\mathbf{M} = diag(m_r)$, $\mathbf{e}_n = \{0, 0, \ldots, 0, 1\}$ and $\mathbf{E}$ is given in equation (2.2.10).

Equations (2.3.2) - (2.3.5) may be combined to give

$$\mathbf{M}\ddot{\mathbf{u}} + \mathbf{K}\mathbf{u} = \phi_n\mathbf{e}_n + \ell_n^{-1}\tau_n\mathbf{E}\mathbf{e}_n, \tag{2.3.6}$$

where

$$\mathbf{K} = \mathbf{E}\mathbf{L}^{-1}\mathbf{E}\hat{\mathbf{K}}\mathbf{E}^T\mathbf{L}^{-1}\mathbf{E}^T. \tag{2.3.7}$$

This equation has the same general form as equation (2.2.8). We note that $\mathbf{M}$ and $\mathbf{K}$ are again symmetric and positive-definite, $\mathbf{M}$ being diagonal, and $\mathbf{K}$ being *pentadiagonal.*

2.4 Generalised coordinates and Lagrange's equations: *the rod*

The idea that a discrete system is one composed of a finite number of masses connected by springs is unnecessarily restrictive. The general concept is that of a system whose motion is specified by n *generalised coordinates* $(q_r)_1^n$ that are functions of time t alone. The systems considered in Sections 2.2, 2.3 are indeed discrete in this sense and the generalised coordinates corresponding to the system in Figure 2.2.1 are $(u_r)_1^n$. However, the more general concept would also cover, for instance, a model of a non-uniform longitudinally vibrating rod constructed by using the *finite element method* (see for example, Zienkiewicz (1971) [343]), Strang and Fix (1973) [311].

In such a model, shown in Figure 2.4.1, the rod is first divided into $n+1$ elements. In the rth element, shown in Figure 2.4.2., the longitudinal displacement $y(x,t)$ is taken to have a simple linear form.

$$y(x,t) = y_r(t)(1-\xi) + y_{r+1}(t)\xi\ , \quad x_r \le x \le x_{r+1}, \tag{2.4.1}$$

where

$$\xi = (x - x_r)/\ell_r,$$

runs from 0 at the left hand end of the element to 1 at the right. Equations (2.4.1) with $r = 0, 1, \ldots, n$ express the displacement at every point of the rod in terms of the $n+2$ generalised coordinates $(y_r)_0^{n+1}$. When the end conditions are imposed there will be, as before, only n coordinates $(y_r)_1^n$.

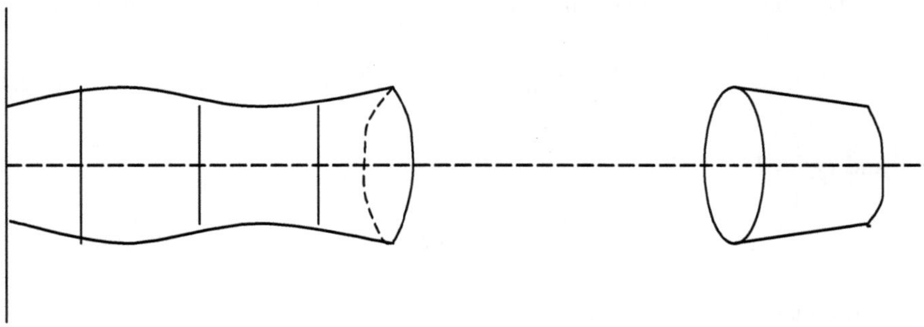

Figure 2.4.1 - A rod divided into elements

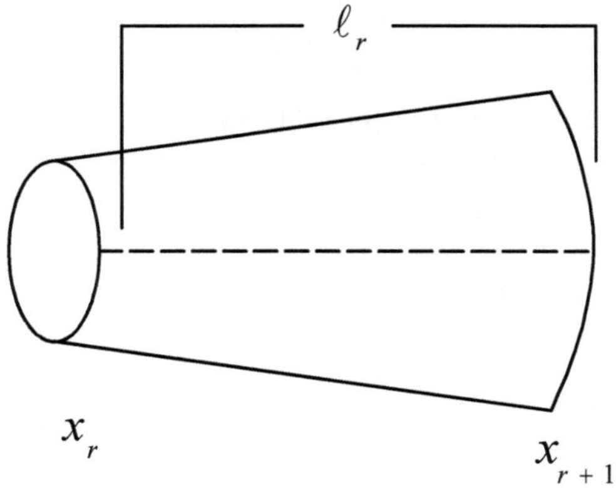

Figure 2.4.2 - One element of the rod

When the finite element method is used, it is not possible to set up the equations of motion by using Newton's equations of motion, for there is no actual 'mass' to which forces are applied. Instead we may use Lagrange's equations. For a conservative system with *kinetic energy* T and *potential* or *strain energy* V, which are functions of n coordinates $(q_r)_1^n$, Lagrange's equations state that

$$\frac{d}{dt}\left(\frac{\partial T}{\partial \dot{q}_r}\right) - \frac{\partial T}{\partial q_r} + \frac{\partial V}{\partial q_r} = Q_r \ , \quad (r = 1, 2, \ldots, n). \tag{2.4.2}$$

Here Q_r is the generalised force corresponding to q_r in the sense that the work done by external forces acting on the system when the system is displaced from a configuration specified by $(q_r)_1^n$ to one specified by $(q_r + \delta q_r)_1^n$, is

$$\delta W_e = \sum_{r=1}^{n} Q_r \delta q_r.$$

For the system shown in Figure 2.2.1 the kinetic and potential energies are

$$T = \frac{1}{2}\sum_{r=1}^{n} m_r \dot{y}_r^2 , \qquad V = \frac{1}{2}\sum_{r=1}^{n} k_r (y_{r+1} - y_r)^2, \tag{2.4.3}$$

and $Q_r = F_r(t)$. Thus

$$\frac{\partial T}{\partial \dot{y}_r} = m_r \dot{y}_r , \quad \frac{\partial V}{\partial y_r} = -k_r(y_{r+1} - y_r) + k_{r-1}(y_r - y_{r-1}),$$

and equation (2.4.3) yields (2.2.1).

For the finite element model of Figure 2.4.1, the kinetic and potential energies of the system will be

$$T = \frac{1}{2}\int_0^{\ell} S\rho[\dot{y}(x,t))]^2 dx,$$

$$V = \frac{1}{2}\int_0^{\ell} SE[\frac{\partial y}{\partial x}(x,t)]^2 dx,$$

where $S(x), \rho(x), E(x)$ are the (possibly variable) cross-sectional area, density and Young's modulus of the rod. On inserting the assumed form of $y(x,t)$ given in (2.4.1) we find

$$T = \frac{1}{2}\sum_{r=0}^{n}\int_0^1 S(x_r + \ell_r\xi)\rho(x_r + \ell_r\xi)[\dot{y}_r(1-\xi) + \dot{y}_{r+1}\xi]^2 \ell_r d\xi, \tag{2.4.4}$$

$$V = \frac{1}{2}\sum_{r=1}^{n}\int_0^1 S(x_r + \ell_r\xi)E(x_r + \ell_r\xi)[y_{r+1} - y_r]^2 \ell_r^{-1} d\xi. \tag{2.4.5}$$

On carrying out the integrations, perhaps numerically if $S(x), \rho(x), E(x)$ are variable, we may write

$$T = \frac{1}{2}\sum_{r=0}^{n+1}\sum_{s=0}^{n+1} m_{rs}\dot{y}_r\dot{y}_s, \tag{2.4.6}$$

$$V = \frac{1}{2}\sum_{r=0}^{n+1}\sum_{s=0}^{n+1} k_{rs}y_r y_s. \tag{2.4.7}$$

If the rod is fixed at both ends, then

$$y_0 = 0 = y_{n+1}, \tag{2.4.8}$$

so that all the sums in (2.4.6), (2.4.7) run from 1 to n. In this case

$$\frac{\partial T}{\partial \dot{y}_r} = \sum_{s=1}^{n} m_{rs}\dot{y}_s, \quad \frac{\partial V}{\partial y_r} = \sum_{s=1}^{n} k_{rs}y_s,$$

and equation (2.4.2) yields the following equation for free vibration:

$$\sum_{s=1}^{n} m_{rs}\ddot{y}_s + \sum_{s=1}^{n} k_{rs}y_s = 0, \quad (r = 1, 2, \ldots, n).$$

This equation may, as before, be condensed into the matrix equation

$$\mathbf{M}\ddot{\mathbf{y}} + \mathbf{K}\mathbf{y} = \mathbf{0} \tag{2.4.9}$$

We note that, for the rod with the kinetic and potential energies given by (2.4.6), (2.4.7), the matrices $\mathbf{M}, \mathbf{K}$ are symmetric, *tridiagonal* matrices with sign properties. They are tridiagonal because m_{rs}, k_{rs} are zero unless $r = s$ or $r = s \pm 1$. The sign properties may be deduced from (2.4.4), (2.4.5): the codiagonal elements $m_{r,r+1}, m_{r,r-1}$ of $\mathbf{M}$ are positive, while $k_{r,r+1}, k_{r,r-1}$ are negative. Thus

$$\mathbf{M} = \begin{bmatrix} a_1 & b_1 & & \\ b_1 & a_2 & \ddots & \\ & \ddots & \ddots & b_{n-1} \\ & & b_{n-1} & a_n \end{bmatrix}, \mathbf{K} = \begin{bmatrix} c_1 & -d_1 & & \\ -d_1 & c_2 & \ddots & \\ & \ddots & \ddots & -d_{n-1} \\ & & -d_{n-1} & c_n \end{bmatrix}. \tag{2.4.10}$$

These sign properties of $\mathbf{M}, \mathbf{K}$ will later be shown to have important implications for the qualitative properties of a vibrating rod.

On the basis of these examples we now pass to the general case. For a conservative system with generalised coordinates $(q_r)_1^n$ which specify small displacements from a position of stable equilibrium, the kinetic and potential energies will have the form

$$T = \frac{1}{2}\sum_{r=1}^{n}\sum_{s=1}^{n} m_{rs}\dot{q}_r\dot{q}_s, \tag{2.4.11}$$

$$V = \frac{1}{2}\sum_{r=1}^{n}\sum_{s=1}^{n} k_{rs}q_r q_s, \tag{2.4.12}$$

where the matrices $\mathbf{M} = (m_{rs})$ and $\mathbf{K} = (k_{rs})$ are *symmetric*, in that

$$m_{sr} = m_{rs}, \quad k_{sr} = k_{rs}.$$

The equations governing free vibration may be written

$$\mathbf{M}\ddot{\mathbf{q}} + \mathbf{K}\mathbf{q} = \mathbf{0}. \tag{2.4.13}$$

We note that equations (2.4.11), (2.4.12) may be written

$$T = \frac{1}{2}\dot{\mathbf{q}}^T\mathbf{M}\dot{\mathbf{q}}, \quad V = \frac{1}{2}\mathbf{q}^T\mathbf{K}\mathbf{q}. \tag{2.4.14}$$

It is not possible for any arbitrarily chosen symmetric matrix $\mathbf{M}$ to be an inertia matrix, because the kinetic energy T is an essentially positive quantity, i.e., it is always positive except when each of the $\dot{q}_r$ is zero, in which case it is zero. Thus $\mathbf{M}$ must be *positive definite* (see Section 1.4).

The restrictions on the matrix $\mathbf{K}$ are slightly less severe since, although the strain energy will always be positive or zero, it will actually be zero if the system has a rigid-body displacement. Notice, for example, that the V of (2.4.5) will be zero if $\mathbf{y}$ is the rigid body displacement

$$y_0 = y_1 = \cdots = y_n = y_{n+1}.$$

This will be a possible displacement of the system in Figure 2.2.1 only if both ends are *free*. We conclude that if the system is not constrained so that one point is fixed, then $\mathbf{K}$ is *positive semi-definite.*

Exercises 2.4

1. Use equations (2.4.4), (2.4.5) to evaluate the mass and stiffness matrices for a *uniform* rod in longitudinal vibration subject to the end conditions (2.4.8).

2. Use the form (2.4.5) of the strain energy of the rod to show that the stiffness matrix $\mathbf{K}$ for a rod fixed at the left and free at the right has the form

$$K = \begin{bmatrix} k_1+k_2 & -k_2 & & & & \\ -k_2 & k_2+k_3 & -k_3 & & & \\ \ddots & \ddots & \ddots & & & \\ & \ddots & \ddots & \ddots & & \\ & & \ddots & \ddots & \ddots & \\ & & & \ddots & \ddots & -k_{n-1} \\ & & & & -k_{n-1} & k_n \end{bmatrix}.$$

2.5 Vibration of a membrane and an acoustic cavity

Over the last three or four decades, computational vibration analysis has developed to such an extent that it can analyse the vibration of almost anything: rods, beams, plates, trusses, steel and concrete buildings, bridges, aircraft, and so on. Inverse vibration analysis in the strict form we consider in this book can hope to encompass only comparatively simple structures: strings, rods, beams, membranes and acoustic cavities and, even now, inverse problems for membranes and cavities are still open; all we can do is find some qualitative properties of the vibration. The vibrations of a membrane and of an acoustic cavity are mathematically similar: both involve just one scalar quantity, the transverse displacement $u(x,y)$, for the membrane under unit tension; the excess pressure $p(x,y,z)$, for the acoustic cavity. Both are governed by a wave equation

$$\triangle u = \rho\frac{\partial^2 u}{\partial t^2}, \quad \triangle = \frac{\partial^2}{\partial x^2} + \frac{\partial^2}{\partial y^2}, \tag{2.5.1}$$

for a membrane with mass density $\rho(x,y)$, and

$$\triangle p = \rho\frac{\partial^2 p}{\partial t^2}, \quad \triangle = \frac{\partial^2}{\partial x^2} + \frac{\partial^2}{\partial y^2} + \frac{\partial^2}{\partial z^2}, \tag{2.5.2}$$

for the acoustic cavity.

To set up the finite element model FEM of a membrane we consider the energies

$$T = \frac{1}{2}\int\int_D \rho\dot{u}^2 dxdy, \tag{2.5.3}$$

$$V = \frac{1}{2}\int\int_D (\nabla u)^2 dxdy. \tag{2.5.4}$$

The simplest FEM is based on triangulation. For an arbitrary triangular element P_1, P_2, P_3 as shown in Figure 2.5.1, we take

$$u(x,y) = a + bx + cy \tag{2.5.5}$$

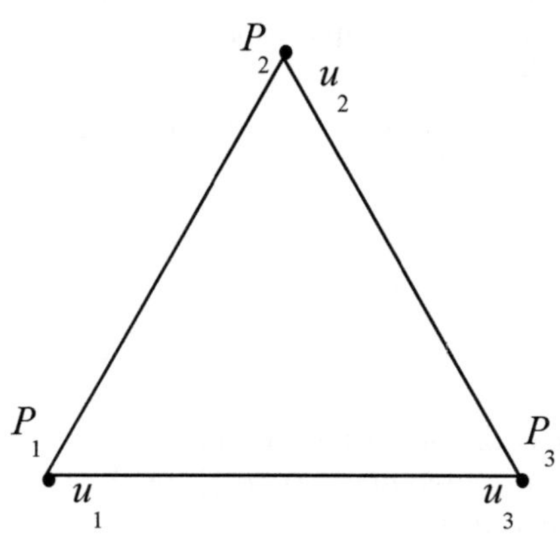

Figure 2.5.1 - A triangular finite element

If u takes the values u_1, u_2, u_3 at the vertices P_1, P_2, P_3 respectively, then

$$u_i = a + bx_i + cy_i, \quad i = 1,2,3. \tag{2.5.6}$$

We can solve these equations for a, b, c and hence express T, V for one element, i.e., T_e, V_e, as quadratic forms

$$T_e = \frac{1}{2}\dot{\mathbf{u}}_e^T \mathbf{M}_e \dot{\mathbf{u}}_e \tag{2.5.7}$$

$$V_e = \frac{1}{2}\mathbf{u}_e^T \mathbf{K}_e \mathbf{u}_e \tag{2.5.8}$$

with coefficients which are functions of the coordinates $(x_i, y_i), i = 1, 2, 3$. We are not particularly interested in the *magnitudes* of the coefficients; we are more interested in their *signs.*

First we investigate the elements of $\mathbf{K}_e$. Equation (2.5.8) give

$$\begin{aligned} b\triangle &= u_1(y_2 - y_3) + u_2(y_3 - y_1) + u_3(y_1 - y_2) \\ c\triangle &= u_1(x_2 - x_3) + u_2(x_3 - x_1) + u_3(x_1 - x_2) \end{aligned}$$

where

$$\triangle = \begin{vmatrix} 1 & x_1 & y_1 \\ 1 & x_2 & y_2 \\ 1 & x_3 & y_3 \end{vmatrix} = 2 \text{ area}(P_1P_2P_3).$$

Since $(\nabla u)^2 = b^2 + c^2$, the coefficient of, say, u_1u_2 in V_e is

$$-\{(x_3 - x_1)(x_3 - x_2) + (y_3 - y_1)(y_3 - y_2)\}/|\triangle| = -|P_1P_3|.|P_2P_3| \cos\gamma/|\triangle|.$$

Users of finite element methods have found that compact, i.e., acute angled, triangles give more accurate computational results than elongated triangles that have an obtuse angle.

If the triangle has all its angles *acute*, then $k_{12,e}$ and $k_{23,e}, k_{31,e}$ are all *negative*: $\mathbf{K}_e$ has the sign pattern

$$\mathbf{K}_e = \begin{bmatrix} + & - & - \\ - & + & - \\ - & - & + \end{bmatrix}. \tag{2.5.9}$$

To find the signs of the coefficients in T_e, it is convenient to write (2.5.7) in terms of the *areal coordinates* , $\phi_i(x, y)$, of the triangle; if P is an arbitrary point of the triangle, then

$$u(x, y) = u_1\phi_1(x, y) + u_2\phi_2(x, y) + u_3\phi_3(x, y)$$

where

$$\phi_1 = \frac{\text{area}(PP_2P_3)}{\text{area}(P_1P_2P_3)}, \quad \phi_2 = \frac{\text{area}(PP_3P_1)}{\text{area}(P_1P_2P_3)}, \quad \phi_3 = \frac{\text{area}(PP_1P_2)}{\text{area}(P_1P_2P_3)}$$

as shown in Figure 2.5.2.

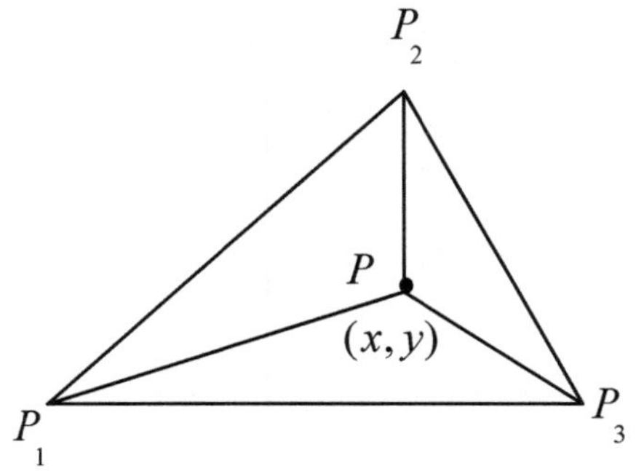

Figure 2.5.2 - $P_1P_2P_3$ is split into three triangles

Since ϕ_1, ϕ_2, ϕ_3 are all positive when P is inside the triangle $P_1P_2P_3$, all the coefficients in T_e are positive: $\mathbf{M}_e$ has the form

$$\mathbf{M}_e = \begin{bmatrix} + & + & + \\ + & + & + \\ + & + & + \end{bmatrix}. \tag{2.5.10}$$

Now we assemble the element matrices to form the global mass and stiffness matrices. The membrane is replaced by an assembly of triangles $\triangle_i$ with vertices P_i and edges P_iP_j as shown in Figure 2.5.3. The boundary condition $u = 0$ is imposed on the outer vertices labelled '0'.

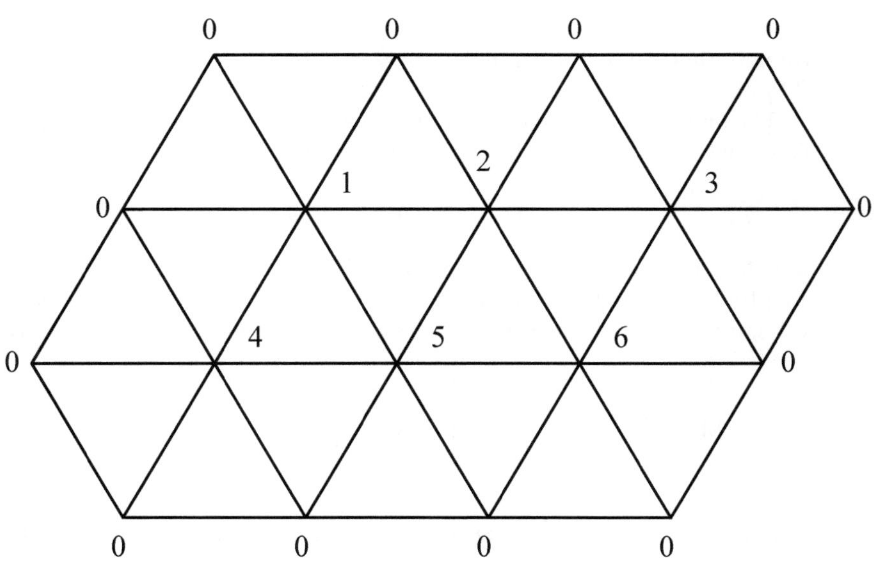

Figure 2.5.3 - An assembly of triangular elements

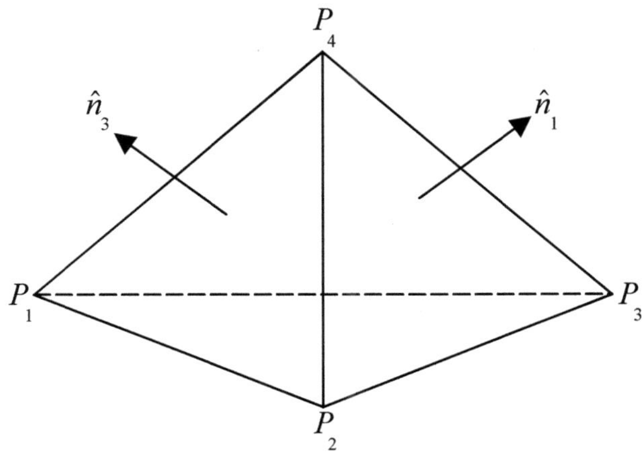

Figure 2.5.4 - The angles between outward drawn normals to the faces are all obtuse

For this particular configuration, the matrices **A**, **C** have the sign patterns

$$\mathbf{M}=\begin{bmatrix} + & + & & + & + & \\ + & + & + & & + & + \\ & + & + & & & + \\ + & & & + & + & \\ + & + & & + & + & + \\ & + & + & & + & + \end{bmatrix}, \quad \mathbf{K}=\begin{bmatrix} + & - & & - & - & \\ - & + & - & & - & - \\ & - & + & & & - \\ - & & & + & - & \\ - & - & & - & + & - \\ & - & - & & - & + \end{bmatrix}. \tag{2.5.11}$$

We note that if $i \neq j$, then $m_{ij} > 0$, $k_{ij} < 0$ iff P_i, P_j are the ends of an edge P_iP_j of the mesh.

The finite element analysis of a 3D- acoustic cavity proceeds in a similar way. The elements are taken to be tetrahedra, and the pressure $p(x, y, z)$ is taken as

$$p(x, y, z) = a + bx + cy + dz \tag{2.5.12}$$

in each tetrahedron. Now it is found Zhu (2000) [342], Gladwell and Zhu (2002) [131] that if the angles between the normals to the faces are all *obtuse*, as shown in Figure 2.5.4, then the element mass and stiffness matrices have the form

$$\mathbf{M}_e=\begin{bmatrix} + & + & + & + \\ + & + & + & + \\ + & + & + & + \\ + & + & + & + \end{bmatrix}, \quad \mathbf{K}_e=\begin{bmatrix} + & - & - & - \\ - & + & - & - \\ - & - & + & - \\ - & - & - & + \end{bmatrix}. \tag{2.5.13}$$

This means that when the matrices are assembled they have the same kind of sign pattern as before: if $i \neq j$ then $m_{ij} > 0$, $k_{ij} < 0$ iff P_iP_j is an edge of the mesh.

Applying Lagrange's equation to the energies

$$T = \frac{1}{2}\dot{\mathbf{u}}^T\mathbf{M}\dot{\mathbf{u}}, \quad V = \frac{1}{2}\mathbf{u}^T\mathbf{K}\mathbf{u}$$

we find the equation governing the vibration as

$$\mathbf{M}\ddot{\mathbf{u}} + \mathbf{K}\mathbf{u} = \mathbf{0}. \tag{2.5.14}$$

2.6 Natural frequencies and normal modes

The matrix equation (2.4.13) represents a set of second order equations with constant coefficients. Following usual practice we seek the solution in the form

$$\mathbf{q} = \begin{bmatrix} q_1 \\ q_2 \\ \cdot \\ q_n \end{bmatrix} = \begin{bmatrix} x_1 \\ x_2 \\ \cdot \\ x_n \end{bmatrix} \sin(\omega t + \phi), \tag{2.6.1}$$

where the constants x_r, frequency ω and phase angle ϕ are to be determined. When $\mathbf{q}$ has the form (2.6.1), then

$$\ddot{\mathbf{q}} = -\omega^2\mathbf{q} = -\omega^2\mathbf{x}\sin(\omega t + \phi), \tag{2.6.2}$$

so that equation (2.4.13) demands that

$$(\mathbf{K} - \lambda\mathbf{M})\mathbf{x} = \mathbf{0}, \quad \lambda = \omega^2. \tag{2.6.3}$$

This is the eigenvalue equation (1.4.1) and, since $\mathbf{M}$ is positive-definite and $\mathbf{K}$ is either positive semi-definite or positive-definite, the whole of the analysis developed in Section 1.4 can be used here. Thus the equation has n eigenvalues $(\lambda_i)_1^n$ satisfying

$$0 \le \lambda_1 \le \lambda_2 < \cdots \le \lambda_n, \tag{2.6.4}$$

and n corresponding eigenvectors $(\mathbf{x}_i)_1^n$ satisfying

$$(\mathbf{K} - \lambda_i\mathbf{M})\mathbf{x}_i = \mathbf{0}. \tag{2.6.5}$$

The frequencies $\omega_i = (\lambda_i)^{\frac{1}{2}}$ are called the *natural* frequencies of the system, and the eigenvectors are called the *normal* or *principal* modes. Note the distinction between x_i, a scalar, and $\mathbf{x}_i$, a vector.

In order to become acquainted with the properties of natural frequencies and normal modes we shall consider the system specified by equation (2.2.7) and, to simplify the algebra, shall assume that

$$(m_r)_1^n = m, \quad (k_r)_1^n = k. \tag{2.6.6}$$

In this case the eigenvalue equation may be written

$$\begin{bmatrix} 2-\lambda & -1 & 0 & \cdots & 0 \\ -1 & 2-\lambda & -1 & \cdots & 0 \\ \cdot & \cdot & \cdot & \cdots & \cdot \\ 0 & \cdots & & 2-\lambda & -1 \\ 0 & \cdots & & -1 & 1-\lambda \end{bmatrix} \begin{bmatrix} x_1 \\ x_2 \\ \cdot \\ x_{n-1} \\ x_n \end{bmatrix} = 0, \tag{2.6.7}$$

where

$$\lambda = m\omega^2/k. \tag{2.6.8}$$

To solve for the x_r we use the idea suggested in Exercise 1.4.4, namely to write (2.6.7) as the *recurrence relation*

$$-x_{r-1} + (2-\lambda)x_r - x_{r+1} = 0, \ (r = 1, 2, \ldots, n). \tag{2.6.9}$$

The first of equations (2.6.7) may be written in this form if x_0 is taken to be zero; this may be interpreted as stating that the left hand mass (m_0) is fixed. On the other hand, the last of equations (2.6.7) may be written in the form (2.6.9) if x_{n+1} is taken to be equal to x_n. Thus the end conditions for the recurrence (2.6.9) are

$$x_0 = 0 = x_{n+1} - x_n. \tag{2.6.10}$$

The recurrence relation has the general solution

$$x_r = A \ \cos r\theta + B \ \sin r\theta, \tag{2.6.11}$$

where, on substitution into (2.6.9) we find that θ must satisfy

$$\begin{aligned} \cos(r-1)\theta + \cos(r+1)\theta &= 2\cos\theta\cos r\theta = (2-\lambda)\cos r\theta, \\ \sin(r-1)\theta + \sin(r+1)\theta &= 2\cos\theta\sin r\theta = (2-\lambda)\sin r\theta, \end{aligned}$$

i.e.,

$$2\cos\theta = 2 - \lambda.$$

The end conditions will be satisfied if and only if

$$A = 0 = \sin(n+1)\theta - \sin n\theta = 2\cos[(n+1/2)\theta]\sin\theta/2,$$

so that the possible values of θ are

$$\theta = \theta_i = \frac{(2i-1)\pi}{2n+1}, \ (i = 1, 2, \ldots, n),$$

while the corresponding values of λ are

$$\lambda_i = 2 - 2\cos\theta_i = 4\sin^2[\frac{(2i-1)\pi}{2(2n+1)}]. \tag{2.6.12}$$

Thus, in the ith mode, the displacement amplitude of the rth mass is

$$x_r = \sin r\theta_i = \sin[\frac{(2i-1)r\pi}{(2n+1)}]. \tag{2.6.13}$$

The modes for the case $n = 4$, which are shown in Figure 2.6.1, exhibit properties that are held by all eigenvectors of a *tridiagonal* matrix (such as that in (2.6.7)), namely

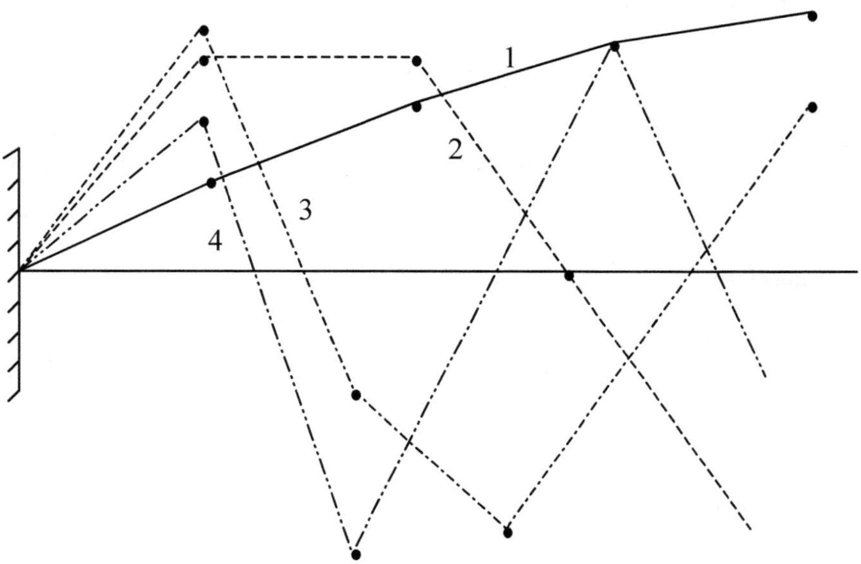

Figure 2.6.1 - The modes of the spring-mass system for $n = 4$

For a proof of the convergence of this class of discrete models to the continuous beam, and for an estimate of the discretisation error on frequencies and mode shapes, see Davini (1996) [74].

(a) the ith mode crosses the axis $(i-1)$ times - the zeros at the ends are not counted;

(b) the *nodes* (points where the mode crosses the axis) of the ith mode *interlace* those of the neighbouring $((i-1)$th and $(i+1)$th) modes.

If instead of being free at the right hand end, the system were pinned there, then the analysis would be unchanged except that the end conditions would be

$$x_0 = 0 = x_n. \tag{2.6.14}$$

In this case θ would have to satisfy

$$\sin n\theta = 0$$

so that

$$\theta = \phi_i = \frac{i\pi}{n}, \quad i = 1, 2, \ldots, n-1$$

and the corresponding eigenvalues, which we will label $(\lambda_i^0)_1^{n-1}$, would be

$$\lambda_i^0 = 4\sin^2(\frac{i\pi}{2n}). \tag{2.6.15}$$

In the ith mode, the rth displacement amplitude is

$$y_r = \sin(r\phi_i) = \sin[\frac{ri\pi}{n}]. \tag{2.6.16}$$

The two sets of eigenvalues $(\lambda_i)_1^n$ and $(\lambda_i^0)_1^{n-1}$ are related in a way which will be found to be general for problems of this type (see equation (2.9.10)), namely

$$0 < \lambda_1 < \lambda_1^0 < \cdots < \lambda_{n-1} < \lambda_{n-1}^0 < \lambda_n. \tag{2.6.17}$$

Exercises 2.6

1. Consider the beam system of Figure 2.3.1 in the case when $(m_i)_{-1}^n = m$, $(k_i)_1^n = k$, $(\ell_i)_0^n = \ell$. Show that the recurrence relation linking the $(u_r)_0^{n+2}$ may be written

   $$u_{r-2} - 4u_{r-1} + (6-\lambda)u_r - 4u_{r+1} + u_{r+2} = 0$$

   where $\lambda = m\omega^2\ell^2/k$. Seek a solution of the recurrence relation of the form

   $$u_r = A\cos r\theta + B\sin r\theta + C\cosh r\phi + D\sinh r\phi$$

   and find θ, ϕ so that the end conditions $u_{-1} = 0 = u_0 = u_{n-1} = u_n$ are satisfied. Hence find the natural frequencies and normal modes of the system; i.e., a clamped-clamped beam. A physically more acceptable discrete approximation of a beam is considered in detail by Gladwell (1962) [103] and Lindberg (1963) [215].

2.7 Principal coordinates and receptances

Theorem 1.4.5 states that the vectors $(\mathbf{x}_i)_1^n$ span the space of n-vectors, so that any arbitrary vector $\mathbf{q}(t)$ may be written

$$\mathbf{q}(t) = p_1\mathbf{x}_1 + p_2\mathbf{x}_2 + \cdots + p_n\mathbf{x}_n. \tag{2.7.1}$$

This may be condensed into the matrix equation

$$\mathbf{q} = \mathbf{X}\mathbf{p}, \tag{2.7.2}$$

where $\mathbf{X}$ is the $n\times n$ matrix with the $\mathbf{x}_i$ as its columns i.e., $\mathbf{x}_i = \{x_{1i}, x_{2i}, \ldots, x_{ni}\}$. The coordinates $p_1, p_2, \ldots, p_n$, called the *principal coordinates*, will in general be functions of t; they indicate the extent to which the various eigenvectors $\mathbf{x}_i$ participate in the vector $\mathbf{q}$. The energies T, V take particularly simple forms when $\mathbf{q}$ is expressed in terms of the principal coordinates. For equation (2.7.2) implies

$$\dot{\mathbf{q}} = \mathbf{X}\dot{\mathbf{p}}, \tag{2.7.3}$$

so that

$$T = \frac{1}{2}(\mathbf{X}\dot{\mathbf{p}})^T\mathbf{M}(\mathbf{X}\dot{\mathbf{p}}) = \frac{1}{2}\dot{\mathbf{p}}^T(\mathbf{X}^T\mathbf{M}\mathbf{X})\dot{\mathbf{p}}. \tag{2.7.4}$$

But the element in row i, column j of the matrix $\mathbf{X}^T\mathbf{M}\mathbf{X}$ is simply $\mathbf{x}_i^T\mathbf{M}\mathbf{x}_j$ and according to (1.4.12) this is zero if $i \neq j$, a_i if $i = j$. Thus

$$\mathbf{X}^T\mathbf{M}\mathbf{X} = diag(a_1, a_2, \ldots, a_n), \tag{2.7.5}$$

so that

$$T = \frac{1}{2}\{a_1\dot{p}_1^2 + a_2\dot{p}_2^2 + \cdots a_n\dot{p}_n^2\}. \tag{2.7.6}$$

Similarly

$$V = \frac{1}{2}\mathbf{p}^T(\mathbf{X}^T\mathbf{K}\mathbf{X})\mathbf{p} \tag{2.7.7}$$

and

$$\mathbf{X}^T\mathbf{K}\mathbf{X} = diag(\lambda_1 a_1, \lambda_2 a_2, \ldots \lambda_n a_n) \tag{2.7.8}$$

so that

$$V = \frac{1}{2}\{\lambda_1 a_1 p_1^2 + \lambda_2 a_2 p_2^2 + \cdots \lambda_n a_n p_n^2\}. \tag{2.7.9}$$

Equations (2.7.6), (2.7.9) show that the search for eigenvalues and eigenvectors for a symmetric matrix pair $(\mathbf{M}, \mathbf{K})$ is equivalent to the search for a coordinate transformation $\mathbf{q} \rightarrow \mathbf{p}$ which will simultaneously convert two quadratic forms $\mathbf{q}^T\mathbf{M}\mathbf{q}$ and $\mathbf{q}^T\mathbf{K}\mathbf{q}$ to sums of squares.

We shall now use the principal coordinates to obtain the response of a system to sinusoidal forces. Equations (2.4.2) and (2.4.14) show that the equation governing the response to generalised forces $(Q_r)_1^n$ is

$$\mathbf{M}\ddot{\mathbf{q}} + \mathbf{K}\mathbf{q} = \mathbf{Q} \tag{2.7.10}$$

where $\mathbf{q} = \{q_1, q_2, \ldots, q_n\}$. If the forces have frequency ω and are all in phase, then $\mathbf{Q}$ and $\mathbf{q}$ may be written

$$\mathbf{Q} = \boldsymbol{\Phi}\sin(\omega t + \phi), \quad \mathbf{q} = \mathbf{x}\sin(\omega t + \phi). \tag{2.7.11}$$

In this case equations (2.6.1) - (2.6.2) yield

$$(\mathbf{K} - \lambda\mathbf{M})\mathbf{x} = \boldsymbol{\Phi}. \tag{2.7.12}$$

To solve this equation we express $\mathbf{x}$ in terms of the eigenvectors $\mathbf{x}_i$, so that

$$\mathbf{x} = \pi_1\mathbf{x}_1 + \pi_2\mathbf{x}_2 + \cdots \pi_n\mathbf{x}_n = \mathbf{X}\boldsymbol{\pi}, \tag{2.7.13}$$

where $\pi_1, \pi_2, \ldots, \pi_n$ are the amplitudes of the principal coordinates $p_1, p_2, \ldots, p_n$. Substitute (2.7.13) into (2.7.12) and multiply the resulting equation by $\mathbf{X}^T$; the result is

$$\mathbf{X}^T(\mathbf{K} - \lambda\mathbf{M})\mathbf{X}\boldsymbol{\pi} = \mathbf{X}^T\boldsymbol{\Phi} = \boldsymbol{\Xi}. \tag{2.7.14}$$

But now the matrix of coefficients of the set of N equations for the unknowns $\pi_1, \pi_2, \ldots, \pi_n$ is *diagonal*, and the ith equation is simply

$$(\lambda_i - \lambda)a_i\pi_i = \Xi_i,$$

so that

$$\pi_i = \frac{\Xi_i}{a_i(\lambda_i - \lambda)}. \tag{2.7.15}$$

In order to interpret this result we consider the response to a single generalised force Q_r. In this case

$$\mathbf{Q} \equiv \quad \boldsymbol{\Phi} = \{0, 0, \ldots, \Phi_r, 0, \ldots, 0\}, \quad \boldsymbol{\Xi} = \Phi_r\{x_{r1}, x_{r2}, \ldots, x_{rn}\}$$
$$\pi_i = \frac{\Phi_r x_{ri}}{a_i(\lambda_i - \lambda)},$$

and the sth displacement amplitude is:

$$x_s = \sum_{i=1}^{n} \pi_i x_{si} = \alpha_{rs}\Phi_r \tag{2.7.16}$$

where

$$\alpha_{rs} = \sum_{i=1}^{n} \frac{x_{ri}x_{si}}{a_i(\lambda_i - \lambda)}. \tag{2.7.17}$$

The quantity $\alpha_{rs}$is the *receptance* Bishop and Johnson (1960) [34] giving the amplitude of response of q_s to a unit amplitude generalised force Q_r. The fact that α_{rs} is symmetric, i.e.,

$$\alpha_{rs} = \alpha_{sr} \tag{2.7.18}$$

is a reflection of the reciprocal theorem which holds for forced harmonic excitation.

Exercises 2.7

1. Use the orthogonality of the $(\mathbf{x}_i)_1^n$ w.r.t the inertia matrix to show that

$$\mathbf{x}_i^T\mathbf{M}\mathbf{q} = p_i a_i.$$

2.8 Rayleigh's Principle

Consider a conservative system with generalised coordinates $(q_r)_1^n$ vibrating with harmonic motion given by (2.6.1). Its kinetic and potential energies will be

$$T = \frac{1}{2}\dot{\mathbf{q}}^T\mathbf{M}\dot{\mathbf{q}} = \omega^2\cos^2\omega t T_0,$$

$$V = \frac{1}{2}\mathbf{q}^T\mathbf{K}\mathbf{q} = \sin^2 \omega t V_0,$$

where

$$T_0 = \frac{1}{2}\mathbf{x}^T\mathbf{M}\mathbf{x}, \quad V_0 = \frac{1}{2}\mathbf{x}^T\mathbf{K}\mathbf{x}. \tag{2.8.1}$$

Since the system is conservative,

$$T + V = const.,$$

so that

$$\omega^2 \cos^2 \omega t T_0 + (1 - \cos^2 \omega t)V_0 = const.,$$

and therefore

$$\omega^2 T_0 = V_0.$$

This we may write as

$$\lambda = \frac{V_0}{T_0} = \frac{\mathbf{x}^T\mathbf{K}\mathbf{x}}{\mathbf{x}^T\mathbf{M}\mathbf{x}}.$$

If the system is vibrating freely at frequency ω, then ω must be one of the natural frequencies and $\mathbf{x}$ the corresponding eigenvector. If $\omega = \omega_i$, then $\lambda = \lambda_i$, $\mathbf{x} = \mathbf{x}_i$ and

$$\lambda_i = \frac{\mathbf{x}_i^T\mathbf{K}\mathbf{x}_i}{\mathbf{x}_i^T\mathbf{M}\mathbf{x}_i} \tag{2.8.2}$$

which agrees with equation (1.4.5).

Rayleigh's Principle states that *the stationary values of the Rayleigh Quotient*

$$\lambda_R = \frac{\mathbf{x}^T\mathbf{K}\mathbf{x}}{\mathbf{x}^T\mathbf{M}\mathbf{x}} \tag{2.8.3}$$

viewed as a function of the components $(x_r)_1^n$, *occur when* $\mathbf{x}$ *is an eigenvector* $\mathbf{x}_i$. *The corresponding stationary value of* λ_R *is* λ_i.

Proof. Rayleigh's Principle has a long history - see for example Temple and Bickley (1933) [322] or Washizu (1982) [330]. We shall state the proof in a number of ways because each is instructive. First consider λ_R as a ratio of V_0 and T_0 and write down the partial derivative of this quotient w.r.t. x_r. We have

$$\frac{\partial T_0}{\partial x_r} = m_{r1}x_1 + m_{r2}x_2 + \cdots + m_{rn}x_n,$$

$$\frac{\partial V_0}{\partial x_r} = k_{r1}x_1 + k_{r2}x_2 + \cdots + k_{rn}x_n,$$

and

$$\frac{\partial}{\partial x_r}\left(\frac{V_0}{T_0}\right) = \frac{1}{T_0}\frac{\partial V_0}{\partial x_r} - \frac{V_0}{T_0^2}\frac{\partial T_0}{\partial x_r} = \frac{1}{T_0}\left\{\frac{\partial V_0}{\partial x_r} - \lambda_R\frac{\partial T_0}{\partial x_r}\right\},$$

so that, on inserting the expressions for $\partial V_0/\partial x_r$ and $\partial T_0/\partial x_r$ we obtain just the rth row of the matrix equation (2.6.3) with λ_R in place of λ. The complete set of n equations which state that V_0/T_0 is stationary w.r.t. all the $(x_r)_1^n$ is the matrix equation (2.6.3) which is satisfied when $\mathbf{x}$ is an eigenvector $\mathbf{x}_i$ and λ is the corresponding eigenvalue λ_i.

Now express the energies in terms of principal coordinates. If

$$p_i = \pi_i \sin(\omega t + \phi),$$

then equations (2.7.6), (2.7.9) show that

$$T_0 = \frac{1}{2}\left\{a_1\pi_1^2 + a_2\pi_2^2 + \cdots + a_n\pi_n^2\right\},$$

$$V_0 = \frac{1}{2}\left\{\lambda_1 a_1\pi_1^2 + \lambda_2 a_2\pi_2^2 + \cdots + \lambda_n a_n\pi_n^2\right\}.$$

Since $\mathbf{M}$ is assumed to be positive definite, there is no loss in generality in taking each $a_i = 1$, then

$$\lambda_R = \frac{\lambda_1\pi_1^2 + \lambda_2\pi_2^2 + \cdots + \lambda_n\pi_n^2}{\pi_1^2 + \pi_2^2 + \cdots + \pi_n^2}, \tag{2.8.4}$$

so that, in particular,

$$\lambda_R - \lambda_1 = \frac{(\lambda_2 - \lambda_1)\pi_2^2 + \cdots + (\lambda_n - \lambda_1)\pi_n^2}{\pi_1^2 + \pi_2^2 + \cdots + \pi_n^2}. \tag{2.8.5}$$

Since the λ_i are labelled in increasing (or non-decreasing) order, the quantities $\lambda_i - \lambda_1, i = 2, 3, \ldots, n$ are non-negative, and so

$$\lambda_R \geq \lambda_1.$$

If λ_1 is strictly less than λ_2 then equality occurs only when $\pi_2 = 0 = \ldots = \pi_n$, i.e., when the system is vibrating in its first principal mode. Equation (2.8.5) states the important property that whenever values are taken for $(x_r)_1^n$, the values of the Rayleigh quotient will always be greater than λ_1 and (when $\lambda_1 < \lambda_2$) will be equal to λ_1 only if the ratios $x_1 : x_2 : \ldots x_n$ correspond to those of the first eigenvector $x_{11}, x_{21}, \ldots x_{n1}$. Equation (2.8.5) shows that λ_1 is the *global minimum* of λ_R, and it may be proved in an exactly similar way that

$$\lambda_R \leq \lambda_n, \tag{2.8.6}$$

so that λ_n is the global *maximum* of λ_R.

If λ_i is an intermediate eigenvalue, so that $\lambda_1 < \lambda_i < \lambda_n$, then

$$\lambda_R - \lambda_i = \frac{-\sum_{j=1}^{i-1}(\lambda_i - \lambda_j)\pi_j^2 + \sum_{j=i+1}^{n}(\lambda_j - \lambda_i)\pi_j^2}{\pi_1^2 + \pi_2^2 + \cdots + \pi_n^2}. \tag{2.8.7}$$

In this case λ_R will not be strictly less nor strictly greater than λ_i for variations of the π_j; λ_R has a saddle point in the ith mode ($\pi_j = 0, j \neq i$). However, for computational purposes it is important that the difference between λ_R and λ_i depends on the **squares** of the quantities π_j. This means that if $\mathbf{x}$ is 'nearly' in the ith mode, so that the π_j with $j \neq i$ are much smaller than π_i, i.e., $\pi_i \approx 1, \pi_j = 0(\varepsilon)$, then $\lambda_R - \lambda_i = 0(\varepsilon^2)$.

Since $\mathbf{M}$ is positive definite, $\mathbf{x}^T\mathbf{M}\mathbf{x} > 0$, and the problem of finding the stationary values of the Rayleigh quotient λ_R given by equation (2.8.3) is equivalent to finding the stationary values of $\mathbf{x}^T\mathbf{K}\mathbf{x}$ subject to the restriction that $\mathbf{x}^T\mathbf{M}\mathbf{x} = 1$. This in turn is equivalent to finding the stationary values of

$$F \equiv \mathbf{x}^T\mathbf{K}\mathbf{x} - \lambda\mathbf{x}^T\mathbf{M}\mathbf{x}, \tag{2.8.8}$$

subject to $\mathbf{x}^T\mathbf{M}\mathbf{x} = 1$. Here λ acts as a Lagrange parameter. Note that

$$\frac{\partial F}{\partial x_r} = 2\sum_{s=1}^{n} k_{rs}x_s - 2\lambda\sum_{s=1}^{n} m_{rs}x_s,$$

so that the set of equations $\partial F/\partial x_r = 0$ yields equation (2.6.3), viz.

$$(\mathbf{K} - \lambda\mathbf{M})\mathbf{x} = \mathbf{0}. \quad \blacksquare$$

2.9 Vibration under constraint

The concept of a system vibrating under constraint is important in the solution of inverse problems. Suppose a system has generalised coordinates $(q_r)_1^n$, but they are constrained to satisfy a relation

$$f(q_1, q_2, \ldots, q_n) = 0.$$

For small vibrations about $q_1 = 0 = \ldots = q_n$, this relation may be replaced by

$$\mathbf{q}^T\mathbf{d} = d_1q_1 + d_2q_2 + \ldots + d_nq_n = 0,$$

where

$$d_r = \frac{\partial f}{\partial q_r}(q_1, q_2, \ldots, q_n)|_{q_1=0=q_2=\ldots=q_n}.$$

Two of the most important constraints will correspond to a certain q_r being zero, or two, q_r and q_s, being equal. Now suppose that the system is vibrating with frequency ω, where $\omega^2 = \lambda$, and

$$\mathbf{q} = \mathbf{x}\sin\omega t.$$

Rayleigh's Principle states that the (natural frequencies)^2 will be the stationary values of F, given in equation (2.8.8) but now subject to the further constraint

$$\mathbf{x}^T\mathbf{d} = 0. \tag{2.9.1}$$

Thus we must find the stationary values of

$$\mathbf{F} = \mathbf{x}^T\mathbf{K}\mathbf{x} - \lambda\mathbf{x}^T\mathbf{M}\mathbf{x} - 2\nu\mathbf{x}^T\mathbf{d}, \tag{2.9.2}$$

where ν is another Lagrange parameter (the 2 is inserted purely for convenience). The equations $\partial F/\partial x_r = 0$ now yield

$$\mathbf{Kx} - \lambda\mathbf{Mx} - \nu\mathbf{d} = \mathbf{0}. \tag{2.9.3}$$

By comparing this with equation (2.7.12) we see that $\nu\mathbf{d}$ is a generalised force; it is the force required to maintain the constraint (2.9.1).

In order to analyse equation (2.9.3) we express $\mathbf{x}$ in terms of principal coordinates, using equation (2.7.13). Then

$$\mathbf{KX}\boldsymbol{\pi} - \lambda\mathbf{MX}\boldsymbol{\pi} - \nu\mathbf{d} = \mathbf{0}. \tag{2.9.4}$$

Multiply throughout by $\mathbf{X}^T$ and use equations (2.7.5) and (2.7.8) which show that both $\mathbf{X}^T\mathbf{MX}$ and $\mathbf{X}^T\mathbf{KX}$ will be diagonal matrices; the rth row of the resulting equation is

$$\lambda_r a_r \pi_r - \lambda a_r \pi_r - \nu b_r = 0, \quad r = 1, 2, \ldots, n, \tag{2.9.5}$$

where

$$\mathbf{b} = \mathbf{X}^T\mathbf{d}. \tag{2.9.6}$$

Equations (2.9.5) yield

$$\pi_r = \frac{\nu b_r}{a_r(\lambda_r - \lambda)}, \tag{2.9.7}$$

which, when substituted in the constraint (2.9.1); i.e.,

$$\mathbf{x}^T\mathbf{d} \equiv \boldsymbol{\pi}^T\mathbf{X}^T\mathbf{d} \equiv \boldsymbol{\pi}^T\mathbf{b} = 0, \tag{2.9.8}$$

yields the frequency equation

$$B(\lambda) \equiv \sum_{i=1}^{n} \frac{b_i^2}{a_i(\lambda_i - \lambda)} = 0. \tag{2.9.9}$$

The form of this equation has important consequences. Consider first the case in which none of the b_i is zero. The coefficients $(b_i^2/a_i)_1^n$ will all be positive and the graph of $B(\lambda)$ against λ will have the form shown in Figure 2.9.1. Since $B(\lambda_i + 0)$ is very large negative, $B(\lambda_{i+1} - 0)$ is very large positive, and $B(\lambda)$ is steadily increasing between λ_i and $\lambda_{i+1}, B(\lambda)$ will have just $n-1$ zeros, $(\lambda_i')_i^{n-1}$, that interlace the λ_i in the sense that

$$\lambda_i < \lambda_i' < \lambda_{i+1}, (i = 1, 2, \ldots, n-1). \tag{2.9.10}$$

This inequality may be interpreted as follows: *if a linear constraint is applied to a system, each natural frequency increases (or, more precisely, does not decrease), but does not exceed the next natural frequency of the original system.*

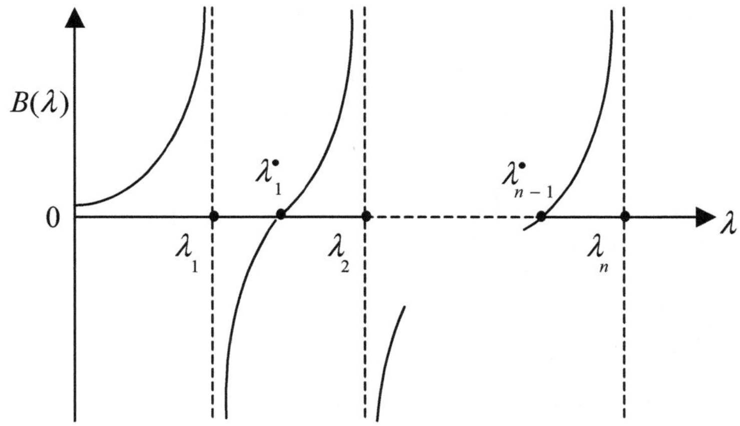

Figure 2.9.1 - The eigenvalues of a constrained system interlace the original eigenvalues

If all the b_i are non-zero then the inequalities in (2.9.10) are strictly obeyed. Now, however, suppose some of the b_i are zero; in particular consider the constraint

$$\pi_1 = 0, \tag{2.9.11}$$

for which $(b_i)_2^n = 0$. In this case $(\pi_i)_2^n$ are the principal coordinates of the system and the corresponding eigenvalues are

$$\lambda_i' = \lambda_{i+1}, (i = 1, 2, \ldots, n-1). \tag{2.9.12}$$

If the constraint is

$$\pi_j = 0, \; 1 < j \leq n,$$

then the principal coordinates are $\pi_1, \pi_2, \ldots, \pi_{j-1}, \pi_{j+1}, \ldots, \pi_n$, so that

$$\lambda_i' = \lambda_i, \quad i = 1, 2, \ldots, j-1; \lambda_i' = \lambda_{i+1}, \quad i = j, j+1, \ldots, n-1.$$

If the constraint is (2.9.8) and some particular b_j is zero, then equation (2.9.5) shows that

$$\pi_i = \delta_{ij} \equiv \begin{cases} 1 & i = j \\ 0 & i \neq j \end{cases}$$

is a solution corresponding to $\lambda = \lambda_j$. This means that a constraint (2.9.8) with $b_j = 0$ does not affect the jth principal mode. Figure 2.9.2 shows the form of $B(\lambda)$ when $b_2 = 0$. The graph may a) pass to the left of λ_2, in which case $\lambda_{1a}' < \lambda_2$, $\lambda_{2a}' = \lambda_2$; or b) pass to the right, in which case $\lambda_{1b}' = \lambda_2$, $\lambda_{2b}' > \lambda_2$.

If *two* constraints are applied, then the constrained system will have $n-2$ eigenvalues $(\lambda_i'')_1^{n-2}$ satisfying

$$\lambda_i' \leq \lambda_i'' \leq \lambda_{i+1}', \quad (i = 1, 2, \ldots, n-2),$$

where λ_i' are the eigenvalues of the system subject to one of the constraints. Thus

$$\lambda_i \le \lambda_i' \le \lambda_i'' \le \lambda_{i+1}' \le \lambda_{i+2}, \quad (i = 1, 2, \ldots, n-2)$$

or

$$\lambda_i \le \lambda_i'' \le \lambda_{i+2}, \quad (i = 1, 2, \ldots, n-2). \tag{2.9.13}$$

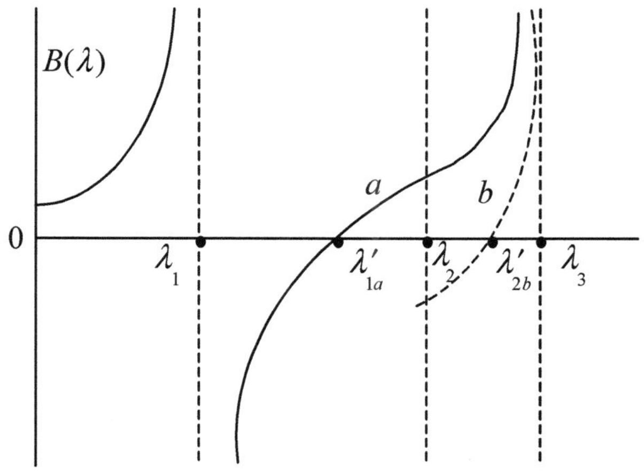

Figure 2.9.2 - The form of $B(\lambda)$ when $b_2 = 0$; either a) $\lambda_1' < \lambda_2, \lambda_2' = \lambda_2$ or b) $\lambda_1' = \lambda_2, \lambda_2' > \lambda_2$.

2.10 Iterative and independent definitions of eigenvalues

In this section we take a closer look at the eigenvalues of (2.6.3) in relation to the Rayleigh Quotient

$$\lambda_R = \frac{\mathbf{x}^T\mathbf{K}\mathbf{x}}{\mathbf{x}^T\mathbf{M}\mathbf{x}}. \tag{2.10.1}$$

We assume that $\mathbf{K}$ is symmetric (it may or may not be positive semi definite) and that $\mathbf{M}$ is positive definite. The importance of the latter assumptions is that the denominator of (2.10.1) is never zero and always positive for all $\mathbf{x} \neq \mathbf{0}$.

First, we note that λ_R is a homogeneous function of $\mathbf{x}$ in the sense that

$$\lambda_R(c\mathbf{x}) = \lambda_R(\mathbf{x}), \; c \neq 0.$$

This means that we can always scale any $\mathbf{x}$ so that the denominator of (2.10.1) is **unity**, i.e.,

$$\mathbf{x}^T\mathbf{M}\mathbf{x} = 1. \tag{2.10.2}$$

The vectors $\mathbf{x}$ with this property constitute a closed and bounded subspace $D_1 \subset V_n$. Now consider the Rayleigh Quotient on D_1; it is

$$\lambda_R = \mathbf{x}^T\mathbf{K}\mathbf{x}. \tag{2.10.3}$$

This is a continuous function of the variables $x_1, x_2, \ldots, x_n$ on the *closed* bounded region D_1 so that, by *Weierstrass' Theorem* on continuous functions, it attains its minimum value **on** D_1, i.e., for some vector $\mathbf{x} \in D_1$. (Recall the definition of a closed set S: if $\{y_i\}$ is a convergent sequence in S then its limit $\lim_{i\to\infty} y_i = y$ is also in S.) There may be more than one such minimizing vector, but there is always at least one, which we denote by $\mathbf{x}_1$. The corresponding minimum value of λ_R we denote by λ_1. We have the result

$$\lambda_1 = \min_{\mathbf{x}\in D_1} \mathbf{x}^T\mathbf{K}\mathbf{x} = \mathbf{x}_1^T\mathbf{K}\mathbf{x}_1. \tag{2.10.4}$$

Having found $\mathbf{x}_1$ and λ_1, we set up a new minimum problem: finding the minimum of $\mathbf{x}^T\mathbf{K}\mathbf{x}$ on the subspace D_2 of D_1 that consists of vectors $\mathbf{x}$ orthogonal to $\mathbf{x}_1$, i.e., $\mathbf{x}$ satisfying $\mathbf{x}^T\mathbf{M}\mathbf{x}_1 = 0$. This subspace is again closed and bounded so that by Weierstrass' Theorem there is a vector $\mathbf{x}_2 \in D_2$ which minimizes $\mathbf{x}^T\mathbf{K}\mathbf{x}$ on D_2; the minimum value is λ_2. We have

$$\lambda_2 = \min_{\mathbf{x}\in D_2} \mathbf{x}^T\mathbf{K}\mathbf{x} = \mathbf{x}_2^T\mathbf{K}\mathbf{x}_2, \tag{2.10.5}$$

and $\mathbf{x}_2^T\mathbf{M}\mathbf{x}_2 = 1$, $\mathbf{x}_2^T\mathbf{M}\mathbf{x}_1 = 0$. Since λ_2 is the minimum of $\mathbf{x}^T\mathbf{K}\mathbf{x}$ on D_2, a subspace of D_1, λ_2 cannot be less than λ_1, i.e., $\lambda_2 \geq \lambda_1$.

Proceeding in this way we find a set of vectors $\mathbf{x}_i$ and numbers $\lambda_i, (i = 1, 2, \ldots, n)$ such that

$$\lambda_i = \min_{\mathbf{x}\in D_i} \mathbf{x}^T\mathbf{K}\mathbf{x} = \mathbf{x}_i^T\mathbf{K}\mathbf{x}_i, \tag{2.10.6}$$

$$\mathbf{x}_i^T\mathbf{M}\mathbf{x}_j = \begin{cases} 1 & i = j, \\ 0 & i \neq j, \end{cases} \tag{2.10.7}$$

and $\lambda_1 \leq \lambda_2 \leq \ldots \leq \lambda_n$.

This procedure is **iterative**: we cannot set up the minimizing problem that gives λ_2 until we have found $\mathbf{x}_1$, and generally we cannot set up the minimizing problem that gives λ_i until we have found $\mathbf{x}_1, \mathbf{x}_2, \ldots, \mathbf{x}_{i-1}$. There is another procedure in which we can find any $\lambda_i, \mathbf{x}_i$ without first finding $\mathbf{x}_1, \mathbf{x}_2, \ldots, \mathbf{x}_{i-1}$; this is called the *independent* or *minimax* procedure.

In the independent procedure we start as before:

$$\lambda_1 = \min_{\mathbf{x}\in D_1} \mathbf{x}^T\mathbf{K}\mathbf{x} = \mathbf{x}_1^T\mathbf{K}\mathbf{x}_1.$$

Now we return to the analysis of Section 2.9 relating to vibration under a constraint. The inequality (2.9.10) shows that if none of the $(b_i)_1^n$ is zero, then the first constrained eigenvalue, λ_1^*, is strictly less than λ_2. Equations (2.9.11), (2.9.12) show that if the constraint is $\pi_1 = 0$, then $\lambda_1^* = \lambda_2$. The quantity π_1 is the amplitude of the component of $\mathbf{x}_1$ in $\mathbf{x}$, and on premultiplying equation (2.7.13) by $\mathbf{x}_1^T\mathbf{K}$ we see that

$$\mathbf{x}_1^T\mathbf{K}\mathbf{x} = \pi_1\mathbf{x}_1^T\mathbf{M}\mathbf{x}_1 = \pi_1. \tag{2.10.8}$$

Thus $\pi_1 = 0$ means that $\mathbf{x}$ is orthogonal to $\mathbf{x}_1$ w.r.t. the matrix $\mathbf{M}$; this is the constraint which yields the **maximum** value of λ_1^*, namely λ_2.

Thus

$$\max_{\mathbf{d}} \min_{\substack{\mathbf{x}\in D_1 \\ \mathbf{x}\perp\mathbf{d}}} \mathbf{x}^T\mathbf{K}\mathbf{x} = \lambda_2 \tag{2.10.9}$$

where $\mathbf{x} \perp \mathbf{d}$ means $\mathbf{x}^T\mathbf{M}\mathbf{d} = 0$; the $\mathbf{d}$ which maximizes the minimum is $\mathbf{x}_1$.

We may now extend this analyses to higher eigenvalues by using (2.9.13); thus

$$\max_{\mathbf{d}_1,\mathbf{d}_2} \min_{\substack{\mathbf{x}\in D_1 \\ \mathbf{x}\perp\mathbf{d}_1,\mathbf{d}_2}} \mathbf{x}^T\mathbf{K}\mathbf{x} = \lambda_3,$$

and generally

$$\max_{\mathbf{d}_1,\mathbf{d}_2,\ldots,\mathbf{d}_i} \min_{\substack{\mathbf{x}\in D_1 \\ \mathbf{x}\perp\mathbf{d}_1,\mathbf{d}_2,\ldots,\mathbf{d}_i}} \mathbf{x}^T\mathbf{K}\mathbf{x} = \lambda_{i+1}. \tag{2.10.10}$$

Again, the $\mathbf{d}$'s that maximize the minimum in the general case are $\mathbf{d}_1 = \mathbf{x}_1, \mathbf{d}_2 = \mathbf{x}_2, \ldots, \mathbf{d}_{n-1} = \mathbf{x}_{n-1}$.

The minimax definition of eigenvalues seems to have been noted first by Fischer (1905) [88]. The iterative and independent definitions of eigenvalues are discussed at length in Courant and Hilbert (1953) [64], and in the more specialised volume Gould (1966) [151]. The motivation for Gould's book was the search for *lower* bounds for eigenvalues; discretising methods like the finite element method almost always lead to *upper* bounds.

Exercises 2.10

1. Examine the arguments in Sections 2.9, 2.10 in the case when two eigenvalues are equal, e.g., $\lambda_1 = \lambda_2$.

2. Use the minimax procedure to show that if stiffness is added to a system, i.e., the stiffness matrix is changed from $\mathbf{K}$ to $\mathbf{K}'$, and $\mathbf{x}^T\mathbf{K}'\mathbf{x} \geq \mathbf{x}^T\mathbf{K}\mathbf{x}$ for all $\mathbf{x} \in V_n$, then none of the eigenvalues of the system decreases. Why can you prove this result only for λ_1 by using the iterative definition?

Chapter 3

Jacobi Matrices

Let no one say that I have said nothing new; the arrangement of the subject is new.

Pascal's *Pensées*, 22

3.1 Sturm sequences

In this Chapter we will analyse the properties of the eigenvalues and eigenvectors of systems with the special *tridiagonal* mass and stiffness matrices met in Chapter 2. We will start by considering systems like that for the system in Figure 2.2.1, for which the mass matrix is diagonal and the stiffness matrix is tridiagonal, with negative codiagonal. At the end of the section we will show that many of the results may be generalised to apply to systems like that in (2.4.10) in which the mass matrix is tridiagonal with positive codiagonal. The most important property of the eigenvalues of such systems is that they are simple, i.e., distinct (Theorem 3.1.3). Thus

$$\lambda_1 < \lambda_2 < \ldots < \lambda_n.$$

If $\mathbf{x}_r$ is the rth eigenvector, then as r increases, the eigenvectors oscillate more and more (Theorem 3.3.1) in such a way that the zeros of $\mathbf{x}_r$ interlace those of the neighbouring $\mathbf{x}_{r-1}$ and $\mathbf{x}_{r+1}$ (3.3.4). We shall now establish these and other results. Throughout the next few Chapters, we redevelop analysis originally established by Gantmacher and Krein (1950) [98]. Their book was republished in 2002..

We start with a definition:

Definition 3.1.1 *A* ***Jacobi*** *matrix is a positive semi-definite symmetric tridiagonal matrix with (strictly) negative codiagonal.*

Note: Different authors define a Jacobi matrix in different ways; some choose the codiagonal to be strictly positive.

Now we consider the equation

$$(\mathbf{K} - \lambda\mathbf{M})\mathbf{x} = \mathbf{0} \tag{3.1.1}$$

where $\mathbf{K}$ is a Jacobi matrix. First, we suppose that $\mathbf{M}$ is a (strictly) positive diagonal matrix, as in (2.2.7), and we reduce (3.1.1) to standard form.

Take

$$\mathbf{M} = diag(m_1, m_2, \ldots, m_n)$$

and write $\mathbf{M} = \mathbf{D}^2$, where

$$\mathbf{D} = diag(d_1, d_2, \ldots, d_n), \; d_i = m_i^{\frac{1}{2}},$$

introduce the vector $\mathbf{u}$ related to $\mathbf{x}$ by

$$\mathbf{u} = \mathbf{D}\mathbf{x}, \quad \mathbf{x} = \mathbf{D}^{-1}\mathbf{u}$$

and premultiply (3.1.1) by $\mathbf{D}^{-1}$ to obtain

$$\mathbf{D}^{-1}(\mathbf{K} - \lambda\mathbf{D}^2)\mathbf{D}^{-1}\mathbf{u} = \mathbf{0},$$

i.e.,

$$(\mathbf{J} - \lambda\mathbf{I})\mathbf{u} = \mathbf{0}, \tag{3.1.2}$$

where

$$\mathbf{J} = \mathbf{D}^{-1}\mathbf{K}\mathbf{D}^{-1}. \tag{3.1.3}$$

The matrix $\mathbf{J}$, like $\mathbf{K}$, is a Jacobi matrix, and has the same eigenvalues as the system (3.1.1). We write

$$\mathbf{J} = \begin{bmatrix} a_1 & -b_1 & 0 & \cdots & 0 \\ -b_1 & a_2 & -b_2 & \cdots & 0 \\ \ddots & \ddots & \ddots & \ddots & \\ & & \ddots & \ddots & -b_{n-1} \\ & & & -b_{n-1} & a_n \end{bmatrix}. \tag{3.1.4}$$

The analysis now centres on the leading *principal minors* (see (1.4.6)) of the matrix $\mathbf{J} - \lambda\mathbf{I}$. We define

$$P_0 = 1, \; P_1(\lambda) = a_1 - \lambda, \; P_2(\lambda) = \begin{vmatrix} a_1 - \lambda & -b_1 \\ -b_1 & a_2 - \lambda \end{vmatrix}, \text{etc.} \tag{3.1.5}$$

so that finally

$$P_n(\lambda) = \det(\mathbf{J} - \lambda\mathbf{I}). \tag{3.1.6}$$

The minors satisfy the three-term recurrence relation

$$P_{r+1}(\lambda) = (a_{r+1} - \lambda)P_r(\lambda) - b_r^2 P_{r-1}(\lambda), \; r = 1, 2, \ldots, n-1, \tag{3.1.7}$$

which enables us to calculate $P_2, P_3, \ldots, P_n$ successively from P_0, P_1. Since the zeros of any $P_r(\lambda)$ are the eigenvalues of the truncated symmetric matrix obtained by retaining just the first r rows and columns of $\mathbf{J}$, they are all real.

We now prove:

Theorem 3.1.1 *If $b_r^2 > 0$ $(r = 1, 2, \ldots, n-1)$, then the $(P_r(\lambda))_0^n$ form a* ***Sturm*** *sequence, the defining properties of which are*

1. $P_0(\lambda)$ has everywhere the same sign ($P_0(\lambda) \equiv 1$).
2. When $P_r(\lambda)$ vanishes, $P_{r+1}(\lambda)$ and $P_{r-1}(\lambda)$ are non-zero and have opposite signs.

Proof. In order to establish property 2 we note first that two successive P_r cannot be simultaneously zero - i.e., for the same $\lambda = \lambda^0$. For if $P_{s+1}(\lambda^0) = 0 = P_s(\lambda^0)$ then equation (3.1.7) shows that $P_{s-1}(\lambda^0) = 0$, so that finally we must have P_1 and P_0 zero; but $P_0(\lambda^0) = 1$, which yields a contradiction.

The latter part of property 2 now follows directly from (3.1.7). ∎

Before proceeding further we must define the *sign change function* $s_r(\lambda)$. This is the integer-valued function equal to the cumulative number of sign-changes in the sequence $P_0, P_1(\lambda), P_2(\lambda), \ldots, P_r(\lambda)$. Thus if

$$\mathbf{J} = \begin{bmatrix} 2 & -1 & 0 \\ -1 & 3 & -2 \\ 0 & -2 & 4 \end{bmatrix},$$

then,

$$\begin{aligned} P_0 &= 1, \; P_1(\lambda) = -\lambda + 2, \\ P_2(\lambda) &= \lambda^2 - 5\lambda + 5, \; P_3(\lambda) = -\lambda^3 + 9\lambda^2 - 21\lambda + 12. \end{aligned}$$

For $\lambda = 0$ the sequence of values is 1, 2, 5, 12. Since there is no change of sign in the sequence, each $s_r(0) = 0$. For $\lambda = 3$ the sequence is 1, -1, -1, 3, so that $s_1(3) = s_2(3) = 1, \; s_3(3) = 2$.

Theorem 3.1.2 *$s_r(\lambda)$ changes only when λ passes through a zero of the last polynomial, $P_r(\lambda)$.*

Proof. Clearly, $s_r(\lambda)$ can change only when λ passes through a zero of one of the $P_s(\lambda)$, $(s \le r)$; it therefore suffices to prove that $s_r(\lambda)$ does not change at all when λ passes through a zero of an intermediate $P_s(\lambda)$, $(s < r)$. Suppose $P_s(\lambda^0) = 0$, where $1 \le s < r$, then $P_{s-1}(\lambda^0)$ and $P_{s+1}(\lambda^0)$ will be both non-zero and have opposite signs. The signs of the triad $P_{s-1}(\lambda^0)$, $P_s(\lambda^0)$, $P_{s+1}(\lambda^0)$ are therefore +0 - or - 0+. Suppose the first to be the case, so that $P_s(\lambda)$ increases as λ passes through λ^0, (the other possibility may be handled similarly). Then for values of λ sufficiently close to λ^0 and less than λ^0 the signs are + - -, while for values of λ sufficiently close and greater than λ^0 the signs are + + -. Thus, whether λ is greater than or less than λ^0 there is just one change of sign in the triad of values of $P_{s-1}(\lambda)$, $P_s(\lambda)$, $P_{s+1}(\lambda)$. In other words the triad of polynomials $P_{s-1}(\lambda)$, $P_s(\lambda)$, $P_{s+1}(\lambda)$ will not contribute any change to $s_r(\lambda)$ as λ passes through λ^0. But no other members of the sequence will contribute any change to $s_r(\lambda)$ as λ passes through λ^0 (unless λ^0 is a zero of another $P_t(\lambda)$, $|t - s| \ge 2$, in which case again there will be no change in $s_r(\lambda)$) so that $s_r(\lambda)$ will not change at all. ∎

Clearly, $s_r(\lambda)$ is not well defined when $P_r(\lambda) = 0$.

Theorem 3.1.3 *The zeros of $P_r(\lambda)$, are simple, i.e., distinct. In addition, if $P_r(\lambda^0) \neq 0$ and $s_r(\lambda^0) = k$, then $P_r(\lambda)$ has k zeros less than λ^0.*

Proof. Since $P_s(\lambda) = (-)^s\lambda^s + \cdots$, all $P_s(\lambda)$ will be positive for sufficiently large negative λ, i.e., $\lambda \leq \alpha$, so that $s_r(\alpha) = 0$: α may be taken to be zero if $\mathbf{J}$ is positive definite. On the other hand, for sufficiently large positive λ, i.e., $\lambda \geq \beta$, the $P_s(\lambda)$ will alternate in sign, so that $s_r(\beta) = r$. Now since $s_r(\lambda)$ can increase only when λ passes through a zero of $P_r(\lambda)$, all the zeros of $P_r(\lambda)$ must be distinct. For if λ^0 were a zero of even multiplicity then $s_r(\lambda)$ would not increase at all as λ passed through λ^0, while $s_r(\lambda)$ would increase only by unity if λ^0 were a zero of odd multiplicity. The second part of the theorem now follows immediately. ■

Corollary 3.1.1 *The eigenvalues of a Jacobi matrix are **distinct**.*

Corollary 3.1.2 *The number of zeros of $P_r(\lambda)$ satisfying $\alpha < \lambda < \beta$ is equal to $s_r(\beta) - s_r(\alpha)$.*

Corollary 3.1.3 *If λ^0 is a zero of $P_r(\lambda)$ then, as λ passes from λ^0- to λ^0+ the sign of $P_{r-1}(\lambda)P_r(\lambda)$ changes from + to -, and $s_r(\lambda)$ increases by unity.*

Theorem 3.1.4 *Between any two neighbouring zeros of $P_r(\lambda)$ there lies one and only one zero of $P_{r-1}(\lambda)$, and one and only one zero of $P_{r+1}(\lambda)$.*

Proof. Let μ_1, μ_2 be the two neighbouring zeros. Suppose, for the sake of argument that $P_r(\mu_1-) > 0$, then $P_r(\mu_1+) < 0$ and $P_r(\mu_2-) < 0$. By Corollary 3.1.3, $P_{r-1}(\mu_1+) > 0$ and $P_{r-1}(\mu_2-) < 0$, so that $P_{r-1}(\lambda)$ changes sign between μ_1+ and μ_2-, and therefore has at least one zero in (μ_1, μ_2).

Now property 2 of Sturm sequences shows that $P_{r+1}(\mu_i)$ and $P_{r-1}(\mu_i)$, $(i = 1, 2)$ have opposite signs. Thus $P_{r+1}(\mu_1+) < 0$, $P_{r+1}(\mu_2-) > 0$ so that $P_{r+1}(\lambda)$ has *at least one* zero in (μ_1, μ_2). Now suppose, if possible, that $P_{r-1}(\lambda)$ (or $P_{r+1}(\lambda)$) had two (or more) zeros in (μ_1, μ_2) then $P_r(\lambda)$ would have a zero in (μ_1, μ_2), contrary to the hypothesis that μ_1, μ_2 are neighbouring zeros. ■

This theorem is usually stated in the form: *the eigenvalues of successive principal minors interlace each other.*

3.2 Orthogonal polynomials

There is an intimate connection between Jacobi matrices and orthogonal polynomials. In this section we outline some of the basic properties of orthogonal polynomials.

Two polynomials $p(x)$, $q(x)$ are said to be *orthogonal* w.r.t. the *weight function* $w(x) > 0$ over an interval (a, b) if

$$(p, q) \equiv \int_a^b w(x)p(x)q(x)dx = 0. \tag{3.2.1}$$

A familiar example is provided by the Laguerre polynomials $(L_n(x))_0^\infty$, i.e.,

$$L_0(x) = 1,\ L_1(x) = x - 1,\ L_2(x) = x^2 - 4x + 2, \ldots$$

which are orthogonal w.r.t. the weight function e^{-x} over $(0, \infty)$, i.e.,

$$\int_0^\infty e^{-x} L_n(x) L_m(x) dx = 0,\ m \neq n.$$

One of the important properties of such polynomials is that they satisfy a *three-term recurrence relation.* The relation for the $L_r(x)$, for example, is

$$L_{r+1}(x) = (x - 2r - 1)L_r(x) - r^2 L_{r-1}(x).$$

In this section we shall be concerned, not with a *continuous* orthogonality relation of the form (3.2.1), but with a *discrete* orthogonality relation

$$(p, q) \equiv \sum_{i=1}^{n} w_i p(\xi_i) q(\xi_i) = 0;\ (w_i)_1^n > 0. \tag{3.2.2}$$

where $(\xi_i)_1^n$ are n points, satisfying $\xi_1 < \xi_2 < \cdots < \xi_n$.

To introduce the concept formally we let $\mathbf{P}_n$ denote the linear space of polynomials of order n, i.e., the set of all polynomials $p(x)$ with degree $k \leq n$, with real coefficients. On this space (,) acts as an *inner product* since it is *positive definite*, *bilinear* and *symmetric*, i.e.,

1. $(p, p) \equiv ||p||^2 > 0$ if $p(x) \neq 0$
2. $(\alpha p, q) = \alpha(p, q),\ (p + q, r) = (p, r) + (q, r)$
3. $(p, q) = (q, p)$

In addition

4. $(xp, q) = (p, xq)$

We now prove

Theorem 3.2.1 *There is a unique sequence of monic polynomials, i.e., $(q_i(x))_0^n$ such that $q_i(x)$ has degree i and leading coefficient (of x^i) unity, which are orthogonal with respect to the inner product (,), i.e., for which*

$$(q_i, q_j) = 0,\ i \neq j.$$

Proof. The $q_i(x)$ may be constructed by applying the familiar Gram-Schmidt orthogonalisation procedure to the linearly independent polynomials $(x^i)_0^{n-1}$. Thus

$$q_0 = 1,\ q_i(x) = x^i - \sum_{j=0}^{i-1} \alpha_{ij} q_j(x),\ (i = 1, 2, \ldots, n-1),$$

where

$$(q_i, q_j) = (x^i, q_j) - \alpha_{ij}(q_j, q_j) = 0$$

so that

$$\alpha_{ij} = (x^i, q_j)/||q_j||^2, \; j = 0, 1, \ldots, i-1. \; \blacksquare$$

We note that the polynomial

$$q_n(x) = \prod_{i=1}^{n}(x - \xi_i)$$

is the monic polynomial of degree n in the sequence. It is orthogonal to $(q_i)_0^{n-1}$; in fact it is orthogonal to all functions, since $q_n(\xi_i) = 0, \; i = 1, 2, \ldots, n$.

The Gram-Schmidt procedure does not provide a computationally convenient means for computing the q_i; instead we use Forsythe (1957) [90].

Theorem 3.2.2 *The monic polynomials $(q_i)_0^n$ satisfy a three-term recurrence relation of the form*

$$q_i(x) = (x - \alpha_i)q_{i-1}(x) - \beta_{i-1}^2 q_{i-2}(x), \; i = 1, 2, \ldots, n, \tag{3.2.3}$$

with the initial values

$$q_{-1}(x) = 0, \; q_0(x) = 1. \tag{3.2.4}$$

Proof. $q_i(x) - xq_{i-1}(x)$ is a polynomial of degree $(i-1)$. It may therefore be expressed in terms of (the linearly independent - see Ex. 3.2.1) $q_0, q_1, \ldots, q_{i-1}$. Thus

$$q_i(x) - xq_{i-1}(x) = c_0 q_0 + c_1 q_1 + \cdots + c_{i-1}q_{i-1}. \tag{3.2.5}$$

Take the inner product of this equation with $q_j(x)$, $(j = 0, 1, \ldots, i-1)$ thus

$$(q_i, q_j) - (q_{i-1}, xq_j) = \sum_{k=0}^{i-1} c_k(q_k, q_j) = c_j||q_j||^2, \tag{3.2.6}$$

where the second term on the left has been rewritten by using property 4, above. But if $j = 0, 1, \ldots, i-1$, then the first term on the left is zero, and if $j = 0, 1, \ldots, i-3$, then xq_j has degree at most $i-2$ and so is orthogonal to q_{i-1}. Thus $c_j = 0$ if $j = 0, 1, 2, \ldots, i-3$ and there only *two* terms c_{i-1} and c_{i-2} on the right of (3.2.5) i.e.,

$$q_i(x) - xq_{i-1}(x) = c_{i-2}q_{i-2}(x) + c_{i-1}q_{i-1}(x). \tag{3.2.7}$$

Moreover equation (3.2.6) gives

$$\alpha_i = -c_{i-1} = (q_{i-1}, xq_{i-1})/||q_{i-1}||^2 \tag{3.2.8}$$

$$c_{i-2} = -(q_{i-1}, xq_{i-2})/||q_{i-2}||^2.$$

But xq_{i-2} is a monic polynomial of degree $i-1$; it may therefore be expressed in the form

$$xq_{i-2}(x) = q_{i-1}(x) + \sum_{j=0}^{i-2} d_j q_j(x)$$

so that

$$(q_{i-1}, xq_{i-2}) = ||q_{i-1}||^2$$

and thus c_{i-2} is negative and equal to $-\beta_{i-1}^2$, where

$$\beta_i = ||q_i||/||q_{i-1}||. \quad \blacksquare \tag{3.2.9}$$

Equations (3.2.3), (3.2.4) with (3.2.8), (3.2.9) enable us to compute the polynomials $\{q_i\}_1^{n-1}$ step by step. Thus with $q_{-1} = 0$, $q_0 = 1$ we first compute α_1 from (3.2.8); this substituted into (3.2.3) gives q_1. Now we compute α_2, β_1 and find q_2, etc.

In *inverse* problems we will need to express the weights w_i in terms of the polynomials q_{n-1} and q_n. For this we note that if $f(x)$ is any polynomial in $\mathbf{P}_{n-1}$, i.e., of degree $n-2$ or less, then

$$(q_{n-1}, f) \equiv \sum_{i=1}^{n} w_i q_{n-1}(\xi_i) f(\xi_i) = 0.$$

But if such a combination

$$\sum_{i=1}^{n} m_i f(\xi_i), \quad m_i = w_i q_{n-1}(\xi_i)$$

is zero for any $f(x)$ in $\mathbf{P}_{n-1}$ then

$$\sum_{i=1}^{n} m_i \xi_i^k = 0, \ (k = 0, 1, \ldots, n-2),$$

since each x^k is in $\mathbf{P}_{n-1}$, i.e.,

$$\mathbf{Bm} = \begin{bmatrix} 1 & 1 & 1 & \cdots & 1 \\ \xi_1 & \xi_2 & \xi_3 & \cdots & \xi_n \\ \xi_1^2 & \xi_2^2 & \xi_3^2 & \cdots & \xi_n^2 \\ \cdot & \cdot & \cdot & \cdot & \cdot \\ \xi_1^{n-2} & \xi_2^{n-2} & \xi_3^{n-2} & \cdots & \xi_n^{n-2} \end{bmatrix} \begin{bmatrix} m_1 \\ m_2 \\ m_3 \\ \cdots \\ \cdot \\ m_n \end{bmatrix} = \mathbf{0}. \tag{3.2.10}$$

It is shown in Ex. 3.2.2 that this equation has the solution

$$m_i = \gamma / \prod_{j=1}^{n}{}'(\xi_i - \xi_j). \tag{3.2.11}$$

Apart from the arbitrariness of the factor γ, this is the unique solution. The prime means that the term $j = i$ is omitted. Now since

$$q_n(\xi) = \prod_{j=1}^{n} (\xi - \xi_j), \tag{3.2.12}$$

we have

$$q_n'(\xi_i) = \prod_{j=1}^{n}{}' (\xi_i - \xi_j),$$

where the prime on the left denotes differentiation!

Returning to equation (3.2.11) we can deduce that, for some γ,

$$m_i \equiv w_i q_{n-1}(\xi_i) = \gamma / q_n'(\xi_i), \; i = 1, 2, \ldots, n.$$

Since the $\{q_i\}_0^n$ satisfy the three-term recurrence relation (3.2.3) it follows, by the arguments used in Section 3.1, that the zeros of $q_n(x)$ and $q_{n-1}(x)$ must interlace and therefore (Ex. 3.2.3) $q_{n-1}(\xi_i) q_n'(\xi_i) > 0$. This means that the weights

$$w_i = \gamma / \{q_{n-1}(\xi_i) q_n'(\xi_i)\} \tag{3.2.13}$$

are *positive*.

This equation is important: *it means that if the monic polynomials* $q_n(\xi)$, $q_{n-1}(\xi)$ *are given, and if their zeros interlace, then they may be viewed as the nth and* $(n-1)$*th members, respectively, of a sequence of monic polynomials orthogonal w.r.t. the weights* w_i *given by (3.2.13), and the points* $\{\xi_i\}_1^n$.

Exercises 3.2

1. Show that if the polynomials $\{q_i\}_0^k$, $k < n$, are orthogonal w.r.t. the inner-product (3.2.2), then they are linearly independent. Hence deduce that any polynomial $p(x)$ of degree $k-1$ may be expressed uniquely in the form

$$p(x) = \sum_{j=0}^{k-1} c_j q_j(x),$$

and that $q_k(x)$ is orthogonal to each polynomial of degree $k-1$.

2. Show that if the Vandermonde determinant $\triangle$ is defined by

$$\triangle = \begin{vmatrix} 1 & 1 & \cdots & 1 \\ \xi_1 & \xi_2 & \cdots & \xi_{n-1} \\ \xi_1^2 & \xi_2^2 & \cdots & \xi_{n-1}^2 \\ \cdots & \cdots & \cdots & \cdots \\ \xi_1^{n-2} & \xi_2^{n-2} & \cdots & \xi_{n-1}^{n-2} \end{vmatrix},$$

then

$$\triangle = \prod_{j=2}^{n-1} \prod_{k=1}^{j-1} (\xi_j - \xi_k) = \Gamma / \prod_{j=1}^{n-1} (\xi_n - \xi_j),$$

where

$$\Gamma = \prod_{j=2}^{n} \prod_{k=1}^{j-1} (\xi_j - \xi_k).$$

Hence deduce (3.2.11).

3. The zeros $\{\xi_i\}_1^n$ of $q_n(x)$, and $\{\xi_i^*\}_1^{n-1}$ of $q_{n-1}(x)$ must satisfy $\xi_1 < \xi_1^* < \xi_2 < \cdots < \xi_{n-1}^* < \xi_n$. Show that $(-)^{n-i} q_n'(\xi_i) > 0$, $(-)^{n-i} q_{n-1}(\xi_i) > 0$ and hence $q_n'(\xi_i) q_{n-1}(\xi_i) > 0$.

3.3 Eigenvectors of Jacobi matrices

In this section we establish some properties of the eigenvectors of Jacobi matrices, in preparation for the solution of 'inverse mode problems'. We return to the analysis of Section 3.1 and prove

Theorem 3.3.1 *The sequence $(u_{r,j})_{r=1}^n$ for the jth eigenvector has exactly $j-1$ sign reversals.*

Proof. The $u_{r,j}$ are determined from equation (3.1.2) for $\lambda = \lambda_j$; this may be written

$$-b_{r-1} u_{r-1,j} + (a_r - \lambda_j) u_{r,j} - b_r u_{r+1,j} = 0, \ (r = 1, 2, \ldots, n) \tag{3.3.1}$$

where $u_{0,j}, u_{n+1,j}$ are interpreted as zero, i.e.,

$$u_{0,j} = 0 = u_{n+1,j}. \tag{3.3.2}$$

Choose an arbitrary $b_n > 0$ and put

$$v_1 = u_{1,j}, \ v_2 = b_1 u_{2,j}, \ldots, v_{n+1} = b_1 b_2 \cdots b_n u_{n+1,j}$$

and multiply equation (3.3.1) by $b_1 b_2 \cdots b_{r-1}$ to obtain

$$-b_{r-1}^2 v_{r-1} + (a_r - \lambda_j) v_r - v_{r+1} = 0, \ (r = 1, 2, \ldots, n). \tag{3.3.3}$$

On comparing this equation with (3.1.7) we see that it has the solution

$$v_0 = 0, \ v_1 = 1, \ v_r = P_{r-1}(\lambda_j), \ (r = 1, 2, \ldots, n+1)$$

which because of $P_n(\lambda_j) = 0$, satisfies the end-condition $v_{n+1} = 0$.
Thus,

$$u_{r,j} = (b_1 b_2 \cdots b_{r-1})^{-1} P_{r-1}(\lambda_j), \tag{3.3.4}$$

and since λ_j lies between the $(j-1)$th and jth zeros of $P_{n-1}(\lambda)$, $s_{n-1}(\lambda_j) = j-1$. ∎

Before establishing further properties of the eigenvectors we introduce the concept of a u-line.

Definition 3.3.1 *Let* $\mathbf{u} = \{u_1, u_2, \ldots, u_{n+1}\}$ *be a vector. We shall define the **u-line** as the broken line in the plane joining the points with coordinates*

$$x_r = r, \; y_r = u_r, \; (r = 1, 2, \ldots, n+1).$$

Thus, between (x_r, y_r) and (x_{r+1}, y_{r+1}), $y(x)$ is defined by

$$y(x) = (r + 1 - x)u_r + (x - r)u_{r+1}, \; (r = 1, 2, \ldots, n),$$

as shown in Figure 3.3.1.

Now return to Theorem 3.3.1. For arbitrary (real) λ the sequence given by

$$u_0 = 0, \; u_r(\lambda) = u_1(b_1 b_2 \cdots b_{r-1})^{-1} P_{r-1}(\lambda), \; (r = 1, 2, \ldots, n+1),$$

satisfies the recurrence (3.3.1) for $r = 1, 2, \ldots, n$. (It will satisfy the last equation with $u_{n+1} = 0$ iff $P_n(\lambda) = 0$.) For arbitrary λ, the vector $\mathbf{u}(\lambda) = \{u_1(\lambda), \ldots, u_{n+1}(\lambda)\}$ defines a $u(\lambda)-$ line. We now investigate the *nodes* of this line, i.e., the points x at which $y(x) = 0$. First we note that if $u_r(\lambda) = 0$, i.e., $P_{r-1}(\lambda) = 0$, then $P_r(\lambda)$ and $P_{r-2}(\lambda)$, i.e., u_{r+1} and u_{r-1}, will have opposite signs, so that the $u(\lambda)-$ line will cross the x-axis at $x = r$. Secondly, if u_r and u_{r+1} have opposite signs, then $y(x)$ has a node between r and $r + 1$. This implies that the $u(\lambda_j)-$ line has exactly j nodes, *excluding* the left hand end where $u_0 = 0$, but *including* the right hand end. Moreover, if $\lambda_j \leq \lambda < \lambda_{j+1}$, then the $u(\lambda)-$ line will have exactly j nodes, again excluding the left hand end where $u_0 = 0$. Table 3.3.1 shows the signs of u_r for the whole range of λ-values, for the case $n = 3$. The last line in the table shows the number of nodes in the $u(\lambda)$. Figure 3.3.1 shows the form of the $u(\lambda)$ for the starred values of λ. We now establish an identity which will enable us to prove further results concerning the eigenvectors.

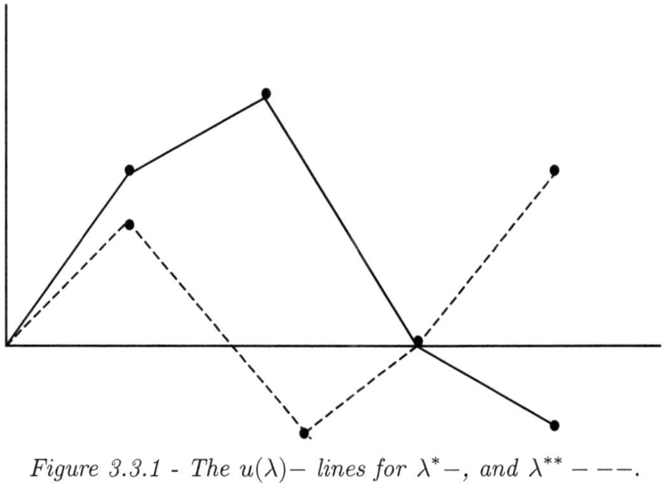

Figure 3.3.1 - The $u(\lambda)-$ *lines for* λ^*-, *and* λ^{**} $---$.

Table 3.3.1 -The signs of u_r for different values of λ

λ	0	λ_1	λ^*	λ	λ_2	λ^{**}	λ_3
u_1	+	+	+	+	+	+	+
u_2	+	+	+	0	-	-	-
u_3	+	+	0	-	-	0	+
u_4	+	0	-	-	0	+	0
	0	1	1	1	2	2	3

Consider the solutions $\mathbf{u}, \mathbf{v}$ of the equations (3.3.1) corresponding to λ, μ respectively. Suppose that $u_0 = 0 = v_0$ and that some positive value has been assigned to b_n. Then

$$\begin{aligned} -b_{r-1}u_{r-1} + a_r u_r - b_r u_{r+1} &= \lambda u_r, (r = 1, 2, \ldots, n) \\ -b_{r-1}v_{r-1} + a_r v_r - b_r v_{r+1} &= \mu v_r, (r = 1, 2, \ldots, n). \end{aligned}$$

Eliminating a_r from these equations, we find

$$b_r t_r - b_{r-1} t_{r-1} = (\mu - \lambda)(s_r - s_{r-1}) \tag{3.3.5}$$

where

$$t_r = u_{r+1} v_r - u_r v_{r+1}, \quad s_r = \sum_{i=1}^{r} u_i v_i \tag{3.3.6}$$

so that on summing over $r = p, p+1, \ldots, q$ $(1 \leq p \leq q \leq n)$, we obtain

$$b_q t_q - b_{p-1} t_{p-1} = (\mu - \lambda)(s_q - s_{p-1}). \tag{3.3.7}$$

In particular, if $p = 1$, so that $u_0 = 0 = v_0 = t_0 = s_0$,

$$b_q t_q = (\mu - \lambda) s_q. \tag{3.3.8}$$

We now prove

Theorem 3.3.2 *If $\lambda < \mu$, then between any two nodes of the $u(\lambda)$ - line there is at least one node of the $u(\mu)$ - line.*

Proof. Let $\alpha, \beta (\alpha < \beta)$ be two neighbouring nodes of the u-line and suppose that

$$p - 1 \leq \alpha < p, \quad q < \beta \leq q + 1, \quad (p \leq q),$$

so that

$$y(\alpha, \lambda) \equiv (p - \alpha) u_{p-1} + (\alpha - p + 1) u_p = 0, \tag{3.3.9}$$

$$y(\beta, \lambda) \equiv (q + 1 - \beta) u_q + (\beta - q) u_{q+1} = 0 \tag{3.3.10}$$

and $y(x, \lambda) \neq 0$ for $\alpha < x < \beta$. For the sake of definiteness suppose that $y(x, \lambda) > 0$ for $\alpha < x < \beta$, then $u_p, u_{p+1}, \ldots, u_q$ are all positive. We now need to prove that $y(x, \mu)$ has a zero between α and β. Suppose $y(x, \mu)$ has no such zero, that is, it has the same sign for $\alpha < x < \beta$. Without loss of generality we can assume that

$$y(x, \mu) > 0 \text{ for } \alpha < x < \beta.$$

that is $y(\alpha, \mu) \geq 0, y(\beta, \mu) \geq 0$ and $v_p, v_{p+1}, \ldots, v_q$ are all positive. Thus,

$$(p - \alpha)v_{p-1} + (\alpha - p + 1)v_p \geq 0, \tag{3.3.11}$$

$$(q + 1 - \beta)v_q + (\beta - q)v_{q+1} \geq 0, \tag{3.3.12}$$

and on eliminating α between (3.3.9), (3.3.11), and β between (3.3.10), (3.3.12) we deduce that $t_{p-1} \geq 0$, $t_q \leq 0$. On the other hand, $s_q - s_{p-1} = \sum_{i=p}^{q} u_i v_i > 0$, so that the LHS of (3.3.7) is non-positive, while the RHS is positive, providing a contradiction. If we had assumed $y(x, \mu) < 0$ for $\alpha < x < \beta$, the we would have found the LHS of (3.3.5) non-negative and the RHS negative. ∎

Theorem 3.3.3 *As λ increases continuously, then the nodes of the $u(\lambda)$ - line shift continuously to the left.*

Proof. Let $\alpha_1(\lambda), \alpha_2(\lambda), \ldots$ be the nodes of the $u(\lambda)$ - line, and suppose $0 < \alpha_1(\mu), \alpha_2(\mu), \ldots$ are the nodes of the $u(\mu)$ - line. We need to prove that

$$\alpha_r(\mu) < \alpha_r(\lambda)$$

for all those values of r corresponding to the $u(\lambda)$ - line. Since, by Theorem 3.3.2, there is a least one of the $\alpha_r(\mu)$ between any two of the $\alpha_r(\lambda)$, it is sufficient to prove that

$$\alpha_1(\mu) < \alpha_1(\lambda) = x.$$

Suppose if possible that $\alpha_1(\mu) \geq x$ and that

$$q < x \leq q + 1 \quad (1 \leq q \leq n)$$

then all $u_1, u_2, \ldots, u_q$ and $v_1, v_2, \ldots, v_q$ will be positive while

$$\begin{aligned} (q + 1 - x)u_q + (x - q)u_{q+1} &= 0 \\ (q + 1 - x)v_q + (x - q)v_{q+1} &\geq 0 \end{aligned}$$

which imply $t_q \leq 0$. On the other hand $s_q > 0$, which, when used with (3.3.8), provides a contradiction. ∎

Table 3.3.1 shows how the first node of $u(\lambda)$ appears at the right hand end $(n+1)$ when $\lambda = \lambda_1$ and gradually shifts to the left, how the second zero appears when $\lambda = \lambda_2$, etc.

Theorem 3.3.4 *The nodes of two successive eigenvectors interlace.*

Proof. Let the eigenvectors correspond to λ_j and λ_{j+1}. The nodes of the $u(\lambda_j)$ and $u(\lambda_{j+1})$ - lines are $(\alpha_r(\lambda_j))_{r=1}^{j}$ and $(\alpha_r(\lambda_{j+1}))_{r=1}^{j+1}$ respectively; and $\alpha_j(\lambda_j) = \alpha_{j+1}(\lambda_{j+1}) = n + 1$. Theorem 3.3.3 shows that $\alpha_1(\lambda_{j+1}) < \alpha_1(\lambda_j)$, while Theorem 3.3.2 applied to the two zeros $\alpha_{j-1}(\lambda_j)$ and $\alpha_j(\lambda_j) \equiv n+1$ shows that $\alpha_j(\lambda_{j+1}) > \alpha_{j-1}(\lambda_j)$. These two inequalities imply that the only possible ordering of the nodes is

$$\begin{aligned} 0 &< \alpha_1(\lambda_{j+1}) < \alpha_1(\lambda_j) < \alpha_2(\lambda_{j+1}) < \cdots < \alpha_{j-1}(\lambda_j) \\ &< \alpha_j(\lambda_{j+1}), < \alpha_j(\lambda_j) = \alpha_{j+1}(\lambda_{j+1}) = n+1. \quad \blacksquare \end{aligned} \tag{3.3.13}$$

The derivation of certain other important properties of the eigenmodes will be deferred until Section 5.7, where properties of an oscillatory matrix will be used. See Gladwell (1991a) [119] for some related results.

Exercises 3.3

1. Show that the first and last components of any eigenvector of a Jacobi matrix must be non-zero.

2. Show that if the matrix $\mathbf{J}$ of (3.1.4), with negative off-diagonal elements has an eigenpair $\lambda_i, \mathbf{u}_i$, then the corresponding matrix $\mathbf{J}^*$ with positive off-diagonal elements, has eigenpair $\lambda_i, \mathbf{Z}\mathbf{u}_i$ where $\mathbf{Z}$ is given by $\mathbf{Z} = diag$ $(1, -1, 1, \cdots (-)^{n-1})$. This means that the eigenvector corresponding to the smallest eigenvalue, λ_1, has $n-1$ sign changes, while that corresponding to λ_n has none. Show that if the eigenvalues of $\mathbf{J}^*$ are numbered *in reverse*, i.e., $\lambda_1 > \lambda_2 > \cdots > \lambda_n > 0$, then Theorem 3.3.1 remains valid.

3.4 Generalised eigenvalue problems

In Section 2.4 we showed that the eigenvalue problem for a finite element model of a vibrating rod could be reduced to a generalised eigenvalue problem

$$(\mathbf{K} - \lambda\mathbf{M})\mathbf{u} = \mathbf{0} \tag{3.4.1}$$

where $\mathbf{K}, \mathbf{M}$ were both symmetric tridiagonal matrices, $\mathbf{K}$ having negative codiagonal and $\mathbf{M}$ having positive codiagonal. If $\mathbf{M}$ is positive definite, and $\mathbf{K}$ is positive semi-definite (i.e., $\mathbf{K}$ is a Jacobi matrix), then the analysis of Chapter 1 shows that the eigenvalues are non-negative. Under these conditions we may prove that the solutions of (3.4.1) share the properties of the eigenvalue problem in normal form i.e., equation (3.1.7). In particular, we can show that the eigenvalues of (3.4.1) are distinct and that the sequence $(u_{r,j})_{r=1}^n$ for the jth eigenvector has exactly $j-1$ sign reversals. To obtain these results we need to return to the analysis in Section 3.1 onwards and see what changes have to be made.

We start with the principal minors of the matrix $\mathbf{K} - \lambda\mathbf{M}$, using the notation of (2.4.10):

$$P_0(\lambda) = 1, \quad P_1(\lambda) = c_1 - \lambda a_1, \quad P_2(\lambda) = \begin{vmatrix} c_1 - \lambda a_1 & -d_1 - \lambda b_1 \\ -d_1 - \lambda b_1 & c_2 - \lambda a_2 \end{vmatrix}, \ldots \tag{3.4.2}$$

so that finally

$$P_n(\lambda) = \det(\mathbf{K} - \lambda\mathbf{M}).$$

The minors satisfy the three-term recurrence relation

$$P_{r+1}(\lambda) = (c_{r+1} - \lambda a_{r+1})P_r(\lambda) - (d_r + \lambda b_r)^2 P_{r-1}(\lambda). \tag{3.4.3}$$

The argument used in Section 3.1, 3.3 was based on the fact that the sequence of principal minors defined by (3.1.5), (3.1.7) was a Sturm sequence. The sequence defined by (3.4.2), (3.4.3) however, is not a Sturm sequence. For if $P_r(\lambda) = 0$, then $P_{r+1}(\lambda) = -(d_r + \lambda b_r)^2 P_{r-1}(\lambda)$, and if it happens that $d_r + \lambda b_r = 0$, then $P_{r+1}(\lambda)$ would be zero, and not, as required by condition 2 of Theorem 3.1.1, of opposite sign to $P_{r-1}(\lambda)$. Now we make the crucial observation, that if we restrict attention to $\lambda \geq 0$, then the $P_r(\lambda)$ do form a Sturm sequence because b_r, d_r being positive, eliminates the possibility that $d_r + \lambda b_r = 0$. If we assume that $\mathbf{M}$ is positive definite and $\mathbf{K}$ is positive semi-definite, then all the eigenvalues λ_r will be non-negative and we may proceed as before. Thus Theorem 3.1.1 holds provided that $\lambda \geq 0$, and Theorem 3.1.2 holds. The proof of Theorem 3.1.3 must be slightly changed. In the expansion of $P_s(\lambda)$ in powers of λ we have

$$P_s(\lambda) = \alpha_{s,0} + \alpha_{s,1}\lambda + \ldots + \alpha_{s,s}\lambda^s. \tag{3.4.4}$$

The first term, $\alpha_{s,0}$, is the sth principal minor of $\mathbf{K}$ and, since $\mathbf{K}$ is positive semi-definite, $\alpha_{s,0} > 0$ for $s = 1, 2, \ldots, n-1$ and $\alpha_{n,0} \geq 0$; since $P_s(0) = \alpha_{s,0}$ we have $s_r(0) = 0$. The last term in (3.4.4), is $\alpha_{s,s} = (-)^s *$ (the sth principal minor of $\mathbf{M}$), so that for sufficiently large λ, i.e., $\lambda \geq \beta, s_r(\beta) = r$. The remainder of the proof of Theorem 3.1.1, the corollaries 1-3 and Theorem 3.1.4 follow as before.

We need to make small changes in the proof of Theorem 3.3.1. The $u_{r,j}$ are determined from the equations

$$-(d_{r-1} + \lambda_j b_{r-1})u_{r-1,j} + (c_r - \lambda_j a_r)u_{r,j} - (d_r + \lambda_j b_r)u_{r+1,j} = 0 \tag{3.4.5}$$

for $r = 1, 2, \ldots, n$, where $u_{0,j} = 0 = u_{n+1,j}$.

Put $d_r + \lambda_j b_r = e_r$, choose an arbitrary $e_n > 0$, and put

$$v_1 = u_{1,j} \quad v_2 = e_1, u_{2,j}, \ldots \quad v_{n+1} = e_1 e_2 \ldots e_n u_{n+1,j}$$

and multiply equation (3.4.5) by $e_1 e_2 \ldots e_{r-1}$ to obtain

$$-e_{r-1}^2 v_{r-1} + (c_r - \lambda_j a_r)v_r - v_{r+1} = 0 \quad r = 1, 2, \ldots, n.$$

On comparing this with (3.4.3) we see that it has the solution

$$v_0 = 0, v_1 = 1, v_r = P_{r-1}(\lambda_j), (r = 1, 2, \ldots, n+1).$$

Again, we conclude that $s_{n-1}(\lambda_j) = j - 1$.

We may make similar changes to the proofs of Theorems 3.3.2-3.3.4.

Exercises 3.4

1. Make appropriate changes in the proofs of Theorems 3.3.2-3.3.4.

Chapter 4

Inverse Problems for Jacobi Systems

People are generally better persuaded by the reasons which they themselves have discovered than by those which have come into the minds of others.
Pascal's *Pensées*, 10

4.1 Introduction

Research on these inverse problems began in the former Soviet Union, with the work of M.G. Krein. It appears that his primary interest was in the qualitative properties of the solutions of, and the inverse problems for, the Sturm-Liouville equation (see Chapter 10), and the discrete problems were studied because such problems were met in any approximate analysis of Sturm-Liouville problems. Krein's early papers Krein (1933) [198], Krein (1934) [199] concern the theory of Sturm sequences, while the Supplement to Gantmacher and Krein (1950) [98], Gantmacher and Krein (2002) and Krein (1952) [202] make use of the theory of continued fractions developed by Stieltjes (1918) [310]. Krein sees his results as giving mechanical interpretations of Stieltjes' analysis.

Consider the simple system shown in Figure 4.1.1a.

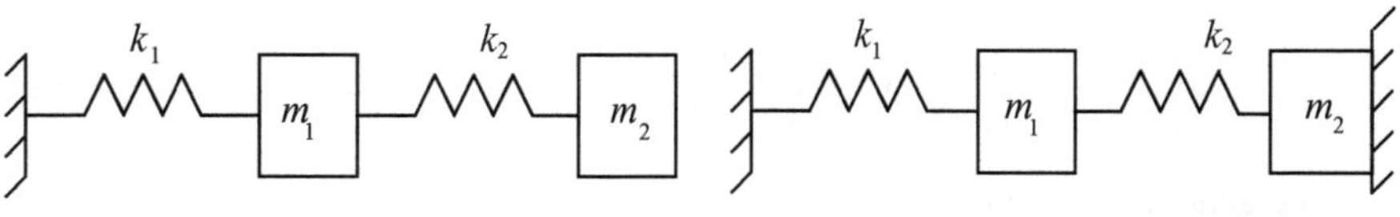

Figure 4.1.1 - The system is a) free and b) fixed at the right hand end

If m_1, m_2, k_1, k_2 are given, then the analysis of Chapter 2 shows how we can find the two natural frequencies, ω_1, ω_2 of the system: $\lambda_1 = \omega_1^2, \lambda_2 = \omega_2^2$ are the eigenvalues of the equation

$$\begin{bmatrix} k_1 + k_2 - \lambda m_1 & -k_2 \\ -k_2 & k_2 - \lambda m_2 \end{bmatrix} \begin{bmatrix} u_1 \\ u_2 \end{bmatrix} = \mathbf{0}. \tag{4.1.1}$$

The eigenvalues are the roots of the determinant:

$$\triangle(\lambda) \equiv m_1 m_2 \lambda^2 - \{k_2 m_1 + (k_1 + k_2) m_2\}\lambda + k_1 k_2 = 0. \tag{4.1.2}$$

Now consider the inverse problem. First it is clear that if one set of values m_1, m_2, k_1, k_2 has been found that yield specified eigenvalues λ_1, λ_2, and if $a > 0$, then am_1, am_2, ak_1, ak_2 will be another set yielding the same eigenvalues: there are not four quantities to be found, only three ratios $m_1 : m_2 : k_1 : k_2$. Knowing these ratios, we would need one more quantity, for instance the total mass $m = m_1 + m_2$, or the total stiffness k given by $1/k = 1/k_1 + 1/k_2$, to find the absolute values of the four quantities m_1, m_2, k_1, k_2.

But even knowing two eigenvalues λ_1, λ_2, we cannot find the three ratios; we need one more piece of information. One possible piece is the single eigenvalue $\lambda^* = {\omega^*}^2$ of the system obtained by fixing m_2, as shown in Figure 4.1.1b. This is

$$\lambda^* = {\omega^*}^2 = \frac{k_1 + k_2}{m_1}. \tag{4.1.3}$$

The sum and product of the roots λ_1, λ_2 of equation (4.1.2) are

$$\lambda_1 + \lambda_2 = \frac{k_2 m_1 + (k_1 + k_2) m_2}{m_1 m_2} = \frac{k_2}{m_2} + \frac{(k_1 + k_2)}{m_1} \tag{4.1.4}$$

$$\lambda_1 \lambda_2 = \frac{k_1 k_2}{m_1 m_2}. \tag{4.1.5}$$

Subtracting (4.1.3) from (4.1.4) we obtain

$$\frac{k_2}{m_2} = \lambda_1 + \lambda_2 - \lambda^*, \tag{4.1.6}$$

and then (4.1.5) gives

$$\frac{k_1}{m_1} = \frac{\lambda_1 \lambda_2}{\lambda_1 + \lambda_2 - \lambda^*}, \tag{4.1.7}$$

and finally (4.1.3) gives

$$\frac{k_2}{m_1} = \lambda^* - \frac{k_1}{m_1} = \lambda^* - \frac{\lambda_1 \lambda_2}{\lambda_1 + \lambda_2 - \lambda^*} = \frac{(\lambda^* - \lambda_1)(\lambda_2 - \lambda^*)}{\lambda_1 + \lambda_2 - \lambda^*}. \tag{4.1.8}$$

The general theory of vibration under constraint (Section 2.9) states that $\lambda_1 < \lambda^* < \lambda_2$, so that all the quantities on the right hand sides of (4.1.6)-(4.1.8) are positive: the solution is realistic. The theory presented in this Chapter provides various generalisations of this analysis to a lumped-mass system made up of n masses. The Chapter falls into three parts: a discussion of inverse problems for a Jacobi matrix; mass-spring realisations of these problems; generalisations and variants of these problems.

Exercises 4.1

1. Show that if $\mathbf{u}_1, \mathbf{u}_2$ are the eigenvectors of (4.1.1), normalised so that $\mathbf{u}_i^T \mathbf{M} \mathbf{u}_j = \delta_{ij}$, then the equation giving the eigenvalue λ^* is
$$\frac{u_{2,1}^2}{\lambda_1 - \lambda} + \frac{u_{2,2}^2}{\lambda_2 - \lambda} = 0$$
so that knowing λ^* is equivalent to knowing $u_{2,2} : u_{2,1}$.

2. Show that for the system of Figure 4.1.1, the system of given stiffness k, $(1/k = 1/k_1 + 1/k_2)$, and least mass $m = m_1 + m_2$, is found for $\lambda^* = \lambda_1 + \lambda_2 - \sqrt{\lambda_1 \lambda_2}$.

3. Show that for a taut string with tension T and unit length with just one concentrated mass m located at a distance ℓ_1 from the left hand end, ℓ_2 from the right, the frequency ω is given by
$$k_1 + k_2 - \lambda m = 0$$
where
$$k_i = \frac{T}{\ell_i}, \quad \ell_1 + \ell_2 = 1.$$
Hence find the system of *least mass* having a given frequency $\omega = \sqrt{\lambda}$. This suggests the problem of finding a string of *least mass* having concentrated masses $(m_i)_1^n$ separated by distances $\ell_1, \ell_2, \ldots, \ell_{n+1}$, where $\sum_{i=1}^{n+1} \ell_i = 1$. Barcilon and Turchetti (1980) [23] considered this problem in a wider context, but did not find a closed form solution for the discrete problem.

4.2 An inverse problem for a Jacobi matrix

It was shown in Section 3.1 that the (natural frequencies)^2 of a lumped mass system may be obtained as the eigenvalues of a Jacobi matrix

$$\mathbf{J} = \begin{bmatrix} a_1 & -b_1 & & & \\ -b_1 & a_2 & -b_2 & & \\ \cdot & \cdot & \cdot & \cdot & \cdot \\ 0 & 0 & & a_{n-1} & -b_{n-1} \\ 0 & 0 & & -b_{n-1} & a_n \end{bmatrix}. \tag{4.2.1}$$

If the system is connected, i.e., the stiffnesses between masses are strictly positive, then the codiagonal elements $-b_i$ are strictly negative.

The basic theorem is

Theorem 4.2.1 *There is a unique Jacobi matrix* $\mathbf{J}$ *having specified eigenvalues* $(\lambda_i)_1^n$, *where*

$$0 \le \lambda_1 < \lambda_2 < \cdots < \lambda_n \tag{4.2.2}$$

and with normalised eigenvectors $(\mathbf{u}_i)_1^n$ *having non-zero specified values* $(u_{1i})_1^n$ *or* $(u_{ni})_1^n$ *of their first or last components respectively; recall that* $\mathbf{u}_i = \{u_{1i}, u_{2i}, \ldots, u_{ni}\}$.

(We recall Ex. 3.3.1, that the first and last components of an eigenvector of a Jacobi matrix are both non-zero.)

Proof. The theorem is at once an existence (there is ...) and a uniqueness (... a unique) theorem. We shall prove existence by actually constructing a matrix, and will do so by using the so-called Lanczos algorithm; the algorithm demonstrates that $\mathbf{J}$ is unique. This algorithm has the advantage that numerically it is well conditioned. An independent proof that the matrix is unique is left to Ex. 4.2.2. The proof will be presented for the case in which $(u_{1j})_1^n$ are specified.

The eigenvectors $\mathbf{u}_i$ satisfy

$$\mathbf{J}\mathbf{u}_i = \lambda_i \mathbf{u}_i \tag{4.2.3}$$

Use the column vectors $(\mathbf{u}_i)_1^n$ to construct a square matrix $\mathbf{U} : \mathbf{U} = [\mathbf{u}_1, \mathbf{u}_2, \ldots, \mathbf{u}_n]$. The orthonormality conditions $\mathbf{u}_i^T\mathbf{u}_j = \delta_{ij}$ yield

$$\mathbf{U}^T\mathbf{U} = \mathbf{I}.$$

This means that $\mathbf{U}^T$ is the inverse of $\mathbf{U}$: $\mathbf{U}$ is an *orthogonal* matrix. But if $\mathbf{U}^T\mathbf{U} = \mathbf{I}$, then Theorem 1.3.6 states that $\mathbf{U}\mathbf{U}^T = \mathbf{I}$ also. Now put $\mathbf{U}^T = \mathbf{X}$, then $\mathbf{U}\mathbf{U}^T = \mathbf{X}^T\mathbf{X} = \mathbf{I}$. But this means that the columns of $\mathbf{X}$, like the columns of $\mathbf{U}$, are orthonormal. Call the columns $(\mathbf{x}_i)_1^n$, so that $\mathbf{X} = [\mathbf{x}_1, \mathbf{x}_2, \ldots, \mathbf{x}_n]$, then

$$\mathbf{x}_i^T\mathbf{x}_j = \delta_{ij}.$$

The reason why we have introduced the vectors $\mathbf{x}_i$ is that

$$\mathbf{x}_1 = \{x_{11}, x_{21}, \ldots, x_{n1}\} = \{u_{11}, u_{12}, \ldots, u_{1n}\} \tag{4.2.4}$$

is given as part of the data.

Now we proceed to rewrite the eigenvalue equations (4.2.3) as equations for the $\mathbf{x}_i$. The set of equations (4.2.3) for $i = 1, 2, \ldots, n$, may be written

$$\mathbf{J}\mathbf{U} = \mathbf{U}\wedge. \tag{4.2.5}$$

Thus, on transposing we find

$$\mathbf{X}\mathbf{J} = \wedge\mathbf{X}. \tag{4.2.6}$$

Written in full, this equation is

$$[\mathbf{x}_1, \mathbf{x}_2, \ldots, \mathbf{x}_n]\begin{bmatrix} a_1 & -b_1 & & \\ -b_1 & a_2 & -b_2 & \\ \cdot & \cdot & \cdot & \cdot \\ & & -b_{n-1} & a_n \end{bmatrix} = \wedge\,[\mathbf{x}_1, \mathbf{x}_2, \ldots, \mathbf{x}_n]. \tag{4.2.7}$$

Take this equation column by column. The first column is

$$a_1\mathbf{x}_1 - b_1\mathbf{x}_2 = \wedge\mathbf{x}_1. \tag{4.2.8}$$

Premultiply this by $\mathbf{x}_1^T$, using $\mathbf{x}_1^T\mathbf{x}_1 = 1,\ \mathbf{x}_1^T\mathbf{x}_2 = 0$;

$$a_1\mathbf{x}_1^T\mathbf{x}_1 = a_1 = \mathbf{x}_1^T\wedge\mathbf{x}_1.$$

Now rewrite equation (4.2.8) as

$$b_1\mathbf{x}_2 = a_1\mathbf{x}_1 - \wedge\mathbf{x}_1 = \mathbf{z}_2.$$

The vector $\mathbf{z}_2$ is known, because $a_1, \mathbf{x}_1, \wedge$ are all known. The vector $\mathbf{x}_2$ is to be a unit vector, so that

$$b_1||\mathbf{x}_2|| = b_1 = ||\mathbf{z}_2||.$$

and $\mathbf{x}_2 = \mathbf{z}_2/b_1$. Having found $a_1, b_1, \mathbf{x}_2$ we proceed to the next column of (4.2.7):

$$-b_1\mathbf{x}_1 + a_2\mathbf{x}_2 - b_2\mathbf{x}_3 = \wedge\mathbf{x}_2$$

Again, premultiplying by $\mathbf{x}_2^T$ we find $a_2 = \mathbf{x}_2^T\wedge\mathbf{x}_2$, and then

$$b_2\mathbf{x}_3 = a_2\mathbf{x}_2 - b_1\mathbf{x}_1 - \wedge\mathbf{x}_2 = \mathbf{z}_3$$

so that

$$b_2||\mathbf{x}_3|| = b_2 = ||\mathbf{z}_3||, \quad \mathbf{x}_3 = \mathbf{z}_3/b_3$$

and so on. This procedure is called the Lanczos algorithm; see Lanczos (1950) [203], Golub (1973) [132], Golub and Van Loan (1983) [135] and Kautsky and Golub (1983) [192]. It produces a matrix $\mathbf{J}$ and at the same time constructs the columns $(\mathbf{x}_i)_1^n$ which yield $\mathbf{X} = \mathbf{U}^T$.

Actually, what we have described is an inverse version of the original Lanczos algorithm. This original algorithm solved the following problem: Given a symmetric matrix $\mathbf{A}$ and a vector $\mathbf{x}_1$ such that $\mathbf{x}_1^T\mathbf{x}_1 = 1$, compute a symmetric tridiagonal matrix $\mathbf{J}$ and an orthogonal matrix $\mathbf{X} = [\mathbf{x}_1, \mathbf{x}_2, \ldots, \mathbf{x}_n]$ such that $\mathbf{A} = \mathbf{X}\mathbf{J}\mathbf{X}^T$. In our use of the algorithm, we start with $\mathbf{A} = \wedge$.

We have defined a Jacobi matrix as a *positive semi-definite* symmetric tridiagonal matrix with strictly negative codiagonal. If the spectrum $(\lambda_i)_1^n$ satisfies the inequalities (4.2.2), so that $\lambda_1 \geq 0$, then the $\mathbf{J}$ constructed by the Lanczos algorithm from $\mathbf{A} \equiv \wedge$ will be a Jacobi matrix. ■

Exercises 4.2

1. Show that the vectors $\mathbf{x}_i$ constructed in the Lanczos algorithm satisfy

$$\mathbf{x}_i^T\mathbf{x}_j = \delta_{ij} \quad i, j = 1, 2, \ldots, n$$

even though this orthogonality is apparently established only for $|i-j| \leq 1$.

2. Show that there cannot be two distinct Jacobi matrices $\mathbf{J}$ and $\mathbf{J}'$ with $\sigma(\mathbf{J}) = \sigma(\mathbf{J}')$ and with the same values of the first components $(u_{1i})_1^n$ of their normalised eigenvectors.

3. Rewrite the procedure described in equation (4.2.5) on, to solve the original Lanczos problem.

4.3 Variants of the inverse problem for a Jacobi matrix

First, we introduce some notation. Suppose $\mathbf{A} \in \mathbf{M}_n$. The set of eigenvalues of $\mathbf{A}$, the *spectrum* of $\mathbf{A}$ is denoted by $\sigma(\mathbf{A})$. If $\mathbf{A}$ is symmetric, i.e., $\mathbf{A} \in S_n$, then $\sigma(\mathbf{A})$ is a sequence of real numbers $(\lambda_i)_1^n$, where $\lambda_1 \leq \lambda_2 \leq \lambda_3 \cdots \leq \lambda_n$. If $\mathbf{M}, \mathbf{K} \in S_n$, then the set of eigenvalues of equation (3.1.1) is denoted by $\sigma(\mathbf{M}, \mathbf{K})$; again it is a sequence of real numbers $(\lambda_i)_1^n$ satisfying $\lambda_1 \leq \lambda_2 \leq \cdots \leq \lambda_n$.

See Kautsky and Golub (1983) [192], deBoor and Saff (1986) [76] for a discussion that places the Jacobi matrix problem in a wider context. Friedland and Melkman (1979) [94] discuss the inverse eigenvalue problem in the context of non-negative matrices.

If $\mathbf{A} \in S_n$, the matrix obtained by deleting the ith row and column of $\mathbf{A}$ is called a *truncated* matrix. It will sometimes be denoted by $\mathbf{A}_i$; its eigenvalues will be denoted $\sigma(\mathbf{A}_i)$.

Now suppose that $\mathbf{A} \in S_n$ is a Jacobi matrix $\mathbf{J}$, then its eigenvalues will be distinct, and the eigenvalues $\sigma(\mathbf{J}_1) = (\mu_i)_1^{n-1}$ will strictly interlace $(\lambda_i)_1^n$, i.e.,

$$0 \leq \lambda_1 < \mu_1 < \lambda_2 < \cdots < \mu_{n-1} < \lambda_n. \qquad (4.3.1)$$

The problem of reconstructing $\mathbf{J}$ from $\sigma(\mathbf{J})$ and $\sigma(\mathbf{J}_1)$ seems to have been studied first by Hochstadt (1967) [173]. He proved that there is *at most* one matrix $\mathbf{J}$ with the required property. Hochstadt (1973) [176] attempted to construct this unique Jacobi matrix, but he did not show that his method would always lead to real values of the codiagonal elements b_i. Hald (1976) [160] presented another construction and showed that, in theory, it would always work provided that the eigenvalues satisfied the interlacing condition (4.3.1). In practice, however, the construction was found to break down due to loss of significant figures. Hald also showed that Hochstadt's construction will always lead to real b_i provided that (4.3.1) holds. Gray and Wilson (1976) [154] presented an alternative, inductive construction of $\mathbf{J}$. An independent uniqueness proof was given by Hald (1976) [160].

In this section we shall present two methods for constructing $\mathbf{J}$. The first relies on the theory of orthogonal polynomials described in Section 3.2. The second, which will later be generalised to inverse problems for band matrices, relies on the Lanczos algorithm described in Section 4.2.

Note that we have chosen to define a Jacobi matrix so that it is *positive semi-definite.* Many of the results require only the interlacing of the λ's and the μ's, without any restriction on positivity.

The first method is best described by supposing that $\sigma(\mathbf{J}) = (\lambda_i)_1^n$ and $\sigma(\mathbf{J}_n) = (\mu_i)_1^{n-1}$ are known, and satisfy (4.3.1). Remember that $\mathbf{J}_n$ is obtained by deleting the nth row and column of $\mathbf{J}$. Now we form monic polynomials $p_i(\lambda)$, rather than the polynomials $P_i(\lambda)$ in equation (3.1.5). We form two polynomials

$$p_n(\lambda) = \prod_{i=1}^{n}(\lambda - \lambda_i), \quad p_{n-1}(\lambda) = \prod_{i=1}^{n-1}(\lambda - \mu_i). \tag{4.3.2}$$

The polynomials are the nth and $(n-1)$th monic polynomials of the sequence of monic polynomials with weights given by equation (3.2.13), i.e.,

$$w_i = \gamma/\{p_{n-1}(\lambda_i)p_n'(\lambda_i)\} \tag{4.3.3}$$

and points $(\lambda_i)_1^n$. In addition, they are the nth and $(n-1)$th principal minors of the matrix $(\lambda\mathbf{I} - \mathbf{J})$. The polynomials $p_r(\lambda)$ therefore satisfy

$$p_r(\lambda) = (\lambda - a_r)p_{r-1}(\lambda) - b_{r-1}^2 p_{r-2}(\lambda). \tag{4.3.4}$$

Hald's method of reconstructing $\mathbf{J}$ is as follows: he starts from $p_n(\lambda), p_{n-1}(\lambda)$ and constructs $p_{n-2}(\lambda)$, and in the process finds a_n and b_{n-1}, by synthetic division. Then from $p_{n-1}(\lambda), p_{n-2}(\lambda)$ he constructs $p_{n-3}(\lambda)$ and finds a_{n-1} and b_{n-2}, and so on. The process is inherently unstable because the polynomials $p_{n-2}, p_{n-3}, \ldots, p_1$ are found by successively cancelling the leading terms in the preceding pair of polynomials; the process becomes unstable because of cancellation of leading digits.

de Boor and Golub (1978) [75] proceed quite differently. Having found the weights w_i by using (4.3.3), they construct the polynomials in the natural order by using the analysis of Section 3.2, i.e.,

$$p_{-1}(\lambda) = 0, \quad p_0(\lambda) = 1 \tag{4.3.5}$$

$$p_r(\lambda) = (\lambda - a_r)p_{r-1}(\lambda) - b_{r-1}^2\, p_{r-2}(\lambda), \tag{4.3.6}$$

with the numbers a_r, b_r computed by

$$a_r = \frac{(p_{r-1}, \lambda p_{r-1})}{||p_{r-1}||^2}, \quad b_r = \frac{||p_r||}{||p_{r-1}||}, \quad r = 1, 2, \ldots, n-1. \tag{4.3.7}$$

This process is numerically stable.

The only major difficulty encountered by de Boor and Golub lay in the computation of the weights w_i. In seeking to overcome this difficulty, they used the reflection of $\mathbf{J}$ about its second diagonal. The matrix

$$\mathbf{T} = \begin{bmatrix} 0 & 0 & \ldots & 1 \\ & & 1 & \\ & \cdot & & \\ 1 & & & 0 \end{bmatrix} \tag{4.3.8}$$

is orthogonal and symmetric, so that $\mathbf{T}^2 = \mathbf{I}$. It reverses the order of the rows and the columns of $\mathbf{J}$, i.e., it transforms $\mathbf{J}$ into

$$\bar{\mathbf{J}} = \mathbf{TJT} = \begin{bmatrix} a_n & -b_{n-1} & & \\ -b_{n-1} & a_{n-1} & -b_{n-2} & \\ \ddots & \ddots & \ddots & \\ & \ddots & \ddots & -b_1 \\ & & -b_1 & a_1 \end{bmatrix}. \tag{4.3.9}$$

If, therefore, the elements of $\bar{\mathbf{J}}$ are denoted by $\bar{a}_r, \bar{b}_r$ then

$$\bar{a}_r = a_{n+1-r}, \quad \bar{b}_r = b_{n-r}.$$

The leading principal minors of $\lambda\mathbf{I} - \bar{\mathbf{J}}$ are the trailing principal minors of $\lambda\mathbf{I} - \mathbf{J}$; we denote them by $\bar{p}_r(\lambda)$. We prove

Theorem 4.3.1 *For* $i = 1, 2, \ldots, n$

$$p_{n-1}(\lambda_i)\bar{p}_{n-1}(\lambda_i) = (b_1 b_2 \ldots b_{n-1})^2 = b^2.$$

Proof. For once we step out of sequence, and use the notation we will introduce in Section 6.2. Let α denote the sequence $\{2, 3, \ldots, n-1\}$, then

$$p_{n-1}(\lambda_i) = B(\alpha \cup 1), \quad \bar{p}_{n-1}(\lambda_i) = B(\alpha \cup n).$$

Using Sylvester's theorem (Corollary 2 of Theorem 6.2.2), with $B(\alpha)$ as pivotal block, we obtain

$$0 = B(\alpha)\det(\mathbf{B}) = \begin{vmatrix} B(\alpha \cup 1) & B(\alpha \cup 1; \alpha \cup n) \\ B(\alpha \cup 1; \alpha \cup n) & B(\alpha \cup n) \end{vmatrix}$$

i.e.,

$$0 = p_{n-1}(\lambda_i)\bar{p}_{n-1}(\lambda_i) - (b_1 b_2 \ldots b_{n-1})^2 \quad \blacksquare$$

This result means that the polynomials $\bar{p}_n(\lambda), \bar{p}_{n-1}(\lambda), \ldots, \bar{p}_1(\lambda), \bar{p}_0(\lambda)$ are the monic polynomials related to the weights

$$w_i = \frac{b^2}{\bar{p}_{n-1}(\lambda_i)\bar{p}'_n(\lambda_i)} = \frac{p_{n-1}(\lambda_i)}{p'_n(\lambda_i)}. \tag{4.3.10}$$

These weights are more easily constructed than those in (4.3.3). In this procedure, the terms in the matrix $\mathbf{J}$ are computed in the order $\bar{a}_1, \bar{b}_1, \bar{a}_2, \ldots, \bar{b}_{n-1}, \bar{a}_n$, i.e., in the order $a_n, b_{n-1}, a_{n-1}, \ldots, b_1, a_1$.

The second method of constructing $\mathbf{J}$ is due to Golub and Boley (1977) [133]. See also de Boor and Saff (1986) [76]. It relies on the fact that, once we know $\sigma(\mathbf{J})$ and $\sigma(\mathbf{J}_1)$ we may compute the vector $\mathbf{x}_1$ of first components of the eigenvectors of $\mathbf{J}$; these are the data needed for construction by the Lanczos algorithm of Section 4.2. We can carry out the analysis for an arbitrary symmetric matrix

$\mathbf{A} \in S_n$, rather than a Jacobi matrix. Barcilon (1978) [19] concentrated on the eigen*vectors* corresponding to λ_i and μ_i, rather than using the μ_i to find the quantities x_{i1}; his subsequent analysis did not lend itself to computation.

If $\mathbf{A} \in S_n$, then the eigenvalues of $\mathbf{A}_1$ are the stationary values of $\mathbf{u}^T\mathbf{A}\mathbf{u}$ subject to $\mathbf{u}^T\mathbf{u} = 1$ and the constraint $u_1 = 0$, i.e., $\mathbf{u}^T\mathbf{e}_1 = 0$. Thus they are the stationary values of

$$f = \mathbf{u}^T\mathbf{A}\mathbf{u} - \lambda\mathbf{u}^T\mathbf{u} - 2\nu\mathbf{u}^T\mathbf{e}_1, \tag{4.3.11}$$

where $\mathbf{e}_1 = \{1, 0, \ldots, 0\}$ and λ, ν are Lagrange parameters. The condition that f be stationary yields

$$\mathbf{A}\mathbf{u} - \lambda\mathbf{u} - \nu\mathbf{e}_1 = \mathbf{0}. \tag{4.3.12}$$

Since the eigenvectors $\mathbf{u}_i$ of $\mathbf{A}$ span V_n, we may write

$$\mathbf{u} = \sum_{i=1}^{n} \alpha_i \mathbf{u}_i, \tag{4.3.13}$$

and then

$$\mathbf{A}\mathbf{u} = \sum_{i=1}^{n} \alpha_i \mathbf{A}\mathbf{u}_i = \sum_{i=1}^{n} \lambda_i \alpha_i \mathbf{u}_i,$$

so that (4.3.12) becomes

$$\sum_{i=1}^{n} (\lambda_i - \lambda)\alpha_i \mathbf{u}_i = \nu\mathbf{e}_1,$$

and the orthogonality condition $\mathbf{u}_j^T\mathbf{u}_i = \delta_{ij}$ gives

$$(\lambda_j - \lambda)\alpha_j = \nu\mathbf{u}_j^T e_1 = \nu u_{1j} = \nu x_{j1},$$

where we have used (4.2.4). Substituting for α_i in (4.3.13) we find

$$\mathbf{u} = \nu \sum_{i=1}^{n} \frac{x_{i1}\mathbf{u}_i}{\lambda_i - \lambda}, \tag{4.3.14}$$

and the condition $u_1 = 0$, and $u_{1i} = x_{i1}$, yields the eigenvalue equation

$$\sum_{i=1}^{n} \frac{(x_{i1})^2}{\lambda_i - \lambda} = 0. \tag{4.3.15}$$

We note that if $\mathbf{A}$ is a Jacobi matrix, none of the coefficients x_{i1} will be zero (Ex. 3.3.1). The analysis of Section 2.9 shows that the roots $(\mu_i)_1^{n-1}$ of this equation will then strictly interlace the $(\lambda_i)_1^n$, as in (4.3.1).

Since $\mathbf{x}_1 = \{x_{11}, x_{21}, \ldots, x_{n1}\}$ and $\mathbf{x}_1$ is the first column of the orthogonal matrix $\mathbf{X} = \mathbf{U}^T$, we have $||\mathbf{x}_1||^2 = 1 = \sum_{i=1}^{n}(x_{i1})^2$, so that we have the identity

$$\sum_{i=1}^{n} \frac{(x_{i1})^2}{\lambda_i - \lambda} = \frac{\prod_{i=1}^{n-1}(\mu_i - \lambda)}{\prod_{i=1}^{n}(\lambda_i - \lambda)} \tag{4.3.16}$$

(Note that, for large λ, both sides approach $-1/\lambda$.) On multiplying (4.3.16) through by $(\lambda_j - \lambda)$ and then putting $\lambda = \lambda_j$ we find

$$(x_{i1})^2 = \frac{\prod_{j=1}^{n-1}(\mu_j - \lambda_i)}{\prod_{j=1}^{n\prime}(\lambda_j - \lambda_i)}, \quad i = 1, 2, \ldots, n \tag{4.3.17}$$

where $'$ indicates that the term $j = i$ has been omitted. The interlacing condition ensures that the right hand side of (4.3.17) is strictly positive for each $i = 1, 2, \ldots, n$. This equation thus yields $\mathbf{x}_1$.

We stress the importance of the analysis in equations (4.3.11)-(4.3.17). It shows that if $\mathbf{A}$ is an arbitrary symmetric matrix, then $\sigma(\mathbf{A})$ and $\sigma(\mathbf{A}_1)$ determine the vector $\mathbf{x}_1$ of first components of the normalised eigenvectors of $\mathbf{A}$. Conversely, $\sigma(\mathbf{A})$ and $\mathbf{x}_1$ determine $\sigma(\mathbf{A}_1)$.

There is a third inverse problem which appears in a number of contexts. Given two strictly increasing sequences $(\lambda_i)_1^n$ and $(\lambda_i^*)_1^n$ with

$$0 \le \lambda_1 < \lambda_1^* < \lambda_2 < \lambda_2^* < \cdots < \lambda_n < \lambda_n^* \tag{4.3.18}$$

determine $\mathbf{J} \in S_n$ such that $\sigma(\mathbf{J}) = (\lambda_i)_1^n$, and $\sigma(\mathbf{J}^*) = (\lambda_i^*)_1^n$, where $\mathbf{J}^* = (a_1^* - a_1)\mathbf{E}_{1,1} + \mathbf{J}$. (The matrix $\mathbf{J}^*$ differs from $\mathbf{J}$ only in the 1,1 position.)

Suppose $\mathbf{A} \in S_n$ is an arbitrary symmetric matrix, and that $\mathbf{A}^*$ differs from $\mathbf{A}$ only in the 1,1 position, i.e., $\mathbf{A}^* = \mathbf{A} + (a_{1,1}^* - a_{1,1})\mathbf{E}_{1,1}$. We will show that $\sigma(\mathbf{A})$ and $\sigma(A^*)$ determine $\mathbf{x}_1$. The eigenvalue equation for $\mathbf{A}^*$ is

$$\mathbf{A}^*\mathbf{u} = \lambda\mathbf{u} \tag{4.3.19}$$

which we write

$$\mathbf{A}\mathbf{u} + (a_{1,1}^* - a_{1,1})u_1\mathbf{e}_1 = \lambda\mathbf{u}.$$

Write

$$\mathbf{u} = \sum_{i=1}^n \alpha_i \mathbf{u}_i, \tag{4.3.20}$$

so that equation (4.3.19) becomes

$$\sum_{i=1}^n \lambda_i\alpha_i\mathbf{u}_i + (a_{1,1}^* - a_{1,1})u_1\mathbf{e}_1 = \lambda\sum_{i=1}^n \alpha_i\mathbf{u}_i,$$

and therefore,

$$(\lambda - \lambda_i)\alpha_i = (a_{1,1}^* - a_{1,1})u_1u_{1i},$$

which when substituted into (4.3.20), yields

$$\mathbf{u} = (a_{1,1}^* - a_{1,1})u_1\sum_{i=1}^n \frac{u_{1,i}\mathbf{u}_i}{\lambda - \lambda_i}.$$

Equating the first components on each side of their equation, we have

$$1 = (a_{1,1}^* - a_{1,1})\sum_{i=1}^n \frac{x_{i1}^2}{\lambda - \lambda_i},$$

where $x_{i1} = u_{1i}$. The roots of this equation are $(\lambda_i^*)_1^n$, so that

$$1 - (a_{1,1}^* - a_{1,1})\sum_{i=1}^n \frac{x_{i1}^2}{\lambda - \lambda_i} = \prod_{i=1}^n \left(\frac{\lambda - \lambda_i^*}{\lambda - \lambda_i}\right) \tag{4.3.21}$$

and therefore

$$(a_{1,1}^* - a_{1,1})x_{i1}^2 = (\lambda_i^* - \lambda_i)\prod_{j=1}^n{}' \left(\frac{\lambda_i - \lambda_j^*}{\lambda_i - \lambda_j}\right). \tag{4.3.22}$$

By comparing the traces of $\mathbf{A}$ and $\mathbf{A}^*$ we see that

$$a_{1,1}^* - a_{1,1} = \sum_{j=1}^n (\lambda_j^* - \lambda_j) > 0. \tag{4.3.23}$$

Thus, equation (4.3.22) expresses $(x_{i1})^2$ in terms of $\sigma(\mathbf{A})$ and $\sigma(\mathbf{A}^*)$, and the interlacing condition (4.3.18) insures that $(x_{i1})^2$ will be positive. If we know that $\mathbf{A}$ is a Jacobi matrix then, of course, we can use the Lanczos algorithm to determine it. Note that nowhere in the analysis do we need the restriction that λ_1 is non-negative; only the strict interlacing is needed.

A matrix $\mathbf{A}$ is said to be *persymmetric* if it is symmetric, and also symmetric about the second diagonal, the one going from top right to bottom left. Thus $\mathbf{A}$ is persymmetric if $\bar{\mathbf{A}}$ given by (4.3.9) satisfies

$$\bar{\mathbf{A}} = \mathbf{A}. \tag{4.3.24}$$

If $\mathbf{A}$ is tridiagonal and persymmetric, then

$$a_r = a_{n+1-r}, \quad b_r = b_{n-r}. \tag{4.3.25}$$

The final inverse problem considered here concerns the reconstruction of a persymmetric Jacobian matrix. Now we need only one spectrum, not two. We prove

Theorem 4.3.2 *There is a unique persymmetric Jacobi matrix* $\mathbf{J}$ *with* $\sigma(\mathbf{J}) = (\lambda_i)_1^n$, *satisfying* $0 \le \lambda_1 < \lambda_2 < \cdots < \lambda_n$.

Proof. The simplest proof is perhaps to show that if the eigenvalues $(\lambda_i)_1^n$ are known, then it is possible to find the weights for the construction of the orthogonal polynomials $p_r(\lambda)$. Indeed if $\mathbf{J}$ is persymmetric then the minor $p_r(\lambda)$ is equal to $\bar{p}_r(\lambda)$. But then Theorem 4.3.1 shows that

$$[p_{n-1}(\lambda_i)]^2 = b^2, \quad \text{i.e., } p_{n-1}(\lambda_i) = \pm b$$

so that equation (4.3.10) yields

$$w_i = \pm b / p_n'(\lambda_i). \tag{4.3.26}$$

Since the signs of $p'_n(\lambda_i)$ will alternate with i, then so must the signs in (4.3.26) if the w_i are to be positive. The magnitude of b is irrelevant to the construction of the $p_r(\lambda)$. See Hochstadt (1979) [182] for another variant of this inverse eigenvalue problem. ■

Exercises 4.3

1. Show that if $\mathbf{B} = \lambda_i\mathbf{I} - \mathbf{J}$, then

$$B(1, 2, \ldots, n-1;\ 2, 3, \ldots, n) = (-)^{n-1} b_1 b_2 \ldots b_{n-1}.$$

2. Show that the x_{i1} computed from (4.3.22) do satisfy

$$\sum_{i=1}^{n} x_{i1}^2 = 1.$$

3. If you like using a computer, then try to reconstruct a Jacobi matrix using Hald's method, or that of de Boor and Golub. Start with the matrix $\mathbf{J}$ with

$$a_i = 2, \quad b_i = 1, \quad i = 1, 2, \ldots, n-1; \quad a_n = 2.$$

Set up recurrence relations to give $(\lambda_i)_1^n$ and $(\mu_i)_1^{n-1}$ and use these as data to reconstruct $\mathbf{J}$.

4.4 Reconstructing a spring-mass system; by end constraint

We may divide the problem of reconstructing an in-line spring-mass system into three stages:

i) Formulate the problem as an inverse eigenvalue problem for a Jacobi matrix $\mathbf{J}$.

ii) Solve this problem and find $\mathbf{J}$.

iii) Recover the mass and stiffness matrices $\mathbf{M}$ and $\mathbf{K}$ from $\mathbf{J}$.

Stage i) was discussed in Section 3.1; we repeat the analysis here. For an in-line system, the frequency equation governing free vibration is

$$(\mathbf{K} - \lambda\mathbf{M})\mathbf{y} = \mathbf{0}. \tag{4.4.1}$$

For the system shown in Figure 4.4.1 the matrices $\mathbf{K}$ and $\mathbf{M}$ are given explicitly in (2.2.7). We write $\mathbf{M} = \mathbf{D}^2$, where $\mathbf{D} = diag(d_1, d_2, \ldots, d_n)$, put $\mathbf{Dy} = \mathbf{u}$ and reduce (4.4.1) to

$$(\mathbf{J} - \lambda\mathbf{I})\mathbf{u} = \mathbf{0}, \tag{4.4.2}$$

where

$$\mathbf{J} = \mathbf{D}^{-1}\mathbf{K}\mathbf{D}^{-1}. \tag{4.4.3}$$

Stage ii) was the subject of Section 4.3. Given the spectra of the systems in Figure 4.4.1a) and b), i.e., $\sigma(\mathbf{J}) = (\lambda_i)_1^n$ and $\sigma(\mathbf{J}_1) = (\mu_i)_1^{n-1}$, we construct $\mathbf{x}_1$, the vector of first components of the eigenvectors $\mathbf{u}_i$ of (4.4.2), and then construct $\mathbf{J}$ by using the Lanczos algorithm of Section 4.2.

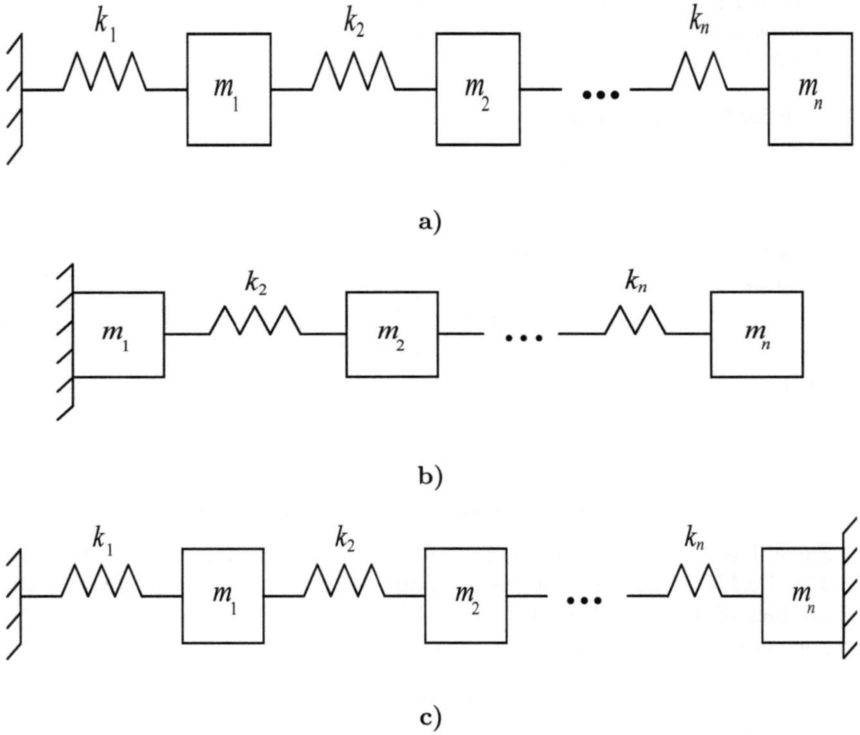

Figure 4.4.1 - Two possible ways of constraining the end of a fixed-free system

It remains to consider Stage iii). By using the explicit form of $\mathbf{K}$ in equation (2.2.7) we can verify that if

$$\mathbf{e} = \{1, 1, \ldots, 1\}, \tag{4.4.4}$$

then

$$\mathbf{Ke} = \{k_1, 0, 0, \ldots, 0\}. \tag{4.4.5}$$

Physically, this equation states that a static force k_1 applied to mass m_1 will extend the first spring by unit amount and at the same time displace all the remaining masses $m_2, m_3, \ldots, m_n$ by unit amount to the right, as if everything to the right of m_1 were a rigid body. Since $\mathbf{K} = \mathbf{DJD}$ we have

$$\mathbf{DJDe} = \mathbf{DJD}\{1, 1, \ldots, 1\} = \{k_1, 0, 0, \ldots, 0\},$$

i.e.,

$$\mathbf{J}\mathbf{d} = \mathbf{J}\{d_1, d_2, \ldots, d_n\} = \{k_1/d_1, 0, 0, \ldots, 0\}. \tag{4.4.6}$$

(Note that $\mathbf{D} = diag(d_1, d_2, \ldots, d_n)$, while $\mathbf{d} = \{d_1, d_2, \ldots, d_n\}$.) We need to be sure that $\mathbf{d}$ so calculated will be a strictly positive vector. We prove

Theorem 4.4.1 *If $\mathbf{J} \in S_n$ is a non-singular Jacobi matrix, then $\mathbf{J}^{-1}$ is a strictly positive matrix, meaning that each element of $\mathbf{J}^{-1}$ is strictly positive; we write this $\mathbf{J}^{-1} > \mathbf{0}$.*

Proof. We use induction. Write

$$\mathbf{J} = \begin{bmatrix} a_1 & -\mathbf{b}^T \\ -\mathbf{b} & \mathbf{J}_1 \end{bmatrix}, \quad \mathbf{b} = \{b_1, 0, \ldots, 0\}.$$

We will have achieved our goal if we can show that *if* $\mathbf{J}_1^{-1} > 0$ *then* $\mathbf{J}^{-1} > 0$. Suppose

$$\mathbf{J}^{-1} = \begin{bmatrix} h_1 & \mathbf{k}^T \\ \mathbf{k} & \mathbf{H} \end{bmatrix}$$

then

$$\mathbf{J}\mathbf{J}^{-1} = \begin{bmatrix} a_1 & -\mathbf{b}^T \\ -\mathbf{b} & \mathbf{J}_1 \end{bmatrix} \begin{bmatrix} h_1 & \mathbf{k}^T \\ \mathbf{k} & \mathbf{H} \end{bmatrix} = \begin{bmatrix} 1 & \mathbf{0} \\ \mathbf{0} & \mathbf{I} \end{bmatrix}$$

so that

$$-\mathbf{b}\mathbf{k}^T + \mathbf{J}_1\mathbf{H} = \mathbf{I}, \quad -\mathbf{b}h_1 + \mathbf{J}_1\mathbf{k} = \mathbf{0}.$$

Since $\mathbf{J}$ is a non-singular Jacobi matrix, it is positive definite; so therefore is $\mathbf{J}^{-1}$, by Ex. 1.4.2; therefore $h_1 > 0$, and so $\mathbf{k} = \mathbf{J}_1^{-1}\mathbf{b}h_1 > 0$. (Note that the product of $\mathbf{J}_1^{-1}$, which is strictly positive by hypothesis, and the non-negative non-zero vector $\mathbf{b}$, is strictly positive.) Therefore,

$$\mathbf{H} = \mathbf{J}_1^{-1}\mathbf{b}\mathbf{k}^T + \mathbf{J}_1^{-1} > \mathbf{0}.$$

(Note that since $\mathbf{J}_1^{-1} > 0$, all we need in order to prove that $\mathbf{H} > 0$, is that $\mathbf{b}\mathbf{k}^T \geq 0$, i.e., $\mathbf{k} \geq 0$; actually though, $\mathbf{k} > \mathbf{0}$.) Thus $\mathbf{H} > \mathbf{0}, \mathbf{k} > \mathbf{0}$ and $h_1 > 0$ so that $\mathbf{J}^{-1} > \mathbf{0}$. ■

We may now return to equation (4.4.6). Take the unique reconstructed non-singular $\mathbf{J}$ and solve

$$\mathbf{J}\mathbf{x} = \mathbf{J}\{x_1, x_2, \ldots, x_n\} = \{1, 0, \ldots, 0\} = \mathbf{e}_1.$$

The solution $\mathbf{x}$ is strictly positive: $\mathbf{x} > \mathbf{0}$. Thus the solution of equation (4.4.6) is

$$\mathbf{d} = c\mathbf{x},$$

for some as yet unknown $c > 0$. The total mass of the system is

$$m = \sum_{i=1}^{n} m_i = \sum_{i=1}^{n} d_i^2 = ||\mathbf{d}||^2 = c^2||\mathbf{x}||^2.$$

Thus, knowing m and $||\mathbf{x}||^2$, we can find $c > 0$ and $\mathbf{d}$, and thus $\mathbf{D}$. Then $\mathbf{K} = \mathbf{DJD}$, and because $\mathbf{K}$ satisfies (4.4.5), it necessarily (Ex. 4.4.1) has the form $\mathbf{K} = \mathbf{E}\hat{\mathbf{K}}\mathbf{E}^T$ given in equation (2.2.12), where $\hat{\mathbf{K}} = diag(k_1, k_2, \ldots, k_n)$. This completes the reconstruction.

The reconstruction from the spectra of a) and c) proceeds along similar lines; we merely renumber the masses starting from the right (Ex. 4.4.2).

This reconstruction may be used in a reversed situation: it shows that any non-singular Jacobi matrix $\mathbf{J}$ may be expressed uniquely as

$$\mathbf{J} = \mathbf{D}^{-1}\mathbf{E}\hat{\mathbf{K}}\mathbf{E}^T\mathbf{D}^{-1}, \tag{4.4.7}$$

where $\mathbf{D}, \hat{\mathbf{K}}$ are strictly positive diagonal matrices and $||\mathbf{D}|| = 1$; this corresponds to $m = 1$ in equation (4.4.6).

Now we consider the fixed-fixed case shown in Figure 4.4.2a; there is essentially only one constraint we can apply, to m_n, as shown in Figure 4.4.2b).

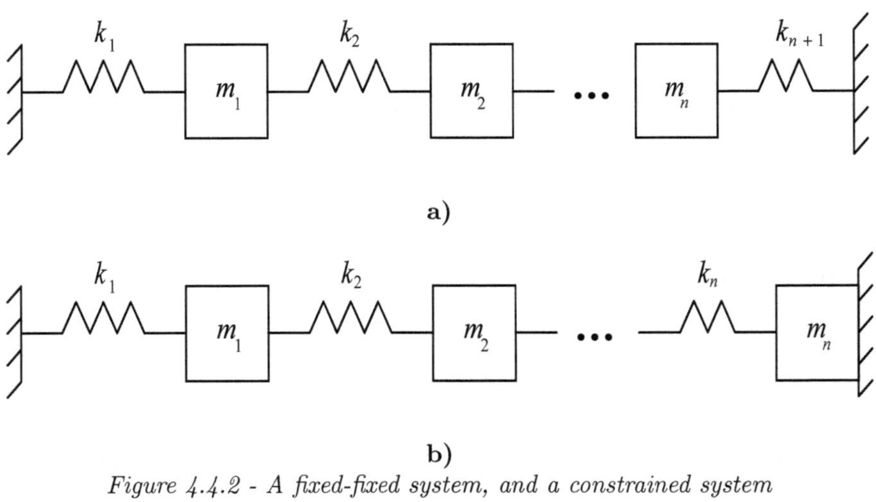

Figure 4.4.2 - A fixed-fixed system, and a constrained system

We start our analysis as before. The stiffness matrix for the system in a) is

$$\mathbf{K} = \begin{bmatrix} k_1 + k_2 & -k_2 & & \\ -k_2 & k_2 + k_3 & -k_3 & \\ \cdot & \cdot & \cdot & \cdot \\ & & & -k_n \\ & & -k_n & k_n + k_{n+1} \end{bmatrix}. \tag{4.4.8}$$

Knowing the spectra $(\lambda_i)_1^n$ and $(\mu_i)_1^{n-1}$ of the systems a) and b) we can construct $\mathbf{J} = \mathbf{D}^{-1}\mathbf{K}\mathbf{D}^{-1}$ where again $\mathbf{M} = \mathbf{D}^2$. Now however

$$\mathbf{K}\{1, 1, \ldots, 1\} = \{k_1, 0, 0, \ldots, 0, k_{n+1}\} : \tag{4.4.9}$$

this states that in order to produce unit static displacements of the masses, we must apply two forces, k_1 at m_1 and k_{n+1} at m_n. Thus

$$\mathbf{DJD}\{1,1,\ldots,1\} = k_1\mathbf{e}_1 + k_{n+1}\mathbf{e}_n$$

so that

$$\mathbf{Jd} = \mathbf{J}\{d_1,d_2,\ldots,d_n\} = (k_1/d_1)\mathbf{e}_1 + (k_{n+1}/d_n)\mathbf{e}_n. \tag{4.4.10}$$

First, consider the equation

$$\mathbf{Jy} = \mathbf{e}_n. \tag{4.4.11}$$

simple algebra shows that the solution is

$$y_i = b_i b_{i+1}\ldots b_{n-1}P_{i-1}/P_i, \tag{4.4.12}$$

where P_i is the ith leading principal minor of $\mathbf{J}$ (see equation (1.4.6). Since $\mathbf{J}$ is positive definite, equation (4.4.12) confirms that the solution $\mathbf{y}$ is positive, as predicted by Theorem 4.4.1. We can find the solution of

$$\mathbf{Jx} = \mathbf{e}_1 \tag{4.4.13}$$

in a similar way (Ex. 4.4.3); all we need here is that, according to Theorem 4.4.1, $\mathbf{x} > \mathbf{0}$.

Using $\mathbf{x}$ and $\mathbf{y}$ we may write the solution of (4.4.10) as

$$\mathbf{d} = (k_1/d_1)\mathbf{x} + (k_{n+1}/d_n)\mathbf{y}. \tag{4.4.14}$$

In particular,

$$\begin{aligned} d_n &= (k_1/d_1)x_n + (k_{n+1}/d_n)y_n \\ &= \frac{k_1}{d_1}x_n + \frac{k_{n+1}}{d_n}\frac{P_{n-1}}{P_n}. \end{aligned} \tag{4.4.15}$$

But $P_n = \prod_{i=1}^n \lambda_i$ and $P_{n-1} = \prod_{i=1}^{n-1}\mu_i$, so that we can write equation (4.4.15) as

$$m_n - k_{n+1}\frac{\prod_{i=1}^{n-1}\mu_i}{\prod_{i=1}^{n}\lambda_i} = \frac{k_1 d_n x_n}{d_1}. \tag{4.4.16}$$

Now consider this equation. The system in Figure 4.4.2a) has $2n+1$ parameters. Choose one of the parameters, and divide the remaining $2n$ parameters by it; we obtain $2n$ ratios. The two spectra $(\lambda_i)_1^n$ and $(\mu_i)_1^{n-1}$ provide $2n-1$ ratios, so one more ratio is needed. The chosen parameter is merely a scaling factor; the total mass, or alternatively one individual mass, say m_n, would determine it. If we take m_n as known, then equation (4.4.16) states that the required $2n$th ratio, k_{n+1}/m_n must be chosen so that

$$0 < \frac{k_{n+1}}{m_n} < \frac{\prod_{i=1}^n \lambda_i}{\prod_{i=1}^{n-1}\mu_i}. \tag{4.4.17}$$

This inequality was first pointed out by Nylen and Uhlig (1997a) [253]. Once we have chosen k_{n+1}/m_n satisfying this inequality, then equation (4.4.16) determines k_1/d_1, since x_n is known, and $d_n = m_n^{\frac{1}{2}}$. With k_{n+1}/d_n and k_1/d_1 known, equation (4.4.14) gives $\mathbf{d}$ and hence $\mathbf{D}$ and $\mathbf{K} = \mathbf{D}^{-1}\mathbf{J}\mathbf{D}^{-1}$. The reconstruction is complete.

The third system is free-free, as shown in Figure 4.4.3a); constraining m_1 we obtain the fixed-free system in Figure 4.4.3b).

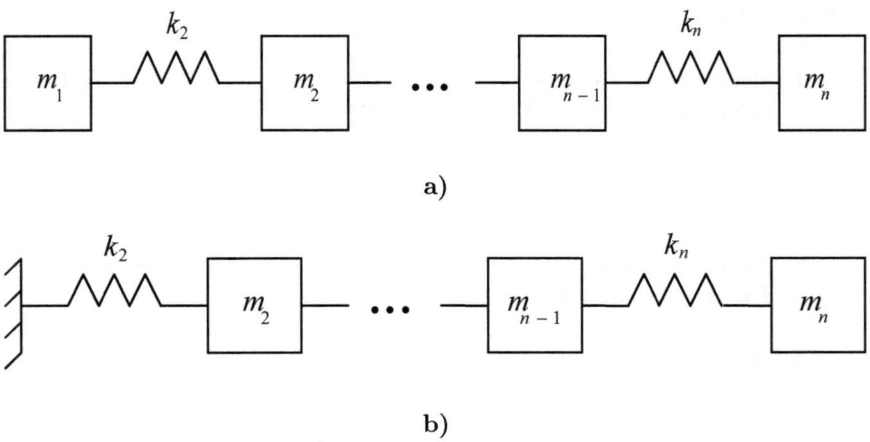

Figure 4.4.3 - A constraint is applied to a free-free system

The pair is essentially the same as the pair in Figure 4.4.1, with $k_1 = 0$. The analysis starts as before; the only difference is that the lowest frequency of a) is $\lambda_1 = 0$. Still, from $\sigma(\mathbf{J})$ and $\sigma(\mathbf{J}_1)$ we can construct $\mathbf{J}$ uniquely, but now $\mathbf{J}$ will be singular, i.e., positive semi-definite.

The stiffness matrix $\mathbf{K}$ of system a) will satisfy

$$\mathbf{K}\{1, 1, \ldots, 1\} = 0. \tag{4.4.18}$$

Now we need a result like Theorem 4.4.1 which covers the case when $\mathbf{J}$ is singular. It is

Theorem 4.4.2 *If $\mathbf{J}$ is a singular Jacobi matrix then the equation $\mathbf{Jx} = \mathbf{0}$ has a unique strictly positive solution $\mathbf{x}$ satisfying $||\mathbf{x}|| = 1$.*

The proof is straightforward; see Ex. 4.4.4.

Now we may complete the reconstruction. We take $\mathbf{J}$ and write $\mathbf{K} = \mathbf{DJD}$, then (4.4.18) becomes

$$\mathbf{Ke} = \mathbf{DJDe} = \mathbf{DJd} = \mathbf{0}.$$

Thus $\mathbf{d} = \mathbf{cx}$ where $\mathbf{x}$ is governed by Theorem 4.4.2, and if the total mass $m = 1$, then $c = 1$. This gives $\mathbf{d}$ and hence $\mathbf{D}$ and

$$\mathbf{K} = \mathbf{DJD} = \mathbf{E}\hat{\mathbf{K}}\mathbf{E}^T$$

where $\hat{\mathbf{K}} = diag(k_1, k_2, \ldots, k_n)$. Again, we can use this result to show that an arbitrary singular Jacobi matrix may be written

$$\mathbf{J} = \mathbf{D}^{-1}\mathbf{E}\hat{\mathbf{K}}\mathbf{E}^T\mathbf{D}^{-1} \tag{4.4.19}$$

where now $\hat{\mathbf{K}}$ has first diagonal entry zero.

Exercises 4.4

1. Show that if $\mathbf{Ke} = k_1\mathbf{e}_1$, and $\mathbf{E}^{-1}$ is given by (2.2.10), then $\mathbf{KE}^{-T}$ is bidiagonal and $\mathbf{E}^{-1}\mathbf{KE}^{-T}$ is diagonal.

2. Reconstruct the system of Figure 4.4.1a) from the spectra $(\lambda_i)_1^n$ and $(\mu_i)_1^{n-1}$ of a) and c) respectively.

3. Use the solution (4.4.12) of equation (4.4.11), and the transformation from $\mathbf{J}$ to $\bar{\mathbf{J}}$ given in (4.3.9) to find the solution to equation (4.4.13).

4. Provide a constructive proof of Theorem 4.4.2, by writing $\mathbf{x}$ in terms of the principal minors of $\mathbf{J}$.

5. Suppose that the eigenvalues $(\lambda_i)_1^n$ of the system in Figure 4.4.2a are known, as are the eigenvalues $(\lambda_i^*)_1^n$ when the stiffness k_{n+1} is replaced by some unknown stiffness k_{n+1}^*. Show that there is a one-parameter family of systems, each member of which has the stated eigenvalues.

6. Show that if $\mathbf{J}$ is a non-singular Jacobi matrix, then its inverse $\mathbf{C} = \mathbf{J}^{-1}$ has the form

$$\mathbf{C} = \begin{bmatrix} u_1v_1 & u_1v_2 & \ldots & u_1v_n \\ u_1v_2 & u_2v_2 & \ldots & u_2v_n \\ \cdot & \cdot & \ldots & \cdot \\ u_1v_n & u_2v_n & \ldots & u_nv_n \end{bmatrix}$$

i.e.,

$$c_{ij} = \begin{cases} u_iv_j & i \leq j \\ u_jv_i & i \geq j \end{cases}$$

and that $(u_i)_1^n$, $(v_i)_1^n$ are strictly positive, and satisfy

$$\frac{u_1}{v_1} \leq \frac{u_2}{v_2} \leq \cdots \frac{u_n}{v_n}.$$

This result is quoted in Gantmacher and Krein (1950) [98], but may have been known earlier.

4.5 Reconstruction by using modification

The simplest way to modify a system is to attach a spring at a free end, thus going from the system in Figure 4.5.1a) to that in Figure 4.5.1b). (We have renumbered the masses so that the spring is attached at m_1.)

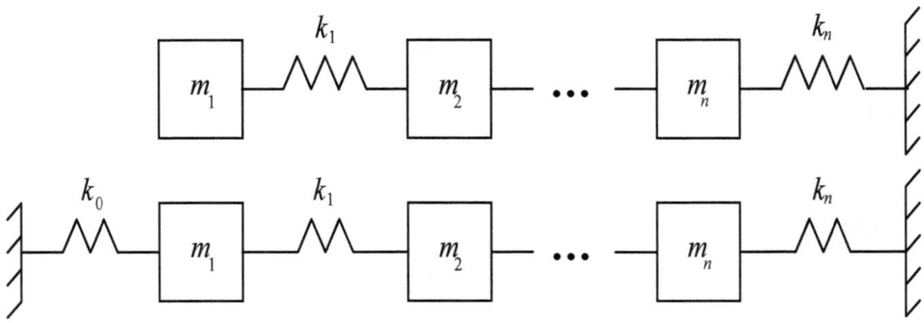

Figure 4.5.1 - A spring is added to the system

This is an example of the analysis of Section 4.3. The spectra for a) and b) are $\sigma(\mathbf{J}) = (\lambda_i)_1^n$ and $\sigma(\mathbf{J}^*) = (\lambda_i^*)_1^n$ respectively. Because we have added stiffness to the system, we have $\lambda_i^* > \lambda_i$, as in (4.3.21).

i) Use the trace condition to find

$$a_1^* - a_1 = \sum_{i=1}^{n} (\lambda_i^* - \lambda_i)$$

ii) Use $a_1 = k_1/m_1$ and $a_1^* = (k_1 + k_0)/m_1$ to find $k_0/m_1 = a_1^* - a_1$.

iii) Use $a_1^* - a_1$ and equation (4.3.22) to find x_{i1}^2, and hence $\mathbf{x}_1 = \{x_{11}, x_{21}, \ldots, x_{n1}\}$.

iv) Use the Lanczos algorithm to find $\mathbf{J}$.

v) Use a variant of the analysis given in Section 4.4 to untangle $\mathbf{K}$ and $\mathbf{M}$ from $\mathbf{J}$.

As an alternative modification we may add mass to the system, specifically a mass m_1^* to m_1. In this case it is easier to work initially with the original equation (4.4.1) than with the reduced equation (4.4.2). Again, we start with the free-fixed system of Figure 4.5.1a. The eigenvalue problem for a) is

$$\mathbf{K}\mathbf{y}_i = \lambda_i \mathbf{M}\mathbf{y}_i.$$

That for the modified system is

$$\mathbf{K}\mathbf{y} = \lambda \mathbf{M}^* \mathbf{y}, \tag{4.5.1}$$

where $\mathbf{M}^* = \mathbf{M} + m_1^*\mathbf{E}_{1,1}$. Since we have added mass to the system, the eigenvalues must satisfy

$$0 < \lambda_1^* < \lambda_1 < \cdots < \lambda_n^* < \lambda_n. \tag{4.5.2}$$

Express $\mathbf{y}$ as a combination of the $\mathbf{y}_i$:

$$\mathbf{y} = \sum_{i=1}^{n} \alpha_i \mathbf{y}_i, \tag{4.5.3}$$

then

$$\mathbf{Ky} = \sum_{i=1}^{n} \alpha_i \mathbf{K}\mathbf{y}_i = \sum_{i=1}^{n} \lambda_i \alpha_i \mathbf{M}\mathbf{y}_i,$$

and

$$\mathbf{M}^*\mathbf{y} = \sum_{i=1}^{n} \alpha_i \mathbf{M}\mathbf{y}_i + m_1^*\mathbf{E}_{1,1}\mathbf{y},$$

so that equation (4.5.1) becomes

$$\sum_{i=1}^{n} \lambda_i \alpha_i \mathbf{M}\mathbf{y}_i = \lambda \sum_{i=1}^{n} \alpha_i \mathbf{M}\mathbf{y}_i + \lambda m_1^* \mathbf{E}_{1,1}\mathbf{y}.$$

Premultiply both sides by $\mathbf{y}_j^T$, using the orthonormality condition $\mathbf{y}_j^T\mathbf{M}\mathbf{y}_i = \delta_{ij}$:

$$\alpha_j \lambda_j = \alpha_j \lambda + \lambda m_1^* y_{1j} y_1,$$

and on substituting for α_j in (4.5.3) and equating the first elements of the vectors on each side, we find

$$1 = \lambda m_1^* \sum_{i=1}^{n} \frac{(y_{1i})^2}{\lambda_i - \lambda}. \tag{4.5.4}$$

In order to use this equation to obtain the first components u_{1i} of the eigenvectors $\mathbf{u}_i$ of the reduced equation, for use in the Lanczos equation, we need to express y_{1i} in terms of u_{1i}. The equation $\mathbf{Dy} = \mathbf{u}$ gives $d_1 y_{1i} = u_{1i} = x_{i1}$ that we may write (4.5.4) as

$$1 = \lambda \alpha \sum_{i=1}^{n} \frac{(x_{i1})^2}{\lambda_i - \lambda}, \quad \alpha = \frac{m_1^*}{m_1}.$$

Since the roots of the equation are $(\lambda_i^*)_1^n$ we have

$$1 - \lambda \alpha \sum_{i=1}^{n} \frac{(x_{i1})^2}{\lambda_i - \lambda} = c \frac{\prod_{i=1}^{n}(\lambda_i^* - \lambda)}{\prod_{i=1}^{n}(\lambda_i - \lambda)}. \tag{4.5.5}$$

Equating both sides for $\lambda = 0$ and $\lambda \to \infty$, we have

$$1 = c \prod_{i=1}^{n} (\lambda_i^*/\lambda_i), \quad 1 + \alpha \sum_{i=1}^{n} x_{i1}^2 = c.$$

The interlacing condition (4.5.2) gives

$$c = \prod_{i=1}^{n} (\lambda_i/\lambda_i^*) > 1.$$

The orthonormality condition gives $\sum_{i=1}^{n} x_{i1}^2 = 1$, so that $m_1^*/m_1 = \alpha = c - 1 > 0$. Finally, multiplying (4.5.5) throughout by $\lambda_i - \lambda$ and then putting $\lambda = \lambda_i$ we find

$$-\lambda_i \alpha x_{i1}^2 = c \prod_{j=1}^{n} (\lambda_j^* - \lambda_i) / \prod_{j=1}^{n}{}' (\lambda_j - \lambda_i). \tag{4.5.6}$$

The interlacing condition (4.5.2) ensures that $x_{i1}^2 > 0$. Now we use $\mathbf{x}_1$ in the Lanczos algorithm, and the untangling procedure as before.

There are still more ways in which to obtain second spectrum, for which see Nylen and Uhlig (1997a) [253], Nylen and Uhlig (1997b) [254]. Ram (1993) [276] supposes that the system of Figure 4.5.1 is modified by adding both a mass m to m_1 and a spring k_0. He makes use of some simple but powerful results found in Ram and Blech (1991) [277].

We close this section by supposing that an oscillating force $F \sin \omega t$ is applied to the free end of the spring-mass system of Figure 4.5.1a). The matrix equation governing the response $\mathbf{y} \sin \omega t$ is

$$(\mathbf{K} - \lambda \mathbf{M})\mathbf{y} = F\mathbf{e}_1.$$

Write

$$\mathbf{y} = \sum_{i=1}^{n} \alpha_i \mathbf{y}_i,$$

where $\mathbf{y}_i$ is the ith eigenvector, normalised so that

$$\mathbf{y}_i^T \mathbf{M} \mathbf{y}_j = \delta_{ij}.$$

We obtain

$$(\lambda_i - \lambda)\alpha_i = F y_{1i},$$

and hence

$$\mathbf{y} = F \sum_{i=1}^{n} \frac{y_{1i}\mathbf{y}_i}{\lambda_i - \lambda},$$

so that

$$y_1 = F \sum_{i=1}^{n} \frac{y_{1i}^2}{\lambda_i - \lambda}. \tag{4.5.7}$$

When the eigenvalue problem is reduced to standard, $\mathbf{J}$, form, then $\mathbf{D}\mathbf{y} = \mathbf{u}$, so that $d_1 y_{1i} = u_{1i} = x_{i1}$ so that we may write

$$y_1 = \frac{F}{m_1} \sum_{i=1}^{n} \frac{x_{i1}^2}{\lambda_i - \lambda}, \tag{4.5.8}$$

where, as usual, $\mathbf{x}_1 = \{x_{11}, x_{21}, \ldots, x_{n1}\}$ is the vector of first components of the eigenvectors of $\mathbf{J}$.

The quantity y_1/F is called the *frequency response function*, specifically the frequency response function for the displacement y_1 due to a unit force applied at y_1. This function may also be identified as a direct *receptance* for y_1, as described, for instance, in Bishop and Johnson (1960) [34]. The two spectra $\sigma(\mathbf{J}) = (\lambda_i)_1^n$ and $\sigma(\mathbf{J}_1) = (\mu_i)_1^{n-1}$ are the poles and zeros of the response function. The interlacing of these two spectra may thus be interpreted as the interlacing of the poles and zeros of the response function, a result which is well known in control theory. The result of Section 4.3 may thus be stated as follows: the response function, and specifically its poles and zeros, uniquely determines the matrix $\mathbf{J}$. As we have seen, once we know $\mathbf{J}$ and the *form* of the stiffness matrix $\mathbf{K}$, we may untangle $\mathbf{M}$ and $\mathbf{K}$ from $\mathbf{J}$. See Gladwell and Gbadeyan (1985) [106] for an alternative treatment.

An experimental - theory study of the problem of reconstructing a spring-mass system from frequency response data for an actual system may be found in Gladwell and Movahhedy (1995) [123] and Movahhedy, Ismail and Gladwell (1995) [242].

4.6 Persymmetric systems

It was shown in Section 4.3 that a persymmetric Jacobi matrix $\mathbf{J}$ can be reconstructed uniquely from its eigenvalues. We shall now consider some physical problems relating to persymmetric matrices. Figure 4.6.1 shows a system of $2n$ masses connected by $(2n+1)$ springs and fixed at each end. Suppose that the system is symmetrical about the mid point, so that

$$m_r = m_{2n+1-r}, \quad k_r = k_{2n+1-r}, \quad (r = 1, 2, \ldots, n). \tag{4.6.1}$$

The odd numbered principal modes of the system will be symmetrical about the mid-point; they will thus be the principal modes of one half (say the left-hand half) of the system with the mid-point of the system free, as in Figure 4.6.2(a). Thus the odd numbered eigenvalues $\wedge_1, \wedge_3, \ldots, \wedge_{2n-1}$ of the complete system will be the eigenvalues of the left-hand half under the conditions fixed-free, i.e.,

$$\wedge_{2i-1} = \lambda_i, \quad i = 1, 2, \ldots, n. \tag{4.6.2}$$

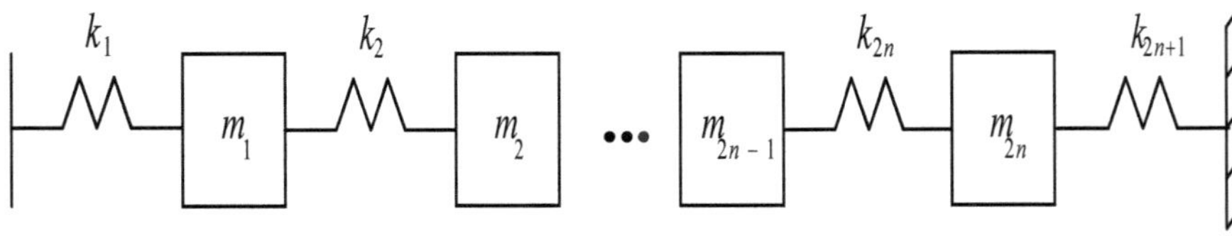

Figure 4.6.1 - A symmetrical system with 2n-masses

On the other hand, the even-numbered principal modes of the system will be antisymmetrical about the mid-point so that the even-numbered eigenvalues $\wedge_2, \wedge_4, \ldots, \wedge_{2n}$ will be the eigenvalues of the left-hand half under the condition fixed-fixed, as in Figure 4.6.2(b).

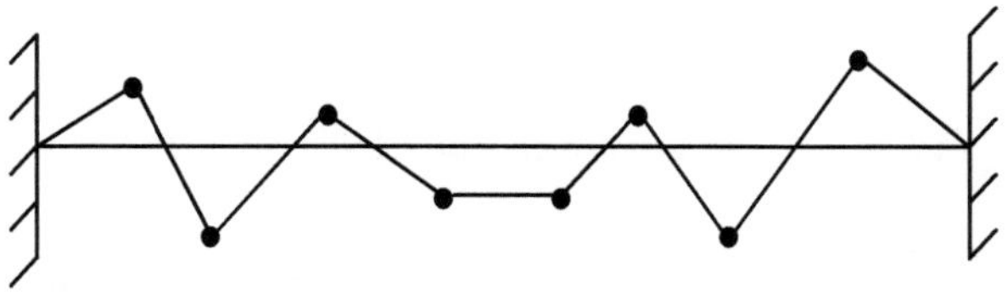

(a)

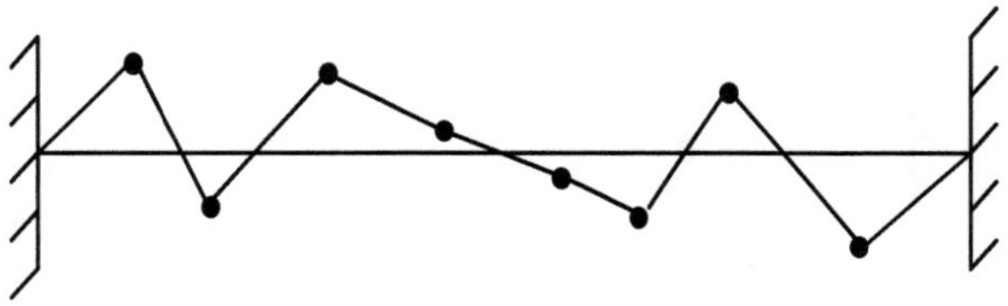

(b)

Figure 4.6.2(a) The odd numbered modes are symmetrical, (b) The even numbered ones are antisymmetrical.

Thus

$$\wedge_{2i} = \lambda_i^*, \quad i = 1, 2, \ldots, n. \tag{4.6.3}$$

This means that the left-hand half, and hence the whole system may be uniquely constructed, using the analysis of Section 4.4 from the eigenvalues $\wedge_1, \ldots, \wedge_{2n}$ and the total mass.

Figure 4.6.3 shows a symmetrical system with $2n-1$ masses and $2n$ springs. Now the odd-numbered symmetrical modes will be the modes of the left-hand half with $(m_n/2)$ at the end and free there, as in Figure 4.6.4(a). On the other hand, the even-numbered, antisymmetrical modes will be the modes of left-hand half with m_n fixed as in Figure 4.6.4(b). Thus

$$\wedge_{2i-1} = \lambda_i, \quad i = 1, 2, \ldots, n, \tag{4.6.4}$$

$$\wedge_{2i} = \mu_i, \quad i = 1, 2, \ldots, n-1. \tag{4.6.5}$$

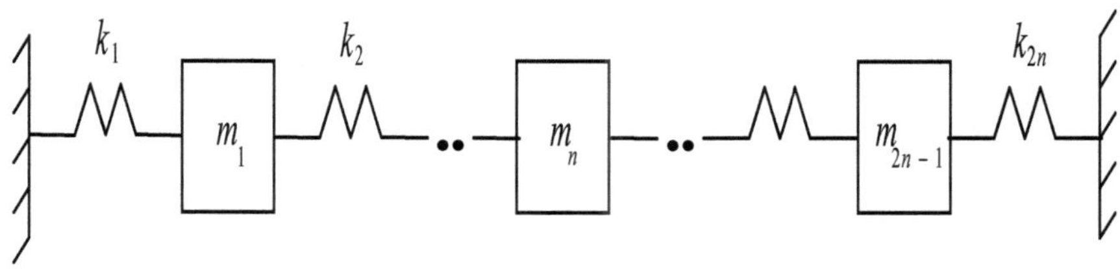

Figure 4.6.3 - A symmetrical system with $2n-1$ masses.

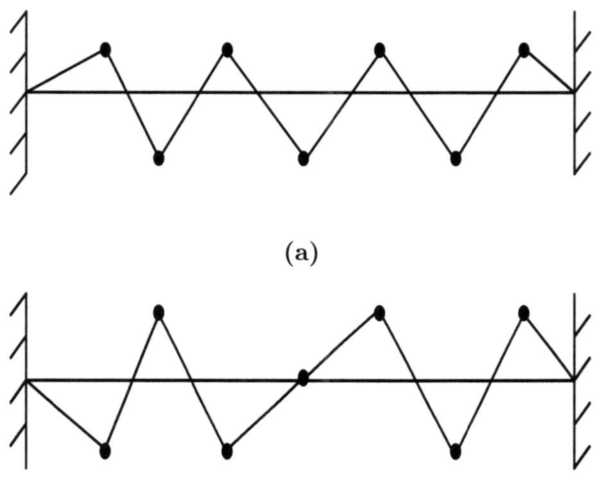

Figure 4.6.4 - (a) The odd numbered modes are symmetrical. (b) The even numbered modes are antisymmetrical.

4.7 Inverse generalised eigenvalue problems

In this section we consider how we can reconstruct a finite element model from spectral data.

The eigenvalue problem is

$$(\mathbf{K} - \lambda\mathbf{M})\mathbf{y} = \mathbf{0}, \tag{4.7.1}$$

where as in (2.4.10), *both* $\mathbf{K}$ and $\mathbf{M}$ are symmetric tridiagonal, $\mathbf{K}$ with negative codiagonal, $\mathbf{M}$ with positive codiagonal. Since one spectrum is insufficient even to reconstruct one tridiagonal matrix, it is certainly insufficient to reconstruct two. We therefore assume Gladwell (1999) [127] that $\mathbf{M}$ can be written in terms of $\mathbf{K}$:

$$\mathbf{M} = \mathbf{D}^2 - c\mathbf{K}, \quad c > 0, \tag{4.7.2}$$

where $\mathbf{D}$ is an as yet undetermined diagonal matrix with positive entries, and c is an arbitrary positive number. Since $\mathbf{K}$ has negative codiagonal, $\mathbf{M}$ will have positive codiagonal. Now

$$\begin{aligned}\mathbf{K} - \lambda\mathbf{M} &= \mathbf{K} - \lambda(\mathbf{D}^2 - c\mathbf{K}) = (1 + c\lambda)\mathbf{K} - \lambda\mathbf{D}^2 \\ &= (1 + c\lambda)\{\mathbf{K} - \nu\mathbf{D}^2\}, \quad \nu = \lambda/(1 + c\lambda).\end{aligned}$$

Thus (4.7.1) reduces to

$$(\mathbf{K} - \nu\mathbf{D}^2)\mathbf{y} = \mathbf{0}, \tag{4.7.3}$$

which, as in Section 3.1, we can reduce to

$$(\mathbf{J} - \nu\mathbf{I})\mathbf{u} = \mathbf{0}, \tag{4.7.4}$$

where $\mathbf{J} = \mathbf{D}^{-1}\mathbf{K}\mathbf{D}^{-1}$ and $\mathbf{u} = \mathbf{D}^{-1}\mathbf{y}$.

Suppose that (4.7.1) has specified eigenvalues $(\lambda_i)_1^n$, where $\lambda_i \geq 0$, then $\mathbf{J}$ has eigenvalues $(\nu_i)_1^n$ where $\nu_i = \lambda_i/(1 + c\lambda_i) \geq 0$, showing that $\mathbf{J}$, and thus $\mathbf{K}$, is positive semi-definite. The matrix $\mathbf{M}$ can be written

$$\mathbf{M} = \mathbf{D}(\mathbf{I} - c\mathbf{J})\mathbf{D} \tag{4.7.5}$$

and the matrix $\mathbf{I} - c\mathbf{J}$ has eigenvalues $1 - c\nu_i = 1/(1 + c\lambda_i) > 0$, showing that $\mathbf{M}$ is positive definite.

To reconstruct $\mathbf{J}$ we need a second spectrum. If the eigenvalues of (4.7.1) under the constraint $u_n = 0$ are $(\mu_i)_1^{n-1}$, then the eigenvalues of (4.7.4) under the same constraint will be $\sigma_i = \mu_i/(1 + c\mu_i)$. We note that the interlacing

$$\lambda_1 < \mu_1 < \lambda_2 < \cdots < \mu_{n-1} < \lambda_n \tag{4.7.6}$$

yields the interlacing

$$\nu_1 < \sigma_1 < \nu_2 < \cdots < \sigma_{n-1} < \nu_n. \tag{4.7.7}$$

Having found $\mathbf{J}$, we need to find $\mathbf{D}$ so that $\mathbf{K} = \mathbf{DJD}$ satisfies the characteristic stiffness equation (4.4.9). This can be done exactly as in Section 4.4. Gladwell (1999) [127] finds wider families of systems with the given spectra. See Ram and Gladwell (1994) [289] for a different approach to reconstructing a finite element model of a rod.

4.8 Interior point reconstruction

Suppose, following Gladwell and Willms (1988) [113], we have a spring-mass system with n masses, under some end conditions, as in Figure 4.8.1(a). (We exclude the free-free condition at this stage.) If a sinusoidal force $F \sin \omega t$ is applied to mass m_{m+1}, where $0 < m < n - 1$, then the response at mass m_{m+1} may be calculated as in equation (4.5.7):

$$x_{m+1}/F = \sum_{i=1}^{n} \frac{(x_{m+1,i})^2}{\lambda_i - \lambda}.$$

The poles of this response function are the eigenvalues $(\lambda_i)_1^n$ of the whole system, A. The zeros of the response function will be the eigenvalues of the system constrained so that $x_{m+1} = 0$, i.e., they will be eigenvalues of the systems, B, on the left, or C, on the right, of x_{m+1}, as shown in Figure 4.8.1(b). Different ways of assigning the eigenvalues of the constrained system to the two subsystems B and C will lead to different reconstructed systems. When this assignment has been made, then we know the eigenvalues $(\lambda_i)_1^n, (\mu_i)_1^m$ and $(\nu_i)_1^p$ of systems A, B, C respectively; $p = n - m - 1$. Within themselves these sets of eigenvalues must be distinct. There are two cases.

a) The constrained system has no double eigenvalues. That is, *all* the $(\mu_i)_1^m$and $(\nu_i)_1^p$ are distinct; if they are arranged in ascending order and relabelled $(\tilde{\mu}_i)_1^{n-1}$, they will satisfy

$$\lambda_1 < \tilde{\mu}_1 < \lambda_2 < \cdots < \tilde{\mu}_{n-1} < \lambda_n;$$

this is equivalent to the statement that no eigenvector $\mathbf{x}_i$ of $\mathbf{J}$ has a node at x_{m+1}, i.e., $x_{m+1,i} \neq 0$ for all $i = 1, 2, \ldots, n$.

b) Two members of a pair (μ_j, ν_k) are identical; now there is an i such that $\lambda_i = \mu_j = \nu_k$; this will occur iff $x_{m+1,i} = 0$. There can be more than one such pair.

To analyse the situation we suppose that the eigenvalue equation (4.4.1) has been reduced to normal form, (4.4.2), and we partition $\mathbf{J}$ as

$$\mathbf{J} = \begin{matrix} \\ m \\ 1 \\ p \end{matrix} \begin{matrix} \begin{matrix} m & \quad 1 & \quad p \end{matrix} \\ \begin{bmatrix} \mathbf{B} & -\mathbf{b}_m & \mathbf{0} \\ -\mathbf{b}_m^T & a_{m+1} & -\mathbf{c}_{m+1}^T \\ \mathbf{0} & -\mathbf{c}_{m+1} & \mathbf{C} \end{bmatrix} \end{matrix}, \tag{4.8.1}$$

where $\mathbf{b}_m^T = \{0, 0, \ldots, b_m\}$, $\quad \mathbf{c}_{m+1}^T = \{b_{m+1}, 0, 0, \ldots, 0\}$.

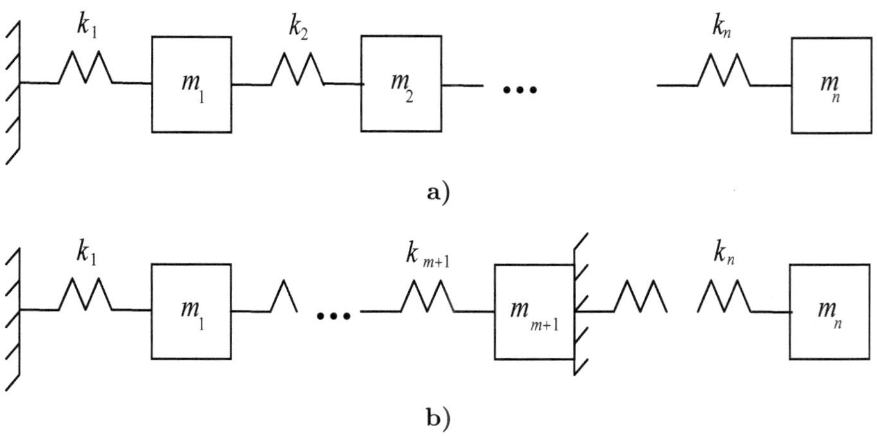

Figure 4.8.1 - The mass m_{m+1} is constrained.

Now we consider the principal minors of $\lambda\mathbf{I} - \mathbf{J}$. We denote the leading principal minors by $p_i(\lambda)$ and the trailing principal minors by $q_i(\lambda)$. The Laplace expansions of $p_n(\lambda) = \det(\lambda\mathbf{I} - \mathbf{J})$ using the first m and first $m+1$ rows are

$$p_n(\lambda) = p_m(\lambda)q_{p+1}(\lambda) - b_m^2 p_{m-1}(\lambda)q_p(\lambda), \tag{4.8.2}$$

$$= p_{m+1}(\lambda)q_p(\lambda) - b_{m+1}^2 p_m(\lambda)q_{p-1}(\lambda). \tag{4.8.3}$$

We know that

$$p_n(\lambda) = \prod_{i=1}^{n}(\lambda - \lambda_i), \quad p_m(\lambda) = \prod_{j=1}^{m}(\lambda - \mu_j), \quad q_p(\lambda) = \prod_{k=1}^{p}(\lambda - \nu_k),$$

and thus equation (4.8.2), (4.8.3) give

$$p_n(\mu_j) = -b_m^2 p_{m-1}(\mu_j)q_p(\mu_j), \tag{4.8.4}$$

$$p_n(\nu_k) = -b_{m+1}^2 p_m(\nu_k)q_{p-1}(\nu_k). \tag{4.8.5}$$

In case (a), *all* the quantities appearing in the latter equations are non-zero, so that, apart from the factors b_m^2 and b_{m+1}^2, these equations yield $p_{m-1}(\mu_j)$ and $q_{p-1}(\nu_k)$, respectively. These quantities are just what is needed to compute the matrices $\mathbf{B}$ and $\mathbf{C}$, respectively, using Forsythe's algorithm in Section 3.2. The weights $(w_j)_b$ for $\mathbf{B}$ are given by

$$\begin{aligned} b_m^2(w_j)_b &= b_m^2 p_{m-1}(\mu_j)/p_m'(\mu_j), \\ &= -p_n(\mu_j)/[p_m'(\mu_j)q_p(\mu_j)], \end{aligned} \tag{4.8.6}$$

while those for $\mathbf{C}$ are

$$\begin{aligned} b_{m+1}^2(w_k)_c &= b_{m+1}^2 q_{p-1}(\nu_k)/q_p'(\nu_k), \\ &= -p_n(\nu_k)/[q_p'(\nu_k)p_m(\nu_k)]. \end{aligned} \tag{4.8.7}$$

To verify that the weights $(w_i)_b$ are positive, we suppose that μ_j has s ν's to its left, and $p - s$ to its right; then $\nu_{j+s} < \mu_j < \nu_{j+s+1}$. If a number x may be written $x = (-1)^n c$, where $c > 0$, then we say $sgn(x) = n$. Now we can easily verify that

$$sgn[p_n(\mu_j),\ p_m'(\mu_j),\ q_p(\mu_j)] = [n - j - s, m - j, p - s]$$

so that

$$\begin{aligned} sgn(w_j)_b &= 1 + (n - j - s) + (m - j) + p - s \\ &= 2n - 2j - 2s = even \end{aligned}$$

so that $(w_j)_b > 0$; we may prove similarly that $(w_k)_c > 0$.

Thus $\mathbf{B}$ may be reconstructed uniquely. At the end, $p_{m-1}(\lambda)$ will be known, and so the $p_{m-1}(\mu_j)$ will be known. Any one of these values may be substituted into (4.8.4) to yield b_m^2. The matrix $\mathbf{C}$ and b_{m+1}^2 may be found in a similar manner.

In case (b) there is a common factor $\lambda - \lambda_i \equiv \lambda - \mu_j = \lambda - \nu_k$ in each of the equations (4.8.2) and (4.8.3). Cancel this factor and then put $\lambda = \lambda_i$. Since

$$\begin{aligned} p_{m+1}(\lambda) &= (\lambda - a_{m+1})p_m(\lambda) - b_m^2 p_{m-1}(\lambda), \\ q_{p+1}(\lambda) &= (\lambda - a_{m+1})q_p(\lambda) - b_{m+1}^2 q_{p-1}(\lambda), \end{aligned}$$

we have

$$\begin{aligned} p_{m+1}(\mu_j) &= -b_m^2 p_{m-1}(\mu_j), \\ q_{p+1}(\nu_k) &= -b_{m+1}^2 q_{p-1}(\nu_k), \end{aligned}$$

and thus both equations (4.8.2) and (4.8.3) reduce to

$$p_n'(\lambda_i) = -b_m^2 p_{m-1}(\mu_j) q_p'(\nu_k) - b_{m+1}^2 p_m'(\mu_j) q_{p-1}(\nu_k). \tag{4.8.8}$$

This is the single equation that replaces the pair of equations (4.8.4) and (4.8.5) for the common eigenvalue. Using (4.8.6) and (4.8.7), we may write (4.8.8) as

$$W_i = -p_n'(\lambda_i)/[p_m'(\mu_j) q_p'(\nu_k)] = b_m^2 (w_j)_b + b_{m+1}^2 (w_k)_c, \tag{4.8.9}$$

and we note that W_i is a positive quantity. Now we proceed as follows to find the family of Jacobi matrices having the specified eigenvalues. Choose $\alpha \in (0, \frac{\pi}{2})$ and put

$$b_m^2 (w_j)_b = W_i \cos^2 \alpha, \quad b_m^2 (w_k)_c = W_i \sin^2 \alpha.$$

If there is more than one triple of common eigenvalues, then this procedure may be followed for each. Combine these weights with those corresponding to the distinct eigenvalues, and compute **B** and **C**. At the final stage $p_{m-1}(\lambda)$ and $q_{p-1}(\lambda)$ will be known, so that b_m^2 and b_{m+1}^2 may be found from equations (4.8.4) and (4.8.5) for one of the distinct eigenvalues, and there will be at least one, as before.

There is an alternative procedure which elucidates the situation in which one or more triples λ_i, μ_j, ν_k are equal, and which uses the Lanczos algorithm. Express the eigenvalue problem for **J** in (4.8.1) in terms of the normalised eigenvectors $(\mathbf{y}_j)_1^m$ and $(\mathbf{z}_j)_1^p$ of **B** and **C** respectively. Thus

$$\mathbf{x} = \tilde{\mathbf{I}}_m (\sum_{j=1}^{m} p_j \mathbf{y}_j) + x_{m+1} \mathbf{e}_{m+1} + \tilde{\mathbf{J}}_p (\sum_{k=1}^{p} q_k \mathbf{z}_k)$$

where $\tilde{\mathbf{I}}_m = \begin{bmatrix} \mathbf{I}_m \\ \mathbf{0} \end{bmatrix}$ and $\tilde{\mathbf{J}}_p = \begin{bmatrix} 0 \\ \mathbf{I}_p \end{bmatrix}$. The eigenvalue problem becomes

$$\begin{bmatrix} \lambda - \mu_1 & & & -s_1 & & & \\ & \ddots & & \vdots & & & \\ & & \lambda - \mu_m & -s_m & & & \\ -s_1 & \cdots & -s_m & \lambda - a_{m+1} & -t_1 & \cdots & -t_p \\ & & & -t_1 & \lambda - v_1 & & \\ & & & \vdots & & \ddots & \\ & & & -t_p & & & \lambda - v_p \end{bmatrix} \begin{bmatrix} p_1 \\ \vdots \\ p_m \\ x_{m+1} \\ q_1 \\ \vdots \\ q_p \end{bmatrix} = \mathbf{0} \tag{4.8.10}$$

where

$$s_j = b_m y_{m,j}, \quad t_k = b_{m+1} z_{1,k}. \tag{4.8.11}$$

Thus

$$\begin{aligned}(\lambda - \mu_j)p_j - s_j x_{m+1} &= 0, \quad j = 1, 2, \ldots, m,\\ (\lambda - \nu_k)q_k - t_k x_{m+1} &= 0, \quad k = 1, 2, \ldots, p,\end{aligned}$$

so that

$$\{a_{m+1} - \lambda_i + \sum_{j=1}^{m} \frac{s_j^2}{\lambda_i - \mu_j} + \sum_{k=1}^{p} \frac{t_k^2}{\lambda_i - \nu_k}\} x_{m+1,i} = 0, \quad i = 1, \ldots, n.$$

In case (a) $x_{m+1,i} \neq 0; i = 1, 2, \ldots, n$, so that

$$a_{m+1} - \lambda + \sum_{j=1}^{m} \frac{s_j^2}{\lambda - \mu_j} + \sum_{k=1}^{p} \frac{t_k^2}{\lambda - \nu_k} = \frac{-p_n(\lambda)}{p_m(\lambda) q_p(\lambda)} \tag{4.8.12}$$

which yields

$$s_j^2 = \frac{-p_n(\mu_j)}{p'_m(\mu_j) q_p(\mu_j)}, \quad t_k^2 = \frac{-p_n(\nu_k)}{p_m(\nu_k) q'_p(\nu_k)} \tag{4.8.13}$$

for $j = 1, 2, \ldots, m; k = 1, 2, \ldots, p$, in agreement with (4.8.6), (4.8.7). Now b_m^2, b_{m+1}^2 may be computed from

$$b_m^2 = \sum_{j=1}^{m} s_j^2, \quad b_{m+1}^2 = \sum_{k=1}^{p} t_k^2. \tag{4.8.14}$$

With $(y_{m,j})_1^m$ and $(z_{1,k})_1^p$ known, from equations (4.8.11)-(4.8.14), **B** and **C** may be computed by using the Lanczos algorithm.

In case (b), suppose that there are $r \geq 1$ triples $\{\lambda_{iq}, \mu_{jq}, \nu_{kq}\}, q = 1, 2, \ldots, r$ such that $\lambda_{iq} = \mu_{jq} = \nu_{kq}$, then

$$f(\lambda) \equiv a_{m+1} - \lambda + \sum_{j=1}^{m}{}^{*} \frac{s_j^2}{\lambda - \mu_j} + \sum_{k=1}^{p}{}^{*} \frac{t_k^2}{\lambda - \nu_k} + \sum_{q=1}^{r} \frac{s_{jq}^2 + t_{kq}^2}{\lambda - \lambda_{iq}} \tag{4.8.15}$$

has, as its $m + p - r + 1 = n - 2$ roots, the $n - r$ non-degenerate λ_i. In equation (4.8.15), *means that the degenerate triples are omitted. Now the separate s_j^2 and t_k^2, and the values $W_q = s_{jq}^2 + t_{kq}^2$ for the degenerate modes will be known. Thus as before

$$s_{jq}^2 + t_{kq}^2 = W_q, \quad s_{jq}^2 = W_q \cos^2 \alpha_q, \quad t_{kq}^2 = W_q \sin^2 \alpha_q$$

where W_q is defined as in (4.8.9). With the α_q chosen, the s_{jq}^2 and t_{kq}^2 are all known. Equation (4.8.14) yields b_m^2 and b_{m+1}^2 as functions of the parameters $\{\alpha_q\}_1^r$ (and note that $b_m^2 + b_{m+1}^2$ is invariant) so that the $(y_{m,i})_1^m$ and $(z_{1,j})_1^p$ are known, and **B** and **C** may be calculated from the Lanczos algorithm.

An alternative approach to the interior reconstruction problem may be found in Nylen and Uhlig (1997a) [253].

The mass-spring models considered in this chapter are very similar to the shear building model used extensively by Takewaki and his coworkers. They have formulated various hybrid inverse problems in which part of a structure is given and part is yet to be found in order to yield a structure with specified spectral (eigenvalue or modal) properties. Full, definitive description of these problems and their use in structural design may be found in the monograph Takewaki (2000) [321]. Among the original papers most closely related to the concerns of this chapter are the following: Takewaki and Nakamura (1995) [317], Takewaki, Nakamura and Arita (1996) [318] and Takewaki and Nakamura (1997) [319], Takewari (1999) [320].

Chapter 5

Inverse Problems for Some More General Systems

Words differently arranged have a different meaning, and meanings differently arranged have different effects.
Pascal's *Pensées*, 23

5.1 Introduction: graph theory

The inverse problems considered in Chapter 4 are special, simply because Jacobi matrices are special matrices. In this chapter we will consider some slightly more general problems but must admit that there are still only a few problems that we have been able to solve.

The special feature of a Jacobi matrix is its *structure*: it is tridiagonal, with strictly negative codiagonal. (It is also positive semi-definite, but that is another matter.) The structure of the matrix **J** in equation (4.4.2) is related to the structures of **K** and **M** in (4.4.1); **K** is tridiagonal while **M** is diagonal. The structures of **K** and **M**, in turn, derive from the structure of the system, an in-line mass system, to which they belong. **K**, the stiffness matrix, relates to the stiffnesses, the connectors, between masses. **K** is tridiagonal because each interior mass m_i $2 \le i \le n-1$ is connected only to its immediate neighbours m_{i-1} and m_{i+1}; the end masses m_1 and m_n each have just one neighbour m_2 or m_n respectively. The natural tool for describing and analysing the structure of a system is *graph theory*.

This is not the place to prove any *theorems* in graph theory, but it is useful to introduce some of the basic *concepts*. A *graph* $\mathcal{G}$ is a set of *vertices*, connected by *edges*. The set of vertices is called the *vertex set*, and is denoted by $\mathcal{V}$; the set of edges is called the *edge set*, $\mathcal{E}$. Figure 5.1.1 shows a graph. This is actually an example of a *simple*, *undirected* graph. It is *simple* because there is at most one edge connecting any two vertices; the edge connecting vertices i and j is denoted by (i, j). The graph is *undirected* because there is no preferred direction associated with an edge. Henceforth, the terms *graph* will be used to mean a *simple, undirected graph.*

The *adjacency matrix* $\mathbf{A}$ of a graph $\mathcal{G}$ is the symmetric matrix defined by

$$\left.\begin{array}{lll} a_{ij} & = 1 & \text{iff } i \neq j \text{ and } (i,j) \in \mathcal{E}, \\ & = 0 & \text{otherwise.} \end{array}\right\} \tag{5.1.1}$$

The adjacency matrix for the graph in Figure 5.1.1 is

$$A = \begin{bmatrix} 0 & 1 & 1 & 0 & 0 \\ 1 & 0 & 1 & 0 & 0 \\ 1 & 1 & 0 & 1 & 1 \\ 0 & 0 & 1 & 0 & 1 \\ 0 & 0 & 1 & 1 & 0 \end{bmatrix}.$$

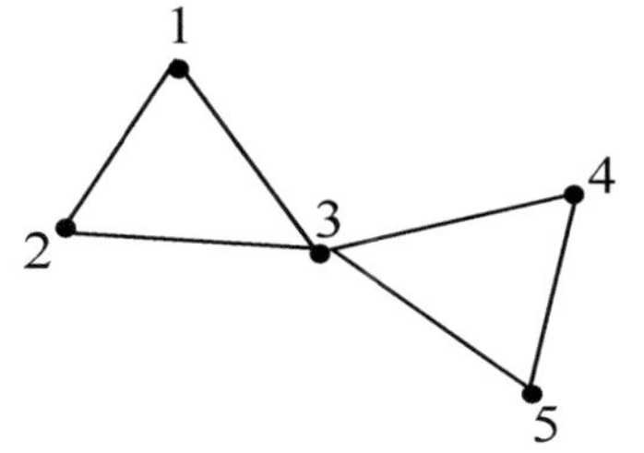

Figure 5.1.1 - A graph.

With any symmetric matrix $\mathbf{A}$ we may associate a graph; the rule is

$$\text{if } i \neq j \text{ then } (i,j) \in \mathcal{E} \text{ iff } a_{ij} \neq 0. \tag{5.1.2}$$

Using this rule we see that the graph associated with a Jacobi matrix is an (unbroken) *path*, as in Figure 5.1.2. The path is clearly one of the simplest graphs.

Figure 5.1.2 - The graph associated with a Jacobi matrix

Another simple graph is a *star* on n vertices, shown in Figure 5.1.3.

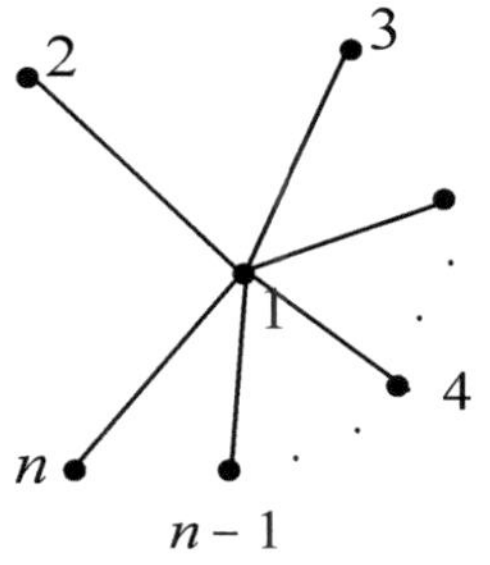

Figure 5.1.3 - A star on n vertices

A (symmetric) bordered diagonal matrix **B** has a star on n vertices as its associated graph.

$$\mathbf{B} = \begin{bmatrix} a_1 & \hat{b}_1 & \ldots & \hat{b}_{n-1} \\ \hat{b}_1 & a_2 & & \\ \cdot & & \ddots & \\ \hat{b}_{n-1} & & & a_n \end{bmatrix} \tag{5.1.3}$$

A *periodic* Jacobi matrix is one of the form

$$\mathbf{J}_{per} = \begin{bmatrix} a_1 & b_1 & & & b_n \\ b_1 & a_2 & b_2 & & \\ & \ddots & \ddots & \ddots & \\ & & \ddots & \ddots & b_{n-1} \\ b_n & & & b_{n-1} & a_n \end{bmatrix}. \tag{5.1.4}$$

It is tridiagonal except for the terms b_n in the top right and bottom left. The underlying matrix is a *ring* on n vertices as shown in Figure 5.1.4.

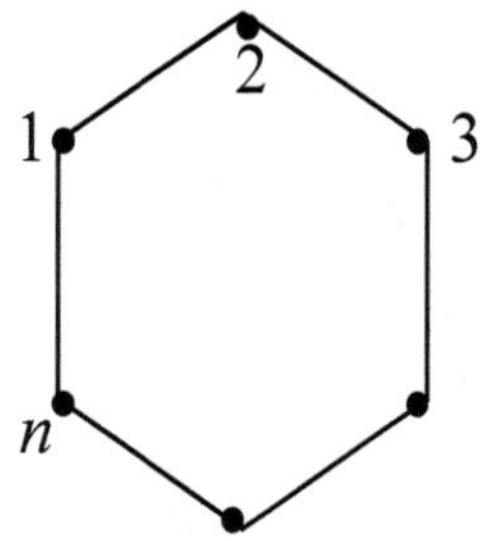

Figure 5.1.4 - A ring on n vertices

The graph associated with a pentadiagonal matrix, such as occurred in Section 2.3 in the analysis of the vibration of a beam, is a *strut*, as shown in Figure 5.1.5.

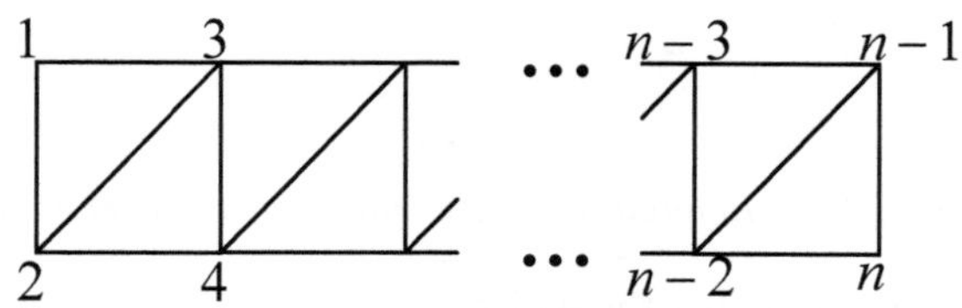

Figure 5.1.5 - The strut on n (even) vertices is the underlying graph of a pentadiagonal matrix

The graph associated with a 2×2 block tridiagonal matrix is also a strut, but now one with double connections, as shown in Figure 5.1.6.

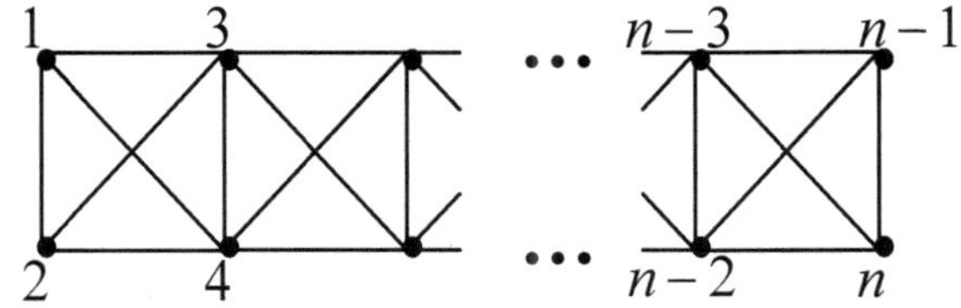

Figure 5.1.6 - The graph underlying a 2×2 block tridiagonal matrix

The graphs shown in Figs. 5.1.1-5.1.6 are all *connected* graphs: there is a chain consisting of a sequence of edges connecting any one vertex to any other vertex. Note that the intersections of the diagonals in Figure 5.1.6 are not vertices of the graph.

The graphs shown in Figure 5.1.7a), b) are *disconnected.*

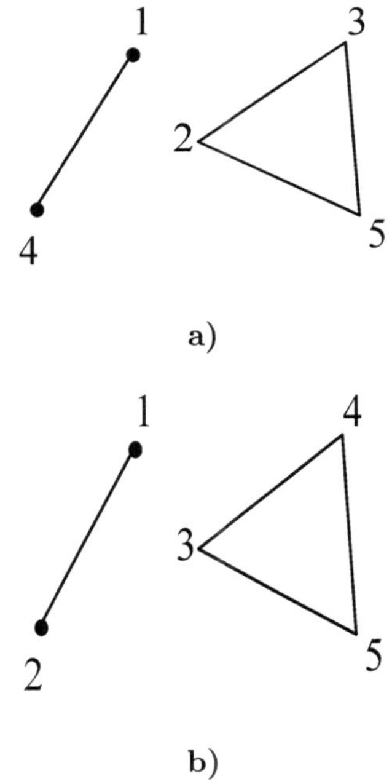

Figure 5.1.7 - Renumbering does not essentially change a graph

In order to test whether the underlying graph of a given (symmetric) matrix is connected or not, we note that *renumbering* the vertices of a graph does not change the essential character of a graph; the graphs a) and b) in Figure 5.1.7 are essentially the same. Renumbering the vertices of a graph leads to a *rearranging* of the rows and of the columns of any (symmetric) matrix based on that graph. When a graph is disconnected, it may be partitioned, as in Figure 5.1.7a) into a set of connected subgraphs. Then we can always rearrange the numbering, as in b) so that vertex numbers in any one connected subgraph form a consecutive sequence. The adjacency matrices of the graphs a) and b) are

$$A_1 = \begin{bmatrix} 0 & 0 & 0 & 1 & 0 \\ 0 & 0 & 1 & 0 & 1 \\ 0 & 1 & 0 & 0 & 1 \\ 1 & 0 & 0 & 0 & 0 \\ 0 & 1 & 1 & 0 & 0 \end{bmatrix}, \quad A_2 = \left[\begin{array}{cc|ccc} 0 & 1 & 0 & 0 & 0 \\ 1 & 0 & 0 & 0 & 0 \\ \hline 0 & 0 & 0 & 1 & 1 \\ 0 & 0 & 1 & 0 & 1 \\ 0 & 0 & 1 & 1 & 0 \end{array}\right].$$

We see, in this example, that when the vertices are renumbered so that each connected subgraph has consecutive numbering, then the adjacency matrix splits into two separate submatrices: such a (symmetric) matrix is said to be *reducible.* A symmetric matrix $\mathbf{A}$ is said to be *irreducible* iff it *cannot* be transformed to the form

$$\mathbf{A} = \left[\begin{array}{c|c} \mathbf{B} & \mathbf{0} \\ \hline \mathbf{0} & \mathbf{C} \end{array}\right] \tag{5.1.5}$$

by any rearrangement of rows and columns. If it is reducible, then it can be transformed to the form (5.1.5), and of course $\mathbf{B}$ and $\mathbf{C}$ may perhaps themselves be reduced further. Note: The concepts of connectedness of a *directed* graph, and the corresponding concept of irreducibility of a *general* (not necessarily symmetric) matrix, are more complex than those described here. See Horn and Johnson (1985) [183] Section 6.2.21.

Now we may state the general result.

Theorem 5.1.1 *The (symmetric) matrix* $\mathbf{A}$ *is irreducible iff its underlying graph is connected.*

It is easy to check that if a spring (other than k_1) is removed from a spring mass system such as that in Figure 4.4.1, then the underlying graph becomes disconnected, and the stiffness matrix becomes reducible.

A *tree* is a special kind of connected graph: one which has no *circuits.* Now there is a *unique* chain of edges connecting any one vertex to any other. The path and the star are both trees, but a ring, see Figure 5.1.4, is not a tree. A connected graph has one or more *spanning* trees. If $\mathcal{G}$ is a connected graph with vertex set $\mathcal{V}$, then a spanning tree $\mathcal{S}$ of $\mathcal{G}$ is a *maximal* tree with the vertex set $\mathcal{V}$; if any more edges in $\mathcal{E}$ were added to $\mathcal{S}$ then it would cease to be a tree: it would have a circuit. Figure 5.1.8 shows three possible spanning trees for the graph $\mathcal{G}$ in Figure 5.1.1.

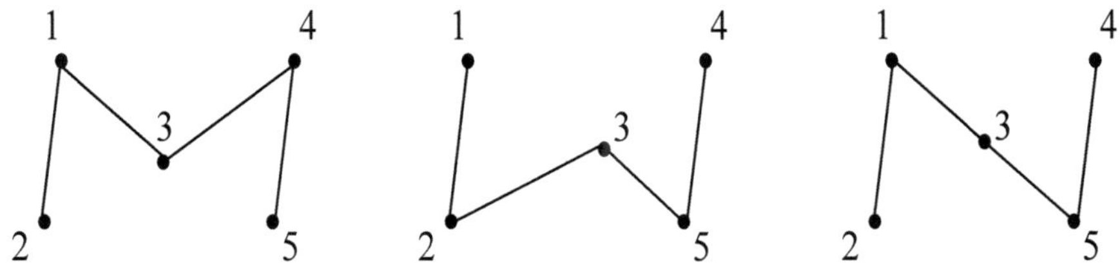

Figure 5.1.8 - Three spanning trees for the graph in Figure 5.1.1.

It may be proved that *all the spanning trees of a given graph $\mathcal{G}$ have the same number of edges.*

Nabben (2001) [243], in a wide ranging paper, discusses Green's matrices for trees.

5.2 Matrix transformations

In the first part of this book we are concerned very largely with matrix eigenvalue problems. One of the basic questions we face is this: 'What operations, i.e., transformations, may we apply to a matrix, or a matrix pair, which will leave its eigenvalues unchanged, i.e., invariant?' We now discuss this question.

Suppose $\mathbf{C}, \mathbf{A} \in M_n$. The set of matrices $\mathbf{C} - \lambda\mathbf{A}$ is called the *matrix pencil* based on the pair $(\mathbf{C}, \mathbf{A})$. As stated in Section 1.4, the eigenvalues of the pair $(\mathbf{C}, \mathbf{A})$ are the values of λ for which the equation

$$(\mathbf{C} - \lambda\mathbf{A})\mathbf{x} = \mathbf{0}$$

has a non-trivial solution $\mathbf{x} \in V_n$. The eigenvalues are the roots of

$$\det(\mathbf{C} - \lambda\mathbf{A}) = 0.$$

Suppose $\mathbf{P}, \mathbf{R} \in M_n$ are constant matrices, i.e., they are independent of λ. Since

$$\det(\mathbf{PCR} - \lambda\mathbf{PAR}) = \det(\mathbf{P}) \cdot \det(\mathbf{C} - \lambda\mathbf{A}) \cdot \det(\mathbf{R})$$

we may deduce that if $\mathbf{P}, \mathbf{R}$ are non-singular, so that $\det(\mathbf{P}) \neq 0$, $\det(\mathbf{R}) \neq 0$, then

$$\det(\mathbf{PCR} - \lambda\mathbf{PAR}) = 0 \text{ iff } \det(\mathbf{C} - \lambda\mathbf{A}) = 0,$$

so that the transformation 'premultiply by $\mathbf{P}$, and postmultiply by $\mathbf{R}$' leaves the eigenvalues invariant. The transformation is called an *equivalence* transformation. It is a special *equivalence relation* (Ex. 5.2.1).

In general, an equivalence (transformation) will transform a symmetric pencil into an unsymmetric pencil. Those which preserve symmetry are characterised by

$$\mathbf{P} = \mathbf{R}^T. \tag{5.2.1}$$

An equivalence changes a pencil $\mathbf{A} - \lambda\mathbf{I}$ into $\mathbf{PAR} - \lambda\mathbf{PR}$. If it is to change $\mathbf{A} - \lambda\mathbf{I}$ into $\mathbf{B} - \lambda\mathbf{I}$, then we must choose $\mathbf{P}, \mathbf{R}$ so that $\mathbf{PR} = \mathbf{I}$, i.e.,

$$\mathbf{P} = \mathbf{R}^{-1}. \tag{5.2.2}$$

An equivalence with this property is called a *similarity* (transformation). An equivalence which satisfies both (5.2.1) and (5.2.2) is called a *rotation* or an *orthogonal transformation.* We reserve the symbol $\mathbf{Q}$ to denote the '$\mathbf{P}$' of such a transformation. Equations (5.2.1), (5.2.2.) show that

$$\mathbf{Q}\mathbf{Q}^T = \mathbf{Q}^T\mathbf{Q} = \mathbf{I} : \tag{5.2.3}$$

$\mathbf{Q}$ is an *orthogonal matrix*; the matrices $\mathbf{U}$ and $\mathbf{X}$ in Section 4.2 were orthogonal matrices. We recall that the columns (rows) of an orthogonal matrix are mutually orthogonal, and each column (row) has norm 1; if $\mathbf{Q} = [\mathbf{q}_1, \mathbf{q}_2, \ldots, \mathbf{q}_n]$, then

$$\mathbf{q}_i^T\mathbf{q}_j = \delta_{ij}. \tag{5.2.4}$$

If $n = 2$, an orthogonal matrix has the form

$$\mathbf{Q} = \begin{bmatrix} \cos\theta & -\sin\theta \\ \sin\theta & \cos\theta \end{bmatrix}. \tag{5.2.5}$$

When $n = 2$, the eigenvalue problem relates to a plane, and this $\mathbf{Q}$ corresponds to a rotation of the x, y axes through an angle θ about the z-axis.

It is difficult to write down the most general expression for an orthogonal matrix in M_n. Instead, we use the fact that a product of orthogonal matrices is itself orthogonal (Ex. 5.2.3).

There is a particularly simple and powerful orthogonal matrix which can be constructed by making a rank-one change to the identity matrix:

$$\mathbf{Q} = \mathbf{I} - 2\mu\mathbf{x}\mathbf{x}^T \tag{5.2.6}$$

will be orthogonal if

$$\begin{aligned} \mathbf{Q}\mathbf{Q}^T &= (\mathbf{I} - 2\mu\mathbf{x}\mathbf{x}^T)(\mathbf{I} - 2\mu\mathbf{x}\mathbf{x}^T) \\ &= \mathbf{I} - 4\mu\mathbf{x}\mathbf{x}^T + 4\mu^2(\mathbf{x}^T\mathbf{x})(\mathbf{x}\mathbf{x}^T) = \mathbf{I}, \end{aligned}$$

i.e., if μ is chosen so that

$$\mu = 1/\mathbf{x}^T\mathbf{x}. \tag{5.2.7}$$

Such a transform is called a *Householder transformation*; note that $\mathbf{Q}$ in (5.2.6) is symmetric, i.e., $\mathbf{Q} = \mathbf{Q}^T$.

Householder transformations are used in various contexts; one is the reduction of a symmetric matrix to tridiagonal form, as we now describe.

Suppose $\mathbf{Q}$ is given by (5.2.6), and $\mathbf{A} \in S_n$. We wish to choose $\mathbf{Q}$, i.e., find $\mathbf{x}$, so that the transformed matrix $\mathbf{QAQ} = \mathbf{B}$ has zero elements in its first row and column, except for the first two, b_{11}, b_{12}. First consider the postmultiplication by $\mathbf{Q}$, and use the abbreviation ρ_1, for row 1 of a matrix. With $\mathbf{Q}$ given by (5.2.6), we have

$$\begin{array}{lllll} \mathbf{C} & = & \mathbf{AQ} & = & \mathbf{A} - 2\mu(\mathbf{Ax})\mathbf{x}^T \\ \rho_1(\mathbf{C}) & = & \rho_1(\mathbf{AQ}) & = & \rho_1(\mathbf{A}) - 2\mu(\rho_1(\mathbf{A})\mathbf{x})\mathbf{x}^T \end{array}$$

Thus $c_{1i} = a_{1i} - 2\mu(\rho_1(\mathbf{A})\mathbf{x})x_i,\ i = 1, 2, \ldots, n.$

We now choose

$$x_i = a_{1i},\ i = 3, 4, \ldots, n. \tag{5.2.8}$$

Then $c_{1i} = 0,\ i = 3, 4, \ldots, n$ if

$$2\mu(\rho_1(\mathbf{A})\mathbf{x}) = 1. \tag{5.2.9}$$

This gives one equation for the remaining unknowns x_1, x_2. Now carry out the premultiplication:

$$\mathbf{QAQ} = \mathbf{B} = \mathbf{QC} = \mathbf{C} - 2\mu\mathbf{x}(\mathbf{x}^T\mathbf{C}),$$

so that

$$\rho_1(\mathbf{B}) = \rho_1(\mathbf{C}) - 2\mu\rho_1(\mathbf{x})(\mathbf{x}^T\mathbf{C}).$$

Thus if the premultiplication is not to change the zero elements in the first row of $\mathbf{C}$, we must choose $x_1 = 0$. Now equations (5.2.7)-(5.2.9), give

$$2(a_{12}x_2 + a_{13}^2 + \cdots + a_{1n}^2) = x_2^2 + (a_{13}^2 + \cdots + a_{1n}^2),$$

which yields

$$x_2 = a_{12} \pm S, \tag{5.2.10}$$

where

$$S^2 = \sum_{i=2}^{n} a_{1i}^2. \tag{5.2.11}$$

Thus the required $\mathbf{x}$ is

$$\mathbf{x} = \{0,\ a_{12} \pm S,\ a_{13}, \ldots, a_{1n}\} \tag{5.2.12}$$

and for numerical purposes we choose the sign of S to be that of a_{12}.

This is the basic Householder transformation; it reduces an arbitrary symmetric $\mathbf{A}$ to a matrix

$$\mathbf{B} = \begin{bmatrix} a_{11} & \mathbf{b}^T \\ \mathbf{b} & \mathbf{B}_1 \end{bmatrix}, \tag{5.2.13}$$

where $\mathbf{b} = b_1\mathbf{e}_1$. This completes the first step in the reduction to tridiagonal form. Now we apply another Householder transformation to the submatrix $\mathbf{B}_1$, using a new $\mathbf{x}$ with $x_1 = 0 = x_2$. This second transformation will leave $a_{11}, \mathbf{b}, \mathbf{b}^T$ unchanged, and will eliminate all but the first two elements of the first row and column of $\mathbf{B}_1$. After $n-2$ applications, the matrix becomes tridiagonal. Once the matrix has been reduced to tridiagonal form, its eigenvalues can easily be located by using the sign count function $s_r(\lambda)$ of Section 3.1. Details on the numerical implementation of this reduction may be found in Bishop, Gladwell and Michaelson (1965) [33] (Chapter 9), Golub and Van Loan (1983) [135] Section 8.2.

We make two comments. Because $x_1 = 0$, the $\mathbf{Q}$ in (5.2.6) may be written

$$\mathbf{Q} = \begin{bmatrix} 1 & 0 \\ 0 & \mathbf{Q}_1 \end{bmatrix}, \tag{5.2.14}$$

where $\mathbf{Q}_1$ is an orthogonal matrix in M_{n-1}. This has an important consequence. Not only does the transformation preserve $\sigma(\mathbf{A})$, i.e., $\sigma(\mathbf{A}) = \sigma(\mathbf{B})$, but also $\sigma(\mathbf{A}_1) = \sigma(\mathbf{B}_1)$.

Secondly, we can use a trivial modification of the Householder transformation to reduce a general symmetric matrix $\mathbf{A}$ to, say, pentadiagonal, form. We take

$$\mathbf{x} = \{0, 0, a_{13} \pm S, a_{14}, \ldots, a_{1n}\} \tag{5.2.15}$$

where $S^2 = \sum_{i=3}^{n} a_{1i}^2$. This transformation preserves $\sigma(\mathbf{A}), \sigma(\mathbf{A}_1), \sigma(\mathbf{A}_{1,2})$, where the last denotes the spectrum of $\mathbf{A}$ with rows and columns 1 and 2 removed.

Exercises 5.2

1. An *equivalence relation*, 'a is related to b', written aRb, has three defining properties:

- *reflexivity*, aRa
- *symmetry*, if aRb then bRa
- *transitivity*, if aRb and bRc, then aRc

A set of elements related by an equivalence relation is called an *equivalence class*. Use the joint operation 'premultiply by $\mathbf{P}$ and postmultiply by $\mathbf{R}$' (with $\mathbf{P}, \mathbf{R}$ non-singular) to define an equivalence relation and an equivalence class for matrix pairs $(\mathbf{C}, \mathbf{A})$.

2. Show that the transformation $\mathbf{B} = \mathbf{Q}\mathbf{A}\mathbf{Q}^T$ defines an equivalence relation and a corresponding equivalence class.

3. Show that if $\mathbf{Q}_1, \mathbf{Q}_2$ are orthogonal, then so is $\mathbf{Q}_1\mathbf{Q}_2$. Show by counterexample that if $\mathbf{Q}_1, \mathbf{Q}_2$ are symmetric, then $\mathbf{Q}_1\mathbf{Q}_2$ is not necessarily so.

4. Show that if $\mathbf{x}$ is given by (5.2.11), then μ in (5.2.7) is given by $2S(S + a_{12})\mu = 1$.

5. Verify that the $\mathbf{Q}$ obtained as a result of $n-2$ successive Householder transformations has the form (5.2.14).

5.3 The star and the path

In Section 5.1 we noted that the graph associated with a bordered diagonal matrix (5.1.3) is a star on n vertices, as in Figure 5.1.3. There is a particularly simple inverse eigenvalue problem connected with a bordered diagonal matrix $\mathbf{B}$: construct $\mathbf{B}$ so that $\sigma(\mathbf{B}) = (\lambda_i)_1^n$, $\sigma(\mathbf{B}_1) = (\mu_i)_1^{n-1}$. The usual variational arguments show that the two spectra must interlace, at least in a loose sense:

$$\lambda_1 \le \mu_1 \le \lambda_2 \le \cdots \le \mu_{n-1} \le \lambda_n. \tag{5.3.1}$$

For simplicity we assume that the $(\mu_i)_1^{n-1}$ are distinct:

$$\mu_1 < \mu_2 < \cdots < \mu_{n-1}. \tag{5.3.2}$$

We write $\mathbf{B}$ in the form (5.1.3), i.e.,

$$\mathbf{B} = \begin{bmatrix} a_1 & \hat{\mathbf{b}}^T \\ \hat{\mathbf{b}} & \mathbf{M} \end{bmatrix}, \tag{5.3.3}$$

where $\mathbf{M}$ is diagonal, and $\hat{\mathbf{b}} = \{\hat{b}_1, \hat{b}_2, \ldots, \hat{b}_{n-1}\}$. Clearly, we can make $\sigma(\mathbf{B}_1) = (\mu_i)_1^{n-1}$ by taking $\mathbf{M} = diag(\mu_1, \mu_2, \ldots, \mu_{n-1})$. The trace condition gives

$$a_1 = \sum_{i=1}^{n} \lambda_i - \sum_{i=1}^{n-1} \mu_i. \tag{5.3.4}$$

Now consider the eigenvector equations for $\mathbf{B}$:

$$\begin{aligned} \hat{b}_i v_1 \quad &+ \quad (\mu_i - \lambda) v_{i+1} &= 0, \quad i = 1, 2, \ldots, n-1, \\ (a_1 - \lambda) v_1 \quad &+ \quad \textstyle\sum_{i=1}^{n-1} \hat{b}_i v_{i+1} &= 0, \end{aligned}$$

which give the eigenvalue equation

$$\lambda - a_1 - \sum_{i=1}^{n-1} \frac{\hat{b}_i^2}{\lambda - \mu_i} = 0.$$

This is to have roots $(\lambda_i)_1^n$, so that

$$\lambda - a_1 - \sum_{i=1}^{n-1} \frac{\hat{b}_i^2}{\lambda - \mu_i} = \frac{\prod_{i=1}^{n} (\lambda - \lambda_i)}{\prod_{i=1}^{n-1} (\lambda - \mu_i)} \tag{5.3.5}$$

and hence

$$\hat{b}_i^2 = \frac{-\prod_{j=1}^{n} (\mu_i - \lambda_j)}{\prod_{j=1}^{\prime n-1} (\mu_i - \mu_j)}, \quad i = 1, 2, \ldots, n-1, \tag{5.3.6}$$

where, as usual, $'$ denotes $i \ne j$; the interlacing condition (5.3.1) yields $\hat{b}_i^2 \ge 0$. We can choose the sign of $\hat{b}_i$ to be + or -. Because we have assumed that the μ_i are distinct, a given μ_i can coincide only with its neighbours λ_i or λ_{i+1}.

Equation (5.3.6) shows that $\hat{b}_i = 0$ iff μ_i coincides with either of these two λ's. If $\hat{b}_i = 0$, then the edge $(1, i+1)$ is absent from the underlying graph.

Having constructed the bordered diagonal matrix $\mathbf{B}$, we have a new way to construct a tridiagonal $\mathbf{J}$ such that $\sigma(\mathbf{J}) = (\lambda_i)_1^n$, $\sigma(\mathbf{J}_1) = (\mu_i)_1^{n-1}$: we can apply Householder transformations to $\mathbf{B}$ to get $\mathbf{J}$. On account of Ex. 5.2.5, the transformation will have the form

$$\begin{bmatrix} 1 & \mathbf{0} \\ \mathbf{0} & \mathbf{Q}_1 \end{bmatrix} \begin{bmatrix} a_1 & \hat{\mathbf{b}}^T \\ \hat{\mathbf{b}} & \mathbf{M} \end{bmatrix} \begin{bmatrix} 1 & \mathbf{0} \\ \mathbf{0} & \mathbf{Q}_1^T \end{bmatrix} = \begin{bmatrix} a_1 & b_1\mathbf{e}_1^T \\ b_1\mathbf{e}_1 & \mathbf{J}_1 \end{bmatrix}, \tag{5.3.7}$$

or equivalently

$$\begin{bmatrix} 1 & \mathbf{0} \\ \mathbf{0} & \mathbf{Q}_1^T \end{bmatrix} \begin{bmatrix} a_1 & b_1\mathbf{e}_1^T \\ b_1\mathbf{e}_1 & \mathbf{J}_1 \end{bmatrix} \begin{bmatrix} 1 & \mathbf{0} \\ \mathbf{0} & \mathbf{Q}_1 \end{bmatrix} = \begin{bmatrix} a_1 & \hat{\mathbf{b}}^T \\ \hat{\mathbf{b}} & \mathbf{M} \end{bmatrix}. \tag{5.3.8}$$

On carrying out the multiplication, we find

$$\mathbf{Q}_1^T\mathbf{J}_1\mathbf{Q}_1 = \mathbf{M}, \quad \mathbf{Q}_1^T b_1\mathbf{e}_1 = \hat{\mathbf{b}}. \tag{5.3.9}$$

The first equation shows that the eigenvectors of $\mathbf{J}_1$ are the columns of $\mathbf{Q}_1$: the ith eigenvector is

$$\mathbf{q}_i = \{q_{1i}, q_{2i}, \ldots, q_{(n-1),i}\}. \tag{5.3.10}$$

The second equation shows that, apart from the factor b_1, the vector $\hat{b}$ is the vector of first components of the eigenvectors of $\mathbf{J}_1$:

$$\hat{\mathbf{b}} = b_1\{q_{11}, q_{12}, \ldots, q_{1,n-1}\}. \tag{5.3.11}$$

Thus, apart from the factor b_1, $\hat{\mathbf{b}}$ is the vector $\mathbf{x}_1$ needed for the construction of $\mathbf{J}_1$ from the Lanczos algorithm of Section 4.2. The factor b_1 is given by $b_1 = ||\hat{\mathbf{b}}||$.

Note the difference between (5.3.6) and (4.3.17): the former, according to (5.3.11), gives the first components of the eigenvectors of $\mathbf{J}_1$; the latter gives the first components of the eigenvectors of $\mathbf{J}$.

Sussman-Fort (1982) [312] discusses connections between the inverse eigenvalue problems for Jacobi and bordered matrices.

Exercises 5.3

1. Explore what happens to $\mathbf{J}$ when one or more of the λ's coincides with a μ.

5.4 Periodic Jacobi matrices

In Section 5.1 we showed that the graph underlying a periodic Jacobi matrix is a *ring* on n vertices.The following analysis is due to Ferguson (1980) [87] and Boley and Golub (1984) [35], Boley and Golub (1987) [36].

A periodic Jacobi matrix $\mathbf{J}_{per}$ has $2n$ terms, $(a_i, b_i)_1^n$. We show how to construct $\mathbf{J}_{per}$ from $\sigma(\mathbf{J}_{per}) = (\lambda_i)_1^n$, $\sigma(\mathbf{J}_{per,1}) = (\mu_i)_1^{n-1}$ and one extra piece of data:

$$\beta = b_1 b_2 \dots b_n. \tag{5.4.1}$$

It is convenient to consider two matrices, the original matrix $\mathbf{J}_{per}$ of (5.1.4), and another matrix $\mathbf{J}_{per}^-$ with b_n replaced by $-b_n$. We suppose $\sigma(\mathbf{J}_{per}^-) = (\lambda_i^-)_1^n$; clearly there are relations between the λ_i^- and the λ_i. The λ_i and μ_i will again interlace as in (5.3.1), as will the λ_i^- and μ_i; again we suppose that the $(\mu_i)_1^{n-1}$ are distinct, i.e., (5.3.2) holds.

We start by constructing two bordered diagonal matrices, $\mathbf{B}$ from $(\lambda_i)_1^n$ and $(\mu_i)_1^{n-1}$, $\mathbf{B}^-$ from $(\lambda_i^-)_1^n$ and $(\mu_i)_1^{n-1}$. They will have the form

$$\mathbf{B} = \begin{bmatrix} a_1 & \hat{\mathbf{b}}^T \\ \hat{\mathbf{b}} & \mathbf{M} \end{bmatrix}, \quad \mathbf{B}^- = \begin{bmatrix} a_1^- & \hat{\mathbf{b}}^{-T} \\ \hat{\mathbf{b}}^- & \mathbf{M} \end{bmatrix}. \tag{5.4.2}$$

Here $a_1, \hat{\mathbf{b}}$ will be given by (5.3.4), (5.3.6), and $a_1^-, \hat{\mathbf{b}}^-$ will be obtained from (5.3.4), (5.3.6) by replacing λ_i by λ_i^-.

Since $\sigma(\mathbf{J}_{per}) = \sigma(\mathbf{B})$ and $\sigma(\mathbf{J}_{per,1}) = \sigma(\mathbf{B}_1)$, $\mathbf{J}_{per}$ and $\mathbf{B}$ are related by an orthogonal transformation of the form

$$\mathbf{B} = \begin{bmatrix} a_1 & \hat{\mathbf{b}}^T \\ \hat{\mathbf{b}} & \mathbf{M} \end{bmatrix} = \begin{bmatrix} 1 & 0 \\ 0 & \mathbf{Q}_1^T \end{bmatrix} \begin{bmatrix} a_1 & b_1\mathbf{e}_1^T + b_n\mathbf{e}_{n-1}^T \\ b_1\mathbf{e}_1 + b_n\mathbf{e}_{n-1} & \mathbf{A}_1 \end{bmatrix} \begin{bmatrix} 1 & \mathbf{0} \\ \mathbf{0} & \mathbf{Q}_1 \end{bmatrix}$$

where $\mathbf{e}_1 = \{1, 0, \dots, 0\}$, $\mathbf{e}_{n-1} = \{0, 0, \dots, 1\}$ are in V_{n-1} and similarly

$$\mathbf{B}^- = \begin{bmatrix} a_1^- & \hat{\mathbf{b}}^{-T} \\ \hat{\mathbf{b}}^- & \mathbf{M} \end{bmatrix} = \begin{bmatrix} 1 & \mathbf{0} \\ \mathbf{0} & \mathbf{Q}_1^T \end{bmatrix} \begin{bmatrix} a_1^- & b_1\mathbf{e}_1^T - b_n\mathbf{e}_{n-1}^T \\ b_1\mathbf{e}_1 - b_n\mathbf{e}_{n-1} & \mathbf{A}_1 \end{bmatrix} \begin{bmatrix} 1 & \mathbf{0} \\ \mathbf{0} & \mathbf{Q}_1 \end{bmatrix}.$$

The subblocks of these equations corresponding to $\hat{\mathbf{b}}$ and $\hat{\mathbf{b}}^-$ are

$$\begin{aligned} \hat{\mathbf{b}} &= \mathbf{Q}_1^T(b_1\mathbf{e}_1 + b_n\mathbf{e}_{n-1}) \\ \hat{\mathbf{b}}^- &= \mathbf{Q}_1^T(b_1\mathbf{e}_1 - b_n\mathbf{e}_{n-1}) \end{aligned}$$

which on addition and subtraction give

$$\begin{aligned} \hat{\mathbf{b}} + \hat{\mathbf{b}}^- &= 2b_1\mathbf{Q}_1^T\mathbf{e}_1 = 2b_1\mathbf{x}_1 \\ \hat{\mathbf{b}} - \hat{\mathbf{b}}^- &= 2b_n\mathbf{Q}_1^T\mathbf{e}_{n-1} = 2b_n\mathbf{x}_{n-1} \end{aligned}$$

where, as in (5.3.9), $\mathbf{x}_1$ is the first column of $\mathbf{Q}_1^T$ and $\mathbf{x}_{n-1}$ is the $(n-1)$th column. If we know $\hat{\mathbf{b}}$ and $\hat{\mathbf{b}}^-$, then these equations give b_1 and b_n (up to sign) since $||\mathbf{x}_1|| = 1 = ||\mathbf{x}_{n-1}||$. Once we have a_1, b_1 and $\mathbf{x}_1$ we may compute $\mathbf{J}_{per,1}$ from the Lanczos algorithm as before.

However, in finding $\mathbf{B}$ and $\mathbf{B}^-$, specifically in finding $\hat{\mathbf{b}}$ and $\hat{\mathbf{b}}^-$, we assumed that we knew both the $(\lambda_i)_1^n$ and the $(\lambda_i^-)_1^n$. We complete the analysis by showing that we can in fact find $\hat{\mathbf{b}}^-$ from the $(\lambda_i)_1^n$ and β in (5.4.1).

The periodic Jacobi matrix $\mathbf{J}_{per}$ differs from a regular Jacobi matrix only in the presence of the entries b_n in the corners. This means that $\det(\lambda\mathbf{I}-\mathbf{J}_{per})$ and $\det(\lambda\mathbf{I}-\mathbf{J}_{per}^-)$ will differ from the nth principal minor $p_n(\lambda)$ only by quadratic terms in b_n. In fact

$$\begin{aligned}\det(\lambda\mathbf{I}-\mathbf{J}_{per}) &= \prod_{i=1}^n(\lambda-\lambda_i) &= p_n(\lambda)-b_n^2r_{n-2}(\lambda)-2\beta\\ \det(\lambda\mathbf{I}-\mathbf{J}_{per}^-) &= \prod_{i=1}^n(\lambda-\lambda_i^-) &= p_n(\lambda)-b_n^2r_{n-2}(\lambda)+2\beta\end{aligned}$$

where r_{n-2} is the principal minor taken from rows and columns $2,3,\ldots,n-1$. Subtacting these two equations, we find

$$\prod_{i=1}^n(\lambda-\lambda_i^-)=\prod_{i=1}^n(\lambda-\lambda_i)+4\beta.$$

This means that we can express $(\hat{b}_j^-)^2$ in terms of $(\lambda_i)_1^n$ and $(\mu_i)_1^{n-1}$:

$$(\hat{b}_j^-)^2=-\frac{\prod_{i=1}^n(\mu_j-\lambda_i^-)}{\prod_{i=1}^{n-1}{}'(\mu_j-\mu_i)}=-\frac{\prod_{i=1}^n(\mu_j-\lambda_i)+4\beta}{\prod_{i=1}^{n-1}{}'(\mu_j-\mu_i)}.$$

But this expression is not automatically non-negative if the $(\lambda_i)_1^n$ and $(\mu_i)_1^{n-1}$ satisfy the interlacing condition. We must examine this more closely. Suppose first that $\beta=0$. The expression is certainly non-negative, and actually positive if the λ's and μ's strictly interlace. If they strictly interlace then, from continuity considerations we can conclude that, for each value of j, the expression will be non-negative for β lying in a closed interval $[-e_j,f_j]$ around zero, $e_j>0,\ f_j>0$. This means that all the $(b_j^-)^2$ will actually be non-negative in the intersection of these closed intervals. For β in this intersection the problem as posed, has a solution; for β outside this interval it has no (real) solution. Boley and Golub (1987) [36] present an algorithm to compute $\mathbf{J}_{per}$ in this way. See also Boley and Golub (1984) [35]. Xu (1998) [339] provides a detailed analysis of the problem and shows (Theorem 2.8.3) that there is a solution iff

$$\sum_{k=1}^n|\mu_j-\lambda_k|\geq 2\beta(1+(-1)^{n-j+1}) \tag{5.4.3}$$

for all $j=1,2,\ldots,n-1$. Note that if $(\lambda_i)_1^n$ and $(\mu_i)_1^{n-1}$ are given, then the inequality (5.4.3) provides an upper bound for β. Andrea and Berry (1992) [9] provide a completely different approach to the problem via continued fractions.

5.5 The block Lanczos algorithm

In Section 5.1, we exhibited Figs. 5.1.5 and 5.1.6, and showed that the matrices underlying these graphs were pentadiagonal or block tridiagonal. In order to develop methods for solving inverse problems for such systems, we need a block version of the fundamental Lanczos algorithm described in Section 4.2.

First we recall the original scalar version: Given a symmetric matrix $\mathbf{A}$, and a vector $\mathbf{x}_1$, compute a Jacobi matrix $\mathbf{J}$ as in equation (4.2.1) and an orthogonal matrix $\mathbf{X} = [\mathbf{x}_1, \mathbf{x}_2, \ldots, \mathbf{x}_n]$ such that $\mathbf{A} = \mathbf{XJX}^T$.

The algorithm proceeds by using the two equations

$$\mathbf{J} = \mathbf{X}^T\mathbf{AX}, \quad \mathbf{AX} = \mathbf{XJ}, \tag{5.5.1}$$

alternately. Thus the (1,1) term in (5.5.1a) gives

$$a_1 = \mathbf{x}_1^T\mathbf{A}\mathbf{x}_1$$

and the first column of (5.5.1b) gives

$$\mathbf{A}\mathbf{x}_1 = a_1\mathbf{x}_1 - b_1\mathbf{x}_2,$$

which we rewrite as

$$b_1\mathbf{x}_2 = a_1\mathbf{x}_1 - \mathbf{A}\mathbf{x}_1 = \mathbf{z}_2, \tag{5.5.2}$$

which gives

$$b_1 = ||\mathbf{z}_2||, \quad \mathbf{x}_2 = \mathbf{z}_2/b_1.$$

Now the (2,2) term in $\mathbf{J}$ gives $a_2 = \mathbf{x}_2^T\mathbf{A}\mathbf{x}_2$, and the second column of (5.5.1b) is

$$\mathbf{A}\mathbf{x}_2 = -b_1\mathbf{x}_1 + a_2\mathbf{x}_2 - b_2\mathbf{x}_3$$

which we rewrite as

$$b_2\mathbf{x}_3 = -b_1\mathbf{x}_1 + a_2\mathbf{x}_2 - \mathbf{A}\mathbf{x}_2 = \mathbf{z}_3$$

which gives

$$b_2 = ||\mathbf{z}_3||, \quad \mathbf{x}_3 = \mathbf{z}_3/b_2$$

and so on.

We now construct a block version of these equations, following Boley and Golub (1987) [36]. We start with a symmetric matrix $\mathbf{A} \in S_n$ and suppose $n = ps$ for some integer s. We will reduce $\mathbf{A}$ to a block tridiagonal matrix $\mathbf{J}$, where

$$\mathbf{J} = \begin{bmatrix} \mathbf{A}_1 & -\mathbf{B}_1^T & & \\ -\mathbf{B}_1 & \mathbf{A}_2 & -\mathbf{B}_2^T & \\ & \ddots & \ddots & \\ & & -\mathbf{B}_{s-1} & \mathbf{A}_s \end{bmatrix}. \tag{5.5.3}$$

Here $\mathbf{A}_1, \ldots, \mathbf{A}_s$ are symmetric, i.e., in S_p, and the $\mathbf{B}_i$ are upper triangular matrices in M_p. We assume that in addition to $\mathbf{A}$, we are given p orthonormal vectors $(\mathbf{x}_i)_1^p \in V_n$ which form the columns of $\mathbf{X}_1 = [\mathbf{x}_1, \mathbf{x}_2, \ldots, \mathbf{x}_p] \in M_{n,p}$. The matrix $\mathbf{X}_1$ therefore satisfies $\mathbf{X}_1^T\mathbf{X}_1 = \mathbf{I}_p$.

The aim of the procedure is to construct $\mathbf{J}$ and an orthogonal matrix $\mathbf{X} = [\mathbf{X}_1, \mathbf{X}_2, \ldots, \mathbf{X}_s]$ such that

$$\mathbf{A} = \mathbf{XJX}^T.$$

Just as in the scalar Lanczos process, we consider the two equations

$$\mathbf{J} = \mathbf{X}^T\mathbf{A}\mathbf{X}, \quad \mathbf{A}\mathbf{X} = \mathbf{X}\mathbf{J}. \tag{5.5.4}$$

The first $p \times p$ block of the first equation gives

$$\mathbf{A}_1 = \mathbf{X}_1^T\mathbf{A}\mathbf{X}_1$$

while the first $n \times p$ block of the second gives

$$\mathbf{A}\mathbf{X}_1 = \mathbf{X}_1\mathbf{A}_1 - \mathbf{X}_2\mathbf{B}_1$$

which we rewrite as

$$\mathbf{X}_2\mathbf{B}_1 = \mathbf{X}_1\mathbf{A}_1 - \mathbf{A}\mathbf{X}_1 = \mathbf{Z}_2.$$

In the scalar version we had $b_1\mathbf{x}_2 = \mathbf{z}_2$, from which we immediately concluded that $b_1 = ||\mathbf{z}_2||$, and hence $\mathbf{x}_2 = \mathbf{z}_2/b_1$. In the block version we have constructed $\mathbf{Z}_2 \in M_p$ and we wish to write it as $\mathbf{X}_2\mathbf{B}_1$. Write $\mathbf{X}_2 = [\mathbf{y}_1, \mathbf{y}_2, \ldots, \mathbf{y}_p]$, $\mathbf{Z}_2 = [\mathbf{z}_1, \mathbf{z}_2, \ldots, \mathbf{z}_p]$ and

$$\mathbf{B}_1 = \begin{bmatrix} b_{11} & b_{12} & \ldots & b_{1p} \\ & b_{22} & \ldots & b_{2p} \\ & & & \\ & & & b_{pp} \end{bmatrix}$$

then finding $(\mathbf{y}_i)_1^p$ and the elements of $\mathbf{B}_1$ is essentially a Gram-Schmidt process: finding orthonormal combinations of the vectors $(\mathbf{z}_i)_1^p$. Thus

$$b_{11}\mathbf{y}_1 = \mathbf{z}_1 \text{ implies } b_{11} = \pm||\mathbf{z}_1||, \quad \mathbf{y}_1 = \mathbf{z}_1/b_{11} \tag{5.5.5}$$

and then

$$b_{12}\mathbf{y}_1 + b_{22}\mathbf{y}_2 = \mathbf{z}_2$$

gives

$$b_{12} = \mathbf{y}_1^T\mathbf{z}_2, \quad b_{22}\mathbf{y}_2 = \mathbf{z}_2 - b_{12}\mathbf{y}_1 = \mathbf{w}_2$$

so that

$$b_{22} = \pm||\mathbf{w}_2||, \quad \mathbf{y}_2 = \mathbf{w}_2/b_{22} \text{ etc.} \tag{5.5.6}$$

The Gram-Schmidt process is closely related to the QR algorithm. The decomposition $\mathbf{X}_2\mathbf{B}_1 = \mathbf{Z}_2$ involves writing $\mathbf{Z}_2$ as the product of $\mathbf{X}_2$ which is in $M_{n,p}$, but which satisfies $\mathbf{X}_2^T\mathbf{X}_2 = \mathbf{I}_p$, and an upper triangular matrix $\mathbf{B}_1 \in M_p$. Because $\mathbf{X}_2$ is not simply an orthogonal matrix in M_p, the usual QR algorithm has to be modified to effect the decomposition.

Now we can proceed as before. We have found $\mathbf{X}_2$, so that

$$\mathbf{A}_2 = \mathbf{X}_2^T\mathbf{A}\mathbf{X}_2$$

and

$$\mathbf{A}\mathbf{X}_2 = -\mathbf{X}_1\mathbf{B}_1^T + \mathbf{X}_2\mathbf{A}_2 - \mathbf{X}_3\mathbf{B}_2$$

so that

$$\mathbf{X}_3\mathbf{B}_2 = -\mathbf{X}_1\mathbf{B}_1^T + \mathbf{X}_2\mathbf{A}_2 - \mathbf{A}\mathbf{X}_2 = \mathbf{Z}_3$$

from which $\mathbf{X}_3, \mathbf{B}_2$ may be found, as before, by Gram-Schmidt. Note that different choices for the square roots, as in (5.5.5) and (5.5.6) will lead to different matrices $\mathbf{J}$. Boley and Golub (1987) [36] present a detailed algorithm for the process.

Further studies on the block-Lanczos algorithm have been carried out by Underwood (1975) [325] and Golub and Underwood (1977) [134]. See also Mattis and Hochstadt (1981) [222]. A completely different and highly efficient procedure for the solution of band matrix inverse problems has been developed by Biegler-König (1980) [28], Biegler-König (1981a) [29], Biegler-König (1981b) [30], Biegler-König (1981c) [31]. See also Gragg and Harrod (1984) [153] for a procedure based on Rutishauser's algorithm; they explore the connections to a number of other problems. See also Gladwell and Willms (1989) [114] and Friedland (1977) [92], Friedland (1979) [93], and particularly, Chu (1998) [58].

5.6 Inverse problems for pentadiagonal matrices

We could pose an inverse eigenvalue problem for a general symmetric matrix with $2p+1$ bands, as in Boley and Golub (1987) [36]. Instead, we will confine ourselves to the case $p = 2$, a pentadiagonal matrix $\mathbf{A}$. The pentadiagonal case occurs in the inverse problem for a vibrating beam, but we shall defer considering the beam until we have discussed positivity in Chapter 6; the pentadiagonal matrix giving the stiffness matrix of the beam has a very special form; certain terms in it must be positive, and others must be negative. In this section we will not be concerned with these matters of sign.

Suppose we are given

$$\sigma(\mathbf{A}) = (\lambda_i)_1^n, \ \sigma(\mathbf{A}_1) = (\mu_i)_1^{n-1}, \ \sigma(\mathbf{A}_{1,2}) = (\nu_i)_1^{n-2}, \tag{5.6.1}$$

where, as before, $\sigma(\mathbf{A}_{1,2})$ denotes the spectrum of $\mathbf{A}_{1,2}$ when its first two rows and columns are removed. Clearly the eigenvalues must interlace; and for simplicity we assume that the interlacing is strict.

$$\lambda_1 < \mu_1 < \lambda_2 < \cdots \quad < \mu_{n-1} < \lambda_n, \tag{5.6.2}$$

$$\mu_1 < \nu_1 < \mu_2 < \cdots \quad < \nu_{n-2} < \mu_{n-1}. \tag{5.6.3}$$

Our aim is to construct $\mathbf{A}$ such that (5.6.1) holds. We write

$$\mathbf{A} = \begin{bmatrix} a_1 & \mathbf{b}^T \\ \mathbf{b} & \mathbf{A}_1 \end{bmatrix}, \tag{5.6.4}$$

where only the first two components of the vector $\mathbf{b}$ are non-zero. We denote the eigenvector matrix of $\mathbf{A}$ by $\mathbf{Q}$, and of $\mathbf{A}_1$ by $\mathbf{Q}^{(1)}$ so that

$$\mathbf{Q}^T\mathbf{A}\mathbf{Q} = \wedge, \quad \mathbf{Q}_1^{(1)T}\mathbf{A}_1\mathbf{Q}^{(1)} = \mathbf{M}. \tag{5.6.5}$$

The eigenvectors of $\mathbf{A}$ are therefore $\mathbf{q}_i$, where $\mathbf{Q} = [\mathbf{q}_1, \mathbf{q}_2, \ldots, \mathbf{q}_n]$ while those of $\mathbf{A}_1$ are $\mathbf{q}_i^{(1)}$, where $\mathbf{Q}^{(1)} = [\mathbf{q}_1^{(1)}, \mathbf{q}_2^{(1)}, \ldots, \mathbf{q}_{n-1}^{(1)}]$.

We start by constructing a bordered diagonal matrix, as in Section 5.3:

$$\mathbf{B} = \begin{bmatrix} a_1 & \hat{\mathbf{b}}^T \\ \hat{\mathbf{b}} & \mathbf{M} \end{bmatrix} \tag{5.6.6}$$

such that $\sigma(\mathbf{B}) = (\lambda_i)_1^n$, and $\sigma(\mathbf{M}) = (\mu_i)_1^{n-1}$. The term a_1 is given by the trace:

$$a_1 = \sum_{i=1}^{n} \lambda_i - \sum_{i=1}^{n-1} \mu_i, \tag{5.6.7}$$

while $\hat{\mathbf{b}}$ is given by (5.3.6):

$$(\hat{b}_i)^2 = -\frac{\prod_{j=1}^{n}(\mu_i - \lambda_j)}{\prod_{j=1}^{n-1}{}'(\mu_i - \mu_j)}. \tag{5.6.8}$$

Now, following equation (5.3.8) we relate $\mathbf{A}$ to

$$\mathbf{B} = \begin{bmatrix} a_1 & \hat{\mathbf{b}}^T \\ \hat{\mathbf{b}} & \mathbf{M} \end{bmatrix} = \begin{bmatrix} 1 & 0 \\ 0 & \mathbf{Q}^{(1)T} \end{bmatrix} \begin{bmatrix} a_1 & \mathbf{b}^T \\ \mathbf{b} & \mathbf{A}_1 \end{bmatrix} \begin{bmatrix} 1 & 0 \\ 0 & \mathbf{Q}^{(1)} \end{bmatrix}. \tag{5.6.9}$$

As in (5.3.9), we have

$$\mathbf{Q}^{(1)T}\mathbf{b} = \hat{\mathbf{b}} \tag{5.6.10}$$

Now however, in contrast to the situation in Section 5.3, $\mathbf{b}$ is not just a multiple of $\mathbf{e}_1$, so that $\hat{\mathbf{b}}$ does not give the vector of first components of the eigenvectors of $\mathbf{A}_1$. But we can use the analysis of Section 4.3 to obtain the first components of the eigenvectors of $\mathbf{A}$ and $\mathbf{A}_1$:

$$q_{1i}^2 = \frac{\prod_{j=1}^{n-1}(\mu_j - \lambda_i)}{\prod_{j=1}^{n}{}'(\lambda_j - \lambda_i)}, \quad (q_{1i}^{(1)})^2 = \frac{\prod_{j=1}^{n-2}(\nu_j - \mu_i)}{\prod_{j=1}^{n-1}{}'(\mu_j - \mu_i)}. \tag{5.6.11}$$

To apply the block Lanczos algorithm to construct $\mathbf{A}$ we need not just the vector $\mathbf{x}_1$ of first components of eigenvectors of $\mathbf{A}$, but also $\mathbf{x}_2$ of second components, making up $\mathbf{X}_1 = [\mathbf{x}_1, \mathbf{x}_2] \in M_{n,2}$. Partition the vector $\mathbf{q}_i$:

$$\mathbf{q}_i = \begin{bmatrix} q_{1i} \\ \mathbf{y}_i \end{bmatrix}, \quad \mathbf{y}_i \in V_{n-1}. \tag{5.6.12}$$

Since $\mathbf{q}_i$ is the ith eigenvector of $\mathbf{A}$, and $\mathbf{A}$ is given by (5.6.4), we may write

$$\begin{bmatrix} a_1 & \mathbf{b}^T \\ \mathbf{b} & \mathbf{A}_1 \end{bmatrix} \begin{bmatrix} q_{1i} \\ \mathbf{y}_i \end{bmatrix} = \lambda_i \begin{bmatrix} q_{1i} \\ \mathbf{y}_i \end{bmatrix}, \tag{5.6.13}$$

so that

$$q_{1i}\mathbf{b} + \mathbf{A}_1\mathbf{y}_i = \lambda_i\mathbf{y}_i.$$

Now premultiply by $\mathbf{Q}^{(1)T}$ to obtain

$$q_{1i}\mathbf{Q}^{(1)T}\mathbf{b} + \mathbf{Q}_1^{(1)T}\mathbf{A}_1\mathbf{y}_i = \lambda_i\mathbf{Q}^{(1)T}\mathbf{y}_i. \tag{5.6.14}$$

But equation (5.6.10) gives $\mathbf{Q}^{(1)T}\mathbf{b} = \hat{\mathbf{b}}$, and equation (5.6.5b) gives $\mathbf{Q}_1^{(1)T}\mathbf{A}_1 = \mathbf{M}\mathbf{Q}^{(1)T}$, so that equation (5.6.14) gives

$$q_{1i}\hat{\mathbf{b}} = -(\mathbf{M} - \lambda_i\mathbf{I})\mathbf{Q}^{(1)T}\mathbf{y}_i$$

and hence

$$\mathbf{y}_i = -q_{1i}\mathbf{Q}^{(1)}(\mathbf{M} - \lambda_i\mathbf{I})^{-1}\hat{\mathbf{b}}.$$

We need just the first term in $\mathbf{y}_i$; it is

$$y_{1i} = -q_{1i}\sum_{j=1}^{n-1}\frac{q_{1j}^{(1)}\hat{b}_j}{\mu_j - \lambda_i}, \quad i = 1, 2, \ldots, n. \tag{5.6.15}$$

Since $\hat{b}_j$ is given by (5.6.8), and $q_{1i}, q_{1j}^{(1)}$ are given by (5.6.11), this equation yields y_{1i}, and hence $\mathbf{x}_2 = \{y_{11}, y_{12}, \ldots, y_{1n}\}$.

Exercises 5.6

1. Verify that the vector $\mathbf{x}_2$ given by (5.6.15) is indeed orthogonal to $\mathbf{x}_1$, as required.
2. Extend the procedure described in this section to the general case of a $2p + 1$ band matrix.

5.7 Inverse eigenvalue problems for a tree

The inverse eigenvalue problems for a path and a star are particular examples of a general problem. Both the path, as shown in Figure 5.1.2, and the star, in Figure 5.1.3, are *trees*, as defined in Section 5.1. The matrices corresponding to these trees are a Jacobi matrix $\mathbf{J}$, or, as we will choose here, a sign-reversed Jacobi matrix $\mathbf{A} = \tilde{\mathbf{J}}$, and a bordered diagonal matrix respectively. In both problems, two spectra were specified, namely $\sigma(\mathbf{A}) = (\lambda_i)_1^n$, and $\sigma(\mathbf{A}_{,1}) = (\mu_i)_1^{n-1}$; the second spectrum corresponded to the eigenvectors $\mathbf{u}$ set to zero at a prescribed vertex, vertex 1. In both cases the two spectra had to satisfy the Cauchy interlacing inequalities

$$\lambda_1 \le \mu_1 \le \lambda_2 \le \cdots \le \mu_{n-1} \le \lambda_n. \tag{5.7.1}$$

In both cases also, if the inequalities (5.7.1) were all strict, the matrix $\mathbf{A}$ was irreducible, and the corresponding graph $\mathcal{G}$ was connected.

The purpose of this section is to serve as an introduction to an important paper by Duarte (1989) [81]. This paper reviews the history of inverse eigenvalue problems for trees, and establishes a general result. We will present analysis covering the simpler parts of the general case. As we will do sometimes in Chapter 6, Duarte labels eigenvalues in decreasing order, and we do the same. Specifically, we will show that if $\mathcal{G}$ is a tree on n vertices $\mathcal{V}$, and if two spectra $(\lambda_i)_1^n$,$(\mu_i)_1^{n-1}$ are given, satisfying

$$\lambda_1 > \mu_1 > \lambda_2 \cdots > \mu_{n-1} > \lambda_n > 0, \tag{5.7.2}$$

then we can find a symmetric matrix $\mathbf{A} \in S_n$ on $\mathcal{G}$ such that $\sigma \equiv \sigma(\mathbf{A}) = (\lambda_i)_1^n$, $\sigma_1 \equiv \sigma(\mathbf{A}_{,1}) = (\mu_i)_1^{n-1}$. We take the strict interlacing and the positivity condition for simplicity; Duarte relaxes these conditions.

We start by observing that the two cases that we have considered so far, the path (Jacobi), and star (bordered diagonal), have common features. First, we note that the entries of the constructed matrices may be considered as functions of the data σ, σ_1. Secondly, we note that in both matrices there are $n^2 - n - 2(n-1) = n^2 - 3n + 2$ *constant* functions, which in fact are all zero. This suggests the following questions:

1. Can the constant functions appearing in $\mathbf{A}$ be other than the zero function?

2. Can the number of these constant functions be increased?

The answer to the first question is NO. For if $\mathbf{A} \in S_n$ has eigenvalues $(\lambda_i)_1^n$ with maximum modulus λ, then (Ex. 5.7.1) $|a_{ij}| \leq \lambda$, so that $\mathbf{A}$ can have no fixed entry, independent of the eigenvalues, other than zero. To answer the second question we note that if the inequalities (5.7.2) hold, then $\mathbf{A}$ must be irreducible. For if $\mathbf{A}$ were reducible, i.e., after possibly renumbering the vertices, it could be written

$$\mathbf{A} = \begin{bmatrix} \mathbf{B} & 0 \\ 0 & \mathbf{C} \end{bmatrix},$$

then $\mathbf{A}$ and $\mathbf{A}_{,1}$ (which after renumbering, would be $\mathbf{A}_{,i}$) would have a common eigenvalue, a situation that is precluded by (5.7.2). Thus $\mathbf{A}$ is irreducible and $\mathcal{G}$ is connected. Now we note that $\mathbf{A}$ must be positive definite so that no diagonal term a_{ii} can be zero. The maximum number of zero entries will be attained for matrices whose graph is a tree, and this number is precisely $n^2 - 3n + 2$ (Ex. 5.7.2). Thus the answer to the second question is NO also.

Having answered these questions, we proceed to the analysis. We start by considering a tree $\mathcal{G}$, choose a vertex of $\mathcal{V}$, label it 1, and see the effect of deleting vertex 1 - this is the graph corresponding to deleting row 1 and column 1 of $\mathbf{A}$. First, we need a symbol, $\mathcal{N}$, to denote the set of m vertices j of $\mathcal{G}$ which are connected to vertex 1. Now we use $\mathcal{G}'$ to denote the graph obtained from $\mathcal{G}$ by deleting vertex 1. Figure 5.7.1 shows two examples. In Figure 5.7.1a, where vertex 1 is at the end of a path, $\mathcal{N} = \{2\}$ and $\mathcal{G}'$ is the connected graph with vertices $\{2,3,4,5,6\}$. In Figure 5.7.1b, $\mathcal{N} = \{2,4\}$ and $\mathcal{G}'$ has two connected components, one on either side of vertex 1; we call these $\mathcal{G}'_2, \mathcal{G}'_4$ respectively. In general, $\mathcal{G}'$ will have m connected components which we label $\mathcal{G}'_j, j \in \mathcal{N}$; the corresponding matrix $\mathbf{A}_{,1}$ will have m irreducible components.

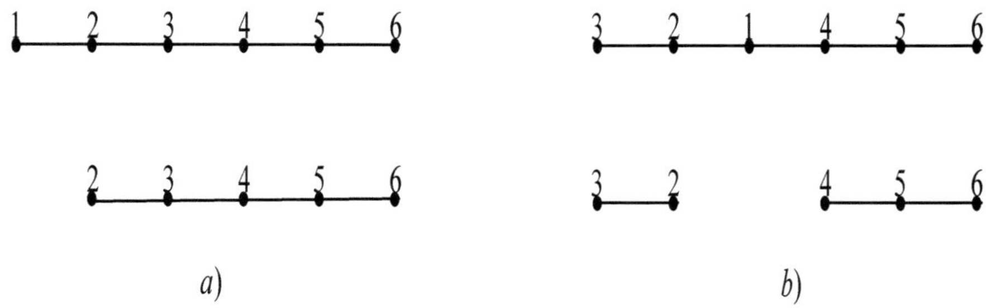

Figure 5.7.1 - Deleting vertex 1 from a path.

Figure 5.7.2 shows another example, a star. Now $\mathcal{N} = \{2, 3, 4, 5\}$ and $\mathcal{G}'$ has 4 connected components: $\mathcal{G}'_j = \{j\}, \quad j = 2, 3, 4, 5$.

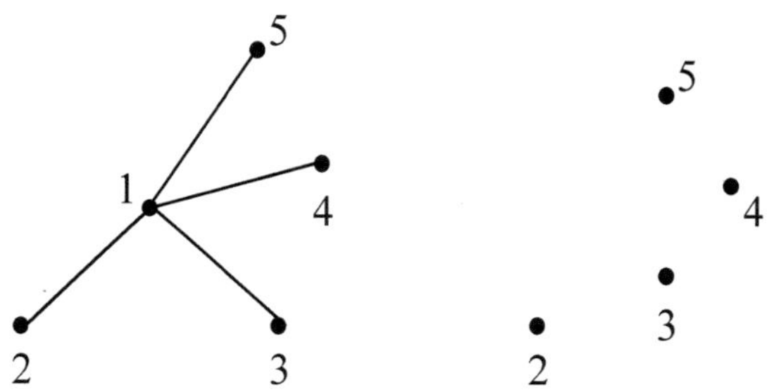

Figure 5.7.2 - Deleting the centre of a star.

Finally, we need a symbol for the graph obtained by deleting vertex $j \in \mathcal{N}$ from $\mathcal{G}'_j$; we call it $\mathcal{G}''_j$. Figure 5.7.3 shows these subgraphs for the graphs $\mathcal{G}'$ in Figure 5.7.1.

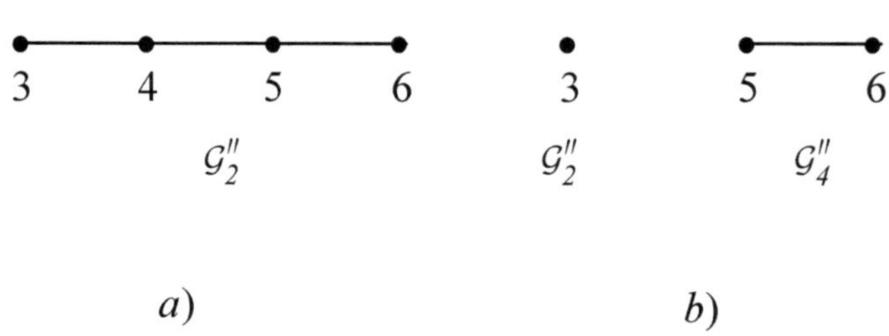

Figure 5.7.3 - The subgraphs $\mathcal{G}''_j$ for the graphs $\mathcal{G}'$.

Note that for the star, the vertex set of $\mathcal{G}''_j$ is empty because the vertex set of $\mathcal{G}'_j$ is $\{j\}$.

Having established notation that allows us to see what happens when we delete a vertex of a graph, we need to consider the two sets of eigenvalues, and how these relate to the matrix $\mathbf{A}$. To do this we first return to two examples we have already treated, those corresponding to deleting an end vertex of a path, and the centre of a star.

First, the path with end vertex 1. The eigenvalues $(\lambda_i)_1^n$ and $(\mu_i)_1^{n-1}$ are the zeros of the trailing monic principal minors, $P'_n(\lambda), P'_{n-1}(\lambda)$ respectively, and, in the notation of equation (4.3.4), these are linked by

$$P'_n(\lambda) = (\lambda - a_1)P'_{n-1}(\lambda) - b_1^2 P'_{n-2}(\lambda). \tag{5.7.3}$$

We note that the graphs corresponding to P'_n, P'_{n-1}, P'_{n-2} are precisely $\mathcal{G}, \mathcal{G}' \equiv \mathcal{G}'_2$ and $\mathcal{G}'' \equiv \mathcal{G}''_2$; in fact P'_n, P'_{n-1}, P'_{n-2} are the characteristic polynomials $\triangle$ of $\mathbf{A}$, and of the submatrices of $\mathbf{A}$ on $\mathcal{G}'$ and $\mathcal{G}''_2$:

$$P'_n(\lambda) = \triangle(\mathbf{A}), \quad P'_{n-1}(\lambda) = \triangle(\mathbf{A}(\mathcal{G}')), \quad P'_{n-2}(\lambda) = \triangle(\mathbf{A}(\mathcal{G}''_2)).$$

We note also that $a_1 = a_{11}, b_1 = a_{12}$ and $\mathcal{N} = \{2\}$. This means that we can write (5.7.3) as

$$\triangle(\mathbf{A}) = (\lambda - a_{11}) \triangle (\mathbf{A}(\mathcal{G}')) - \sum_{j \in \mathcal{N}} a_{1j}^2 \triangle (\mathbf{A}(\mathcal{G}''_j)). \tag{5.7.4}$$

Now consider the star. The equation corresponding to (5.7.3) is equation (5.3.5):

$$\lambda - a_1 - \sum_{i=1}^{n-1} \frac{\hat{b}_i^2}{\lambda - \mu_i} = \frac{\prod_{i=1}^{n}(\lambda - \lambda_i)}{\prod_{i=1}^{n-1}(\lambda - \mu_j)}. \tag{5.7.5}$$

To rewrite this in the same notation as (5.7.4), we note that for a star on $m + 1 = n$ vertices, with the centre labelled 1, $\mathcal{N} = \{2, 3, \ldots, m+1\}$, $a_1 = a_{11}, \quad \hat{b}_i = a_{1,i+1}$, so that

$$\triangle(\mathbf{A}) = (\lambda - a_{11}) \triangle (\mathbf{A}(\mathcal{G}')) - \sum_{j \in \mathcal{N}} a_{1j}^2 \prod_{k \in \mathcal{N} \backslash j} \triangle(\mathbf{A}(\mathcal{G}'_k)). \tag{5.7.6}$$

Note that we have assigned the $m = n - 1$ μ's to the m connected components of $\mathcal{G}'$ so that μ_i is assigned to $\mathcal{G}'_{i+1}, \quad i = 1, 2, \ldots, m$. Note also that although the first terms on the right of (5.7.4), (5.7.6) are identical, the second terms are different. For the star, the vertex set of $\mathcal{G}''_j$ is empty.

Parter (1960) [265] obtained a general result which embraces the particular cases (5.7.4) and (5.7.6):

Lemma 5.7.1 $\triangle(\mathbf{A}) = (\lambda - a_{11})\triangle(\mathbf{A}(\mathcal{G}')) - \sum_{j\in\mathcal{N}} a_{1j}^2 \triangle(\mathbf{A}(\mathcal{G}''_j)).\prod_{k\in\mathcal{N}\backslash j} \triangle(\mathbf{A}(\mathcal{G}'_k))$ *with the convention that* $\triangle(\mathbf{A}(\mathcal{G}''_j)) = 1$ *if, as for the star, the vertex set of* $\mathcal{G}''_j$ *is empty.*

Lemma 5.7.1, like the corresponding result (5.7.5) for the star, is effectively a partial fraction expansion. In the general case, it is

$$\frac{\triangle(\mathbf{A})}{\triangle(\mathbf{A}(\mathcal{G}'))} = \lambda - a_{11} - \sum_{j \in \mathcal{N}} a_{1j}^2 \frac{\triangle(\mathbf{A}(\mathcal{G}''_j))}{\triangle(\mathbf{A}(\mathcal{G}'_j))}, \tag{5.7.7}$$

where we have used the fact that $\mathcal{G}'$ has m separate connected components $\mathcal{G}'_j$, so that

$$\triangle(\mathbf{A}(\mathcal{G}')) = \prod_{j\in\mathcal{N}} \triangle(\mathbf{A}(\mathcal{G}'_j)).$$

Equation (5.7.7) provides the basis for an inductive argument: deleting vertex 1 of $\mathcal{G}$ splits $\mathcal{G}'$ into m components $\mathcal{G}'_j$, and $\mathcal{G}''_j$ bears the same relation to $\mathcal{G}'_j$ as $\mathcal{G}'$ does to $\mathcal{G}$. This means that if we can effect the reconstruction of $\mathbf{A}$ on the components $\mathcal{G}'_j$ of $\mathcal{G}'$ from data referring to $\mathcal{G}''_j$ and $\mathcal{G}'_j$, then we can reconstruct the whole of $\mathbf{A}$.

Now since $\mathcal{G}'_j$ is itself a tree, and $\mathcal{G}''_j$ is obtained by deleting vertex j from $\mathcal{G}'_j$, the roots of $\triangle(\mathbf{A}(\mathcal{G}''_j))$ should interlace the roots of $\triangle(\mathbf{A}(\mathcal{G}'_j))$, just as the μ_i interlace the λ_i, i.e., (5.7.2). But equation (5.7.7) gives $\triangle(\mathbf{A}(\mathcal{G}''_j))$ as a result of the partial fraction expansion. We are given

$$\triangle(\mathbf{A}) = f(\lambda) = \prod_{i=1}^{n}(\lambda - \lambda_i), \tag{5.7.8}$$

$$\triangle(\mathbf{A}(\mathcal{G}')) = g(\lambda) = \prod_{i=1}^{n-1}(\lambda - \mu_i). \tag{5.7.9}$$

Now we must assign the $n-1$ μ's among the m components $\mathcal{G}'_j$. Suppose $\mathcal{G}'_j$ has v_j vertices then we must split the indices $\{1, 2, \ldots, n-1\}$ into m sets, so that if $j \in \mathcal{N}$ then $\mathcal{G}'_j$ is assigned v_j indices. This is equivalent to grouping the terms in $g(\lambda)$ into m terms $g_j(\lambda)$, where $g_j(\lambda)$ has degree v_j:

$$g(\lambda) = \prod_{j\in\mathcal{N}} g_j(\lambda).$$

This means that we must check to see if, when $f(\lambda)/g(\lambda)$ is expanded into partial fractions, as

$$\frac{f(\lambda)}{g(\lambda)} = \lambda - a - \sum_{j\in\mathcal{N}} y_j \frac{h_j(\lambda)}{g_j(\lambda)}, \tag{5.7.10}$$

where $h_j(\lambda)$ is a monic polynomial with $\deg(h_j) < \deg(g_j)$, and if the λ's and μ's interlace as in (5.7.2), then y_j is positive, and the zeros of $h_j(\lambda)$ and $g_j(\lambda)$ interlace. To do this, it is best to change back into a form like Lemma 5.7.1 by multiplying throughout by $g(\lambda)$:

$$f(\lambda) = (\lambda - a)g(\lambda) - \sum_{j\in\mathcal{N}} y_j h_j(\lambda) u_j(\lambda)$$

where $u_j(\lambda) = g(\lambda)/g_j(\lambda)$. We note that

$$u_j(\lambda) = \prod_{s\in Q}(\lambda - \mu_s)$$

where Q consists of $\{1, 2, \ldots, n-1\}$ less those indices which are assigned to g_j. Choose $j \in \mathcal{N}$ and suppose that μ_r, μ_{r+p} are two successive zeros of $g_j(\lambda)$, then, since $g(\mu_r) = 0 = g(\mu_{r+p})$, we have

$$\begin{aligned} f(\mu_r) &= -y_j h_j(\mu_r) u_j(\mu_r) \\ f(\mu_{r+p}) &= -y_j h_j(\mu_{r+p}) u_j(\mu_{r+p}). \end{aligned}$$

We need to show that $h_j(\lambda)$ has a zero between μ_r and μ_{r+p}, i.e., that $h_j(\mu_r)$, $h_j(\mu_{r+p})$ have opposite signs. The terms $(\mu_r - \mu_s)$ and $(\mu_{r+p} - \mu_s)$ appearing in $u_j(\mu_r)$ and $u_j(\mu_{r+p})$ will have the same sign except for those μ_s lying between μ_{r+p} and μ_r; these are $p-1$ such μ's, with indices $r+p-1, \ldots, r+2, r+1$. Thus

$$\begin{aligned} &p \text{ odd} \quad ; \quad u_j(u_r), u_j(u_{r+p}) \text{ have the same sign} \\ &p \text{ even} \quad ; \quad u_j(u_r), u_j(u_{r+p}) \text{ have opposite signs.} \end{aligned}$$

By assumption $f(\lambda)$ has just one zero between any two successive μ's; thus

$$\begin{aligned} &p \text{ odd} \quad ; \quad f(\mu_r), f(\mu_{r+p}) \text{ have opposite signs} \\ &p \text{ even} \quad ; \quad f(\mu_r), f(\mu_{r+p}) \text{ have the same sign.} \end{aligned}$$

Combining these results, we see that $h_j(\mu_r)$ and $h_j(\mu_{r+p})$ must have opposite signs.

Now we check that y_j is positive. Suppose $v_j = q$, and the roots of $g_j(\lambda)$ and $h_j(\lambda)$ are $(\mu_{\alpha_i})_1^q$ and $(\nu_{\alpha_i})_1^{q-1}$ respectively, where

$$\mu_{\alpha_1} > \nu_{\alpha_1} > \mu_{\alpha_2} > \cdots > \nu_{\alpha_{q-1}} > \mu_{\alpha_q}.$$

Then

$$\begin{aligned} g_j(\lambda) &= \prod_{i=1}^{q} (\lambda - \mu_{\alpha_i}), \quad h_j(\lambda) = \prod_{i=1}^{q-1} (\lambda - \nu_{\alpha_i}) \\ f(\lambda) &= \prod_{i=1}^{n} (\lambda - \lambda_i) \end{aligned}$$

and suppose $\alpha_1 = r$, so that

$$f(\mu_r) = \prod_{i=1}^{n} (\mu_r - \lambda_i).$$

Now $\lambda_1, \lambda_2, \ldots, \lambda_r$ are all greater than μ_r, so that the sign of $f(\mu_r)$ is $(-)^r$. All the ν_{α_i} are smaller than μ_r so that $h_j(\mu_r) > 0$. Finally

$$u_j(\mu_r) = \prod (\mu_r - \mu_i)$$

where the sum is taken over those $i \in \{1, 2, \ldots, n-1\}/\{\alpha_1, \alpha_2, \ldots, \alpha_q\}$. But for the sign we need to consider only those $\mu_i > \mu_r$; there are $r-1$ of these, so that the sign of $u_j(\mu_r)$ is $(-)^{r-1}$. Thus $y_j > 0$.

This yields the first stage in the construction of $\mathbf{A}$: take $f(\lambda), g(\lambda)$ and form the partial fraction expansion (5.7.10); $a_{11} = a, \quad a_{1j} = (y_j)^{1/2}$, while the zeros of $g_j(\lambda)$ and $h_j(\lambda)$ are the eigenvalues of the components of $\mathbf{A}_{,1}$.

Figure 5.7.1 shows an example of a tree.

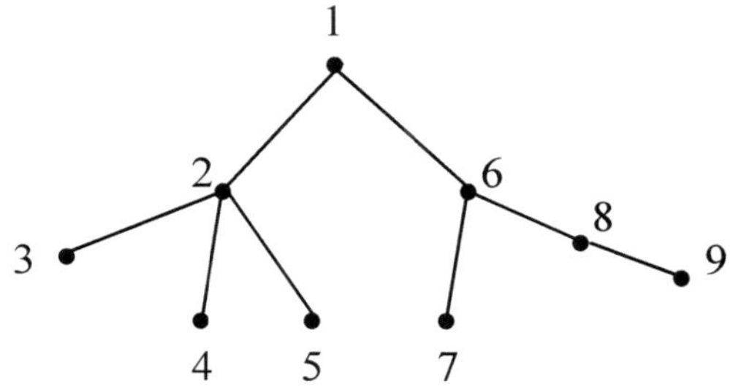

Figure 5.7.4 - A tree on 9 vertices.

The matrix has the form

$$A = \begin{bmatrix} X & X & & & & X & & & \\ X & X & X & X & X & & & & \\ & X & X & & & & & & \\ & X & & X & & & & & \\ & X & & & X & & & & \\ X & & & & & X & X & X & \\ & & & & & X & X & & \\ & & & & & X & & X & X \\ & & & & & & & X & X \end{bmatrix}$$

In stage 1, $\mathbf{A}_{,1}$ has two components; we find a_{11}, a_{12}, a_{16} and we find new data which will allow us to construct the star on vertices $\{2, 3, 4, 5\}$, and the star-path on vertices $\{6, 7, 8, 9\}$. To carry out the second stage we can, if we choose, relabel each of the connected components so that $2 \to 1$ and $6 \to 1$.

We have assumed that the data for constructing $\mathbf{A}$ is two strictly interlacing spectra. However, as with the path and the star, it is possible to use one spectrum $\sigma(\mathbf{A}) = (\lambda_i)_1^n$ and the first coefficients $u_{1i}, \quad i = 1, 2, \ldots, n$ of the normalised eigenvectors of $\mathbf{A}$, instead. We recall the result proved for a general matrix $\mathbf{A} \in S_n$, namely that the eigenvalues of $\mathbf{A}_{,1}$ are the zeros of

$$\sum_{i=1}^{n} \frac{(u_{1i})^2}{\lambda - \lambda_i} = 0.$$

Further discussion of, and reference to, eigenvalue problems related to trees may be found in Nylen and Uhlig (1994) [252].

Further references to the vast literature on inverse eigenvalue problems may be found in Gladwell (1986a) [107], Gladwell (1996) [124], Nocedal and Overton (1983) [251], Friedland, Nocedal and Overton (1987) [95], Ikramov and

Chugunov (2000) [184], Xu (1998) [339], Chu (1998) [58] and Chu and Golub (2001) [59].

Exercises 5.7

1. Show that if the eigenvalues λ_i of $\mathbf{A} \in S_n$ have maximum modulus λ, then $|a_{ij}| \leq \lambda$ for all $i, j = 1, 2, \ldots, n$.

2. Show that if $\mathbf{A} \in S_n$ is a matrix on $\mathcal{G}$, then the maximum number of (non-diagonal) zero entries in $\mathbf{A}$ is attained when $\mathcal{G}$ is a tree, and is $n^2 - 3n + 2$.

3. Construct an algorithm to form $\mathbf{A}$ from $(\lambda_i)_1^n, (\mu_i)_1^n$, given the structure of $\mathcal{G}$. Use it to construct $\mathbf{A}$ on the graph $\mathcal{G}$ of Figure 5.7.4. Take $\{\lambda_i\}_1^9 = \{1, 3, 5, 7, 9, 11, 13, 15, 17\}$, $\{\mu_i\}_1^8 = \{2, 4, 6, 8, 10, 12, 14, 16\}$. As a check, find the eigenvalues of $\mathbf{A}$ and $\mathbf{A}_1$.

Chapter 6

Positivity

> There are then two kinds of intellect: the one able to penetrate acutely and deeply into the conclusions of given premises, and this is the precise intellect; the other able to comprehend a great number of premises without confusing them, and this is the mathematical intellect.
>
> Pascal's *Pensées*, 2

6.1 Introduction

The basic eigenvalue analysis of real symmetric matrices was discussed in Chapter 1. The eigenvalue properties described there are shared by all positive-definite (or semi-definite) matrices. This Chapter, which may be missed on a first reading, provides proofs of some of the results which were used in Chapter 1. Foremost among these are Theorem 6.3.1, that if $\mathbf{A} \in S_n$, then it has n real eigenvectors which are orthonormal, and thus span V_n; and Theorem 6.3.7 that provides necessary and sufficient conditions for the matrix $\mathbf{A}$ to be positive-definite. *Signs*, positive or negative, provide the recurring theme for this Chapter, and hence our choice for the Chapter heading: *positivity.*

In Chapter 3 we focussed our attention on a narrower class, Jacobi matrices, and found that they had additional eigen-properties: they had distinct eigenvalues and, with increasing i, the eigenvector $\mathbf{u}_i$ became increasingly oscillatory, meaning that there was an increasing number of sign changes among the elements $u_{1i}, u_{2i}, \ldots, u_{ni}$. It will be shown in this Chapter that many of the eigen-properties of such matrices are shared by a wider class of so-called *oscillatory* matrices. Actually, there are twin classes of matrices, oscillatory and *sign-oscillatory*, as described in Section 6.5. If $\mathbf{A}$ is oscillatory, and $\mathbf{Z} = diag(1, -1, 1, \ldots, (-)^{n-1})$, then $\tilde{\mathbf{A}} = \mathbf{ZAZ}$ is sign-oscillatory, and *vice versa.* The Jacobi matrix $\mathbf{J}$ of equation (3.1.4) is actually sign-oscillatory. These matrices were introduced and extensively studied by Gantmacher and Krein (1950) [98], see also Gantmacher (1959) [97]. The matrices appearing in lumped-mass or finite element models of strings, rods and beams are all oscillatory or sign-oscillatory; this Chapter serves as reference material for the study of oscillatory matrices.

The theorem upon which the whole of the analysis of oscillatory matrices depends, is Perron's theorem (Theorem 6.5.1). This relates to a strictly positive matrix, one that has all its elements strictly positive, and states that such a matrix has one eigenvalue, the greatest in magnitude, that is real and positive; the corresponding eigenvector has all its coefficients strictly positive.

The matrices appearing in mechanics are usually not strictly positive; such matrices appear in Economics and Operational Analysis. Instead, the matrices are oscillatory. (See the precise definition in Section 6.6.1.) In order to apply Perron's theorem to such matrices, we need two essential steps. First, if $\mathbf{A}$ is oscillatory, then $\mathbf{B} = \mathbf{A}^{n-1}$ is totally positive (TP). This term, which is introduced in Section 6.6.1, means that not only all the elements of $\mathbf{B}$ are strictly positive, but also all the *minors* (Section 6.2) of $\mathbf{B}$. Note that the eigenvalues of $\mathbf{B}$ are the $(n-1)$th powers, λ_i^{n-1}, of the eigenvalues of $\mathbf{A}$, while its eigenvectors are the eigenvectors of $\mathbf{A}$. The other step that is needed is the introduction of the concept of a *compound* matrix (Section 6.2). The compound matrix $\boldsymbol{\mathcal{A}}_p$ is formed from all the $N = \binom{n}{p}$ pth-order minors of $\mathbf{A}$. The important Binet-Cauchy Theorem, Theorem 6.2.3, shows (Ex. 6.3.1) that the eigenvalues of $\boldsymbol{\mathcal{A}}_p$ are simply products of p eigenvalues of $\mathbf{A}$. The argument then runs as follows. Suppose $\mathbf{A}$ is oscillatory, then $\mathbf{B} = \mathbf{A}^{n-1}$ is TP, and hence, for $p = 1, 2, \ldots, n$, $\boldsymbol{\mathcal{B}}_p$ is strictly positive (not TP). The first conclusion (Theorem 6.10.1) is that the eigenvalues of $\mathbf{A}$ are positive and distinct, like those of $\mathbf{J}$ or $\tilde{\mathbf{J}}$.

Before beginning the analysis proper, we point out a notational matter which must be understood if confusion is to be avoided. In Chapter 3, in dealing with a Jacobi matrix $\mathbf{J}$, a positive semi-definite tridiagonal matrix with *negative* codiagonal, the eigenvalues were labelled in *increasing* order, i.e., $0 \leq \lambda_1 < \lambda_2 < \cdots < \lambda_n$. The eigenvectors then became increasingly oscillatory, as described in Theorem 3.3.1. In Ex. 3.3.2, it was pointed out that if the eigenvalues of $\tilde{\mathbf{J}} = \mathbf{ZJZ}$, a positive semi-definite tridiagonal matrix with *positive* codiagonal (an oscillatory matrix if it is actually non-singular, i.e., positive-definite) are labelled in *decreasing* order, i.e., $\lambda_1 > \lambda_2 > \cdots > \lambda_n \geq 0$, then the eigenvectors still satisfy Theorem 3.3.1. In this Chapter, in dealing with oscillatory matrices, we shall keep the same ordering, i.e., $\lambda_1 > \lambda_2 > \cdots > \lambda_n > 0$. Theorem 6.10.2 is a generalisation of Theorem 3.3.1.

6.2 Minors

Suppose $\mathbf{A} \in M_n$. To gain some insight into the structure of $\mathbf{A}$, and into the relative sizes of its elements, we introduce the concept of a *minor*. A minor of order p of the matrix $\mathbf{A}$ is the determinant constructed from the elements of $\mathbf{A}$ in p different rows and p different columns. Thus, the elements of $\mathbf{A}$ themselves are minors of order 1, while $\det(\mathbf{A})$ is the only minor of order n; a_{13}, $\begin{vmatrix} a_{11} & a_{13} \\ a_{21} & a_{23} \end{vmatrix}$ and $\det(\mathbf{A})$ are all minors of $\mathbf{A}$.

Following Ando (1987) [4] we let $Q_{p,n}$, with $1 \le p \le n$, denote the set of strictly increasing sequences α of p integers $\alpha_1, \alpha_2, \ldots, \alpha_p$ taken from $\omega = \{1, 2, \ldots, n\}$. The *complement* $\alpha'$of α is the increasingly arranged sequence $\{1, 2, \ldots, n\}\backslash\alpha = \omega\backslash\alpha$, so that $\alpha' \in Q_{n-p,n}$. When $\alpha \in Q_{p,n}$, $\beta \in Q_{q,n}$ and $\alpha \cap \beta = 0$, their union, $\alpha \cup \beta$, should always be rearranged increasingly to become an element of $Q_{r,n}(r = p+q)$. We will often use two special sequences: $\theta = \theta(p) = \{1, 2, \ldots, p\}$ and $\phi = \phi(p) = \{n-p+1, \ldots, n\}$, and their complements $\theta' = \theta'(p) = \{p+1, \ldots, n\}$, $\phi' = \phi'(p) = \{1, 2, \ldots, n-p\}$. When the argument is omitted in θ or ϕ, it will be understood to be p.

The submatrix formed from rows α and columns β of $\mathbf{A}$ is denoted by $\mathbf{A}[\alpha|\beta]$; $\mathbf{A}[\alpha|\alpha]$ is written $\mathbf{A}[\alpha]$. The minor of $\mathbf{A}$ taken from rows α and columns β is denoted by $A(\alpha;\beta)$; thus

$$A(\alpha;\beta) = \begin{vmatrix} a_{\alpha_1,\beta_1} & a_{\alpha_1,\beta_2} & \cdots & a_{\alpha_1,\beta_p} \\ a_{\alpha_2,\beta_1} & a_{\alpha_2,\beta_2} & \cdots & a_{\alpha_2,\beta_p} \\ \cdot & \cdot & \cdots & \cdot \\ a_{\alpha_p,\beta_1} & a_{\alpha_p,\beta_2} & \cdots & a_{\alpha_p,\beta_p} \end{vmatrix}. \tag{6.2.1}$$

The minor $A(\alpha;\alpha)$ is abbreviated to $A(\alpha)$.

The cofactor of a_{ij}, introduced in Section 1.3, is a minor with a sign attached to it:

$$A_{ij} = (-)^{i+j}\hat{a}_{ij}, \tag{6.2.2}$$

where

$$\hat{a}_{ij} = A(i';j'), \tag{6.2.3}$$

and $i' = \{1, 2, \ldots, i-1, i+1, \ldots, n\} = \omega\backslash i$, $j' = \{1, 2, \ldots, j-1, j+1, \ldots, n\} = \omega\backslash j$; $\hat{a}_{ij}$ is sometimes called *the* minor of a_{ij}.

If $\mathbf{A} \in M_n$, then we can form a new matrix $\hat{\mathbf{A}} = (\hat{a}_{ij})$ from the minors of elements of $\mathbf{A}$. We may prove

Theorem 6.2.1 *Let $\hat{\mathbf{A}} = (\hat{a}_{ij})$, then the minors of $\hat{\mathbf{A}}$ are given by*

$$\hat{A}(\alpha;\beta) = (\det(\mathbf{A}))^{p-1}A(\alpha';\beta').$$

Proof. Consider the theorem for $\alpha = \beta = \theta$; the general case may be obtained by a suitable rearrangement of the rows and columns. Since $\hat{a}_{ij} = (-)^{i+j}A_{ij}$, we may write

$$B = \hat{A}(\alpha;\beta) = \begin{vmatrix} A_{11} & A_{12} & \cdots & A_{1p} \\ A_{21} & A_{22} & \cdots & A_{2p} \\ \cdot & \cdot & \cdots & \cdot \\ A_{p1} & A_{p2} & \cdots & A_{pp} \end{vmatrix}. \tag{6.2.4}$$

Multiplying this by det(**A**), and writing the determinant in (6.2.4) as that of an $n \times n$ matrix, we find

$$B \cdot \det(\mathbf{A}) = \det\left(\begin{bmatrix} A_{11} & A_{12} & \cdots & A_{1p} & A_{1,p+1} & \cdots & A_{1n} \\ A_{21} & A_{22} & \cdots & A_{2p} & A_{2,p+1} & \cdots & A_{2n} \\ \cdot & \cdot & \cdots & \cdot & \cdot & \cdots & \cdot \\ A_{p1} & A_{p2} & \cdots & A_{pp} & A_{p,p+1} & \cdots & A_{pn} \\ 0 & 0 & \cdots & 0 & 1 & \cdots & 0 \\ & & & & & \ddots & \\ 0 & 0 & \cdots & 0 & 0 & \cdots & 1 \end{bmatrix} \cdot \begin{bmatrix} a_{11} & a_{12} & \cdots & a_{1n} \\ a_{21} & a_{22} & \cdots & a_{2n} \\ \cdot & \cdot & \cdots & \cdot \\ \\ a_{n1} & a_{n2} & \cdots & a_{nn} \end{bmatrix}\right),$$

so that on using equation (1.3.10) we obtain

$$\begin{aligned} B \cdot \det(\mathbf{A}) &= \begin{vmatrix} \det(\mathbf{A}) & 0 & \cdots & 0 & \cdots & 0 \\ 0 & \det(\mathbf{A}) & \cdots & 0 & \cdots & 0 \\ & & \det(\mathbf{A}) & 0 & \cdots & 0 \\ a_{1,p+1} & a_{2,p+1} & \cdots & a_{p+1,p+1} & \cdots & a_{n,p+1} \\ a_{1n} & a_{2n} & \cdots & a_{p+1,n} & \cdots & a_{nn} \end{vmatrix} \\ &= (\det(\mathbf{A}))^p A(\alpha';\beta') \end{aligned}$$

so that the theorem holds when $\det(\mathbf{A}) \neq 0$. Continuity considerations show that the theorem also holds when $\det(\mathbf{A}) = 0$. ■

One of the implications of this theorem is that when $\det(\mathbf{A}) = 0$, the rank of $\hat{\mathbf{A}}$ is at most 1, meaning that all the rows of $\hat{\mathbf{A}}$ are multiples of each other, as are all the columns. There is another corollary

Corollary 6.2.1

$$\det(\hat{\mathbf{A}}) = (\det(\mathbf{A}))^{n-1}.$$

There is another way to form a matrix from minors of a given matrix. Suppose $\mathbf{A} \in M_n$ and $1 \leq p < n$, and put $\alpha = \theta(p) := \{1, 2, \ldots, p\}$. We can define b_{ij} by

$$b_{ij} = A(\theta \cup i,\ \theta \cup j), \quad i, j = p+1, p+2, \ldots, n.$$

The matrix $\mathbf{B} \in M_{n-p}$. Thus, if $p = 2$ and

$$\mathbf{A} = \begin{bmatrix} 1 & 2 & 3 & 4 \\ 0 & 1 & -1 & 2 \\ 2 & 1 & 4 & 1 \\ 1 & 0 & 3 & 2 \end{bmatrix}, \text{ then } \mathbf{B} = \begin{bmatrix} -5 & -1 \\ -2 & 2 \end{bmatrix}.$$

The matrix **B** is called a **bordered** matrix; the indices i, j *border* θ.

Sylvester's Theorem on bordered determinants is

Theorem 6.2.2 *Suppose* $\mathbf{A} \in M_n$, $1 \leq p < n$, $b_{ij} = A(\theta \cup i,\ \theta \cup j)$ *for* $i, j = p+1, p+2, \ldots, n$, *and* $\mathbf{B} = (b_{ij})$, *then*

$$\det(\mathbf{B}) = B(\theta';\theta') = (A(\theta;\theta))^{n-p-1} \det(\mathbf{A}).$$

Proof. Theorem 6.2.1, with p replaced by $n-p-1$, and $\alpha = \{p+1, \ldots, r-1, r+1, \ldots, n\} = \theta' \backslash r$, $\beta = \{p+1, \ldots, s-1, s+1, \ldots, n\} = \theta' \backslash s$, shows that

$$\begin{aligned} c_{rs} &= \hat{A}(\alpha; \beta) = (\det(\mathbf{A}))^{n-p-2} A(\theta \cup r; \theta \cup s) \\ &= (\det(\mathbf{A}))^{n-p-2} b_{rs}. \end{aligned} \tag{6.2.5}$$

The Corollary of Theorem 6.2.1 shows that if $\mathbf{C} = (c_{rs})_{p+1}^{n}$, then

$$\det(\mathbf{C}) = (\hat{A}(\theta'; \theta'))^{n-p-1}. \tag{6.2.6}$$

But according to (6.2.5),

$$\begin{aligned} \det(\mathbf{C}) &= B(\theta'; \theta')(\det(\mathbf{A}))^{(n-p-2)(n-p)} \\ &= \det(\mathbf{B})(\det(\mathbf{A}))^{(n-p-2)(n-p)} \end{aligned} \tag{6.2.7}$$

and from Theorem 6.2.1

$$\hat{A}(\theta'; \theta') = (\det(\mathbf{A}))^{n-p-1} A(\theta; \theta) \tag{6.2.8}$$

so that on substituting (6.2.8) into (6.2.6) we find

$$\det(\mathbf{C}) = (\det(\mathbf{A}))^{(n-p-1)^2} (A(\theta; \theta))^{n-p-1}$$

which, on comparison with (6.2.7), yields the required result when $\det(\mathbf{A}) \neq 0$. Continuity considerations show that the theorem still holds when $\det(\mathbf{A}) = 0$. ■

Corollary 6.2.2 *If* $\alpha, \beta \in Q_{s,n}$, $\theta \cap \alpha = 0$, $\theta \cap \beta = 0$, *then*

$$B(\alpha; \beta) = (A(\theta; \theta))^{s-1} A(\theta \cup \alpha; \theta \cup \beta).$$

Corollary 6.2.3 *Suppose* $\alpha, \beta \in Q_{p,n}$ *and*

$$b_{ij} = A(\alpha \cup i; \beta \cup j)$$

for $i = \alpha_p + 1, \ldots, n$; $j = \beta_p + 1, \ldots, n$ *and* $\gamma, \delta \in Q_{q,n}$ *with* $\gamma_1 > \alpha_p$, $\delta_1 > \beta_p$, *then*

$$B(\gamma; \delta) = (A(\alpha; \beta))^{q-1} A(\alpha \cup \gamma; \beta \cup \delta).$$

This is the general form of Sylvester's Theorem. For a proof, see Gantmacher (1959) [97], Vol. I, p. 32.

We now introduce the powerful *Binet-Cauchy Theorem.*

Theorem 6.2.3 *If* $\mathbf{A} \in M_{m,k}$, $\mathbf{B} \in M_{k,n}$ *and* $\mathbf{C} = \mathbf{AB}$, $\alpha \in Q_{p,m}$, $\beta \in Q_{p,n}$ *then*

$$C(\alpha; \beta) = \sum A(\alpha; \gamma) B(\gamma; \beta) \tag{6.2.9}$$

where the sum extends to all $\gamma \in Q_{p,k}$.

The theorem is a generalisation of the formula for c_{ij}, namely

$$c_{ij} = \sum_{p=1}^{k} a_{ip} b_{pj}.$$

The proof may be found in Gantmacher (1959) [97], Vol. I, p. 9.

The importance of the Binet-Cauchy Theorem lies in its application to *compound matrices*, which we now define.

Suppose first that $\mathbf{A}$ is square, i.e., $\mathbf{A} \in M_n$. We shall define the compound matrix $\boldsymbol{\mathcal{A}}_p$. Consider all the sequences $\alpha \in Q_{p,n}$; there are

$$N = \binom{n}{p} = \frac{n!}{p!(n-p)!}$$

such sequences. For given n, p, the N sequences may be arranged in ascending order $1, 2, \ldots, N$. This may be done by associating with the sequence $\alpha = \{\alpha_1, \alpha_2, \ldots, \alpha_p\}$ the number with digits $\alpha_1, \alpha_2, \ldots, \alpha_p$ in the base of $d = N + 1$. This procedure associates a specific index $s = s(\alpha)$ with each sequence α; s lies in the range $1 \leq s \leq N$. Thus, when $n = 5$, $p = 3$, we have $N = 10$, and the combinations are 123, 124, 125, 134, 135, 145, 234, 235, 245, 345. Thus $s(124) = 2$, while $s(245) = 9$. The element a_{st} of $\mathcal{A}_p$ is then given by

$$a_{st} = A(\alpha; \beta)$$

where $s = s(\alpha)$, $t = s(\beta)$.

Although we shall not need it in this book, a compound matrix can be defined for a rectangular matrix $\mathbf{A} \in M_{m,n}$. Now

$$\mathcal{A}_p \in M_{M,N}, \quad M = \binom{m}{p}, \quad N = \binom{n}{p}$$

and a_{st} is given by (6.2.9) for $\alpha \in Q_{p,m}$, $\beta \in Q_{p,n}$. The Binet-Cauchy Theorem now states

Theorem 6.2.4 *If* $\mathbf{A} \in M_{m,k}$, $\mathbf{B} \in M_{k,n}$, $\mathbf{C} = \mathbf{AB}$ *and* $p \leq \min(m, k, n)$, *then*

$$\boldsymbol{\mathcal{C}}_p = \boldsymbol{\mathcal{A}}_p \boldsymbol{\mathcal{B}}_p.$$

Proof. The equation (6.2.9) may be written

$$c_{rs} = \sum_{t=1}^{k} a_{rt} b_{ts}$$

where $r = s(\alpha)$, $s = s(\beta)$, $t = s(\gamma)$. ■

Corollary 6.2.4 *If* $\mathbf{A} \in M_n$ *is non-singular then the pth order compound matrix of* $\mathbf{A}^{-1}$ *is the inverse of the pth order compound matrix of* $\mathbf{A}$.

Proof. Let $\mathbf{B} = \mathbf{A}^{-1}$, then $\mathbf{AB} = \mathbf{I}$ implies $\boldsymbol{\mathcal{A}}_p \boldsymbol{\mathcal{B}}_p = \boldsymbol{\mathcal{I}}_p$ so that $\mathbf{B}_p = (\boldsymbol{\mathcal{A}}_p)^{-1}$. ■

Exercises 6.2

1. If $\mathbf{A} \in M_n$ is non-singular, then equation (1.3.20) shows that its inverse $\mathbf{R} = \mathbf{A}^{-1}$ has elements

$$r_{ij} = A_{ji}/\det(\mathbf{A}) = (-)^{i+j}\hat{a}_{ji}/\det(\mathbf{A}).$$

Use Theorem 6.2.1 to show that if $\alpha, \beta \in Q_{p,n}$ then

$$\det(\mathbf{A})R(\alpha;\beta) = (-)^t A(\alpha';\beta')$$

where

$$t = \sum_{m=1}^{p}(\alpha_m + \beta_m).$$

2. If $\theta = \{1, 2, \ldots, p\}$ and $\phi = \{n-p+1, \ldots, n\}$ then $A(\theta;\phi)$ and $A(\phi;\theta)$ are called the pth order *corner* minors of $\mathbf{A}$. Use Ex. 6.2.1 to show that the corner minors of $\mathbf{R}$ are given by

$$\det(\mathbf{A}) \cdot R(\theta, \phi) = (-)^t A(\theta';\phi'),$$

where $t = r(p+1)$. Note that $A(\theta';\phi')$ is an $(n-p)$th order corner minor of $\mathbf{A}$.

3. Equations (1.3.7), (1.3.8) are a particular case of *Laplace's expansion* of a determinant,

$$\det(\mathbf{A}) = \sum(-)^t A(\alpha;\beta)A(\alpha';\beta')$$

where $\alpha \in Q_{p,n}$ is fixed, the sum is taken over all $\beta \in Q_{p,n}$ and $t = \sum_{m=1}^{p}(\alpha_m + \beta_m)$. Establish this result and show that there is a similar expansion with β fixed and α varying over $Q_{p,n}$.

4. Suppose $\mathbf{A} \in M_n$. Use the Binet-Cauchy theorem to show that the pth compound matrix of $\mathbf{A}^m$ is $(\mathcal{A}_p)^m$, i.e.,

$$(\mathcal{A}^m)_p = (\mathcal{A}_p)^m = \mathcal{A}_p^m.$$

5. Use the Binet-Cauchy theorem to show that if $\mathbf{Q}$ is an orthogonal matrix, then so is $\mathcal{Q}_p$, the pth compound matrix of $\mathbf{Q}$.

6. If $\mathbf{B} = \mathbf{A}^m$, write the minors of $\mathbf{B}$ in terms of the minors of $\mathbf{A}$; use the notation (6.2.9) to show that

$$B(\alpha;\beta) = \sum A(\alpha;\gamma)A(\gamma;\gamma')\ldots A(\gamma'^{(m-1)};\beta)$$

where the sum is over all $\gamma;\gamma'\ldots\gamma'^{(m-1)} \in Q_{p,n}$.

6.3 A general representation of a symmetric matrix

We begin with two theorems.

Theorem 6.3.1 *If* $\mathbf{A} \in S_n$*, then* $\mathbf{A}$ *has* n *real eigenvectors forming an orthonormal system.*

Theorem 6.3.2 *To each* m*-fold eigenvalue* λ_0 *of* $\mathbf{A} \in S_n$*, there correspond* m *linearly independent eigenvectors.*

In Section 1.4 we showed that the eigenvectors corresponding to *distinct* eigenvalues are orthogonal. This means that if all the eigenvalues of $\mathbf{A}$ are distinct, then it has n orthogonal eigenvectors which may be scaled so that they are orthonormal. It sufficies to prove Theorem 6.3.2.

Proof. Suppose that λ_0 is an m-fold eigenvalue of $\mathbf{A}$, i.e., $\triangle(\lambda) = \det(\mathbf{A}-\lambda\mathbf{I})$ has λ_0 as an m-fold root, and that $\mathbf{B} = \mathbf{A} - \lambda_0\mathbf{I}$ has rank p, so that the equation

$$\mathbf{Bu} \equiv (\mathbf{A} - \lambda_0\mathbf{I})\mathbf{u} = \mathbf{0} \tag{6.3.1}$$

has $r = n - p$ linearly independent solutions. We need to prove that $r = m$. Now

$$\begin{aligned} \triangle(\lambda) &= \det(\mathbf{A} - \lambda\mathbf{I}) = \det(\mathbf{B} - (\lambda - \lambda_0)\mathbf{I}) \\ &= \textstyle\sum_{i=0}^{n}(-)^i T_j(\lambda - \lambda_0)^i, \quad j = n - i, \end{aligned}$$

where T_j is the sum of the jth-order principal minors of $\mathbf{B}$, and $T_0 = 1$. But $\mathbf{B}$ has rank p, so that $T_n = 0 = T_{n-1} = \cdots = T_{p+1}$ and therefore

$$\triangle(\lambda) = \pm(\lambda - \lambda_0)^r\{T_p - (\lambda - \lambda_0)T_{p-1} + \cdots \pm (\lambda - \lambda_0)^p T_0\},$$

so that $m \geq r$. It is sufficient to prove that $T_p \neq 0$, for then $\triangle(\lambda)$ will have a r-fold root, i.e., $m = r$. Without loss of generality we may assume that the *first* p rows of $\mathbf{B}$ are linearly independent, so that any row of $\mathbf{B}$ may be expressed as a linear combination of the first p rows, i.e.,

$$b_{ij} = \sum_{k=1}^{p} c_{ik}b_{kj}, \quad i, j = 1, 2, \ldots, n,$$

which may be written

$$\mathbf{B} = \mathbf{C}\mathbf{B}_0, \tag{6.3.2}$$

where $\mathbf{C} \in M_{n,p}$, and $\mathbf{B}_0 \in M_{p,n}$ is formed from the first p rows of $\mathbf{B}$. Now apply the Binet-Cauchy Theorem 6.2.3 to (6.3.2):

$$\mathbf{B}(\alpha;\beta) = \sum \mathbf{C}(\alpha;\gamma)\mathbf{B}_0(\gamma;\beta). \tag{6.3.3}$$

But $\mathbf{C}$ has only p columns, and similarly $\mathbf{B}_0$ has only p rows, and they are the rows $1, 2, \ldots, p$ of $\mathbf{B}$. Thus, there is only one term in the expansion (6.3.3):

$$\mathbf{B}(\alpha;\beta) = \mathbf{C}(\alpha;\theta)\mathbf{B}(\theta;\beta), \tag{6.3.4}$$

where $\theta = \{1, 2, \ldots, p\}$. Similarly

$$\mathbf{B}(\beta; \theta) = \mathbf{C}(\beta; \theta)\mathbf{B}(\theta; \theta). \tag{6.3.5}$$

But $\mathbf{B}$ is symmetric, so that $\mathbf{B}(\beta; \theta) = \mathbf{B}(\theta; \beta)$, and thus, on combining (6.3.4), (6.3.5), we have

$$\mathbf{B}(\alpha; \beta) = \mathbf{C}(\alpha; \theta)\mathbf{C}(\beta; \theta)\mathbf{B}(\theta; \theta). \tag{6.3.6}$$

All the minors on the left cannot vanish, since then $\mathbf{B}$ would have rank *less* than p; we must have $\mathbf{B}(\theta; \theta) \neq 0$. But then (6.3.6) with $\beta = \alpha$ gives

$$\mathbf{B}(\alpha; \alpha) = (\mathbf{C}(\alpha; \theta))^2\mathbf{B}(\theta; \theta).$$

This means that all the pth order principal minors of $\mathbf{B}$ have the same sign, and one at least, $\mathbf{B}(\theta; \theta)$ is non-zero. Thus T_p, their sum, must be non-zero. Hence $m = r$. ■

We may now assert that if $\mathbf{A} \in S_n$, then it *has* n eigenvalues $(\lambda_i)_1^n$ and n orthonormal eigenvectors $(\mathbf{u}_i)_1^n$. This means that

$$\mathbf{A}\mathbf{u}_i = \lambda_i \mathbf{u}_i, \quad i = 1, 2, \ldots, n,$$

which may be combined to yield

$$\mathbf{AU} = \mathbf{U}\wedge, \quad \mathbf{UU}^T = \mathbf{U}^T\mathbf{U} = \mathbf{I} \tag{6.3.7}$$

and this may be transformed to

$$\mathbf{A} = \mathbf{U}\wedge\mathbf{U}^T. \tag{6.3.8}$$

This is a most important representation of a symmetric matrix.

Exercises 6.3

1. Apply the Binet-Cauchy Theorem, in the form of Theorem 6.2.4, to equation (6.3.7) to show that the eigenvalues of $\boldsymbol{\mathcal{A}}_p$ are all the products $\lambda_{i_1}\lambda_{i_2} \ldots \lambda_{i_p}$.
2. Show that the eigenvectors of $\boldsymbol{\mathcal{A}}_p$ are the columns of the compound matrix $\boldsymbol{\mathcal{U}}_p$.

6.4 Quadratic forms

Suppose $\mathbf{A} \in S_n$, then

$$A(\mathbf{x}, \mathbf{x}) \equiv \mathbf{x}^T\mathbf{A}\mathbf{x} = a_{11}x_1^2 + 2a_{12}x_1x_2 + \cdots + 2a_{n,n-1}x_{n-1}x_n + a_{nn}x_n^2 \tag{6.4.1}$$

is called the *quadratic form* associated with $\mathbf{A}$. One of our aims in this section is to find necessary and sufficient conditions for $\mathbf{A}$ to be positive definite (PD), i.e., $A(\mathbf{x}, \mathbf{x}) > 0$ for all $\mathbf{x} \neq \mathbf{0}$.

First, we consider a number of different ways of expressing $A(\mathbf{x},\mathbf{x})$. Let

$$A_i(\mathbf{x}) = \sum_{j=1}^{n} a_{ij}x_j, \quad i = 1, 2, \ldots, n, \tag{6.4.2}$$

then

$$A(\mathbf{x},\mathbf{x}) = \sum_{j=1}^{n} x_j A_j(\mathbf{x}). \tag{6.4.3}$$

This yields

$$\begin{vmatrix} a_{11} & a_{12} & \cdots & a_{1n} & A_1(\mathbf{x}) \\ a_{21} & a_{22} & \cdots & a_{2n} & A_2(\mathbf{x}) \\ \cdot & \cdot & \cdots & \cdot & \cdot \\ a_{n1} & a_{n2} & \cdots & a_{nn} & A_n(\mathbf{x}) \\ A_1(\mathbf{x}) & A_2(\mathbf{x}) & \cdots & A_n(\mathbf{x}) & A(\mathbf{x},\mathbf{x}) \end{vmatrix} = 0, \tag{6.4.4}$$

since the last column is a combination of the first n columns, and

Theorem 6.4.1 *If* $\det(\mathbf{A}) \neq 0$, *then*

$$A(\mathbf{x},\mathbf{x}) = \frac{-1}{\det(\mathbf{A})} \begin{vmatrix} a_{11} & a_{12} & \cdots & a_{1n} & A_1(\mathbf{x}) \\ a_{21} & a_{22} & \cdots & a_{2n} & A_2(\mathbf{x}) \\ \cdot & \cdot & \cdots & \cdot & \cdot \\ a_{n1} & a_{n2} & \cdots & a_{nn} & A_n(\mathbf{x}) \\ A_1(\mathbf{x}) & A_2(\mathbf{x}) & \cdots & A_n(\mathbf{x}) & 0 \end{vmatrix}. \tag{6.4.5}$$

Proof. Expand the zero determinant (6.4.4) along its last row. ■
Now we introduce the quantities

$$X_1(\mathbf{x}) = A_1(\mathbf{x}), X_2(\mathbf{x}) = \begin{vmatrix} a_{11} & A_1(\mathbf{x}) \\ a_{21} & A_2(\mathbf{x}) \end{vmatrix}, X_3(\mathbf{x}) = \begin{vmatrix} a_{11} & a_{12} & A_1(\mathbf{x}) \\ a_{21} & a_{22} & A_2(\mathbf{x}) \\ a_{31} & a_{32} & A_3(\mathbf{x}) \end{vmatrix} \tag{6.4.6}$$

etc., up to $X_n(\mathbf{x})$, and prove

Theorem 6.4.2 *If* $\theta = \{1, 2, \ldots, p\}$ *and* $D_p = A(\theta;\theta) \neq 0, \quad p = 1, 2, \ldots, n$ *then the* $(X_p(\mathbf{x}))_1^n$ *are linearly independent.*

Proof. We see that $X_1(\mathbf{x}) = \sum_{j=1}^{n} a_{1j}x_j$, while $X_2(\mathbf{x}) = \sum_{j=2}^{n}(a_{11}a_{2j} - a_{21}a_{1j})x_j$, and generally

$$X_p(\mathbf{x}) = D_p x_p + \text{ terms in } x_{p+1}, \ldots, x_n.$$

Thus we see that in the reversed sequence $X_n(\mathbf{x}), X_{n-1}(\mathbf{x}), \ldots, X_1(\mathbf{x})$, each term involves one more x_j than the previous one. This means that the $(X_i(\mathbf{x}))_1^n$ can all be simultaneously zero iff all the $(x_i)_1^n$ are zero. ■

This leads us to an important expression for $A(\mathbf{x},\mathbf{x})$ given by

Theorem 6.4.3 *(Jacobi). If $D_0 = 1$, $\theta = \{1, 2, \ldots, p\}$ and $D_p = A(\theta;\theta) \neq 0$, $p = 1, 2, \ldots, n$ then*

$$A(\mathbf{x},\mathbf{x}) = \sum_{p=1}^{n} \frac{(X_p(\mathbf{x}))^2}{D_p D_{p-1}}. \tag{6.4.7}$$

Note that, on account of Theorem 6.4.2, this equation expresses $A(\mathbf{x},\mathbf{x})$ as a sum of multiples of squares of linearly independent combinations of the $(x_i)_1^n$.

Proof. Put $P_0 = 0$, and

$$P_p(\mathbf{x},\mathbf{x}) = \begin{vmatrix} a_{11} & a_{12} & \cdots & a_{1p} & A_1(\mathbf{x}) \\ a_{21} & a_{22} & \cdots & a_{2p} & A_2(\mathbf{x}) \\ \cdot & \cdot & \cdots & \cdot & \cdot \\ a_{p1} & a_{p2} & \cdots & a_{pp} & A_p(\mathbf{x}) \\ A_1(\mathbf{x}) & A_2(\mathbf{x}) & \cdots & A_p(\mathbf{x}) & 0 \end{vmatrix} \tag{6.4.8}$$

and find the recurrence relation linking P_p and P_{p-1}. $P_p(\mathbf{x},\mathbf{x})$ is the determinant of a symmetric matrix $\mathbf{C} \in S_{p+1}$, i.e.,

$$P_p(\mathbf{x},\mathbf{x}) = C(\theta(p+1);\theta(p+1)).$$

Apply Theorem 6.2.2 to this, letting

$$b_{ij} = C(\theta(p-1) \cup i;\ \theta(p-1) \cup j),\ \ i,j = p, p+1$$

then

$$\begin{aligned} \det(\mathbf{B}) &= b_{pp}b_{p+1,p+1} - b_{p,p+1}b_{p+1,p} \\ &= C(\theta(p-1);\theta(p-1)).\ C(\theta(p+1);\theta(p+1)) \end{aligned} \tag{6.4.9}$$

But $b_{pp} = D_p$, $b_{p+1,p+1} = P_{p-1}(\mathbf{x},\mathbf{x})$, $b_{p,p-1} = b_{p-1,p} = X_p(\mathbf{x})$ while $C(\theta(p-1);\theta(p-1)) = D_{p-1}$, $C(\theta(p+1);\theta(p+1)) = P_p(\mathbf{x},\mathbf{x})$. Thus, equation (6.4.9) gives

$$D_p P_{p-1}(\mathbf{x},\mathbf{x}) - X_p^2(\mathbf{x}) = D_{p-1}P_p(\mathbf{x},\mathbf{x})$$

or, since the D_p are non-zero

$$\frac{-P_p(\mathbf{x},\mathbf{x})}{D_p} = \frac{-P_{p-1}(\mathbf{x},\mathbf{x})}{D_{p-1}} + \frac{X_p^2(\mathbf{x})}{D_p D_{p-1}},\ p = 1, 2, \ldots, n. \tag{6.4.10}$$

Now Theorem (6.4.1) states that

$$A(\mathbf{x},\mathbf{x}) = \frac{-P_n(\mathbf{x},\mathbf{x})}{D_n}$$

so that on summing equation (6.4.10) from 1 to n and using $P_0 = 0$ we find the required sum (6.4.7). ■

Theorem 6.4.4 *Suppose $\mathbf{A} \in S_n$, then $\mathbf{A}$ is PD iff $D_i > 0$, $i = 1, 2, \ldots, n$.*

Proof. First we prove that *if* $\mathbf{A}$ is PD, then $\det(\mathbf{A}) > 0$. Since $\mathbf{A} \in S_n$, it has, by the Corollary to Theorem 6.3.2, n eigenvalues $(\lambda_i)_1^n$ and n orthonormal eigenvectors $(\mathbf{u}_i)_1^n$ such that $\mathbf{A}\mathbf{u}_i = \lambda_i \mathbf{u}_i$. Thus $\lambda_i = \mathbf{u}_i^T \mathbf{A} \mathbf{u}_i = A(\mathbf{u}_i, \mathbf{u}_i) > 0$ and therefore $\det(\mathbf{A}) = \prod_{i=1}^n \lambda_i > 0$, i.e., $D_n > 0$.

If $\mathbf{A}$ is PD, then the matrix obtained by deleting the last j rows and columns of $\mathbf{A}$ is PD, for $j = 1, 1, \ldots, n-1$. Therefore, their determinants are positive, i.e., $(D_{n-j})_1^{n-1} > 0$. We have proved that *if* $\mathbf{A}$ is PD, then $(D_i)_1^n > 0$.

Now suppose that $(D_i)_1^n > 0$, then equation (6.4.7) shows that $\mathbf{A}(\mathbf{x}, \mathbf{x}) > 0$, for the $(X_i(\mathbf{x}))_1^n$ can be simultaneously zero only if $\mathbf{x} = \mathbf{0}$. Thus $\mathbf{A}$ is PD. ∎

Corollary 6.4.1 *If* $\mathbf{A} \in S_n$ *is PD, then all the principal minors* $A(\alpha;\alpha) = A(\alpha)$, $\alpha \in Q_{p,n}$, $p = 1, 2, \ldots, n$, *are positive.*

If $\mathbf{A} \in S_n$ is merely positive semi-definite (PSD), then the leading principal minors, and indeed all the principal minors are non-negative. We prove

Theorem 6.4.5 *If* $\mathbf{A} \in S_n$ *is PSD and, for some* p *satisfying* $1 \le p < n$, $D_p = A(\theta;\theta) = 0$, *then every principal minor bordering* D_p *is zero. In particular, the leading principal minors* D_q, $p \le q \le n$, *are zero.*

We prove that the D_q are zero, and leave the remaining result to an Exercise.

Proof. There are two cases:

i) $p = 1$, then $D_1 = a_{11} = 0$, and

$$\begin{vmatrix} a_{11} & a_{1j} \\ a_{j1} & a_{jj} \end{vmatrix} = -a_{1j}^2 \ge 0$$

implies $(a_{1j})_1^n = 0$, so that $(D_q)_1^n = 0$; in this case $\mathbf{x}_1$ does not appear in $A(\mathbf{x}, \mathbf{x})$ at all.

ii) $a_{11} \ne 0$ and, for some p, $1 \le p \le n-1$, $D_p \ne 0$, $D_{p+1} = 0$. (If $p = n-1$, there is nothing further to prove; we may therefore take $p < n-1$.)

We introduce bordered determinants

$$b_{ij} = A(\theta \cup i; \theta \cup j), \;\; i, j = p+1, \ldots, n$$

and form $\mathbf{B} = (b_{ij})_{p+1}^n$. By Sylvester's identity (Corollary to Theorem 6.2.2), if $\alpha_1 > p$ and $\alpha \in Q_{r,n}$, $r \le n-p$, then

$$B(\alpha;\alpha) = (A(\theta;\theta))^{r-1} A(\theta \cup \alpha; \theta \cup \alpha)$$

so that $\mathbf{B}$ is PSD. Since $b_{p+1,p+1} = D_{p+1} = 0$, the matrix falls under case 1 and if $q > p+1$, $r = q-p-1$, $\alpha = \{p+1, \ldots, q\}$, then

$$0 = B(\alpha;\alpha) = \{A(\theta(p);\theta(p))\}^r A(\theta(q);\theta(q))$$

so that $A(\theta(p);\theta(p)) = D_p \ne 0$ implies $A(\theta(q);\theta(q)) = D_q = 0$. ∎

This theorem implies that if $\mathbf{A} \in S_n$ is PSD, then, for some p, $1 \le p < n$, $(D_i)_1^p > 0$, $(D_i)_{p+1}^n = 0$.

Exercises 6.4

1. Show that $\mathbf{A} \in S_n$ is PD iff its eigenvalues $(\lambda_i)_1^n$ are positive; it is PSD iff its eigenvalues are non-negative.

2. Show that if $\mathbf{A} \in S_n$ is PSD then $\mathbf{A}$ is singular, and that $\mathbf{x}^T\mathbf{A}\mathbf{x} = 0$ iff $\mathbf{A}\mathbf{x} = \mathbf{0}$.

3. Show that if $\mathbf{A} \in S_n$ is PSD and if $a_{ii} = 0$ for some i satisfying $1 \le i \le n$, then $a_{ij} = 0$ for $j = 1, 2, \ldots, n$. This means that if $a_{ii} = 0$ then x_i does not appear in $\mathbf{x}^T\mathbf{A}\mathbf{x}$.

4. Show that if $\mathbf{A} \in S_n$ is PSD and has rank r then it has a positive principal minor of order r.

These examples are merely a selection of properties of PD and PSD matrices to be found in Chapter 7 of Horn and Johnson (1985) [183].

6.5 Perron's theorem

Most matrices appearing in classical vibration problems are symmetric. It is therefore known that they have real eigenvalues, and a complete set of orthonormal eigenvectors. Often the matrices are PD, so that their eigenvalues, in addition to being real, are positive. However, the whole theory relating to oscillatory matrices depends on a basic result relating to a class of not necessarily symmetric matrices, as we now describe.

We recall some definitions. If a vector $\mathbf{x}$ has all its elements positive (non-negative) we shall say $\mathbf{x} > \mathbf{0}$ ($\ge \mathbf{0}$) and shall say that $\mathbf{x}$ is *positive* (*non-negative*). If $\mathbf{x}, \mathbf{y}$ are in V_n then $\mathbf{x} \ge \mathbf{y}$ is equivalent to $\mathbf{x} - \mathbf{y} \ge \mathbf{0}$. The matrix $\mathbf{A} \in \mathbf{M}_n$ is said to be *positive* (*non-negative*) if $a_{ij} > 0$ (≥ 0) for all $i, j = 1, 2, \ldots, n$.

Up to this point the only norm we have used for a vector $\mathbf{x} \in V_n$ is the *Euclidean*, or so-called L_2 norm:

$$||\mathbf{x}||_2 = (\sum_{i=1}^{n} |x_i|^2)^{\frac{1}{2}}. \tag{6.5.1}$$

We can define the L_2 norm of a matrix $\mathbf{A} \in \mathbf{M}_n$:

$$||\mathbf{A}||_2 = (\sum_{i,j=1}^{n} |a_{ij}|^2)^{\frac{1}{2}}. \tag{6.5.2}$$

The norm is variously called the *Frobenius* norm, *Schur* norm or *Hilbert-Schmidt* norm.

We will need another norm, the L_1 norm:

$$||\mathbf{x}||_1 = \sum_{i=1}^{n} |x_i|, \tag{6.5.3}$$

$$||\mathbf{A}||_1 = \sum_{i,j=1}^{n} |a_{ij}|. \tag{6.5.4}$$

A norm is like a distance; as such it must satisfy various fundamental conditions, for which see Ex. 6.5.1. For a definitive and extensive study of vector and matrix norms, see Horn and Johnson (1985) [183], Section 5.6.

We may now prove Perron's theorem, following Bellman (1970) [25].

Theorem 6.5.1 *(Perron). Suppose* $\mathbf{A} \in \mathbf{M}_n$ *and* $\mathbf{A} > \mathbf{0}$. *Then* $\mathbf{A}$ *has a unique eigenvalue* ρ *which has greatest absolute value. This eigenvalue is positive and simple, and its associated eigenvector can be taken to be positive. The eigenvalue* ρ *is often called the Perron root of* $\mathbf{A}$.

Proof. Let $S(\lambda)$ be the set of all non-negative λ for which there exist non-negative $\mathbf{x}$ such that

$$\mathbf{A}\mathbf{x} \geq \lambda\mathbf{x}. \tag{6.5.5}$$

We shall consider only L_1-normalised vectors $\mathbf{x}$, i.e., such that $||\mathbf{x}||_1 = \sum_{i=1}^{n} x_i = 1$. (Since $\mathbf{x} \geq \mathbf{0}$, $|x_i| = x_i$.) This therefore excludes the zero vector. If $\mathbf{x}$ satisfies (6.5.5), then Ex. 6.5.2 shows that

$$\lambda||\mathbf{x}||_1 \leq ||\mathbf{A}\mathbf{x}||_1 \leq ||\mathbf{A}||_1.||\mathbf{x}||_1, \tag{6.5.6}$$

so that $0 \leq \lambda \leq ||\mathbf{A}||_1$. This shows that the set $S(\lambda)$ is bounded. It is clearly not empty, because $\mathbf{A}$ is positive. The bounded set $S(\lambda)$ has a least upper bound; let it be λ_0. Let $\lambda_1, \lambda_2, \ldots$ be a sequence of λ's in $S(\lambda)$ converging to λ_0, and $\mathbf{x}_i$ a corresponding sequence of $\mathbf{x}$'s satisfying $\mathbf{A}\mathbf{x}_i \geq \lambda_i\mathbf{x}_i$. The set of all $\mathbf{x}$ such that $||\mathbf{x}||_1 = 1$ is closed and bounded; therefore, the sequence $\{\mathbf{x}_i\}$ contains a convergent sequence $\{\mathbf{x}_{\nu_i}\}$ converging to a non-negative vector $\mathbf{x}_0$ satisfying $||\mathbf{x}_0||_1 = 1$, and (Ex. 6.5.3)

$$\mathbf{A}\mathbf{x}_0 \geq \lambda_0\mathbf{x}_0. \tag{6.5.7}$$

This means that $\lambda_0 \in S(\lambda)$. We shall now show that equality holds in (6.5.7), and we do so by reduction to a contradiction.

Let

$$\mathbf{d} = \mathbf{A}\mathbf{x}_0 - \lambda_0\mathbf{x}_0 \geq \mathbf{0}$$

and suppose one of the d_i, say d_j, is positive. Put

$$y_i = x_{i0} + (d_j/2\lambda_0)\delta_{ij}$$

then the ith row of $\mathbf{A}\mathbf{y} - \lambda_0\mathbf{y}$ is

$$e_i = d_i + a_{ij}d_j/(2\lambda_0) - d_j\delta_{ij}/2 > 0.$$

Now let $\lambda = \lambda_0 + \min_i(e_i/y_i)$, then $\lambda > \lambda_0$, and

$$\begin{aligned} \mathbf{A}\mathbf{y} - \lambda\mathbf{y} &= \mathbf{e} - (\lambda - \lambda_0)\mathbf{y} \\ &= \mathbf{e} - \min_i(e_i/y_i)\mathbf{y} \geq \mathbf{0} \end{aligned}$$

This states that $\lambda \in S(\lambda)$, and that λ is greater than the least upper bound, λ_0, of $S(\lambda)$. This contradiction implies that there is equality in equation (6.5.7), i.e.,

$$\mathbf{A}\mathbf{x}_0 = \lambda_0 \mathbf{x}_0. \tag{6.5.8}$$

Thus λ_0 is an eigenvalue and $\mathbf{x}_0$ is an eigenvector, and $\mathbf{x}_0$ is necessarily positive (Ex. 6.5.4). We will show that λ_0 is the required Perron root.

Suppose that there is another eigenvalue λ, possibly complex, such that $|\lambda| \geq \lambda_0$, with $\mathbf{z} \neq \mathbf{0}$ being an associated eigenvector, so that $\mathbf{A}\mathbf{z} = \lambda\mathbf{z}$. Let $|\mathbf{z}|$ denote the vector with elements $|z_1|, |z_2|, \ldots, |z_n|$, then we deduce that

$$|\lambda|\ |\mathbf{z}| = |\mathbf{A}\mathbf{z}| \leq \mathbf{A}|\mathbf{z}|. \tag{6.5.9}$$

But then the maximum property of λ_0 implies $|\lambda| \leq \lambda_0$, and hence $|\lambda| = \lambda_0$. Now the argument used earlier shows that equality holds in equation (6.5.9), i.e.,

$$\mathbf{A}|\mathbf{z}| = \lambda_0|\mathbf{z}|, \quad |\mathbf{z}| > 0.$$

But then

$$|\mathbf{A}\mathbf{z}| = \mathbf{A}|\mathbf{z}| \tag{6.5.10}$$

and (Ex. 6.5.5) this can hold only if $\mathbf{z} = c\mathbf{w}$, where c is complex and $\mathbf{w}$ is positive; and this implies that λ is positive, i.e., $\lambda = \lambda_0$. We now show that $\mathbf{x}_0$ and $\mathbf{w}$, both positive and both eigenvectors corresponding to λ_0, are equivalent. Put $\mathbf{y} = \mathbf{x}_0 - \varepsilon\mathbf{w}$, and take

$$\varepsilon = \min_i (x_{i0}/w_i) = x_{j0}/w_j,$$

then $\mathbf{y}$ is a non-negative eigenvector corresponding to λ_0 with $y_j = 0$, so that

$$a_{j1}y_1 + a_{j2}y_2 + \cdots + a_{jn}y_n = 0,$$

and since $a_{j1} > 0$ for $i = 1, 2, \ldots, n$ we must have $\mathbf{y} = \mathbf{0}$. Thus $\mathbf{x}_0 = \varepsilon\mathbf{w}$ so that λ_0 is a simple eigenvalue. Thus λ_0 has all the properties asserted for the Perron root ρ. ■

Exercises 6.5

1. A vector norm must satisfy three conditions:

a) $||\mathbf{x}|| \geq 0$, and $||\mathbf{x}|| = 0$ iff $\mathbf{x} = \mathbf{0}$
b) $||c\mathbf{x}|| = |c| \cdot ||\mathbf{x}||$
c) $||\mathbf{x} + \mathbf{y}|| \leq ||\mathbf{x}|| + ||\mathbf{y}||$
Show that both the L_1 and the L_2 norm satisfy these conditions.

2. Show that $\mathbf{A} \in M_n$, $\mathbf{x} \in V_n$, then

$$||\mathbf{A}\mathbf{x}||_1 \leq ||\mathbf{A}||_1 \cdot ||\mathbf{x}||_1.$$

3. Verify that the vector $\mathbf{x}_0$ will in fact satisfy the inequality (6.5.7).

4. Show that if $\mathbf{x}$ is a non-negative eigenvector of a positive matrix $\mathbf{A} \in M_n$, then $\mathbf{x} > \mathbf{0}$. This has the following logical negative consequence:

$$\text{if } \mathbf{A} > \mathbf{0},\ \mathbf{x} \geq \mathbf{0},\ \mathbf{A}\mathbf{x} = \lambda\mathbf{x}$$

and $x_i = 0$ for some $i = 1, \ldots, n$, then $\mathbf{x} = \mathbf{0}$.

5. Show that if $\mathbf{A} \in M_n$ is positive, then equation (6.5.10) can hold only if $\mathbf{z} = c\mathbf{w}$, where c is complex and $\mathbf{w} > \mathbf{0}$.

6.6 Totally non-negative matrices

Suppose $\mathbf{A} \in M_n$. The matrix $\mathbf{A}$ is said to be *positive* (see Section 6.5), written $\mathbf{A} > 0$, if $a_{ij} > 0$ for all $i, j = 1, 2, \ldots, n$. *Total positivity* concerns all the minors of $\mathbf{A}$, (see Section 6.2) not just its elements. If $\mathbf{A} \in M_{m,n}$, we say that $\mathbf{A}$ is

1. *TN (totally non-negative) if all the minors of* $\mathbf{A}$ *are non-negative*;

   If $\mathbf{A} \in M_n$, we say that $\mathbf{A}$ is

2. *NTN (non-singular and totally non-negative) if* $\mathbf{A}$ *is non-singular and TN*;

3. *TP (totally positive) if all the minors are (strictly) positive*;

4. *O (oscillatory) if* $\mathbf{A}$ *is TN, and a power,* $\mathbf{A}^m$, *is TP.*

Note that some authors, including ourselves in Gladwell (1986b) [108], use *totally positive* (TP) instead of *totally non-negative* (TN), and *strictly totally positive* (STP) instead of *totally positive* (TP). Also, in Gladwell (1986b) [108], following Gantmacher and Krein (1950) [98], we used *completely* instead of *totally*; completely positive now has a quite different connotation. Reader, beware of these subtle distinctions!

The concept of an oscillatory (or oscillation) matrix was effectively introduced by Gantmakher and Krein in the 1930's, see Gantmacher and Krein (1950) [98]. It was developed further by Gantmacher (1959) [97]. The concept of total positivity had arisen much earlier than this, e.g., Fekete (1913) [86]; it was first systematically explored by Karlin (1968) [190] in his book *Total Positivity*, Volume 1. (Volume 2 has never appeared!) Ando (1987) [4] reviews its history and proves important new results. All the concepts, total positivity, oscillatory, etc., arise in the study of in-line systems, rods, beams, splines, Sturm-Liouville differential equations, etc.

The study of total positivity involves the delicate treatment of inequalities. Here are two typical examples, which the reader may verify:

i) if $a \geq 0$ or $d \geq 0$; $b \geq 0$ and $c \geq 0$; and

$$\begin{vmatrix} a & b \\ c & d \end{vmatrix} > 0,$$

then $a > 0$ and $d > 0$;

ii) if $a \geq 0$ or $d \geq 0$; $b > 0$ and $c > 0$; and

$$\begin{vmatrix} a & b \\ c & d \end{vmatrix} \geq 0$$

then $a > 0$ and $d > 0$.

The concept of total positivity is similar to positive-definiteness, but there are important differences between the two concepts: positive definiteness applies only to symmetric matrices, TP applies to any matrices in M_n; the condition for positive-definiteness involves only the principal minors, while TP involves all the minors. Clearly, if $\mathbf{A} \in S_n$ is TN then it is PSD; if it is TP then it is PD; but the converses of these results are false. (Ex. 6.6.1). There is a theorem like Theorem 6.4.5 for matrices which are TN:

Theorem 6.6.1 *If* $\mathbf{A} \in M_n$*, and* $\mathbf{A}$ *is TN, and* $\mathbf{A}$ *has a zero principal minor, then every minor bordering it is also zero.*

Proof. For simplicity we confine our attention to the *leading* principal minors; this restriction can be removed at the expense of some complication in the argument. As in Theorem 6.4.5, there are two cases:

1) $D_1 = a_{11} = 0$. We assert that this implies that *either* $(a_{i1})_1^n = 0$ or $(a_{1j})_1^n = 0$. If this were not true, then we could find $a_{i1} > 0$ and $a_{1j} > 0$ for some i, j satisfying $2 \leq i \leq n$, $2 \leq j \leq n$. But then

$$\begin{vmatrix} a_{11} & a_{1j} \\ a_{i1} & a_{ij} \end{vmatrix} = -a_{i1}a_{1j} < 0,$$

which contradicts the statement that $\mathbf{A}$ is TN. Thus if $a_{11} = 0$ then either the first row of $\mathbf{A}$ or the first column of $\mathbf{A}$ must be zero. (See also Ex. 6.6.2.)

2) $a_{11} \neq 0$. Then for some $p (1 \leq p \leq n-1)$ we have $D_p \neq 0$, $D_{p+1} = 0$. (Again, if $p = n-1$, there is nothing further to prove.) We introduce bordered determinants

$$b_{ij} = A(\theta \cup i; \theta \cup j) \quad i, j = p+1, \ldots, n$$

and form the matrix $\mathbf{B} = (b_{ij})_{p+1}^n$. By Sylvester's identity (Corollary 6.2.3), if $\alpha, \beta \in Q_{r,n}$, $\alpha_1 > p$, $\beta_1 > p$, then

$$B(\alpha; \beta) = (A(\theta; \theta))^{r-1} A(\theta \cup \alpha; \theta \cup \beta)$$

so that $\mathbf{B}$ is TN. Since $b_{p+1,p+1} = D_{p+1} = 0$, the matrix falls under case 1. If $q > p+1$, and $\alpha = \beta = \{p+1, \ldots, q\}$, then

$$B(\alpha; \alpha) = (A(\theta(p); \theta(p)))^{q-p-1} A(\theta(q); \theta(q)) = 0.$$

But since $D_p = A(\theta(p); \theta(p)) \neq 0$, we have $D_q = 0$. ■

Theorem 6.6.2 *If* $\mathbf{A} \in M_n$ *and* $\mathbf{A}$ *is NTN, then all its principal minors are positive.*

Proof. If any principal minor were zero, then, by Theorem 6.6.1, $D_n = \det(\mathbf{A})$ would be zero, but $\mathbf{A}$ is non-singular, so that $\det(\mathbf{A}) \neq 0$ (in fact $\det(\mathbf{A}) > 0$). ■

Corollary 6.6.1 *If* $\mathbf{A} \in S_n$ *is NTN, then* $\mathbf{A}$ *is PD.*

If $\mathbf{A} \in M_n$ is NTN, then we know that some elements of $\mathbf{A}$ are strictly positive, in particular, by Theorem 6.6.2, the diagonal elements a_{ii}. We now prove an important result which shows that $\mathbf{A}$ will have a so-called *staircase* structure. We first introduce some definitions.

Let $p = \{p_1, p_2, \ldots, p_n\}$ be a sequence of integers from $\{1, 2, \ldots, n\}$. Then p is a *staircase sequence* if $p_1 \le p_2 \le \cdots \le p_n$ and $p_i \ge i$ for all $i = 1, 2, \ldots, n$. Thus $p = \{2, 3, 3, 5, 5\}$ is a staircase sequence.

Suppose p and q are staircase sequences. A matrix $\mathbf{A} \in M_n$ is said to be a *p, q-staircase matrix* if $a_{ij} = 0$ when $j > p_i$ or $i > q_j$. Suppose $p = \{2, 3, 3, 5, 5\}$, $q = \{2, 4, 5, 5, 5\}$ then

$$A = \begin{bmatrix} a_{11} & a_{12} & & & \\ a_{21} & a_{22} & a_{23} & & \\ & a_{32} & a_{33} & & \\ & a_{42} & a_{43} & a_{44} & a_{45} \\ & & a_{53} & a_{54} & a_{55} \end{bmatrix}$$

is a p, q-staircase matrix. The characteristic of a staircase matrix is that if an element in the upper (lower) triangle is zero then all elements to the right (left) and above (below) are zero. Clearly, if $\mathbf{A} \in S_n$, then $p = q$; we say $\mathbf{A}$ is a p-staircase matrix. We are now ready for

Theorem 6.6.3 *If* $\mathbf{A} \in M_n$ *is NTN, it is a staircase matrix.*

Proof. The elements in the upper and lower triangles may be dealt with separately; we consider just the upper triangle.

Suppose $j > 1$ and $a_{ij} = 0$. If $k > j$, then

$$\begin{vmatrix} a_{ij} & a_{ik} \\ a_{jj} & a_{jk} \end{vmatrix} \ge 0, \; a_{ij} = 0, \; a_{jj} > 0, \; a_{ik} \ge 0$$

imply $a_{ik} = 0$; all elements to the right of a_{ij} are also zero. Now consider the terms $\{a_{ii}, a_{i,i+1}, \ldots, a_{in}\}$ in the ith row. The first is positive; there is therefore a least index m, with $i \le m \le n$, such that $a_{ij} = 0$ for $j > m$; call this index $p_i; p_i \ge i$.

Now suppose $j > i$ and $a_{ij} = 0$. If $k < i$ then

$$\begin{vmatrix} a_{ki} & a_{kj} \\ a_{ii} & a_{ij} \end{vmatrix} \ge 0, \; a_{ij} = 0, \; a_{ii} > 0, \; a_{kj} \ge 0$$

imply $a_{kj} = 0$. Thus $a_{ij} = 0$ and $k < i$ implies $a_{kj} = 0$. Thus $j > p_i$ implies $a_{kj} = 0$, i.e., $j > p_i$ implies $j > p_k$; $p_k \leq p_i$. Thus p is a staircase sequence, and the upper triangle of $\mathbf{A}$ is a staircase. ■

Theorem 6.6.4 *If* $\mathbf{A} \in M_n$ *is TN and* $1 \leq p < n$, *then* $A(\theta(n);\theta(n)) \leq A(\theta(p);\theta(p)).A(\theta'(p);\theta'(p))$. *Recall that* $\theta(p) = \{1,2,\ldots,p\}$, $\theta'(p) = \{p+1,\ldots,n\}$.

Proof. On account of Theorem 6.6.1 we may assume without loss of generality that all the principal minors are *positive*, for if any were zero, then the inequality would be satisfied trivially because then by Theorem 6.6.1, $\det(\mathbf{A}) = 0$. The theorem is true for $n = 2$ since

$$A(\theta(2);\theta(2)) = a_{11}a_{22} - a_{12}a_{21} \leq a_{11}a_{22}.$$

We prove the theorem by induction, and assume that it holds for matrices of order $n-1$ or less. We introduce the matrix $\mathbf{B}$ of Theorem 6.6.1:

$$b_{ij} = A(\theta \cup i;\theta \cup j),\ i,j = p+1,\ldots,n.$$

and

$$\begin{aligned} B(\theta';\theta') &= (A(\theta;\theta))^{n-p-1}A(\theta(n);\theta(n)) \\ &= (D_p)^{n-p-1}D_n \end{aligned}$$

which we reverse to give

$$D_n = B(\theta';\theta')/(D_p)^{n-p-1}.$$

Since $\mathbf{B}[\theta'|\theta']$ is of order $n-p \leq n-1$, the inductive hypothesis applies to it:

$$B(\theta';\theta') \leq b_{p+1,p+1}B(\theta'(p+1);\theta'(p+1))$$

and thus

$$D_n \leq b_{p+1,p+1}B(\theta'(p+1);\theta'(p+1))/(D_p)^{n-p-1}. \tag{6.6.1}$$

Applying Sylvester's identity again, we have

$$B(\theta'(p+1);\theta'(p+1)) = (A(\theta;\theta))^{n-p-2}A(\alpha;\alpha)$$

where $\alpha = \theta \cup\, \theta'(p+1) = \{1,2,\ldots,p,p+2,\ldots,n\}$ which when combined with (6.6.1) and $b_{p+1,p+1} = D_{p+1}$ gives

$$\begin{aligned} D_n &\leq D_{p+1}(D_p)^{n-p-2}A(\alpha;\alpha)/(D_p)^{n-p-1} \\ &\leq D_{p+1}A(\alpha;\alpha)/D_p \end{aligned}. \tag{6.6.2}$$

Now we use the inductive hypothesis again to give

$$A(\alpha;\alpha) \leq D_p\ A(\theta'(p+1);\theta'(p+1))$$

which, when combined with (6.6.2) gives

$$A(\theta(n);\theta(n)) \leq A(\theta(p+1);\theta(p+1))A(\theta'(p+1);\theta'(p+1))$$

which shows that the result holds for matrices of order n. ■

Corollary 6.6.2 *If* $\mathbf{A} \in M_n$ *is TN then*

$$D_p \leq a_{11}a_{22}\ldots a_{pp},\ 1 \leq p \leq n.$$

Theorem 6.6.4 is expressed as a result concerning *principal* minors of a TN matrix **A**, but since any square matrix taken from a subset of rows and columns of such an **A** is also TN we can state

Corollary 6.6.3 *If* $\mathbf{A} \in M_n$ *is TN,* $\beta, \gamma \in Q_{q,n}$*, and* $\beta, \gamma \in \theta(p) = \theta$ *(i.e.,* $\beta_q, \gamma_q \leq p$*), then*

$$A(\beta \cup \theta'; \gamma \cup \theta') \leq A(\beta; \gamma)A(\theta'; \theta').$$

Similarly, if $\beta, \gamma \in Q_{q,n}$, and $\beta, \gamma \in \theta'(p) = \theta'$ (i.e., $\beta_1, \gamma_1 \geq p+1$), then

$$A(\theta \cup \beta; \theta \cup \gamma) \leq A(\beta; \gamma)A(\theta; \theta).$$

See Ando (1987) [4] for generalisations of this result.

Theorem 6.6.5 *Suppose* $\mathbf{A} \in M_{m,n}$ *is TN. If* **A** *has* p *linearly dependent rows, labelled by* $\alpha \in Q_{p,m}$ *with* $\alpha_1 = 1, \alpha_p = m$*, of which the first* $p-1$*, labelled* $\alpha\backslash\alpha_p$*, and the last* $p-1$*, labelled* $\alpha\backslash\alpha_1$*, are linearly independent (l.i.), then* **A** *has rank* $p-1$*.*

Proof. Clearly, the rank of **A** is at least $p-1$; we show that it is not greater than $p-1$, i.e., it is exactly $p-1$.

The linearly dependent rows are specified by $\alpha = \{\alpha_1, \alpha_2, \ldots, \alpha_p\}$. If $p > n$, then rank $(\mathbf{A}) \leq n < p$, so that rank $(\mathbf{A}) = p-1$. Take $p \leq n$. The row α_p may be expressed in terms of rows $\alpha_1, \alpha_2, \ldots, \alpha_{p-1}$:

$$a_{\alpha_p,j} = \sum_{k=1}^{p-1} c_k a_{\alpha_k,j} \quad j = 1, 2, \ldots, n, \tag{6.6.3}$$

and since rows $\alpha\backslash\alpha_1$ are *l.i.*, $c_1 \neq 0$. Since rows $\alpha\backslash\alpha_p$ are l.i., there is $\beta^0 \in Q_{p-1,n}$ such that $A(\alpha\backslash\alpha_p; \beta^0) > 0$. On substituting for $a_{\alpha_p,j}$ from (6.6.3) we find

$$A(\alpha\backslash\alpha_1; \beta^0) = (-)^p c_1 A(\alpha\backslash\alpha_p; \beta^0).$$

Therefore $A(\alpha\backslash\alpha_1; \beta^0) \neq 0$, but this minor is non-negative and therefore it is strictly positive; therefore $(-)^p c_1 > 0$.

Now suppose $q \in \alpha'$ then $\alpha_r < q < \alpha_{r+1}$ for some index r satisfying $1 \leq r < p-1$. If $\beta \in Q_{p,n}$ then, on substituting for $a_{\alpha_p,j}$, as before, we find

$$A((\alpha\backslash\alpha_1) \cup q; \beta) = (-)^{p+1} c_1 A((\alpha\backslash\alpha_p) \cup q; \beta). \tag{6.6.4}$$

The inequality $(-)^p c_1 > 0$ implies that the minors on either side of (6.6.4) have opposite signs. But both are non-negative so that both are zero. Since β is an arbitrary member of $Q_{p,n}$, this means that any row $q \in \alpha'$ may be expressed as a linear combination of the rows $\alpha\backslash\alpha_1$, or equivalently of $\alpha\backslash\alpha_p$. Thus the rank of **A** is $p-1$. ■

We now prove a corollary of this result, but since its truth is not immediately clear, we state it as

Theorem 6.6.6 *If* $\mathbf{A} \in M_{m,n}$ *is TN and there exist* $\alpha \in Q_{p,m}$, $\beta \in Q_{p,n}$ *such that* $\alpha_1 = 1, \alpha_p = m, \beta_1 = 1, \beta_p = n$ *and*

$$A(\alpha;\beta) = 0,\ A(\alpha\backslash\alpha_p;\beta\backslash\beta_p) > 0,\ A(\alpha\backslash\alpha_1;\beta\backslash\beta_1) > 0$$

then $\mathbf{A}$ *has rank* $p-1$.

Proof. Apply Theorem 6.6.5 to the matrix with rows $\{1,2,\ldots,m\}$ and columns β of $\mathbf{A}$. It has p linearly dependent rows α, of which the first $p-1$, $\alpha\backslash\alpha_p$, and the last $p-1$, $\alpha\backslash\alpha_1$, are linearly independent. Therefore, it has rank $p-1$, so that its p columns are linearly dependent. These columns are columns of $\mathbf{A}$, and so are rows of $\mathbf{A}^T$. Now apply Theorem 6.6.5 to $\mathbf{A}^T$. Its rows β are linearly dependent, while the first $p-1$, $\beta\backslash\beta_p$, and last $p-1$, $\beta\backslash\beta_1$, are linearly independent. Therefore, by Theorem 6.6.5, $\mathbf{A}^T$ has rank $p-1$; A has rank $p-1$. ■

Exercises 6.6

1. Exhibit $\mathbf{A} \in S_2$ which is PD but not TN.

2. Use Theorem 6.6.3 to prove that if $\mathbf{A}$ is NTN and $a_{1n} > 0$, $a_{n1} > 0$, then $\mathbf{A}$ is a (strictly) positive matrix. Markham (1970) [221] stated this result for oscillatory $\mathbf{A}$, but NTN is sufficient. Find even weaker conditions for the result to hold. (See Gladwell (1998) [126].) See Gasca and Pena (1992) [99] for related work.

6.7 Oscillatory matrices

We introduced four terms at the beginning of Section 6.6: TN, NTN, TP and O. In this section we are concerned with the last, *oscillatory*. We note that TN is weaker than NTN, which in turn is weaker than TP. O is by definition stronger than NTN; it is weaker than TP because

$$\mathbf{A} = \begin{bmatrix} 2 & 1 & \\ 1 & 2 & 1 \\ & 1 & 2 \end{bmatrix} \tag{6.7.1}$$

is O because

$$\mathbf{A}^2 = \begin{bmatrix} 5 & 4 & 1 \\ 4 & 6 & 4 \\ 1 & 4 & 5 \end{bmatrix}$$

is TP, but $\mathbf{A}$ itself is not TP. Note that if $\mathbf{A}^m$ is TP, then $\mathbf{A}$ is necessarily non-singular. We can therefore define $\mathbf{A}$ to be O, if $\mathbf{A}$ is TN and $\mathbf{A}^m$ is TP. We will show later that if $\mathbf{A} \in S_n$, $\mathbf{A}$ is PD, and tridiagonal with positive co-diagonal, then $\mathbf{A}$ is O. Clearly though, the class of oscillatory matrices is much larger than this. We will first obtain some preliminary results which will allow us to characterise oscillatory matrices. It is oscillatory matrices, and not TP matrices, which appear in applications to inverse problems.

We have defined an oscillatory (O) matrix as a TN matrix which is such that a power $\mathbf{A}^m$ is TP. Using this definition, we cannot easily check whether a TN matrix is O. Our principal aim in this section is to obtain an easily applicable test for $\mathbf{A}$ to be O. As a first step we prove

Theorem 6.7.1 *If $\mathbf{A} \in M_n$ is O, then any principal submatrix $\mathbf{B} \in M_p$ formed by deleting successive rows and columns of $\mathbf{A}$ is O.*

Proof. Clearly, any principal submatrix is TN; the question is whether it is O. It is sufficient to show that $\mathbf{B} = \mathbf{A}_1$, obtained by deleting the *first* row and column of $\mathbf{A}$ is O.

We use Ex. 6.2.6, deduced from the Binet-Cauchy Theorem (equation (6.2.9)), to obtain the minors of a power of a matrix in terms of the minors of the original matrix. Suppose that $\mathbf{A}^m = \mathbf{C}$ is TP, and consider the minors of $\mathbf{D} = \mathbf{B}^m$. We retain the original numbering of rows and columns, so that $\mathbf{B} = (a_{ij})_2^n$.

Then if $\alpha, \beta \in Q_{p,n}$ and $\alpha_1 \geq 2,\ \beta_1 \geq 2$, we have

$$D(\alpha;\beta) = \sum A(\alpha;\gamma^{(1)})A(\gamma^{(1)};\gamma^{(2)})\ldots A(\gamma^{(m-1)};\beta) \tag{6.7.2}$$

where the sum is taken over all sequences $\gamma^{(1)}, \gamma^{(2)}, \ldots, \gamma^{(m-1)} \in Q_{p,n}$ with $\gamma^{(i)} \geq 2$, $i = 1, 2, \ldots, m-1$.

Now consider the corresponding minors of $\mathbf{C} = \mathbf{A}^m$: $C(1 \cup \alpha; 1 \cup \beta)$; we have

$$C(1 \cup \alpha; 1 \cup \beta) = \sum A(1 \cup \alpha; \delta^{(1)})A(\delta^{(1)};\delta^{(2)})\ldots A(\delta^{(m-1)}; 1 \cup \beta) \tag{6.7.3}$$

where the sum is taken over all sequences $\delta^{(1)}; \delta^{(2)}, \ldots, \delta^{(m-1)} \in Q_{p+1,n}$. Since $\mathbf{C}$ is TP, each of its minors must be positive; this implies that for at least one set $\delta^{(1)}; \delta^{(2)}, \ldots, \delta^{(m-1)}$, the product on the right of (6.7.3) must be positive; this implies that each of the minors entering that product must be strictly positive, for they are all non-negative. Now if $\delta \in Q_{p+1,n}$, it may be written $\delta_0 \cup \gamma$, where $\gamma \in Q_{p,n}$ and $\gamma_1 \geq 2$. This means that with the particular set $\delta^{(1)}, \ldots, \delta^{(m-1)} \in Q_{p+1,n}$ for which all the terms in (6.7.3) are positive, one may associate a set $\gamma^{(1)}, \ldots, \gamma^{(m-1)} \in Q_{p,n}$ which appears in the product (6.7.2). Now we use Corollary 2 of Theorem 6.6.3; it shows that for this particular choice of $(\gamma^{(i)})_1^{m-1}$, all the minors on the right of (6.7.2) must be positive, for if one were zero, say the first, then

$$A(1 \cup \alpha; \delta_0^{(1)} \cup \gamma^{(1)}) \leq a_{1,\delta_0^{(1)}} A(\alpha; \gamma^{(1)}) = 0$$

contrary to the fact that the minor on the left is positive. We conclude that one product in the sum on the right of (6.7.2) is positive; $\mathbf{D} = \mathbf{B}^m$ is TP; $\mathbf{B}$ is O. ■

We defined a principal minor of $\mathbf{A}$ as $A(\alpha;\alpha) \equiv A(\alpha)$. We now define a *quasi-principal* minor. The minor $A(\alpha;\beta)$ is said to be quasi-principal if $\alpha, \beta \in Q_{p,n}$ and

$$1 \leq \alpha_1, \beta_1 < \alpha_2, \beta_2 < \cdots < \alpha_p, \beta_p \leq n \tag{6.7.4}$$

and

$$|\alpha_1 - \beta_1| \le 1, \quad |\alpha_2 - \beta_2| \le 1, \ldots, |\alpha_p - \beta_p| \le 1. \qquad (6.7.5)$$

Thus a principal minor is also a quasi-principal minor.

The statement $\alpha_1, \beta_1 < \alpha_2, \beta_2$ means that each of α_1 and β_1 is less than each of α_2 and β_2, but there is no ordering of α_1 and β_1, nor of α_2 and β_2; thus

$$\alpha_1 < \alpha_2, \ \alpha_1 < \beta_2, \ \beta_1 < \alpha_2, \ \beta_1 < \beta_2.$$

The minors

$$A(1,3;2,3), \ A(1,3;1,3), \ A(1,2;1,3)$$

are all quasi-principal, but $A(1,2;2,3)$ is not.

Note that for $\mathbf{A}$ given in (6.7.1), and this matrix $\mathbf{A}$ is O, all these quasi-principal minors are positive. This is a particular case of

Theorem 6.7.2 *If $\mathbf{A} \in M_n$, $\mathbf{A}$ is NTN and $a_{i,i+1} > 0$, $a_{i+1,i} > 0$ for $i = 1, 2, \ldots, n-1$, then all the quasi-principal minors of $\mathbf{A}$ are positive.*

Proof. We will use induction on the order, p, of the minors. The first order quasi-principal minors are the diagonal terms a_{ii}, which are positive because of Corollary 6.6.2; and $a_{i,i+1}$ and $a_{i+1,i}$, which are positive by the statement of the theorem. Suppose then that all the quasi-principal minors of order $p-1$ are positive. We will prove that all those of order p are positive. For suppose this were not true, so that

$$A(\alpha_1, \alpha_2, \ldots, \alpha_p; \beta_1, \beta_2, \ldots, \beta_p) = 0$$

where the indices satisfy the inequalities (6.7.4), (6.7.5). But then

$$A(\alpha_1, \alpha_2, \ldots, \alpha_{p-1}; \beta_1, \beta_2, \ldots, \beta_{p-1})$$

and

$$A(\alpha_2, \alpha_3, \ldots, \alpha_p; \beta_2, \beta_3, \ldots, \beta_p)$$

will be quasi principal minors of order $p-1$, and so positive. Now Theorem 6.6.6 states that the matrix with rows $\alpha_1, \alpha_i+1, \ldots, \alpha_p$ and columns $\beta_1, \beta_1+1, \ldots, \beta_p$ has rank $p-1$. Let $h = \max(\alpha_1, \beta_1)$, then it follows from the inequalities (6.7.4), (6.7.5), that

$$\alpha_1, \beta_1 \le h; \quad \alpha_p, \beta_p \ge h + p - 1.$$

Therefore, the minor $A(h, h+1, \ldots, h+p-1)$ is a pth order minor of a matrix with rank $p-1$, and so is zero. But this minor is a principal minor of $\mathbf{A}$, and Theorem 6.6.4 shows that $\det(\mathbf{A}) = 0$; but $\mathbf{A}$ is NTN and thus non-singular. This contradiction implies that all the quasi-principal minors of $\mathbf{A}$ are positive. ■

We are now in a position to prove the important

Theorem 6.7.3 *If $\mathbf{A} \in M_n$ is NTN then it is O iff $a_{i,i+1} > 0$, $a_{i+1,i} > 0$ for $i = 1, 2, \ldots, n-1$.*

Proof. We first prove that if it is O, then $a_{i,i+1} > 0,\ a_{i+1,i} > 0$. If it is O then Theorem 6.7.1 states that the matrix

$$\mathbf{B} = \begin{bmatrix} a_{ii} & a_{i,i+1} \\ a_{i+1,i} & a_{i+1,i+1} \end{bmatrix}$$

is O, so that $\mathbf{D} = \mathbf{B}^m$ is TP for some m. But if say $a_{i,i+1} = 0$ then $d_{i,i+1} = 0$, whatever the value of m. Similarly, if $a_{i+1,i} = 0$, then $d_{i+1,i} = 0$. Thus $a_{i,i+1} > 0$ and $a_{i+1,i} > 0$.

We must now prove that if $a_{i,i+1} > 0,\ a_{i+1,i} > 0$ for all $i = 1, 2, \ldots, n-1$, then there is a power of $\mathbf{A}$ which is TP. We shall show that $\mathbf{A}^{n-1}$ is TP. We shall use Theorem 6.7.2, which states that the quasi-principal minors of $\mathbf{A}$ are positive. We recall the result used in Theorem 6.7.1, that a minor of $\mathbf{B} = \mathbf{A}^{n-1}$ is a sum of products of $n-1$ minors of $\mathbf{A}$. We need to show that the sum corresponding to a particular minor $B(\alpha;\beta)$ has at least one positive term in it. First, we note that if $B(\alpha;\beta) > 0$ for one particular value of m, then it will be positive for $m+1$ also, and thus for all subsequent m; for since $\mathbf{C} = \mathbf{A}^{m+1} = \mathbf{A}.\mathbf{A}^m = \mathbf{AB}$, the Binet-Cauchy expansion for the minor $C(\alpha;\beta)$ will contain the term $A(\alpha;\alpha)\ B(\alpha;\beta)$. This is positive because, by Theorem 6.6.2, the principal minors of $\mathbf{A}$ are positive.

This implies that, to show that $B(\alpha;\beta) > 0$ holds for $\mathbf{B} = \mathbf{A}^{n-1}$, it is sufficient to show that for some m satisfying $1 \le m \le n-1$ the expansion for $B(\alpha;\beta)$ will contain one product consisting entirely of quasi-principal minors. The problem is essentially how we can step from the sequence α to the sequence β through intermediate sequences $\gamma^{(1)}, \gamma^{(2)}, \ldots, \gamma^{(m-1)}$ such that $A(\alpha;\gamma^{(1)}), A(\gamma^{(1)};\gamma^{(2)}), \ldots, A(\gamma^{(m-1)};\beta)$ are all quasi-principal. Take an example. Suppose $p = 3,\ \alpha = \{1,2,3\}$ and $\beta = \{3,5,6\}$; we step as follows:

$$\{1,2,3\} \longrightarrow \{2,3,4\} \longrightarrow \{3,4,5\} \longrightarrow \{3,5,6\}.$$

The required exponent m is the number of steps needed to go from α to β, and this is

$$D = \max_{1\le r\le p} |\alpha_r - \beta_r|. \tag{6.7.6}$$

The quantity D (3 in the example) may be viewed as the distance $D(\alpha,\beta)$ between two sequences (see Ex. 6.7.2). If $A(\alpha,\beta)$ is quasi-principal then $D(\alpha,\beta) \le 1$; if $A(\alpha;\beta)$ is quasi-principal but not principal, then $D(\alpha,\beta) = 1$. The greatest distance between two sequences $\alpha, \beta \in Q_{p,n}$ is $n-p$; it occurs for instance when

$$\alpha = \{1,2,\ldots,p\}, \quad \beta = \{n-p+1, n-p+2, \ldots, n\};$$

this in turn is maximized when $p = 1$, i.e., $\alpha = \{1\},\ \beta = \{n\}$. We conclude that if $m = n-1$, then the Binet-Cauchy expansion for any minor of $\mathbf{B}$ will contain one product consisting entirely of quasi-principal minors of $\mathbf{A}$; $\mathbf{B}$ is TP; $\mathbf{A}$ is O. ■

We conclude this section by analyzing how oscillatory matrices relate to the Jacobi matrices which occupied our attention in earlier chapters. We defined a

Jacobi matrix in Section 3.1: $\mathbf{J} \in S_n$, $\mathbf{J}$ is PSD, and $\mathbf{J}$ has negative co-diagonal. $\mathbf{J}$ is clearly not O, but

Theorem 6.7.4 *If* $\mathbf{J}$ *is PD, then* $\mathbf{A} = \tilde{\mathbf{J}} = \mathbf{ZJZ}$ *is O.*

Proof. We recall that $\mathbf{Z} = diag(1, -1, 1, \ldots, (-)^{n-1})$, so that in the notation of equation (3.1.4),

$$a_{i,i+1} = a_{i+1,i} = b_i > 0.$$

According to Theorem 6.7.3 it is sufficient to show that $\mathbf{A} \equiv \tilde{\mathbf{J}}$ is TN. Consider a minor $A(\alpha; \beta)$. There are three cases:

1) $\alpha = \beta$, then $A(\alpha; \alpha) > 0$ since $\mathbf{A}$ is PD.

2) $d(\alpha, \beta) = 1$, i.e., $A(\alpha; \beta)$ is quasi-principal, thus it may be expressed as a product of principal minors and b's; $A(\alpha, \beta) > 0$.

3) $d(\alpha, \beta) > 1$, then $A(\alpha; \beta) = 0$. ∎

For $\mathbf{A} = \tilde{\mathbf{J}}$ only the quasi-principal minors are positive; the others are zero.

If $\mathbf{A} \in M_n$, then $\tilde{\mathbf{A}} = \mathbf{ZAZ}$ is called the *sign-reverse* of $\mathbf{A}$.

Theorem 6.7.5 *Suppose* $\mathbf{A} \in M_n$. $\mathbf{A}$ *is NTN, TP, O iff* $(\tilde{\mathbf{A}})^{-1}$ *is NTN, TP, O respectively.*

Proof. We recall from Section 1.3, that $\mathbf{A}^{-1} = \mathbf{R}$, where $r_{ij} = A_{ji}/\det(\mathbf{A})$, where A_{ij} is given by $A_{ij} = (-)^{i+j}\hat{a}_{ij}$. This means that it is sufficient to show that $\mathbf{A}$ is NTN, TP or O iff $\hat{\mathbf{A}} = (\hat{a}_{ij})$ is NTN, TP, O. But Theorem 6.2.1 shows immediately that $\mathbf{A}$ is NTN or TP iff $\hat{\mathbf{A}}$ is NTN or TP respectively. If $\mathbf{A}$ is O then $\hat{a}_{i,i+1}$ and $\hat{a}_{i+1,i}$ are given by Theorem 6.2.1 as quasi-principal minors of $\mathbf{A}$, and so are positive; $\hat{\mathbf{A}}$ is O; and *vice versa*, if $\hat{\mathbf{A}}$ is O, then so is $\mathbf{A}$. ∎

If $\tilde{\mathbf{A}}$ is oscillatory we shall say that $\mathbf{A}$ is sign-oscillatory (SO). This implies, in particular, that a non-singular Jacobi matrix is SO.

Corollary 6.7.1 *If* $\mathbf{A}$ *is SO, then* $\mathbf{A}^{-1}$ *is O.*

Exercises 6.7

1. Why is it not sufficient to define $\mathbf{A}$ to be O if, for some m, $\mathbf{A}^m$ is TP? Exhibit an example of $\mathbf{A} \in M_2$ such that $\mathbf{A}^2$ is TP but $\mathbf{A}$ is not TN.

2. Show that the distance $D(\alpha, \beta)$ satisfies the basic conditions for a distance:

   $D(\alpha, \beta) \geq 0$; $D(\alpha, \beta) = 0$ iff $\alpha = \beta$;

   $D(\alpha, \gamma) \leq D(\alpha, \beta) + D(\beta, \gamma)$.

3. Show that if $\mathbf{A} \in M_n$ is tridiagonal, then it is O iff

   a) its principal minors are non-negative

   b) $a_{i,i+1} > 0$, $a_{i+1,i} > 0$ for $i = 1, 2, \ldots, n-1$

   c) it is non-singular.

4. We say that a tridiagonal matrix **A** as described in Ex. 6.7.3 has half-bandwidth 1; it has 1 diagonal above, and 1 below, the principal diagonal. Show that if $1 \leq p \leq n-1$ then $\mathbf{A}^p$ has half-bandwidth p.

5. Show that if $a_{i,i+1} \neq 0,\ a_{i+1,i} \neq 0$, then a tridiagonal matrix **A** may be *symmetrized* by using diagonal matrices, i.e., we can find diagonal $\mathbf{C}, \mathbf{D}$ so that **CAD** is symmetric. Show that this means that an oscillatory tridiagonal matrix may be symmetrized to a $\tilde{\mathbf{J}}$ matrix by using *positive* diagonals $\mathbf{C}, \mathbf{D}$, i.e., $\mathbf{CAD} = \tilde{\mathbf{J}}$.

6. Suppose $\mathbf{A}, \mathbf{B} \in M_n$. Show that if $\mathbf{A}, \mathbf{B}$ are TN, then so is $\mathbf{C} \equiv \mathbf{AB}$.

7. Show that if $\mathbf{A}, \mathbf{B}$ are O, then so is $\mathbf{C} \equiv \mathbf{AB}$.

8. Show that if $\mathbf{A} \in M_n$ is O then $\boldsymbol{\mathcal{A}}_{n-1}$ is O.

9. Show that if $\mathbf{A} = \mathbf{J}^{-1}$ then **A**, which is O by Theorem 6.7.5, is actually a (strictly) positive matrix, i.e., it is full. Note that by Ex. 6.6.2, it is sufficient to show that $a_{1n} > 0$.

10. Show that if **A** is O, then the indices of its staircase structure (Section 6.6) satisfy $p_i \geq i+1,\ q_j \geq j+1$.

11. Show that if **A** has eigenvalues λ_l and eigenvectors $\mathbf{u}_i$, then $\mathbf{B} = (\tilde{\mathbf{A}})^{-1}$ has eigenvalues $\mu_k = 1/\lambda_\ell$ and eigenvectors $\mathbf{v}_k = \mathbf{Z}\mathbf{u}_l$, where $l = n+1-k$.

12. Exhibit counterexamples to show that if **A** is one of TN, TP or O, then a compound matrix $\mathcal{A}_p$ need not have the same property.

6.8 Totally positive matrices

The matrix $\mathbf{A} \in M_n$ is TP if all its minors are positive. This is equivalent to the statement that all the compound matrices $\boldsymbol{\mathcal{A}}_p \quad p = 1, 2, \ldots, n$, are (strictly) positive. There are

$$P = n^2 + \binom{n}{2}^2 + \binom{n}{3}^2 + \ldots + \binom{n}{n-1}^2 + 1 \tag{6.8.1}$$

elements to be checked. Using a result due to Fekete (1913) [86], Ando (1987) [4] proved that one need check only a much smaller set of minors.

As in Section 6.2, let $Q_{p,n}$ denote the set of strictly increasing sequences $\alpha = \{\alpha_1, \alpha_2, \ldots, \alpha_p\}$ chosen from $1, 2, \ldots, n$. We write

$$d(\alpha) = \sum_{i=1}^{p-1} (\alpha_{i+1} - \alpha_i - 1)$$

and note that if $\alpha \in Q_{p,n}$, then $d(\alpha) = 0$ iff $\alpha_{i+1} = \alpha_i + 1$ for $i = 1, 2, \ldots, p-1$; i.e., $d(\alpha) = 0$ iff $\alpha_1, \alpha_2, \ldots, \alpha_p$ consists of *consecutive* integers. We define $Q^0_{k,n}$ as the subset of $Q_{k,n}$ consisting of those α with $d(\alpha) = 0$. Following Theorem 2.5 of Ando (1987) [4] we have

Theorem 6.8.1 *$A \in M_n$ is TP if $A(\alpha;\beta) > 0$ for all $\alpha, \beta \in Q^0_{k,n}$, $k = 1, 2, \ldots, n$.*

Proof. Let us prove that

$$A(\alpha;\beta) > 0 \text{ for } \alpha, \beta \in Q_{k,n}, k = 1, 2, \ldots, n \tag{6.8.2}$$

by induction on k. When $k = 1$, this is trivial because $Q_{k,n} = Q^0_{k,n}$. Assume that (6.8.2) is true with $k-1 (k \geq 2)$ in place of k. First fix $\alpha \in Q_{k,n}$ with $d(\alpha) = 0$, i.e., $\alpha \in Q^0_{k,n}$, and let us prove (6.8.2) with this α by induction on $\ell = d(\beta)$. When $\ell = 0$ this follows by the assumption of the theorem. Suppose $A(\alpha;\delta) > 0$ for all minors whenever $\delta \in Q_{k,n}$ and $d(\delta) \leq \ell - 1$, with $\ell \geq 1$. Take $\beta \in Q_{k,n}$ with $d(\beta) = \ell$. Then there is a p such that $\beta_1 < p < \beta_k, d(\tau \cup \{\beta_1, p\}) \leq \ell - 1$ and $d(\tau \cup \{p, \beta_k\}) \leq \ell - 1$ where $\tau = \{\beta_2, \beta_3, \ldots, \beta_{k-1}\}$. Now use the identity ((1.39) of Ando (1987) [4])

$$A(\omega; \tau \cup \{p\}) A(\alpha; \tau \cup \{\beta_1, \beta_k\}) = A(\omega; \tau \cup \{\beta_1\}) A(\alpha; \tau \cup \{p, \beta_k\})$$

for any $\omega \in Q_{k-1,n}$ with $\omega \subset \alpha$. It follows from the induction assumptions that the right hand side is positive, as is $A(\omega; \tau \cup \{p\})$, so that $A(\alpha; \tau \cup \{\beta_1, \beta_k\}) \equiv A(\alpha;\beta) > 0$. This proves (6.8.2) for $\alpha \in Q_{k,n}$ with $d(\alpha) = 0$. Apply the same argument row wise to conclude that (6.8.2) is generally true. ■

We may use precisely the same argument to prove the

Corollary 6.8.1 *Suppose $A \in M_n$. If all minors $A(\alpha;\beta) > 0$ for $\alpha, \beta \in Q_{k-1,n}$, and $A(\alpha;\beta) > 0$ for $\alpha, \beta \in Q^0_{k,n}$, then, $A(\alpha;\beta) > 0$ for all $\alpha, \beta \in Q_{k,n}$.*

This result mirrors the test for a matrix $\mathbf{A} \in S_n$ to be PD; to show that $\mathbf{A}$ is PD, it is sufficient to show that the leading principal minors $D_1, D_2, \ldots, D_n$ are all positive. The importance of the result lies in the fact that, with it, the number of minors to be checked for positivity is much smaller than that given by (6.8.1).

The test in Theorem 6.8.2 determines whether an arbitrary matrix $\mathbf{A} \in M_n$ is TP. If it is known that $\mathbf{A}$ is TN, then one needs to check only a very small number of minors for strict positivity to determine whether $\mathbf{A}$ is TP, as stated in

Theorem 6.8.2 *If $\mathbf{A} \in M_n$ is TN, then it is TP if its corner minors are positive.*

Proof. The corner minors are the minors

$$A(1, 2, \ldots, p; n-p+1, \ldots, n), \quad A(n-p+1, \ldots, n; 1, 2, \ldots, p),$$

$p = 1, 2, \ldots, n$. The result follows immediately from Theorem 6.6.6 and Theorem 6.8.1. Consider a minor $A(\alpha;\beta)$ with $\alpha, \beta \in Q^0_{k,n}$. Suppose $\alpha = \{i, i+1, \ldots, i+k-1\}$, $\beta \in \{j, j+1, \ldots, j+k-1\}$. If $i \geq j$ then $A(\alpha;\beta)$ is a principal minor of the corner submatrix $A(n-p+1, \ldots, n; 1, 2, \ldots, p)$ with $p = n - (i - j)$. This submatrix is NTN, so that, by Theorem 6.6.6, its principal minors are positive. If $i < j$, then $A(\alpha;\beta)$ is a principal minor of $A(1,2,\ldots,p;n-p+1,\ldots,n)$ with $p = n-(j-i)$. Since $A(\alpha;\beta) > 0$ for all $\alpha,\beta \in Q^0_{k,n}$, $k = 1,2,\ldots,n$, Theorem 6.8.1 states that A is TP. ■

Exercises 6.8

1. Show that if **A** is NTN and **B** is TP, then **AB** and **BA** are TP.
2. Show that if $p_{ij} = \exp[-k(i-j)^2]$, then $\mathbf{P} = (p_{ij})$ is TP. See Section 7 of Ando (1987) [4].
3. Use Ex. 6.8.3 to show that a NTN matrix **A** may be approximated arbitrarily closely, in the L_1 norm (see (6.5.4)) by the TP matrix $\mathbf{B} = \mathbf{PAP}$. (Again, see Section 7 of Ando (1987) [4].)

6.9 Oscillatory systems of vectors

Before discussing the eigenproperties of totally positive matrices, we need to analyse some sign properties of vectors.

Let $u_1, u_2, \ldots, u_n$ be a sequence of real numbers. If some of them are zero we may assign them arbitrarily chosen signs. We can then compute the number of sign changes in the sequence. This number may change, depending on the choice of signs for the zero terms. The greatest and least values of this number are denoted by $S^+_{\mathbf{u}}$ and $S^-_{\mathbf{u}}$ respectively, where $\mathbf{u} = \{u_1, u_2, \ldots, u_n\}$.

If $S^+_{\mathbf{u}} = S^-_{\mathbf{u}}$, we speak of an *exact* number of sign changes in the sequence, and denote this by $S_{\mathbf{u}}$. Clearly this case can occur only when

1. $u_1, u_n \neq 0$
2. when $u_i = 0$ for some i satisfying $2 \le i \le n-1$, then $u_{i-1}u_{i+1} < 0$, i.e., u_{i-1} and u_{i+1} are both non-zero, and have opposite signs. In this case $S_{\mathbf{u}}$ is the number of sign changes when the zero terms are removed.

We say that a system of vectors $\mathbf{u}_k = \{u_{1k}, u_{2k}, \ldots, u_{nk}\}$, $k = 1,2,\ldots,p$, is an *oscillatory* system if, for any $(c_k)_1^p$ with

$$\sum_{k=1}^{p} c_k^2 > 0, \tag{6.9.1}$$

the vector

$$\mathbf{u} = \sum_{k=1}^{p} c_k \mathbf{u}_k \tag{6.9.2}$$

satisfies $S^+_{\mathbf{u}} \le p-1$. Clearly, we need only consider $p \le n$. Taking $p = 1$ we see that $S^+_{\mathbf{u}_1} = 0$, i.e., $\mathbf{u}_1 > \mathbf{0}$; for $p = 2$, $S^+_{\mathbf{u}} \le 1$, etc.

Theorem 6.9.1 *The necessary and sufficient condition for the system* $(\mathbf{u}_k)_1^p$ *to be an oscillatory system is that all the minors*

$$U(\alpha;\theta)$$

be different from zero, and have the same sign, for

$$\alpha \in Q_{p,n},\ \ \theta = \{1,2,\ldots,p\}.$$

Proof. The minors in question are

$$U(\alpha_1,\alpha_2,\ldots,\alpha_p;1,2,\ldots,p). \tag{6.9.3}$$

Remember that $\alpha_1,\alpha_2,\ldots,\alpha_p$ refer to components of the vectors, while $1,2,\ldots,p$ refer to the vector index k. The theorem states that when $p=1$, $u_{11},u_{21},\ldots,u_{n1}$ must all be non-zero and have the same sign; this is certainly equivalent to $S_{\mathbf{u}}^+ = 0$. For $p=2$, it states that

$$U(1,2;1,2) = \begin{vmatrix} u_{11} & u_{12} \\ u_{21} & u_{22} \end{vmatrix},\ U(1,3;1,2) = \begin{vmatrix} u_{11} & u_{12} \\ u_{31} & u_{32} \end{vmatrix},$$
$$\ldots U(n-1,n;1,2) = \begin{vmatrix} u_{n-1,1} & u_{n-1,2} \\ u_{n,1} & u_{n,2} \end{vmatrix},$$

are all non-zero and have the same sign.

We first prove the necessity. If a minor (6.9.3) were to vanish, then we could find numbers $(c_k)_1^p$, not all zero, such that

$$\sum_{k=1}^{p} c_k u_{\alpha_j,k} = 0 \quad j=1,2,\ldots,p. \tag{6.9.4}$$

But then the vector $\mathbf{u}$ given by (6.9.2) would have p zero terms

$$u_{\alpha_1} = 0 = u_{\alpha_2} = \cdots = u_{\alpha_p}$$

so that, by Ex. 6.9.1, $S_{\mathbf{u}}^+ \geq p > p-1$.

In order to show that the minors all have the same sign it is sufficient to show that all minors $U(\alpha;\theta)$ for α next to θ in the sense $D(\alpha;\theta) = 1$, (see equation (6.7.6)) all have the same sign. These are the minors $(U_j)_1^p$, where $U_j = U(\alpha^{(j)};\theta)$ and $\alpha^{(1)} = \{2,3,\ldots,p+1\}$, $\alpha^{(j)} = \{1,2,\ldots,j-1,j+1,\ldots,p+1\}$, $j=2,3,\ldots,p$. These must all have the same sign as $U_{p+1} = U(\theta;\theta)$. Introduce a vector $\mathbf{u}_{p+1}$ such that

$$u_{i,p+1} = \begin{cases} (-)^{j+1}U_{p+1}, & i=j \\ (-)^{p+1}U_j, & i=p+1 \\ 0 & \text{otherwise} \end{cases}. \tag{6.9.5}$$

Then

$$U(1,2,\ldots,p{+}1;1,2,\ldots,p{+}1) = (-)^{p+1}\{(-)^{p+1}u_{p+1,p+1}U_{p+1}+(-)^j u_{j,p+1}U_j\} = 0$$

so that we can find $(c_k)_1^{p+1}$, not all zero, such that

$$\sum_{k=1}^{p+1} c_k u_{i,k} = 0, \ i = 1, 2, \ldots, p+1.$$

But then the vector (6.9.2) will have coordinates

$$u_i = -c_{p+1} u_{i,p+1} \quad i = 1, 2, \ldots, p+1.$$

The quality c_{p+1} cannot be zero, for then $\mathbf{u}$ would have $p+1$ zero terms and hence $S_{\mathbf{u}}^+ \geq p$. Choose c_{p+1} so that $c_{p+1}U_{p+1} > 0$, then, according to (6.9.5), $(u_i)_1^{j-1} = 0$, $u_j = (-)^j c_{p+1} U_{p+1}$, $(u_i)_{j+1}^p = 0$, $u_{p+1} = (-)^p c_{p+1} U_j$. If U_j, U_{p+1} had opposite signs, then u_j would have the sign of $(-)^j$, and u_{p+1} would have the sign of $(-)^{p+1}$. This means that we can assign the signs of the zero u_i so that, for all $i = 1, 2, \ldots, p+1$, u_i has the sign of $(-)^i$. But then $S_{\mathbf{u}}^+ = p$. This proves the necessity.

Now we prove the sufficiency. Suppose that all the minors (6.9.3) were non-zero and had the same sign, which we may take to be positive. We will prove $S_{\mathbf{u}}^+ \leq p-1$, by assuming the contrary, i.e., $S_{\mathbf{u}}^+ \geq p$. If that were so we could find $p+1$ components $u_{\alpha_1}, u_{\alpha_2}, \ldots, u_{\alpha_{p+1}}$ such that

$$u_{\alpha_j} u_{\alpha_{j+1}} \leq 0, \ j = 1, 2, \ldots, p. \tag{6.9.6}$$

The $(u_{\alpha_j})_1^p$ cannot be simultaneously zero, for then the $(c_k)_1^p$, not all zero, would satisfy equation (6.9.4), the determinant of which is not zero.

Now consider the zero determinant

$$\begin{vmatrix} u_{\alpha_1,1} & u_{\alpha_1,2} & \cdots & u_{\alpha_1,p} & u_{\alpha_1} \\ u_{\alpha_2,1} & u_{\alpha_2,2} & \cdots & u_{\alpha_2,p} & u_{\alpha_2} \\ \cdot & \cdot & \cdots & \cdot & \cdot \\ u_{\alpha_{p+1},1} & u_{\alpha_{p+2},2} & \cdots & u_{\alpha_{p+1},p} & u_{\alpha_{p+1}} \end{vmatrix} = 0,$$

and expand it along its last column

$$\sum_{k=1}^{p+1} u_{\alpha_k} (-)^{p+1+k} U(\alpha_1, \alpha_2, \ldots, \alpha_{k-1}, \alpha_{k+1}, \ldots, \alpha_{p+1}; 1, 2, \ldots, p) = 0.$$

But this is impossible because the minors are all positive and, by (6.9.6), the quantities $(-)^k u_{\alpha_k}$ all have the same sign, and are not zero. This completes the proof. ■

Exercises 6.9

1. Consider the real sequence $u_1, u_2, \ldots, u_n$. Show that if $(u_i)_1^n = 0$ then $S_{\mathbf{u}}^- = 0, S_{\mathbf{u}}^+ = n-1$. Show also that if $p (0 \leq p < n)$ of the u_i are zero then

$$S_{\mathbf{u}}^- \leq n-p-1 \text{ and } p \leq S_{\mathbf{u}}^+ \leq n-1$$

while if $p (1 < p < n)$ *successive* u_i are zero then $S_{\mathbf{u}}^+ - S_{\mathbf{u}}^- \geq p$.

6.10 Eigenproperties of TN matrices

Since TN matrices are not necessarily symmetric we cannot immediately assume that their eigenvalues are real; to do so we must make use of their special properties.

Theorem 6.10.1 *The eigenvalues of an TP matrix are positive and distinct.*

Proof. Suppose that $\mathbf{A} \in M_n$ has eigenvalues $\lambda_1, \lambda_2, \ldots, \lambda_n$, possibly complex. We order them in decreasing modulus, i.e., so that $|\lambda_1| \geq |\lambda_2| \geq \cdots \geq |\lambda_n|$. Since $\mathbf{A}$ is TP, it is positive; Perron's theorem (Theorem 6.5.1) states that λ_1 is positive and $\lambda_1 > |\lambda_2|$. Since $\mathbf{A}$ is TP, the compound matrix $\mathcal{A}_2$ is positive; its eigenvalues are the products $\lambda_i\lambda_j, i,j = 1,2,\ldots,n$. It too has a positive eigenvalue, greater in magnitude than any other; it must be $\lambda_1\lambda_2$ so that $\lambda_1\lambda_2 > 0$ and $\lambda_1\lambda_2 > |\lambda_1\lambda_3|$. Thus $\lambda_2 > 0$ and $\lambda_2 > |\lambda_3|$. Now we consider $\mathcal{A}_3$ and deduce that $\lambda_1\lambda_2\lambda_3 > 0$ and $\lambda_1\lambda_2\lambda_3 > |\lambda_1\lambda_2\lambda_4|$, i.e., $\lambda_3 > 0$ and $\lambda_3 > |\lambda_4|$, and so on. ■

Corollary 6.10.1 *The eigenvalues of an oscillatory matrix are positive and distinct.*

Proof. For if $\mathbf{A} \in M_n$ is O, then $\mathbf{B} = \mathbf{A}^m$ is TP for all $m \geq n-1$. But if the eigenvalues of $\mathbf{A}$ are $(\lambda_i)_1^n$, those of $\mathbf{B}$ are $\mu_i = \lambda_i^m$; since $\mu_1 > \mu_2 > \cdots > \mu_n > 0$, and $\lambda_i \geq 0$, we have $\lambda_1 > \lambda_2 > \cdots > \lambda_n > 0$. ■

We now show that the eigenvectors of an oscillatory matrix behave exactly like those of a $\tilde{\mathbf{J}}$ matrix, i.e., like those of a Jacobi matrix when the ordering of the eigenvalues is reversed (see the comment at the end of Section 6.1).

Theorem 6.10.2 *Suppose $\mathbf{A} \in M_n$ is O, and has eigenvalues $(\lambda_i)_1^n$ satisfying $\lambda_1 > \lambda_2 > \cdots > \lambda_n > 0$. Let $\mathbf{u}_k = \{u_{1k}, u_{2k}, \ldots, u_{nk}\}$ be an eigenvector corresponding to λ_k; it is unique apart from a factor. Let*

$$\mathbf{u} = \sum_{k=p}^{q} c_k \mathbf{u}_k, \quad \sum_{k=p}^{q} c_k^2 > 0 \tag{6.10.1}$$

then the number of sign changes among the components of $\mathbf{u}$ for differing $(c_k)_p^q$ satisfies

$$p - 1 \leq S_{\mathbf{u}}^- \leq S_{\mathbf{u}}^+ \leq q - 1. \tag{6.10.2}$$

Proof. Since the eigenvectors of $\mathbf{A}$ are also the eigenvectors of $\mathbf{A}^m$, and since $\mathbf{A}^m$ is TP for some $m \leq n-1$, we lose no generality by assuming that $\mathbf{A}$ is TP.

Suppose $1 \leq q \leq n$, $\alpha = \{i_1, i_2, \ldots, i_q\} \in Q_{q,n}$, $\theta = \{1, 2, \ldots, q\}$. Then the minors $U(\alpha;\theta)$ are the coordinates of the eigenvector of the compound matrix $\mathcal{A}_q$ corresponding to the maximum eigenvalue $\lambda_1\lambda_2\ldots\lambda_q$. By Perron's theorem all the components of this eigenvector have the same sign. If the sign of the q-th set of minors is E_q then, by multiplying the vectors $(\mathbf{u}_k)_1^n$ by $E_1, E_2/E_1, \ldots, E_n/E_{n-1}$ respectively, we can make

$$U(\alpha;\theta) > 0 \quad q = 1, 2, \ldots, n.$$

Theorem 6.9.1 now shows that $S_{\mathbf{u}}^+ \leq q-1$.

To prove the second part of the theorem we put $\mathbf{B} = (\tilde{\mathbf{A}})^{-1}$. Theorem 6.7.5 shows that if $\mathbf{A}$ is TP, then so is $\mathbf{B}$, and Ex. 6.7.11 shows that it has eigenvalues $\mu_k = 1/\lambda_l$ and eigenvectors $\mathbf{v}_k = \mathbf{Z}\mathbf{u}_l$, where $l = n+1-k$. Thus

$$\mathbf{v}_k = \{v_{1k}, v_{2k}, \ldots, v_{nk}\} = \{u_{1l}, -u_{2l}, u_{3l}, \ldots, (-)^{n-1}u_{nl}\}.$$

The result already proved shows that the number of sign interchanges in

$$\mathbf{v} = \sum_{k=n+1-q}^{n+1-p} c_{n+1-k}\mathbf{v}_k = \mathbf{Z}\sum_{l=p}^{q} c_l \mathbf{u}_l$$

satisfies $S_{\mathbf{v}}^+ \leq n-1$. But since $\mathbf{v}_i = (-)^{i-1}u_i$ we have $S_{\mathbf{v}}^+ + S_{\mathbf{u}}^- = n-1$ so that $S_{\mathbf{u}}^- \geq p-1$. ■

Corollary 6.10.2 *The vector $\mathbf{u} = \mathbf{u}_k$ has exactly $k-1$ sign changes.* ($S_{\mathbf{u}}^- = S_{\mathbf{u}}^+ = k-1$).

Corollary 6.10.3 *$u_{nk} \neq 0$, so that $\mathbf{u}_k$ may be chosen so that $u_{nk} > 0$.*

The argument used in this theorem leads directly to

Corollary 6.10.4 *For each p such that $1 \leq p \leq n$, the minors $U(\alpha_1, \alpha_2, \ldots, \alpha_p; 1, 2, \ldots, p)$, have the same sign for all $\alpha \in Q_{p,n}$.*

The minors of Corollary 6.10.4 relate to components $\alpha_1, \alpha_2, \ldots, \alpha_p$ of the first p eigenvectors. We now prove a result in which components and eigenvalue indices are reversed; this theorem will play a vital role in the inverse problem for the discrete vibrating beam (Chapter 8). Before stating the theorem we repeat comments we have made on the relation between oscillatory (O) and sign-oscillatory (SO) matrices.

If $\mathbf{A}$ is O, with eigenvalues $(\lambda_i)_1^n$ ordered so that $\lambda_1 > \lambda_2 > \cdots > \lambda_n > 0$, then its eigenvectors $(\mathbf{u}_k)_1^n$ satisfy Theorem 6.10.2 so that, in particular, $\mathbf{u}_k$ has exactly $k-1$ sign changes. If $\mathbf{A}$ is SO and we label its eigenvalues $(\lambda_i)_1^n$ in reverse order, i.e., so that $0 < \lambda_1 < \lambda_2 < \cdots < \lambda_n$, then its eigenvectors $(\mathbf{u}_k)_1^n$ again satisfy Theorem 6.10.2, so that, in particular, $\mathbf{u}_k$ has $k-1$ sign changes. We will phrase the final theorem of this section for an SO matrix.

Theorem 6.10.3 *If $\mathbf{A} \in M_n$ is SO, with eigenvalues $(\lambda_i)_1^n$ satisfying $0 < \lambda_1 < \lambda_2 < \cdots < \lambda_n$, then its eigenvectors $(\mathbf{u}_i)_1^n$ may be chosen so that*

$$U(\phi;\alpha) > 0. \tag{6.10.3}$$

for $\phi = \{n-p+1, n-p+2, \ldots, n\}$ and each $\alpha \in Q_{p,n}$.

Proof. The analysis of Section 6.3 (See Ex. 6.3.2) shows that $U(\phi;\alpha)$ is the last component of the eigenvector of the compound matrix $\boldsymbol{\mathcal{A}}_p$ corresponding to the sth eigenvalue $\lambda_{\alpha_1}, \lambda_{\alpha_2}, \ldots, \lambda_{\alpha_p}$, where $s = s(\alpha_1, \alpha_2, \ldots, \alpha_p)$. The more general statement of Theorem 6.10.3 is that all the elements $U(\phi;\alpha)$ have the *same sign*, which is thus the sign for the case $p = 1$, i.e., for u_{ni}.

The proof is by induction on p. Corollary 6.10.3 shows that $u_{ni} \neq 0$. Choose $u_{ni} > 0$ for $i = 1, 2, \ldots, n$; the theorem then holds for $p = 1$. Suppose the result holds for p. Corollary 6.10.2 shows that $\mathbf{u}_i$ has $i - 1$ sign changes, so that $(-)^{i-1}u_{1i} > 0$. Choose $(c_j)_i^{p+1}$ so that

$$\mathbf{u} = \sum_{j=i}^{i+p} c_j \mathbf{u}_j, \quad \sum_{j=i}^{i+p} c_j^2 > 0$$

and

$$u_{n-p+1} = 0 = u_{n-p+2} = \cdots = u_n,$$

using the choice

$$c_j = (-)^{j-i} U(\phi; \beta \backslash j) \quad j = i, i+1, \ldots, i+p$$

where $\beta = \{i, i+1, \ldots, i+p\}, \phi = \{n-p+1, \ldots, n\}$.

The vector $\mathbf{u}$ has the form

$$\mathbf{u} = \{u_1, u_2, \ldots, u_{n-p}, 0, 0, \ldots, 0\}$$

and has first element

$$u_1 = c_i u_{1,i} + c_{i+1} u_{1,i+1} + \cdots c_{i+p} u_{i,i+p}.$$

Since, by hypothesis, the result is true for p, the coefficients c_j satisfy $(-)^j c_{i+j} > 0$; this and the inequality $(-)^{i+j-1}u_{1,i+j} > 0$, yield $(-)^{i-1}c_{i+j}u_{1,i+j} > 0$, so that $(-)^{i-1}u_1 > 0$. By Theorem 6.10.2,

$$i - 1 \leq S_{\mathbf{u}}^- \leq S_{\mathbf{u}}^+ \leq p + i - 1$$

and since the last p elements of $\mathbf{u}$ are zero, there must be exactly $i - 1$ sign changes in the first $n - p$ elements of $\mathbf{u}$; but $(-)^{i-1}u_1 > 0$, so that the last non-zero element, u_{n-p}, must be positive, i.e.,

$$u_{n-p} = U(n-p, n-p+1, \ldots, n; i, i+1, \ldots, i+p) > 0, \ i + p \leq n.$$

This shows that all $(p+1)$th order minors with *consecutive* indices $i, i+1, \ldots, i+p$ are positive, and Theorem 6.8.1 shows that all $(p+1)$th order minors are positive. ∎

Exercises 6.10

1. Show that if $\mathbf{u}_j, \mathbf{u}_{j+1}$ are eigenvectors of an O or SO matrix $\mathbf{A} \in M_n$, then $u_{n-1,j}u_{n,j+1} - u_{n-1,j+1}u_{n,j}$ is non-zero and has the same sign for $j = 1, 2, \ldots, n-1$.

2. Show that the proof used in Theorem 6.10.3 may be used to show that if $\mathbf{A} \in M_n$ is O with eigenvalues $(\lambda_i)_1^n$ satisfying $0 < \lambda_n < \lambda_{n-1} < \cdots < \lambda_1$, then its eigenvalues $(\mathbf{u}_i)_1^n$ may be chosen so that

$$U(\phi; \alpha) > 0$$

for $\phi = \{n - p + 1, \ldots, n$ and each $\alpha \in Q_{p,n}$.

3. The matrix

$$A = \begin{bmatrix} 2 & 1 & & \\ 1 & 2 & 1 & \\ & 1 & 2 & 1 \\ & & 1 & 2 \end{bmatrix}$$

is O. Use the recurrence method described in Section 2.6 to find its eigenvalues $(\lambda_i)_1^4$, labelled so that $0 < \lambda_4 < \lambda_3 < \lambda_2 < \lambda_1$, and its eigenvectors. [Note: the eigenvectors may be written explicitly in terms of $x = \sin\left(\frac{\pi}{5}\right)$ and $x = \sin\left(\frac{2\pi}{5}\right)$.] Choose the signs of the eigenvectors so that they obey Corollary 6.10.4. Make a different choice so that they obey Ex. 6.10.2.

4. If $\mathbf{u}$ is an eigenvector of $\mathbf{A} \in M_n$, and $\mathbf{T}$ is the reversing matrix given in equation (4.3.8), then $\mathbf{v} = \mathbf{Ty}$ is an eigenvector of $\mathbf{B} = \mathbf{TAT}$ corresponding to the same eigenvalue λ. Use this result, and Ex. 6.10.2, to show that if B is O, then $V(p, p-1, \ldots, 1; \alpha_1, \ldots, \alpha_p) > 0$.

6.11 u-line analysis

We recall the concept of a **u**-line corresponding to the vector $\mathbf{u} = \{u_1, u_2, \ldots, u_n\}$, from Section 3.3: it is the broken line made up on the links joining the points with coordinates $(x, y) = (i, u_i)$, so that

$$u(x) = (i + 1 - x)u_i + (x - i)u_{i+1}, \quad i \leq x \leq i + 1.$$

Theorem 6.11.1 *Let $\mathbf{u}_k$ be an eigenvector corresponding to eigenvalue λ_k of an oscillatory matrix $\mathbf{A}$. The corresponding $\mathbf{u}_k$-line, $\mathbf{u}^{(k)}(x)$ has no links on the x-axis, and has just $k-1$ nodes, i.e., simple zeros where $\mathbf{u}^{(k)}(x)$ changes sign.*

Proof. A link of a **u**-line can lie along the x-axis only if two successive u_i are zero, but this is precluded by the Corollary to Theorem 6.10.2. Since $S_{\mathbf{u}} = k - 1$, the **u**-line has just $k - 1$ nodes. ■

Corollary 6.11.1 *If α, β are successive nodes of a $\mathbf{u}_k$-line, then $|\alpha - \beta| > 1$.*

Theorem 6.11.2 *The* **u**-*lines corresponding to two successive eigenvectors of an oscillatory matrix cannot have a common node.*

Proof. Suppose, if possible, that $u^{(k)}(\alpha) = 0 = u^{(k+1)}(\alpha)$, and put

$$u(x) = cu^{(k)}(x) - u^{(k+1)}(x).$$

Theorem 6.10.2 shows that

$$k - 1 \leq S_{\mathbf{u}}^{-} \leq S_{\mathbf{u}}^{+} \leq k. \tag{6.11.1}$$

The Corollary to Theorem 6.11.1 shows that $u^{(k)}(x)$ and $u^{(k+1)}(x)$ will both be non-zero in $(\alpha, \alpha + 1]$. Choose γ so that $\alpha < \gamma \leq \alpha + 1$, and put $c = u^{(k+1)}(\gamma)/u^{(k)}(\gamma)$. Then $u(x)$ will have two zeros, α, γ such that $\alpha < \gamma \leq \alpha + 1$; it must therefore have a link along the x-axis, means that two successive u_i must vanish. According to Ex. 6.9.1 this means that $S_{\mathbf{u}}^{+} - S_{\mathbf{u}}^{-} \geq 2$, contradicting (6.11.1). ∎

Theorem 6.11.3 *The nodes of* **u**-*lines corresponding to two successive eigenvectors $\mathbf{u}_k, \mathbf{u}_{k+1}$ of an oscillatory matrix interlace.*

Proof. Suppose that α, β are two successive nodes of the $\mathbf{u}_{k+1}$-line, then $u^{(k+1)}(\alpha) = 0 = u^{(k+1)}(\beta)$ and $\beta - \alpha > 1$. Suppose if possible that the $\mathbf{u}_k$-line has no node in (α, β). Without loss of generality we may assume that

$$u^{(k)}(x) > 0 \text{ in } [\alpha, \beta],\ u^{(k+1)}(x) > 0 \text{ in } (\alpha, \beta).$$

Put

$$u(x) = cu^{(k)}(x) - u^{(k+1)}(x)$$

then

$$k - 1 \leq S_{\mathbf{u}}^{-} \leq S_{\mathbf{u}}^{+} \leq k. \tag{6.11.2}$$

For sufficiently large c, $u(x) > 0$ in $[\alpha, \beta]$. Decrease c to a certain value c_0 at which $u(x)$ first vanishes at least once, at a point γ in $[\alpha, \beta]$. Clearly $c_0 > 0$ and

$$u_0(x) = c_0 u^{(k)}(x) - u^{(k+1)}(x)$$

does not vanish at α or β, so that $\alpha < \gamma < \beta$. Thus $u_0(x) \geq 0$ in $[\alpha, \beta]$ and $u_0(\gamma) = 0$. The broken line $u_0(x)$ cannot have a complete link on the x-axis, for then, as in Theorem 6.11.2, it would be zero at two successive $u_0(i)$ and $S_{\mathbf{u}_0}^{+} - S_{\mathbf{u}_0}^{-} \geq 2$, contradicting (6.11.2). Since $u_0(\gamma) = 0$, and $u_0(x)$ is positive on either side of γ, γ must be a break-point of the $u_0(x)$ line, say i, so that

$$u_0(i-1) > 0,\ u_0(i) = 0,\ u_0(i+1) > 0$$

and again $S_{\mathbf{u}_0}^{+} - S_{\mathbf{u}_0}^{-} \geq 2$, contradicting (6.11.2). We conclude that between any two nodes of $u^{(k+1)}(x)$ there must be *at least one* node of $u^{(k)}(x)$. But $u^{(k)}(x)$ has only $k - 1$ nodes, while $u^{(k+1)}(x)$ has k nodes. Thus $u^{(k)}(x)$ has *no more than* one node between two nodes of $u^{(k+1)}(x)$, i.e., it has exactly one node there; the two sets of nodes interlace. ∎

Chapter 7

Isospectral Systems

We view things not only from different sides, but with different eyes; we have no wish to find them alike.
Pascal's *Pensées*, 124

7.1 Introduction

We will say that two systems are *isospectral* if they have the same eigenvalues. (Some authors use the term cospectral.) In our context a 'system' is characterised by a *symmetric* matrix $\mathbf{A} \in S_n$, or perhaps by two symmetric matrices $\mathbf{M}, \mathbf{K} \in S_n$. In the notation of Section 4.3, two matrices $\mathbf{A}, \mathbf{B} \in S_n$ are said to be is isospectral if

$$\sigma(\mathbf{A}) = \boldsymbol{\sigma}(\mathbf{B}) \tag{7.1.1}$$

and two systems $(\mathbf{M}, \mathbf{K})$ and $(\mathbf{M}', \mathbf{K}')$ are said to be isospectral if

$$\sigma(\mathbf{M}, \mathbf{K}) = \sigma(\mathbf{M}', \mathbf{K}'). \tag{7.1.2}$$

We recall that if $\mathbf{M}, \mathbf{M}'$ are positive definite, then we may reduce the problem to (7.1.1).

In Section 5.2, when discussing matrix transformations, we showed that if $\mathbf{Q}$ is an orthogonal matrix, i.e., one satisfying

$$\mathbf{Q}\mathbf{Q}^T = \mathbf{Q}^T\mathbf{Q} = \mathbf{I} \tag{7.1.3}$$

and if

$$\mathbf{B} = \mathbf{Q}\mathbf{A}\mathbf{Q}^T \tag{7.1.4}$$

then $\mathbf{A}$ and $\mathbf{B}$ are isospectral. The converse is true: if $\mathbf{A}$ and $\mathbf{B} \in S_n$ are isospectral, *then* they are related by (7.1.4) for some $\mathbf{Q}$. To prove this, we may use the general representation of a symmetric matrix given in the Corollary to Theorem 6.3.2. Suppose $\mathbf{A}, \mathbf{B} \in S_n$ have the same eigenvalues $(\lambda_i)_1^n$. Put $\wedge = diag(\lambda_1, \lambda_2, \ldots, \lambda_n)$, then

$$\mathbf{A} = \mathbf{U} \wedge \mathbf{U}^T \text{ and } \mathbf{B} = \mathbf{V} \wedge \mathbf{V}^T$$

where both $\mathbf{U}$ and $\mathbf{V}$ are orthogonal. Thus

$$\mathbf{B} = \mathbf{V}\mathbf{U}^T.\mathbf{U}\wedge\mathbf{U}^T.\mathbf{U}\mathbf{V}^T = \mathbf{V}\mathbf{U}^T.\mathbf{A}.\mathbf{U}\mathbf{V}^T.$$

But since $\mathbf{U}$, $\mathbf{V}$ are orthogonal, so is $\mathbf{Q} = \mathbf{V}\mathbf{U}^T$, (Ex. 5.2.2). Thus $\mathbf{B} = \mathbf{Q}\mathbf{A}\mathbf{Q}^T$. We recall Ex. 5.2.2, that this transformation defines an equivalence class, an *isospectral family* of matrices.

This means that, from a purely mathematical viewpoint, the problem of characterizing isospectral systems governed by a single matrix is solved: the matrices $\mathbf{A}$ and $\mathbf{B}$ are linked by some orthogonal matrix $\mathbf{Q}$. However, this result is insufficient for applications to vibrating systems. For there we are concerned with vibrating systems of a *particular type*, as described for instance in Section 5.1. It may easily be verified that if the matrix $\mathbf{A}$ has a particular form, in the sense that it relates to a particular graph $\mathcal{G}$, and if $\mathbf{Q}$ is an arbitary orthogonal matrix, then $\mathbf{B}$ will not necessarily have the same form, i.e., relate to the same graph $\mathcal{G}$. In practice, the conditions on the system matrix are even more stringent; there are conditions on the *signs* of matrix elements.

This is the question we address in this Chapter: given one system, specified by $\mathbf{A}$ or $(\mathbf{M}, \mathbf{K})$, with the matrices having some particular form, specified by a graph $\mathcal{G}$, and perhaps some sign conditions, how can we find other systems $\mathbf{B}$ or $(\mathbf{M}', \mathbf{K}')$ satisfying the same conditions? We do not seek just an isospectral family, but a special isospectral family (i.e., a subfamily), the members of which share certain special characteristics. So far, the results which have been obtained relate to comparatively simple systems. We start our quest by considering the concept of isospectral flow.

7.2 Isospectral flow

Suppose $\mathbf{A} \in S_n$ has eigenvalues $(\lambda_i)_1^n$ and eigenvectors $(\mathbf{u}_i)_1^n$, then equation (6.3.8) states that if $\mathbf{U} = [\mathbf{u}_1, \mathbf{u}_2, \ldots, \mathbf{u}_n]$ and $\wedge = diag(\lambda_1, \lambda_2, \ldots, \lambda_n)$, then

$$\mathbf{A} = \mathbf{U}\wedge\mathbf{U}^T, \quad \mathbf{U}\mathbf{U}^T = \mathbf{U}^T\mathbf{U} = \mathbf{I}. \tag{7.2.1}$$

Now suppose that $\mathbf{U}$ depends on a single parameter t, and that $\mathbf{U}(t)$, and hence $\mathbf{A}(t)$ varies in such a way that the eigenvalues, and hence $\wedge$, remain invariant. Using $\cdot$ to denote d/dt, we have

$$\begin{aligned} \dot{\mathbf{A}} &= \mathbf{U}\wedge\dot{\mathbf{U}}^T + \dot{\mathbf{U}}\wedge\mathbf{U}^T \\ &= (\mathbf{U}\wedge\mathbf{U}^T)(\mathbf{U}\dot{\mathbf{U}}^T) + (\dot{\mathbf{U}}\mathbf{U}^T)(\mathbf{U}\wedge\mathbf{U}^T). \end{aligned} \tag{7.2.2}$$

On differentiating the second equation in (7.2.1) we find

$$\dot{\mathbf{U}}\mathbf{U}^T + \mathbf{U}\dot{\mathbf{U}}^T = \mathbf{0}$$

so that on writing

$$\mathbf{S} = \mathbf{U}\dot{\mathbf{U}}^T \tag{7.2.3}$$

we find

$$\dot{\mathbf{U}}\mathbf{U}^T = -\mathbf{S} = \mathbf{S}^T, \tag{7.2.4}$$

and we can write equation (7.2.2) as

$$\dot{\mathbf{A}} = \mathbf{AS} - \mathbf{SA}. \tag{7.2.5}$$

This is the differential equation governing isospectral flow. We note from equation (7.2.4) that the matrix $\mathbf{S}$ is *skew symmetric.* We note also that the differential equation governing $\mathbf{U}$ is

$$\dot{\mathbf{U}} = -\,\mathbf{SU},$$

and that because $\mathbf{S}$ is skew symmetric and $\mathbf{A}$ is symmetric, $\dot{\mathbf{A}}$, given by (7.2.5) is symmetric.

We may apply this analysis in reverse. Suppose $\mathbf{S}(t)$ is a skew symmetric matrix, and let $\mathbf{U}(t)$ be the solution of the equation

$$\dot{\mathbf{U}}(t) = -\mathbf{S}(t)\mathbf{U}(t), \quad \mathbf{U}(0) = \mathbf{U}_0$$

where $\mathbf{U}_0$ is an orthogonal matrix, then

$$(\mathbf{U}^T\mathbf{U})^\bullet = \dot{\mathbf{U}}^T\mathbf{U} + \mathbf{U}^T\dot{\mathbf{U}} = \mathbf{U}^T\mathbf{SU} - \mathbf{U}^T\mathbf{SU} = \mathbf{0}.$$

But since $\mathbf{U}_0^T\mathbf{U}_0 = \mathbf{I}$, $\mathbf{U}^T(t)\mathbf{U}(t) = \mathbf{I}$ for all t; $\mathbf{U}(t)$ is orthogonal.

Now, with this $\mathbf{S}(t)$ and $\mathbf{U}(t)$ we consider the equation

$$\dot{\mathbf{A}}(t) = \mathbf{A}(t)\mathbf{S}(t) - \mathbf{S}(t)\mathbf{A}(t), \ \mathbf{A}(0) = \mathbf{A}_0$$

where $\mathbf{A}_0 = \mathbf{U}_0 \wedge \mathbf{U}_0^T$. We have

$$\begin{aligned}(\mathbf{U}^T\mathbf{AU})^\bullet &= \dot{\mathbf{U}}^T\mathbf{AU} + \mathbf{U}^T\dot{\mathbf{A}}\mathbf{U} + \mathbf{U}^T\mathbf{A}\dot{\mathbf{U}} \\ &= \mathbf{U}^T\mathbf{SAU} + \mathbf{U}^T(\mathbf{AS} - \mathbf{SA})\mathbf{U} - \mathbf{U}^T\mathbf{ASU} = \mathbf{0}\end{aligned}$$

so that

$$\mathbf{U}^T\mathbf{AU} = \mathbf{U}_0^T\mathbf{A}_0\mathbf{U}_0 = \wedge.$$

Equation (7.2.5) provides a way in which to construct a one-dimensional family, i.e., a trajectory, of isospectral systems, and we will explore its use later. At this point however, we will discuss the connection between equation (7.2.5) and matrix factorisation.

One of the basic procedures of numerical linear algebra is the Gram-Schmidt procedure for orthogonalisation: given a set of vectors $(\mathbf{a}_i)_1^m \in V_n$, construct a set of orthonormal vectors $(\mathbf{q}_i)_1^r \in V_n$ by forming combinations of the $\mathbf{a}_i$. The Gram-Schmidt procedure gives a way to factorise a non-singular matrix $\mathbf{A} \in M_n$. Since $\mathbf{A}$ is non-singular, its columns are linearly independent, and so span V_n; the Gram-Schmidt procedure will yield n orthonormal vectors $(\mathbf{q}_i)_1^n$ spanning V_n; we obtain the factorisation by writing the $\mathbf{a}_i$'s in terms of $\mathbf{q}$'s. Let

$$\mathbf{A} = [\mathbf{a}_1, \mathbf{a}_2, \ldots, \mathbf{a}_n],$$

then we choose $(\mathbf{q}_i)_1^n$ so that

$$\mathbf{a}_m = \sum_{k=1}^{m} r_{km}\mathbf{q}_k, \quad m = 1, 2, \ldots, n,$$

which we may assemble to give

$$\begin{bmatrix} a_{11} & a_{12} & \ldots & a_{1n} \\ a_{21} & a_{22} & \ldots & a_{2n} \\ \cdot & \cdot & \ldots & \cdot \\ a_{n1} & a_{n2} & \ldots & a_{nn} \end{bmatrix} = \begin{bmatrix} q_{11} & q_{12} & \ldots & q_{1n} \\ q_{21} & q_{22} & \ldots & q_{2n} \\ \cdot & \cdot & \ldots & \cdot \\ q_{n1} & q_{n2} & \ldots & q_{nn} \end{bmatrix} \begin{bmatrix} r_{11} & r_{12} & \ldots & r_{1n} \\ & r_{22} & \ldots & r_{2n} \\ \cdot & \cdot & \ldots & \cdot \\ & & & r_{nn} \end{bmatrix},$$

i.e.,

$$\mathbf{A} = \mathbf{QR}. \tag{7.2.6}$$

The $\mathbf{q}_i$ and the r's are found in Theorem 3.2.1:

$$r_{11} = ||\mathbf{a}_1||, \quad \mathbf{q}_1 = \mathbf{a}_1/r_{11},$$
$$r_{12} = \mathbf{q}_1^T\mathbf{a}_2, \quad r_{22} = ||\mathbf{a}_2 - r_{12}\mathbf{q}_1||, \quad \mathbf{q}_2 = (\mathbf{a}_2 - r_{12}\mathbf{q}_1)/r_{22},$$

etc. We note that the diagonal terms r_{ii} are *positive.*

One of the basic results related to the **QR** factorisation is that if

$$\mathbf{A} = \mathbf{QR}, \text{ then } \mathbf{A}' \equiv \mathbf{RQ} = \mathbf{Q}^T(\mathbf{QR})\mathbf{Q} = \mathbf{Q}^T\mathbf{AQ} \tag{7.2.7}$$

which means that $\mathbf{A}'$, obtained by reversing the factors **Q** and **R**, is isospectral to **A**. One of the ways in which **QR** algorithm is used in numerical linear algebra is to use it to form a sequence of matrices $\mathbf{A}, \mathbf{A}', \mathbf{A}'', \ldots$ by continually reversing factors:

$$\mathbf{A} = \mathbf{QR}, \ \mathbf{A}' = \mathbf{RQ} = \mathbf{Q}'\mathbf{R}', \ \mathbf{A}'' = \mathbf{R}'\mathbf{Q}' = \mathbf{Q}''\mathbf{R}'', \ldots \tag{7.2.8}$$

Under certain conditions, the sequence converges to an upper triangular matrix or, if **A** is symmetric, to a diagonal matrix composed of the eigenvalues. We will use the basic reversal (7.2.7) and the sequence (7.2.8), in this book, but we are not interested in the convergence properties of the sequence, for which see Golub and Van Loan (1983) [135].

We now show that, for a special choice of the skew symmetric matrix **S**, we may relate the sequence (7.2.8) to an isospectral flow. In doing so we will have to retrace some of the steps we have already taken.

Suppose $\mathbf{A} \in S_n$, and that $\mathbf{A}^+$ is its *strict* upper triangle, i.e.,

$$\mathbf{A}^+ = \begin{bmatrix} a_{12} & \ldots & \ldots & a_{1n} \\ & a_{23} & \ldots & a_{2n} \\ & & \ddots & \\ & & & a_{n-1,n} \end{bmatrix} \tag{7.2.9}$$

then $\mathbf{A}$ may be written

$$\begin{aligned}\mathbf{A} &= \mathbf{A}^{+} + \mathbf{A}^{+T} + diag(a_{11}, a_{22}, \ldots, a_{nn}) \\ &= \mathbf{A}^{+T} - \mathbf{A}^{+} + 2\mathbf{A}^{+} + diag(a_{11}, a_{22}, \ldots, a_{nn}) \\ &= \mathbf{S} + \mathbf{R}\end{aligned} \tag{7.2.10}$$

where $\mathbf{S} = \mathbf{A}^{+T} - \mathbf{A}^{+}$ is skew-symmetric and $\mathbf{R} = 2\mathbf{A}^{+} + diag(a_{11}, a_{22}, \ldots, a_{nn})$ is upper triangular. We note that any symmetric matrix has this unique decomposition into a skew-symmetric matrix and an upper triangular matrix.

We now start to retrace our steps:

Lemma 7.2.1 *Suppose* $\mathbf{S}$ *is skew symmetric, and let* $\mathbf{Q}$ *be the solution to the problem*

$$\dot{\mathbf{Q}} = \mathbf{QS}, \quad \mathbf{Q}(0) = \mathbf{I} \tag{7.2.11}$$

then $\mathbf{Q}$ *is an orthogonal matrix.*

Proof.

$$\begin{aligned}(\mathbf{Q}\mathbf{Q}^T)^{\bullet} &= \dot{\mathbf{Q}}\mathbf{Q}^T + \mathbf{Q}\dot{\mathbf{Q}}^T \\ &= \mathbf{QSQ} + \mathbf{QS}^T\mathbf{Q}^T = \mathbf{Q}(\mathbf{S} + \mathbf{S}^T)\mathbf{Q}^T = \mathbf{0}.\end{aligned}$$

Since $\mathbf{Q}(0)\mathbf{Q}^T(0) = \mathbf{I}$, we have $\mathbf{Q}(t)\mathbf{Q}^T = \mathbf{I}$. ∎

Lemma 7.2.2 *Let* $\mathbf{A}(t)$ *be the solution of the problem*

$$\dot{\mathbf{A}} = \mathbf{AS} - \mathbf{SA} \quad \mathbf{A}(0) = \mathbf{A}_0, \tag{7.2.12}$$

then $\mathbf{A}(t) = \mathbf{Q}^T(t)\mathbf{A}_0\mathbf{Q}(t)$, *where* $\mathbf{Q}(t)$ *is as in Lemma 7.2.1.*

Proof. Let $\mathbf{Z}(t) = \mathbf{Q}(t)\mathbf{A}(t)\mathbf{Q}^T(t)$, then

$$\begin{aligned}\dot{\mathbf{Z}} &= \dot{\mathbf{Q}}\mathbf{A}\mathbf{Q}^T + \mathbf{Q}\dot{\mathbf{A}}\mathbf{Q}^T + \mathbf{QA}\dot{\mathbf{Q}}^T \\ &= \mathbf{QSAQ}^T + \mathbf{Q}(\mathbf{AS} - \mathbf{SA})\mathbf{Q}^T + \mathbf{QAS}^T\mathbf{Q}^T \\ &= \mathbf{QA}(\mathbf{S} + \mathbf{S}^T)\mathbf{Q}^T = \mathbf{0}.\end{aligned}$$

This shows that

$$\mathbf{Z}(t) = \mathbf{Z}(0) = \mathbf{A}(0) = \mathbf{A}_0$$

so that

$$\mathbf{QAQ}^T = \mathbf{A}_0, \text{ i.e., } \mathbf{A} = \mathbf{Q}^T\mathbf{A}_0\mathbf{Q}. \ ∎$$

The orthogonal matrix $\mathbf{Q}$ was introduced as 'the solution to the differential equation (7.2.11)'. We now show that we may identify it through a $\mathbf{QR}$ factorisation:

Lemma 7.2.3 *If the matrix* $\exp(t\mathbf{A}_0)$ *has the* $\mathbf{QR}$*-decomposition*

$$\exp(t\mathbf{A}_0) = \mathbf{Q}(t)\mathbf{R}(t), \tag{7.2.13}$$

then $\mathbf{Q}(t)$ *satisfies equation (7.2.11), and* $\mathbf{A}(t) = \mathbf{Q}^T(t)\mathbf{A}_0\mathbf{Q}(t)$ *is the solution of (7.2.12).*

Proof. Here

$$\exp(t\mathbf{A}_0) = \mathbf{I} + t\mathbf{A}_0 + \frac{t^2}{2}\mathbf{A}_0^2 + \ldots \tag{7.2.14}$$

is the solution of the equation

$$\dot{\mathbf{X}}(t) = \mathbf{A}_0\mathbf{X}(t), \quad \mathbf{X}(0) = \mathbf{I}.$$

Taking derivatives of both sides of (7.2.13), we find

$$(\mathbf{QR})^{\bullet} = \dot{\mathbf{Q}}\mathbf{R} + \mathbf{Q}\dot{\mathbf{R}} = \mathbf{A}_0 \exp(t\mathbf{A}_0) = \mathbf{A}_0\mathbf{QR}$$

so that

$$\dot{\mathbf{Q}} + \mathbf{Q}\dot{\mathbf{R}}\mathbf{R}^{-1} = \mathbf{A}_0\mathbf{Q},$$

and

$$\mathbf{Q}^T\dot{\mathbf{Q}} + \dot{\mathbf{R}}\mathbf{R}^{-1} = \mathbf{Q}^T\mathbf{A}_0\mathbf{Q} = \hat{\mathbf{A}}(t). \tag{7.2.15}$$

But $\hat{\mathbf{A}}(t)$ is a symmetric matrix, $\mathbf{Q}$ is orthogonal, so that $\mathbf{Q}^T\dot{\mathbf{Q}}$ is skew symmetric, and $\dot{\mathbf{R}}\mathbf{R}^{-1}$ is upper triangular: equation (7.2.15) gives the unique decomposition of $\hat{\mathbf{A}}$ as the sum of a skew-symmetric and an upper triangular matrix, i.e.,

$$\mathbf{Q}^T\dot{\mathbf{Q}} = \hat{\mathbf{A}}^{+T} - \hat{\mathbf{A}}^{+} = \hat{\mathbf{S}}.$$

On the other hand

$$\begin{aligned}\dot{\hat{\mathbf{A}}} &= \mathbf{Q}^T\mathbf{A}_0\dot{\mathbf{Q}} + \dot{\mathbf{Q}}^T\mathbf{A}_0\mathbf{Q} \\ &= (\mathbf{Q}^T\mathbf{A}_0\mathbf{Q})(\mathbf{Q}^T\dot{\mathbf{Q}}) + (\dot{\mathbf{Q}}^T\mathbf{Q})(\mathbf{Q}^T\mathbf{A}_0\mathbf{Q}) \\ &= \hat{\mathbf{A}}\hat{\mathbf{S}} - \hat{\mathbf{S}}\hat{\mathbf{A}}\end{aligned}$$

and $\hat{\mathbf{A}}(0) = \mathbf{A}_0$. But this means that $\hat{\mathbf{A}}$ satisfies the same differential equation as $\mathbf{A}$, and has the same initial value, $\mathbf{A}_0$;

$$\hat{\mathbf{A}}(t) = \mathbf{A}(t) = \mathbf{Q}^T\mathbf{A}_0\mathbf{Q}. \ \blacksquare$$

We may now state

Theorem 7.2.1 *Suppose $\mathbf{A}(t)$ is the solution to the differential equation (7.2.12). For $k = 1, 2, \ldots$ suppose*

$$\exp(\mathbf{A}(k-1)) = \mathbf{Q}_k\mathbf{R}_k$$

then

$$\exp(\mathbf{A}(k)) = \mathbf{R}_k\mathbf{Q}_k$$

where $\mathbf{Q}_k = \mathbf{Q}(k)$, $\mathbf{R}_k = \mathbf{R}(k)$.

Proof. Lemmas 7.2.2 and 7.2.3 show that

$$\mathbf{A}(t) = \mathbf{Q}^T\mathbf{A}_0\mathbf{Q}(t)$$

and

$$\exp(t\mathbf{A}(0)) = \mathbf{Q}(t)\mathbf{R}(t) \tag{7.2.16}$$

so

$$\begin{aligned}\mathbf{R}(t)\mathbf{Q}(t) &= \mathbf{Q}^T(t)(\mathbf{Q}(t)\mathbf{R}(t))\mathbf{Q}(t)\\ &= \mathbf{Q}^T(t)\exp(t\mathbf{A}_0)\mathbf{Q}(t).\end{aligned}$$

Now consider

$$\begin{aligned}\exp(t\mathbf{A}(t)) &= \exp(\mathbf{Q}^T(t)t\mathbf{A}_0\mathbf{Q}(t)) \\ &= \mathbf{I}+\mathbf{Q}^T t\mathbf{A}_0\mathbf{Q}+\frac{(\mathbf{Q}^T t\mathbf{A}_0\mathbf{Q})^2}{2!}+\cdots \\ &= \mathbf{I}+\mathbf{Q}^T t\mathbf{A}_0\mathbf{Q}+\frac{(\mathbf{Q}^T t\mathbf{A}_0\mathbf{Q})(\mathbf{Q}^T t\mathbf{A}_0\mathbf{Q})}{2!}+\cdots \\ &= \mathbf{Q}^T(\mathbf{I}+t\mathbf{A}_0+\frac{t^2\mathbf{A}_0^2}{2!}+\cdots)\mathbf{Q} \\ &= \mathbf{Q}^T\exp(t\mathbf{A}_0)\mathbf{Q}=\mathbf{R}(t)\mathbf{Q}(t).\end{aligned} \tag{7.2.17}$$

This means that, taking $t = 1$ in (7.2.16), we have

$$\exp(\mathbf{A}(0)) = \mathbf{Q}_1\mathbf{R}_1$$

and taking $t = 1$ in (7.2.17),

$$\exp\mathbf{A}(1) = \mathbf{R}_1\mathbf{Q}_1 = \mathbf{Q}_2\mathbf{R}_2 \text{ etc.} \quad \blacksquare$$

We describe this result by saying that the solutions of (7.2.12) at integral times $0, 1, 2, \ldots$ give the iterates in the **QR**-sequence (7.2.8) starting from $\exp(\mathbf{A}(0)) = \exp\mathbf{A}_0 = \mathbf{Q}_1\mathbf{R}_1$.

We conclude this section with a note on the historical development of the theory of isospectral flow.

The analysis had its beginnings in the investigation of the so-called *Toda lattice*, Toda (1970) [324], a set of n particles constrained to move on a line under exponential repulsive forces. Symes (1980) [315], Symes (1982) [316] gives references to the roots of the problem in Physics, and establishes the theory, basically as described above, for the particular case encountered in the Toda lattice, that $\mathbf{A}$ is a Jacobi matrix. The analysis for a Jacobi matrix was developed further by Nanda (1982) [245], Nanda (1985) [246] and by Deift, Nanda and Tomei (1983) [77]. The generalisation of the theory to an arbitrary complex non-symmetric matrix is due to Chu (1984) [57]. Watkins (1984) [331] gives a survey of the general theory, and its extension to other matrix factorisations such as **LR** (lower triangular matrix **L**, multiplied by upper triangular matrix **R**) or the Cholesky factorisation $\mathbf{LL}^T$. Chu and Norris (1988) [60] explore the connection between isospectral flows and abstract matrix factorisations.

Most of this research is concerned with the connection between isospectral flow and the procedures used in numerical linear algebra; this is not our concern in this book. Rather, we are interested in isospectral flow as a way of constructing isospectral systems, as we will show in later sections of this Chapter.

We will take up the topic of isospectral flow in Section 7.6 after we have considered algebraic procedures for obtaining isospectral systems.

7.3 Isospectral Jacobi systems

We follow Gladwell (1995) [121] and start our discussion by considering the particular case of the spring-mass system shown in Figure 4.4.2a and reproduced as Figure 7.3.1.

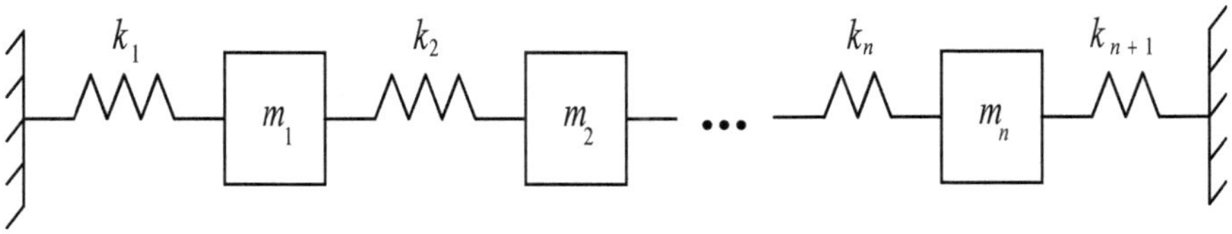

Figure 7.3.1 – An in-line spring-mass system

The governing equation is

$$(\mathbf{K} - \lambda\mathbf{M})\mathbf{y} = \mathbf{0}, \tag{7.3.1}$$

where

$$\mathbf{K} = \begin{bmatrix} k_1+k_2 & -k_2 & 0 & \ldots 0 \\ -k_2 & k_2+k_3 & -k_3 & \ldots 0 \\ \cdot & \cdot & \cdot & \ldots \\ 0 & \ldots & -k_n & k_n+k_{n+1} \end{bmatrix}, \tag{7.3.2}$$

$$\mathbf{M} = diag(m_1, m_2, \ldots, m_n). \tag{7.3.3}$$

We will assume that the chain of masses and springs is unbroken, so that

$$(k_i)_2^n > 0, \quad (m_i)_1^n > 0.$$

There are three particular cases:

(**S**) supported; $k_1 > 0,\ k_{n+1} > 0$

(**C**) cantilever; $k_1 > 0,\ k_{n+1} = 0$

(**F**) free; $k_1 = 0,\ k_{n+1} = 0$

If two systems, 1 and 2, are isospectral then, in the notation of Section 4.3,

$$\sigma(\mathbf{M}_1, \mathbf{K}_1) = \sigma(\mathbf{M}_2, \mathbf{K}_2). \tag{7.3.4}$$

There are two almost trivial ways of obtaining an isospectral pair. First, if $c > 0$, then

$$\sigma(c\mathbf{M}, c\mathbf{K}) = \sigma(\mathbf{M}, \mathbf{K}).$$

Secondly, if we physically turn the system around and renumber the masses and springs from the left, then we will not change the eigenvalues. Renumbering is equivalent to pre- and post-multiplying by the matrix $\mathbf{T}$ of equation (4.3.8). Thus (7.3.4) will hold if

$$\mathbf{M}_2 = \mathbf{T}\mathbf{M}_1\mathbf{T}, \quad \mathbf{K}_2 = \mathbf{T}\mathbf{K}_1\mathbf{T}. \tag{7.3.5}$$

To obtain non-trivial isospectral pairs, we reduce (7.3.1) to standard form. We write

$$\mathbf{M} = \mathbf{D}^2, \quad \mathbf{D}\mathbf{y} = \mathbf{u}, \quad \mathbf{J} = \mathbf{D}^{-1}\mathbf{K}\mathbf{D}^{-1}, \tag{7.3.6}$$

so that

$$(\mathbf{J} - \lambda\mathbf{I})\mathbf{u} = \mathbf{0}. \tag{7.3.7}$$

First, consider a cantilever system. Now as in (4.4.7), $\mathbf{K}$ may be factorised as

$$\mathbf{K} = \mathbf{E}\hat{\mathbf{K}}\mathbf{E}^T, \quad \hat{\mathbf{K}} = diag(k_1, k_2, \ldots, k_n),$$

and

$$\mathbf{J} = \mathbf{D}^{-1}\mathbf{E}\hat{\mathbf{K}}\mathbf{E}^T\mathbf{D}^{-1}. \tag{7.3.8}$$

To obtain an isospectral pair, we need

Lemma 7.3.1 *If* $\mathbf{A}, \mathbf{B} \in M_n$*, then* $\mathbf{AB}$ *and* $\mathbf{BA}$ *have the same eigenvalues, except perhaps for zero.*

Proof. Suppose $\lambda \neq 0$ is an eigenvalue of $\mathbf{AB}$, so that, for some $\mathbf{x} \neq \mathbf{0}$, $\mathbf{ABx} = \lambda\mathbf{x}$. Since $\lambda \neq 0$ and $\mathbf{x} \neq 0$, we have $\mathbf{Bx} \neq 0$. Now $\mathbf{B}(\mathbf{ABx}) = \mathbf{BA}(\mathbf{Bx}) = \lambda\mathbf{Bx}$, so that $\mathbf{Bx}$ is an eigenvector of $\mathbf{BA}$ corresponding to the eigenvalue λ. We have proved that any non-zero eigenvalue of $\mathbf{AB}$ is an eigenvalue of $\mathbf{BA}$. Now reverse the roles of $\mathbf{A}$ and $\mathbf{B}$ to complete the proof. ■

Write $\hat{\mathbf{K}} = \mathbf{F}^2$, so that

$$\mathbf{J} = (\mathbf{D}^{-1}\mathbf{E}\mathbf{F})(\mathbf{F}\mathbf{E}^T\mathbf{D}^{-1}). \tag{7.3.9}$$

Now apply the Lemma: the eigenvalues of $\mathbf{J}$ are non-zero (in fact, positive) so that if

$$\mathbf{J}' = (\mathbf{F}\mathbf{E}^T\mathbf{D}^{-1})(\mathbf{D}^{-1}\mathbf{E}\mathbf{F}), \tag{7.3.10}$$

then

$$\sigma(\mathbf{J}') = \sigma(\mathbf{J}).$$

To form a spring-mass system corresponding to $\mathbf{J}'$ we reverse the reduction to standard form, and write

$$(\mathbf{J}' - \lambda\mathbf{I})\mathbf{u} = \mathbf{0}$$

as

$$(\mathbf{E}^T\mathbf{M}^{-1}\mathbf{E} - \lambda\hat{\mathbf{K}}^{-1})\mathbf{v} = \mathbf{0}, \quad \mathbf{v} = \mathbf{F}\mathbf{u}. \tag{7.3.11}$$

This is the eigenvalue equation for a reversed cantilever, We may verify this by noting that

$$\mathbf{TET} = \mathbf{E}^T, \quad \mathbf{T}^2 = \mathbf{I},$$

and thus

$$\begin{aligned}\mathbf{TE}^T\mathbf{M}^{-1}\mathbf{Ev} &= \mathbf{TE}^T\mathbf{T}\cdot\mathbf{TM}^{-1}\mathbf{T}\cdot\mathbf{TET}\cdot\mathbf{Tv}\\ &= \mathbf{E}\hat{\mathbf{K}}^0\mathbf{E}^T\cdot\mathbf{Tv},\end{aligned}$$

so that we may write equation (7.3.11) as

$$(\mathbf{K}^0-\lambda\mathbf{M}^0)\mathbf{Tv}=0,$$

where

$$\mathbf{K}^0=\mathbf{E}\hat{\mathbf{K}}^0\mathbf{E}^T,\quad \hat{\mathbf{K}}^0=\mathbf{TM}^{-1}\mathbf{T},\quad \mathbf{M}^0=\mathbf{T}\hat{\mathbf{K}}^{-1}\mathbf{T}.$$

This system relates to a cantilever with

$$k_i^0=m_{n-i+1}^{-1},\quad m_i^0=k_{n-i+1}^{-1},\quad i=1,2,\ldots,n,$$

and

$$\sigma(\mathbf{M}^0,\mathbf{K}^0)=\sigma(\mathbf{M},\mathbf{K}).$$

This pair was pointed out by Ram and Elhay (1995a) [285]. See also Ram and Elhay (1998) [287].

In the analysis we have just described, we started with a system specified by $\mathbf{M},\mathbf{K}$ and formed the Jacobi matrix $\mathbf{J}=\mathbf{D}^{-1}\mathbf{KD}^{-1}$. This passage from a spring mass system to a Jacobi matrix is unique, but starting from a given Jacobi matrix we may construct an infinite family of spring mass systems, as we will now show.

The stiffness matrix $\mathbf{K}$ of (7.3.2) has the property

$$\mathbf{K}\{1,1,1,\ldots,1\}=\{k_1,0,\ldots,0,k_{n+1}\};\tag{7.3.12}$$

this equation states that in order to move all the masses statically to the right by unit displacement, we must apply forces k_1 and k_{n+1} to masses m_1 and m_n respectively. We follow the analysis developed in Section 4.4. Since $\mathbf{J}=\mathbf{D}^{-1}\mathbf{KD}^{-1}$ we have $\mathbf{K}=\mathbf{DJD}$ so that equation (7.3.12) yields

$$\mathbf{J}\{d_1,d_2,\ldots,d_n\}=\{k_1d_1^{-1},0,\ldots,k_{n+1}d_n^{-1}\}.$$

Thus in order to find a spring-mass system we must take $\mathbf{J}$ and find a solution to the equation

$$\mathbf{Jd}=\{\alpha,0,\ldots,0,\beta\}\quad \mathbf{d}=\{d_1,d_2,\ldots,d_n\}\tag{7.3.13}$$

where $\alpha\geq 0$, $\beta\geq 0$, $\alpha+\beta>0$. If $\mathbf{J}$ is non-singular, then Theorem 4.4.1 ensures that $\mathbf{d}>\mathbf{0}$. Thus to construct a spring-mass system we may choose α,β to be arbitrary non-negative constants, not both zero. This is equivalent to choosing arbitrary spring stiffnesses k_1 and k_{n+1}; for when we solve equation (7.3.13) we find

$$k_1=d_1\alpha,\quad k_{n+1}=d_n\beta;\tag{7.3.14}$$

we have a two-parameter family of isospectral systems. If we demand that the reconstructed system be a cantilever, so that $\beta=0=k_{n+1}$, then the solution is essentially unique; we can make it unique by taking $m_1=1$ or $\sum_{i=1}^n m_i=1$.

If $\mathbf{J}$ is singular we use Theorem 4.4.2, which ensures that there is a positive solution of

$$\mathbf{Jd} = \mathbf{0} \tag{7.3.15}$$

and then construct $\mathbf{K} = \mathbf{DJD}$, $\mathbf{M} = \mathbf{D}^2$; again the system is essentially unique.

We now discuss two different ways of constructing families of isospectral Jacobi matrices. We let $\mathcal{M}(\lambda_1, \lambda_2, \ldots, \lambda_n)$ denote the set of Jacobi matrices $\mathbf{J}$ such that $\sigma(\mathbf{J}) = (\lambda_i)_1^n$. The first follows directly from the analysis of Section 4.3: we can reconstruct $\mathbf{J}$ uniquely from $\sigma(\mathbf{J}) = (\lambda_i)_1^n$ and the vector $\mathbf{x}_1$ of first components of the normalised eigenvectors $\mathbf{u}_i$ of $\mathbf{J}$. We know that these first components $x_{11}, x_{21}, \ldots, x_{n1}$ are all non-zero, so that we can take them to be all positive, and they satisfy

$$\mathbf{x}_1^T\mathbf{x}_1 = 1 = x_{11}^2 + x_{21}^2 + \cdots + x_{n1}^2. \tag{7.3.16}$$

This means that each $\mathbf{J} \in \mathcal{M}$ may be associated with a point $P = (x_{11}, x_{21}, \ldots, x_{n1})$ in the (strictly) positive orthant of the unit n-sphere. (In more precise terms, $\mathcal{M}$ is a smooth $(n-1)$-dimensional manifold diffeomorphic to the strictly positive orthant of the unit n-sphere.)

The second way uses $\mathbf{QR}$ factorisation, as discussed in Section 7.2. Suppose $\mathbf{A} \in S_n$ and μ is not an eigenvalue of $\mathbf{A}$. Then $\mathbf{A} - \mu\mathbf{I}$ is non-singular, and so may be factorised:

$$\mathbf{A} - \mu\mathbf{I} = \mathbf{QR}. \tag{7.3.17}$$

Here $\mathbf{Q}$ is an orthogonal matrix, and $\mathbf{R}$ is upper triangular with *positive* diagonal terms r_{ii}; this factorisation (7.3.17) is unique. Now form the matrix $\mathbf{A}'$ from the equation

$$\mathbf{A}' - \mu\mathbf{I} = \mathbf{RQ}. \tag{7.3.18}$$

Equations (7.3.17), (7.3.18) define a transformation $\mathcal{G}_\mu : \mathbf{A} \longrightarrow \mathbf{A}'$.

The matrix $\mathbf{A}'$ is symmetrical, and is isospectral to $\mathbf{A}$:

$$\mathbf{A}' = \mu\mathbf{I} + \mathbf{RQ} = \mathbf{Q}^T(\mu\mathbf{I} + \mathbf{QR})\mathbf{Q} = \mathbf{Q}^T\mathbf{AQ}. \tag{7.3.19}$$

We now prove

Theorem 7.3.1 *If* $\mathbf{A}$ *is a Jacobi matrix, then so is* $\mathbf{A}'$.

Proof. We first show that if $\mathbf{A}$ is tridiagonal, then so is $\mathbf{A}'$.

Equations (7.3.17), (7.3.18) give

$$\mathbf{RA} = \mathbf{R}(\mu\mathbf{I} + \mathbf{QR}) = (\mu\mathbf{I} + \mathbf{RQ})\mathbf{R} = \mathbf{A}'\mathbf{R}. \tag{7.3.20}$$

This relation between $\mathbf{A}$ and $\mathbf{A}'$ is fundamental, and is often more instructive than (7.3.17), (7.3.18) or (7.3.19). Consider the i, j term in the products on either side of (7.3.20), and take $i \geq j$:

$$\sum_{k=1}^{n} r_{ik}a_{kj} = \sum_{k=1}^{n} a'_{ik}r_{kj}, \quad j = 1, 2, \ldots, n-1; \ i = j, j+1, \ldots, n. \tag{7.3.21}$$

Since $\mathbf{R}$ is upper triangular, r_{ik} is non-zero only for $k = i, i+1, \ldots, n$. Since $\mathbf{A}$ is tridiagonal, a_{kj} is non-zero only for $k = j-1, j, j+1$. Thus the product on the left is non-zero only for k running from i to $j+1$; it is identically zero if $i \geq j+2$. Since $\mathbf{R}$ is upper triangular, the index k on the right runs from $k = 1, 2, \ldots, j$. Thus

$$\sum_{k=i}^{j+1} r_{ik} a_{kj} = \sum_{k=1}^{j} a'_{ik} r_{kj}. \tag{7.3.22}$$

In particular therefore

$$\sum_{k=1}^{j} a'_{ik} r_{kj} = 0, \quad j = 1, 2, \ldots, n-2; \; i = j+2, \ldots, n. \tag{7.3.23}$$

Taking $j = 1$ we find $a'_{i1} r_{11} = 0$, and since $r_{11} > 0$,

$$a'_{i1} = 0, \quad i = 3, \ldots, n.$$

Now take $j = 2$:

$$a'_{i1} r_{12} + a'_{i2} r_{22} = 0, \quad i = 4, \ldots, n.$$

But $a'_{i1} = 0$ for these values, and $r_{22} > 0$, so that

$$a'_{i2} = 0, \quad i = 4, \ldots, n.$$

Proceeding in this way we find

$$a'_{ij} = 0, \quad j = 1, 2, \ldots, n-2; \; i = j+2, \ldots, n. \tag{7.3.24}$$

Thus $\mathbf{A}'$ has only one non-zero diagonal below the principal diagonal. But $\mathbf{A}'$ is symmetric, so that it is tridiagonal.

To show that if $\mathbf{A}$ is Jacobi, then so is $\mathbf{A}'$ we return to equation (7.3.22). Since $\mathbf{A}'$ is tridiagonal, we can rewrite (7.3.22) as

$$\sum_{k=i}^{j+1} r_{ik} a_{kj} = \sum_{k=i-1}^{j} a'_{ik} r_{kj}. \tag{7.3.25}$$

Take $i = j+1$, then each sum has just one term:

$$r_{ii} a_{i,i-1} = a'_{i,i-1} r_{i-1,i-1}, \quad i = 2, \ldots, n: \tag{7.3.26}$$

if $a_{i,i-1}$ is positive (negative) then so is $a'_{i,i-1}$. ■

We now suppose $\mathbf{A} = \mathbf{J}$, a Jacobi matrix, and prove

Theorem 7.3.2 *The operator $\mathcal{G}_\mu$ is commutative when applied to Jacobi matrices.*

$$\mathcal{G}_\mu \mathcal{G}_\nu \mathbf{J} = \mathcal{G}_\nu \mathcal{G}_\mu \mathbf{J}. \tag{7.3.27}$$

Proof. Consider the relation between the eigenvectors of $\mathbf{J}$ and $\mathbf{J}'$. Suppose $\mathbf{u}$ is a normalised eigenvector of $\mathbf{J}$:

$$\mathbf{J}\mathbf{u} = \lambda\mathbf{u},$$

then

$$\mathbf{J}'\mathbf{Q}^T\mathbf{u} = (\mathbf{Q}^T\mathbf{J}\mathbf{Q})\mathbf{Q}^T\mathbf{u} = \mathbf{Q}^T\mathbf{J}\mathbf{u} = \lambda\mathbf{Q}^T\mathbf{u},$$

so that $\mathbf{u}' = \mathbf{Q}^T\mathbf{u}$ is a normalised eigenvector of $\mathbf{J}'$. We may express this eigenvector in another way. Since

$$\mathbf{J}\mathbf{u} = (\mathbf{Q}\mathbf{R} + \mu\mathbf{I})\mathbf{u} = \lambda\mathbf{u},$$

we have

$$\mathbf{Q}\mathbf{R}\mathbf{u} = (\lambda - \mu)\mathbf{u},$$

or

$$\mathbf{u}' = \mathbf{Q}^T\mathbf{u} = \frac{\mathbf{R}\mathbf{u}}{\lambda - \mu}. \tag{7.3.28}$$

This equation shows that the *last* component of the eigenvector $\mathbf{u}_i'$ may be taken to be

$$u_{ni}' = \frac{r_{nn}(\mu)u_{ni}}{|\lambda_i - \mu|}. \tag{7.3.29}$$

This shows that, under the operation $\mathcal{G}_\mu$, the last components of the eigenvectors are simply multiplied by two terms: one, $r_{nn}(\mu)$, independent of i, and the other $|\lambda_i - \mu|^{-1}$. This means that the last components of the normalised eigenvectors of either of the matrices in (7.3.27) will be proportional to

$$\frac{u_{ni}}{|\lambda_i - \mu||\lambda_i - \nu|}. \tag{7.3.30}$$

Since they are proportional, and the sum of the squares of each set is unity, the two sets must be the same. But a Jacobi matrix is uniquely determined by its eigenvalues and the last components of its normalised eigenvectors. Therefore, (7.3.27) holds, and $\mathcal{G}_\mu$ is commutative. ■

We prove a stronger result in Theorem 7.4.2.

Theorem 7.3.3 *If $\mathbf{A}, \mathbf{B} \in \mathcal{M}$, then we can find a unique set $(\mu_i)_1^{n-1}$ such that $\mu_1 < \mu_2 < \cdots < \mu_{n-1}$ and*

$$\mathcal{G}_{\mu_1}\mathcal{G}_{\mu_2}\ldots\mathcal{G}_{\mu_{n-1}}\mathbf{A} = \mathbf{B}. \tag{7.3.31}$$

Proof. It is sufficient to show that we can pass from one set of last components $(u_{ni})_1^n$ to any other set $(v_{ni})_1^n$ in $n-1$ $\mathcal{G}_\mu$ operations. But equation (7.3.29) shows that this is equivalent to choosing $\mu_1, \mu_2, \ldots, \mu_{n-1}$ such that

$$\prod_{j=1}^{n-1} \frac{u_{ni}}{|\lambda_i - \mu_j|} \propto v_{ni}, \quad i = 1, 2, \ldots, n.$$

This is equivalent to choosing the polynomial

$$P(\lambda) = K \prod_{j=1}^{n-1} (\lambda - \mu_j)$$

such that

$$|P(\lambda_i)| = u_{ni}/v_{ni}, \quad i = 1, 2, \ldots, n.$$

If we choose the $(\mu_i)_1^{n-1}$ so that

$$\lambda_1 < \mu_1 < \lambda_2 < \cdots < \mu_{n-1} < \lambda_n \tag{7.3.32}$$

then

$$P(\lambda_i) = (-)^{n-i} u_{ni}/v_{ni}, \quad i = 1, 2, \ldots, n.$$

But there is a unique such polynomial $P(\lambda)$ of degree $n-1$, taking values of opposite signs at n points λ_i, and it will have $n-1$ roots μ_i satisfying (7.3.32). ■

Corollary 7.3.1 *If* $\mathcal{G}_\mu \mathbf{A} = \mathbf{B}$, *then we can find* $(\mu_i)_1^{n-1}$ *such that* $\mathcal{G}_{\mu_1} \mathcal{G}_{\mu_2} \ldots \mathcal{G}_{\mu_{n-1}}$ $\mathbf{B} = \mathbf{A}$, *and hence find* $\mathcal{G}_\mu^{-1}$.

Corollary 7.3.2 *We can find* $(\mu_i)_1^{n-1}$ *such that*

$$\mathcal{G}_{\mu_1} \mathcal{G}_{\mu_2} \ldots \mathcal{G}_{\mu_{n-1}} \mathbf{A} = \mathbf{A}.$$

Exercises 7.3

1. Consider the case (F), in which $k_1 = 0 = k_{n+1}$. Use Lemma 7.3.1 to obtain a cantilever which has the same eigenvalues as the original system apart from the zero eigenvalue corresponding to the rigid body mode.

2. Construct a formal inductive proof of equation (7.3.24).

7.4 Isospectral oscillatory systems

In Section 7.3 we considered the operator $\mathcal{G}_\mu$ defined by equations (7.3.17) and (7.3.18). We showed, amongst other things, that if $\mathbf{J}$ is tridiagonal, then so is $\mathbf{J}'$; if $a_{i+1,i} < 0$ (> 0), then $a'_{i+1,i} < 0$ (> 0). We recall from Section 6.6 that a positive-definite (symmetric) tridiagonal matrix with positive co-diagonal is a particular example of an *oscillatory* matrix, as defined at the beginning of Section 6.6, and characterised by Theorem 6.7.3. This means that if $\mathbf{A}$ is a symmetric tridiagonal oscillatory matrix, μ is not an eigenvalue of $\mathbf{A}$, and the diagonal elements of $\mathbf{R}$ are positive, then the operations

$$\mathbf{A} - \mu \mathbf{I} = \mathbf{QR} \tag{7.4.1}$$

$$\mathbf{A}' - \mu\mathbf{I} = \mathbf{RQ} \tag{7.4.2}$$

yield a new matrix $\mathbf{A}'$ that is symmetric, tridiagonal and oscillatory. Following Gladwell (1998) [126] we will now state that this is a special case of a general result:

Theorem 7.4.1 *Suppose $\mathbf{A} \in S_n$, let P denote one of the properties NTN, O, TP, let $\mathbf{A}'$ be defined from equations (7.4.1), (7.4.2). $\mathbf{A}'$ has property P iff $\mathbf{A}$ has property P.*

This Theorem states that $\mathbf{A}'$ is NTN iff $\mathbf{A}$ is NTN, $\mathbf{A}'$ is O iff $\mathbf{A}$ is O, and $\mathbf{A}'$ is TP iff $\mathbf{A}$ is TP. Implicit in the theorem is the condition that the diagonal elements of $\mathbf{R}$, which are necessarily non-zero because $\mathbf{A} - \mu\mathbf{I}$ is non-singular, are chosen to be positive.

The two conditions, that $\mathbf{A}$ is symmetric ($\mathbf{A} \in S_n$), and μ is not an eigenvalue of $\mathbf{A}$, are essential, as we now show by counterexamples.

Take $\mu = 0$ and

$$\mathbf{A} = \begin{bmatrix} 2 & a \\ 1 & 2 \end{bmatrix} \tag{7.4.3}$$

then

$$\begin{aligned} \mathbf{Q} &= \frac{1}{\sqrt{5}}\begin{bmatrix} 2 & -1 \\ 1 & 2 \end{bmatrix}, \quad \mathbf{R} = \frac{1}{\sqrt{5}}\begin{bmatrix} 5 & 2a+2 \\ 0 & 4-a \end{bmatrix} \\ \mathbf{A}' &= \frac{1}{5}\begin{bmatrix} 12+2a & 4a-1 \\ 4-a & 2(4-a) \end{bmatrix}. \end{aligned}$$

If $a = \frac{1}{5}$, $\mathbf{A}$ is O and TP, $\mathbf{A}'$ is not even TN; when $a = 0$, $\mathbf{A}$ is NTN and $\mathbf{A}'$ is not TN.

The condition that μ is not an eigenvalue is essential. For when $a = 1$ the matrix $\mathbf{A}$ in (7.4.3) is O and TP, and its eigenvalues are $\lambda_1 = 3$, $\lambda_2 = 1$. (Recall that when we consider oscillatory matrices we label the eigenvalues in decreasing order.) Take $\mu = 1$, then

$$\begin{aligned} \mathbf{A} - \mu\mathbf{I} &= \begin{bmatrix} 1 & 1 \\ 1 & 1 \end{bmatrix} = \begin{bmatrix} c & -c \\ c & c \end{bmatrix}\begin{bmatrix} 2c & 2c \\ 0 & 0 \end{bmatrix}, \quad c = \frac{1}{\sqrt{2}} \\ \mathbf{A}' - \mu\mathbf{I} &= \begin{bmatrix} 2c & 2c \\ 0 & 0 \end{bmatrix}\begin{bmatrix} c & -c \\ c & c \end{bmatrix} = \begin{bmatrix} 2 & 0 \\ 0 & 0 \end{bmatrix}, \quad \mathbf{A}' = \begin{bmatrix} 3 & 0 \\ 0 & 1 \end{bmatrix} \end{aligned} \tag{7.4.4}$$

The matrix $\mathbf{A}'$ in (7.4.4) is not oscillatory.

In general, if $\mathbf{A} \in S_n$ is O then its eigenvalues are distinct (Corollary to Theorem 6.10.1). This means that if $\mu = \lambda_k$ for some k, then $\mathbf{A} - \mu\mathbf{I}$ has rank $n-1$ and $r_{nn} = 0$, but no other r_{ii} is zero. Thus the last row of $\mathbf{A}' - \mu\mathbf{I}$ will be identically zero, in particular $a'_{n,n-1} = 0$, so that, by Theorem 6.7.3, $\mathbf{A}'$ cannot be O.

The proof of Theorem 7.4.1 requires delicate treatment of inequalities. It may be found in Gladwell (1998) [126] and will not be reproduced here. We merely give some hints on the proof. First, it relies on an earlier result of Cryer (1973) [66] for the case $\mu = 0$. See also Cryer (1976) [67] Cryer's results may be used to show that if $\mathbf{A}$ (not necessarily symmetric) is NTN, O or TP, and $\mathbf{A} = \mathbf{LU}$ where $\mathbf{L}(\mathbf{U})$ is lower (upper) triangular, then $\mathbf{A}' = \mathbf{UL}$ is NTN, O or TP respectively. Since $\mathbf{A}$ is PD we may replace $\mathbf{QR}$ factorisation for the case $\mu = 0$ by two successive Cholesky $\mathbf{LL}^T$ factorisations:

$$\mathbf{A} = \mathbf{L}_1\mathbf{L}_1^T, \ \mathbf{B} = \mathbf{L}_1^T\mathbf{L}_1 = \mathbf{L}_2\mathbf{L}_2^T, \ \mathbf{A}' = \mathbf{L}_2^T\mathbf{L}_2.$$

We write

$$\mathbf{Q} = \mathbf{L}_1\mathbf{L}_2^{-T} = \mathbf{L}_1^{-T}\mathbf{L}_2, \ \mathbf{R} = \mathbf{L}_2^T\mathbf{L}_1^T,$$

and note that

$$\mathbf{Q}\mathbf{Q}^T = \mathbf{L}_1\mathbf{L}_2^{-T}(\mathbf{L}_2^T\mathbf{L}_1^{-1}) = \mathbf{I},$$

so that $\mathbf{Q}$ is orthogonal. Now

$$\begin{aligned} \mathbf{A} &= \mathbf{L}_1\mathbf{L}_1^T = (\mathbf{L}_1\mathbf{L}_2^{-T})(\mathbf{L}_2^T\mathbf{L}_1^T) = \mathbf{QR}, \\ \mathbf{A}' &= \mathbf{L}_2^T\mathbf{L}_2 = (\mathbf{L}_2^T\mathbf{L}_1^T)(\mathbf{L}_1^{-T}\mathbf{L}_2) = \mathbf{RQ}. \end{aligned}$$

If $\mathbf{A}$ has property P, then Cryer's result shows that $\mathbf{B}$ has property P, and then again $\mathbf{A}'$ has property P.

The proof also relies on the Binet-Cauchy Theorem. Equation (7.3.20) states that

$$\mathbf{RA} = \mathbf{A}'\mathbf{R}, \tag{7.4.5}$$

so that the Binet-Cauchy Theorem 6.2.4 gives

$$\mathcal{R}_p\mathcal{A}_p = \mathcal{A}'_p\mathcal{R}_p. \tag{7.4.6}$$

We now prove

Lemma 7.4.1

$$\mathcal{R}_p(\mathcal{A}^m)_p = (\mathcal{A}'^m)_p\mathcal{R}_p, \ m = 1, 2, \ldots \tag{7.4.7}$$

Proof. The Binet-Cauchy Theorem gives

$$(\mathcal{A}^m)_p = (\mathcal{A}_p)^m = \mathcal{A}_p^m$$

and similarly $(\mathcal{A}'^m)_p = \mathcal{A}_p'^m$. By equation (7.4.6), the result holds for $m = 1$. Suppose it holds for one value, m, then

$$\begin{aligned} \mathcal{A}_p'^{(m+1)}\mathcal{R}_p &= \mathcal{A}'_p(\mathcal{A}_p'^m\mathcal{R}_p) \\ &= \mathcal{A}'_p(\mathcal{R}_p^m\mathcal{A}_p) = (\mathcal{A}'_p\mathcal{R}_p)\mathcal{A}_p^m \\ &= (\mathcal{R}_p\mathcal{A}_p)\mathcal{A}_p^m = \mathcal{R}_p\mathcal{A}_p^{m+1}; \end{aligned}$$

the result holds for $m + 1$. ■

Equations (7.4.5)-(7.4.7) generally yield complicated relations between the elements of $\mathbf{A}$ and $\mathbf{A}'$, $\mathcal{A}_p$ and $\mathcal{A}'_p$, but for some important special cases the relations are simple. Consider equation (7.4.5) in element form:

$$\sum_{k=1}^{n} r_{ik}a_{kj} = \sum_{k=1}^{j} a'_{ik}r_{kj}. \tag{7.4.8}$$

If $i = n$ and $j = 1$, there is only one term in each sum:

$$r_{nn}a_{n1} = a'_{n1}r_{11}. \tag{7.4.9}$$

The hypothesis of Theorem 7.4.1 is that $\mathbf{A}$ is NTN (at least). Ex. 6.6.1 states that if $\mathbf{A}$ is NTN and $a_{n1} > 0$, then $\mathbf{A}$ is a positive matrix (strictly positive, but not TP!). In fact, $a_{n1} > 0$ is the first of the conditions in Theorem 6.8.2 for a (symmetric) NTN matrix to be TP: a_{n1} is the first of the corner minors of $\mathbf{A}$, as discussed in Theorem 6.8.2. The general corner minor is $\mathbf{A}(\phi;\theta)$ where $\theta = \{1, 2, \ldots, p\}$, $\phi = \{n-p+1, \ldots, n\}$. This is the corner element $N, 1$ in the matrix $\mathcal{A}_p$. Thus equation (7.4.6) gives

$$r_{n-p+1}\ldots r_{nn}A(\phi;\theta) = A'(\phi;\theta)r_{11}\ldots r_{pp} \tag{7.4.10}$$

so that $A'(\phi;\theta) > 0$ iff $A(\phi;\theta) > 0$. This result, combined with some delicate reasoning, shows that $\mathbf{A}'$ is TP iff $\mathbf{A}$ is TP.

To show that $\mathbf{A}'$ is TN iff $\mathbf{A}$ is TN, we use a result due to Ando (1987) [4], that a TN matrix may be approximated arbitrarily closely, in, say, the L_1 norm, by a TP matrix. Finally, to show that $\mathbf{A}'$ is O iff $\mathbf{A}$ is O we use Lemma 7.4.1. That shows that the corner minors of $\mathbf{A}'^m$ are positive iff the corner minors of $\mathbf{A}^m$ are positive. So if $\mathbf{A}$ is O, it is NTN, and therefore, $\mathbf{A}'$ is NTN. Again, if $\mathbf{A}$ is O, $\mathbf{A}^m$ is TP for some $m \leq n-1$, its corner minors are positive, so therefore are those of $\mathbf{A}'^m$; $\mathbf{A}'^m$ is TP; $\mathbf{A}'$ is O.

We conclude from Theorem 7.4.1 that the operator $\mathcal{G}_\mu$ maintains the properties NTN, O, TP (and SO also) invariant, provided of course that $\mathbf{A}$ is symmetric, μ is not an eigenvalue of $\mathbf{A}$, and $\mathbf{R}$ has positive diagonal.

In Section 6.6 we showed (Theorem 6.6.3) that an NTN matrix is a staircase matrix. We now prove

Theorem 7.4.2 *Suppose $\mathbf{A} \in S_n$ is NTN and is a p-staircase matrix, then $\mathbf{A}' = \mathcal{G}_\mu\mathbf{A}$ is also a p-staircase matrix.*

Proof. Since $\mathbf{A}'$ is NTN, it is a staircase matrix, say a p'-staircase. The fundamental relation (7.3.21) gives

$$\sum_{k=1}^{p_j} r_{ik}a_{kj} = \sum_{k=1}^{j} a'_{ik}r_{kj}. \tag{7.4.11}$$

We use induction to prove $p'_j = p_j, j = 1, 2, \ldots, n$. Take $j = 1$. If $i > p_1$, the L.H.S. is zero, so that $a'_{i1}r_{11} = 0$; $p'_1 \leq p_1$. If $i = p_1$ then

$$r_{ii}a_{i1} = a'_{i1}r_{11},$$

so that $a'_{i1} > 0$; $p'_1 = p_1$. Suppose that $p'_j = p_j$ for $j = 1, 2, \ldots, m-1$. If $j = m$ and $i > p_m$ in (7.4.11) then

$$\sum_{k=1}^{m} a'_{ik} r_{kj} = 0. \tag{7.4.12}$$

But since $\mathbf{A}'$ is a staircase, $i > p_m$ implies $i > p_k = p'_k$ for $k = 1, 2, \ldots, m-1$, so that there is only one term, the last, in the sum (7.4.12); $a'_{im} = 0$. Thus $p'_m \leq p_m$. Now take $j = m$, $i = p_m$, then

$$r_{ii} a_{im} = \sum_{k=1}^{m} a'_{ik} r_{km}.$$

If $p_m > p_{m-1}$, then there is only one term, the last, on the right, and

$$r_{ii} a_{im} = a'_{im} r_{mm}$$

so that $p'_m = p_m$. If $p_m = p_{m-1}$, then the inequalities $p'_m \geq p'_{m-1}$, $p'_m \leq p_m$ imply $p'_m = p_m$. Arbenz and Golub (1995) [12] show that staircase patterns are effectively the only ones invariant under the symmetric QR algorithm. ■

In Theorem 7.3.2 we showed that the operator $\mathcal{G}_\mu$ applied to a Jacobi matrix was commutative. We now show a stronger result.

Theorem 7.4.3 *The operator $\mathcal{G}_\mu$ is commutative.*

Proof. We need to show that $\mathcal{G}_{\mu_1}\mathcal{G}_{\mu_2} = \mathcal{G}_{\mu_2}\mathcal{G}_{\mu_1}$. Consider the operations $\mathcal{G}_{\mu_1}\mathbf{A} = \mathbf{A}_1$, $\mathcal{G}_{\mu_2}\mathbf{A}_1 = \mathbf{A}_2$; $\mathcal{G}_{\mu_2}\mathbf{A} = \mathbf{A}_3$, $\mathcal{G}_{\mu_1}\mathbf{A}_3 = \mathbf{A}_4$:

$$\mathbf{A} - \mu_1\mathbf{I} = \mathbf{Q}_1\mathbf{R}_1, \quad \mathbf{A}_1 - \mu_1\mathbf{I} = \mathbf{R}_1\mathbf{Q}_1,$$

$$\mathbf{A}_1 - \mu_2\mathbf{I} = \mathbf{Q}_2\mathbf{R}_2, \quad \mathbf{A}_2 - \mu_2\mathbf{I} = \mathbf{R}_2\mathbf{Q}_2;$$

$$\mathbf{A} - \mu_2\mathbf{I} = \mathbf{Q}_3\mathbf{R}_3, \quad \mathbf{A}_3 - \mu_2\mathbf{I} = \mathbf{R}_3\mathbf{Q}_3,$$

$$\mathbf{A}_3 - \mu_1\mathbf{I} = \mathbf{Q}_4\mathbf{R}_4, \quad \mathbf{A}_4 - \mu_1\mathbf{I} = \mathbf{R}_4\mathbf{Q}_4.$$

These equations give

$$\mathbf{A}_1 - \mu_2\mathbf{I} = \mathbf{Q}_1^T(\mathbf{A} - \mu_2\mathbf{I})\mathbf{Q}_1 = \mathbf{Q}_2\mathbf{R}_2,$$

i.e.,

$$\mathbf{Q}_1^T\mathbf{Q}_3\mathbf{R}_3\mathbf{Q}_1 = \mathbf{Q}_2\mathbf{R}_2, \tag{7.4.13}$$

$$\mathbf{A}_3 - \mu_1\mathbf{I} = \mathbf{Q}_3^T(\mathbf{A} - \mu_1\mathbf{I})\mathbf{Q}_3 = \mathbf{Q}_4\mathbf{R}_4,$$

i.e.,

$$\mathbf{Q}_3^T\mathbf{Q}_1\mathbf{R}_1\mathbf{Q}_3 = \mathbf{Q}_4\mathbf{R}_4. \tag{7.4.14}$$

Equations (7.4.13), (7.4.14) give

$$\begin{aligned}\mathbf{Q}_3\mathbf{R}_3 &= \mathbf{Q}_1\mathbf{Q}_2\mathbf{R}_2\mathbf{Q}_1^T,\\ \mathbf{Q}_1\mathbf{R}_1 &= \mathbf{Q}_3\mathbf{Q}_4\mathbf{R}_4\mathbf{Q}_3^T,\end{aligned}$$

and on multiplying these together, we find

$$(\mathbf{Q}_1\mathbf{Q}_2\mathbf{R}_2\mathbf{Q}_1^T)(\mathbf{Q}_1\mathbf{R}_1) = (\mathbf{Q}_3\mathbf{Q}_4\mathbf{R}_4\mathbf{Q}_3^T)(\mathbf{Q}_3\mathbf{R}_3),$$

or

$$\mathbf{Q}_1\mathbf{Q}_2\mathbf{R}_2\mathbf{R}_1 = \mathbf{Q}_3\mathbf{Q}_4\mathbf{R}_4\mathbf{R}_3.$$

Now $\mathbf{Q}_1\mathbf{Q}_2$, $\mathbf{Q}_3\mathbf{Q}_4$ are orthogonal matrices while $\mathbf{R}_2\mathbf{R}_1$ and $\mathbf{R}_4\mathbf{R}_3$ are upper triangular with positive diagonal. But a non-singular matrix has a unique factorisation $\mathbf{QR}$ (with positive diagonal). Therefore,

$$\mathbf{Q}_1\mathbf{Q}_2 = \mathbf{Q}_3\mathbf{Q}_4, \quad \mathbf{R}_2\mathbf{R}_1 = \mathbf{R}_4\mathbf{R}_3,$$

so that, since

$$\begin{aligned}\mathbf{A}_4 &= \mathbf{Q}_4^T\mathbf{A}_3\mathbf{Q}_4 = \mathbf{Q}_4^T\mathbf{Q}_3^T\mathbf{A}\mathbf{Q}_4\mathbf{Q}_3\\ \mathbf{A}_2 &= \mathbf{Q}_2^T\mathbf{A}_1\mathbf{Q}_2 = \mathbf{Q}_2^T\mathbf{Q}_1^T\mathbf{A}\mathbf{Q}_1\mathbf{Q}_2\end{aligned}$$

we have $\mathbf{A}_4 = \mathbf{A}_2$. ■

7.5 Isospectral beams

We set up the eigenvalue problem for the (cantilever) beam in Section 2.3:

$$\mathbf{Ky} = \lambda\mathbf{My}$$

where

$$\mathbf{K} = \mathbf{EL}^{-1}\mathbf{E}\hat{\mathbf{K}}\mathbf{E}^T\mathbf{L}^{-1}\mathbf{E}^T, \tag{7.5.1}$$

$$\mathbf{M} = \mathbf{D}^2,\ \mathbf{D} = diag(d_1, d_2, \ldots, d_n). \tag{7.5.2}$$

As usual, we reduce the problem to standard form:

$$\mathbf{Au} = \lambda\mathbf{u},$$

where

$$\mathbf{A} = \mathbf{D}^{-1}\mathbf{KD}^{-1}. \tag{7.5.3}$$

First, we obtain a simple isospectral system by using Lemma 7.3.1. Write

$$\hat{\mathbf{K}} = \mathbf{F}^2,\ \mathbf{F} = diag(f_1, f_2, \ldots, f_n)$$

then we may write $\mathbf{A}$ as

$$\mathbf{A} = (\mathbf{D}^{-1}\mathbf{EL}^{-1}\mathbf{EF})\cdot(\mathbf{FE}^T\mathbf{L}^{-1}\mathbf{E}^T\mathbf{D}^{-1}).$$

Now apply Lemma 7.3.1; the eigenvalues of $\mathbf{A}$ are non-zero (in fact, positive) so that if

$$\mathbf{A}' = (\mathbf{F}\mathbf{E}^T\mathbf{L}^{-1}\mathbf{E}^T\mathbf{D}^{-1}) \cdot (\mathbf{D}^{-1}\mathbf{E}\mathbf{L}^{-1}\mathbf{E}\mathbf{F})$$

then

$$\sigma(\mathbf{A}') = \sigma(\mathbf{A}).$$

To form a discrete beam corresponding to $\mathbf{A}'$ we reverse the reduction to standard form, and write

$$\mathbf{A}'\mathbf{u}' = \lambda\mathbf{u}'$$

as

$$\mathbf{K}'\mathbf{y}' = \lambda\mathbf{M}'\mathbf{y}' \tag{7.5.4}$$

where

$$\mathbf{K}' = \mathbf{E}^T\mathbf{L}^{-1}\mathbf{E}^T\hat{\mathbf{K}}'\mathbf{E}\mathbf{L}^{-1}\mathbf{E} \tag{7.5.5}$$

$$\hat{\mathbf{K}}' = \mathbf{M}^{-1}, \ \mathbf{M}' = \hat{\mathbf{K}}^{-1}. \tag{7.5.6}$$

This is the eigenvalue equation for a reversed cantilever, as we may verify just as we did for the spring-mass system in Section 7.3: we operate on (7.5.4) by the reversing matrix $\mathbf{T}$. Thus,

$$\mathbf{T}\mathbf{K}'\mathbf{T} \cdot \mathbf{T}\mathbf{y}' = \lambda\mathbf{T}\mathbf{M}'\mathbf{T} \cdot \mathbf{T}\mathbf{y}',$$

where

$$\begin{aligned} \mathbf{T}\mathbf{K}'\mathbf{T} &= \mathbf{T}\mathbf{E}^T\mathbf{T} \cdot \mathbf{T}\mathbf{L}^{-1}\mathbf{T} \cdot \mathbf{T}\mathbf{E}^T\mathbf{T} \cdot \mathbf{T}\hat{\mathbf{K}}'\mathbf{T} \cdot \mathbf{T}\mathbf{E}\mathbf{T} \cdot \mathbf{T}\mathbf{L}^{-1}\mathbf{T} \cdot \mathbf{T}\mathbf{E}\mathbf{T} \\ &= \mathbf{E}\mathbf{L}^{-0}\mathbf{E}\hat{\mathbf{K}}^0\mathbf{E}^T\mathbf{L}^{-0}\mathbf{E}^T = \mathbf{K}^0. \end{aligned} \tag{7.5.7}$$

The new cantilever is related to the old by

$$k_i^0 = m_{n-i+1}^{-1}, \ l_i^0 = l_{n-i+1}, \ m_i^0 = k_{n-i+1}^{-1}. \tag{7.5.8}$$

To construct a family of isospectral beams, we use the operator $\mathcal{G}_\mu$ defined by equations (7.4.1), (7.4.2). We carry out the following steps:

i) start with a beam, defined by $\hat{\mathbf{K}}, \mathbf{L}, \mathbf{M} = \mathbf{D}^2$.

ii) construct $\mathbf{A}$ as in (7.5.1)-(7.5.3). $\mathbf{A}$ is symmetric, pentadiagonal, and sign-oscillatory.

iii) choose μ, not an eigenvalue of $\mathbf{A}$, and form $\mathbf{A}' = \mathcal{G}_\mu\mathbf{A}$; $\mathbf{A}'$ also is symmetric, pentadiagonal and sign-oscillatory.

iv) factorise $\mathbf{A}' = (\mathbf{D}')^{-1}\mathbf{K}'(\mathbf{D}')^{-1}$ and form $\mathbf{M}' = (\mathbf{D}')^2$, $\mathbf{K}' = \mathbf{E}(\mathbf{L}')^{-1}\mathbf{E}\hat{\mathbf{K}}'\mathbf{E}^T(\mathbf{L}')^{-1}\mathbf{E}^T$.

The only step which needs to be completed is iv). We must show that the new symmetric pentadiagonal sign-oscillatory matrix $\mathbf{A}'$ may be factorised as in (7.5.1)-(7.5.3), with some new positive diagonal matrices $\mathbf{D}', \hat{\mathbf{K}}', \mathbf{L}'$. We first give the gist of the procedure, and afterwards show that it will always work.

The new matrix $\mathbf{A}'$ is related to the new mass and stiffness matrices $\mathbf{K}', \mathbf{M}' = \mathbf{D}'^2$ by equation (7.5.3). We start, as we did with the spring mass system in Section 7.3, by considering simple static deflection of the beam, as shown in Figure 7.5.1. We apply forces $f_1, -f_2$ at masses 1 and 2 so that all the masses have unit deflection. The force-deflection equation is

$$\mathbf{K}'\{1, 1, \ldots, 1\} = \{f_1, -f_2, 0 \ldots, 0\}.$$

But $\mathbf{A}' = \mathbf{D}'^{-1}\mathbf{K}'\mathbf{D}'^{-1}$, so that $\mathbf{K}' = \mathbf{D}'\mathbf{A}'\mathbf{D}'$, and thus

$$\mathbf{D}'\mathbf{A}'\{d'_1, d'_2, \ldots, d'_n\} = \{f_1, -f_2, 0 \ldots, 0\}$$

and

$$\mathbf{A}'\{d'_1, d'_2, \ldots, d'_n\} = \{g_1, -g_2, 0 \ldots, 0\} \tag{7.5.9}$$

where $g_i = f_i/d'_i,\ i = 1, 2$.

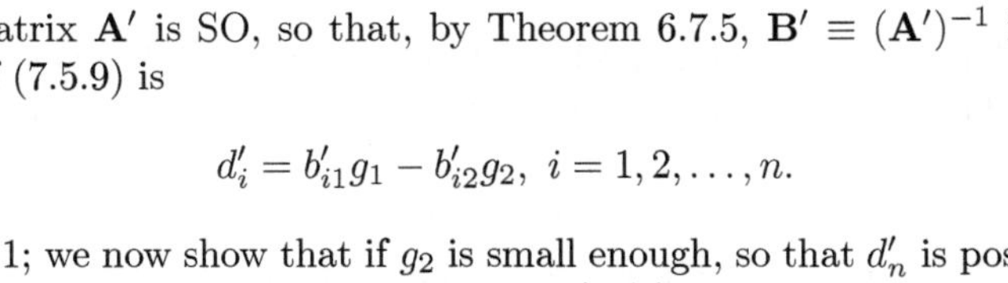

Figure 7.5.1 - Two forces $f_1, -f_2$, are required to produce unit deflections.

The matrix $\mathbf{A}'$ is SO, so that, by Theorem 6.7.5, $\mathbf{B}' \equiv (\mathbf{A}')^{-1}$ is O. The solution of (7.5.9) is

$$d'_i = b'_{i1}g_1 - b'_{i2}g_2,\ i = 1, 2, \ldots, n. \tag{7.5.10}$$

Take $g_1 = 1$; we now show that if g_2 is small enough, so that d'_n is positive, then all the d'_i will be positive. For if $0 < g_2 < b'_{n1}/b'_{n2}$, then

$$\begin{aligned} d'_i &> b'_{i1} - b'_{i2}b'_{n1}/b'_{n2} \\ &= (b'_{i1}b'_{n2} - b'_{i2}b'_{n1})/b'_{n2} \geq 0, \end{aligned}$$

because $\mathbf{B}'$ is O. We will show later that b'_{i1}, b'_{i2} are strictly positive for $i = 1, 2, \ldots, n$, so that the d'_i are strictly positive. Assuming that this is true for the moment, we have now found $\mathbf{d}'$ satisfying (7.5.9) for some $g_1 = 1, g_2 > 0$. The vector $\mathbf{d}'$ is the first column of the matrix $\mathbf{D}'\mathbf{E}^{-T}$; $\mathbf{E}^{-1}$is given in equation (2.2.10).

We now show that the matrix

$$\mathbf{C}' = \mathbf{E}^{-1}\mathbf{D}'\mathbf{A}'\mathbf{D}'\mathbf{E}^{-T} \tag{7.5.11}$$

is a Jacobi matrix. Suppose

$$\mathbf{A}' = \begin{bmatrix} a'_1 & -b'_1 & c'_1 & & & \\ -b'_1 & a'_2 & -b'_2 & c'_2 & & \\ c'_1 & -b'_2 & a'_3 & -b'_3 & c'_3 & \\ & \ddots & \ddots & \ddots & & \\ & & c'_{n-2} & -b'_{n-1} & a'_n & \end{bmatrix}$$

then $\mathbf{A}'\mathbf{D}'\mathbf{E}^{-T}$ has just one diagonal below the principal diagonal; that diagonal has elements $-g_2, -c'_1d'_1, -c'_2d'_2, \ldots, -c'_{n-2}d'_{n-2}$. The matrix $\mathbf{E}^{-1}\mathbf{D}'$ is upper triangular, so that $\mathbf{C}' \equiv \mathbf{E}^{-1}\mathbf{D}'(\mathbf{A}'\mathbf{D}'\mathbf{E}^{-T})$ also will have just one diagonal below the principal diagonal. But $\mathbf{C}'$ is symmetric, so that it will also have just one diagonal above the principal diagonal: it is a symmetric tridiagonal matrix with co-diagonal

$$-d'_2g_2, -c'_1d'_1d'_3, -c'_2d'_2d'_4, \ldots, -c'_{n-2}d'_{n-2}d'_n. \tag{7.5.12}$$

Denote the matrix obtained by deleting rows and columns $1, 2, \ldots, i-1$ of $\mathbf{A}'$ by $\mathbf{A}'_i$ and let $\mathbf{d}'_i = \{0, 0, \ldots, 0, d'_i, d'_{i+1}, \ldots, d'_n\}$, then the diagonal elements of $\mathbf{C}'$ may be written

$$c'_{ii} = \mathbf{d}'^T_i\mathbf{A}'_i\mathbf{d}'_i, \; i = 1, 2, \ldots, n. \tag{7.5.13}$$

To show that $\mathbf{C}'$ is a Jacobi matrix, we need to show that it is PSD. Actually, since the original $\mathbf{A}$ was PD, the new $\mathbf{A}'$ is PD, and so is $\mathbf{C}'$, because

$$\begin{aligned} \mathbf{x}^T\mathbf{C}'\mathbf{x} &= (\mathbf{x}^T\mathbf{E}^{-1}\mathbf{D}')\mathbf{A}'(\mathbf{D}'\mathbf{E}^{-T}\mathbf{x}) \\ &= \mathbf{y}^T\mathbf{A}'\mathbf{y} > 0. \end{aligned}$$

We have constructed a Jacobi matrix $\mathbf{C}'$ from $\mathbf{A}'$. We now use the result obtained in (4.4.7) for the factorisation of a Jacobi matrix. In changed notation we may write

$$\mathbf{C}' = (\mathbf{L}')^{-1}\mathbf{E}\hat{\mathbf{K}}'\mathbf{E}^T(\mathbf{L}')^{-1}, \tag{7.5.14}$$

so that on combining (7.5.11) and (7.5.14) we find

$$\mathbf{A}' = (\mathbf{D}')^{-1}\mathbf{E}(\mathbf{L}')^{-1}\mathbf{E}\hat{\mathbf{K}}'\mathbf{E}^T(\mathbf{L}')^{-1}\mathbf{E}^T(\mathbf{D}')^{-1}, \tag{7.5.15}$$

as required.

We now examine this procedure. We must show that the terms b'_{i1}, b'_{i2} are strictly positive, and that the terms c'_i in the last band of $\mathbf{A}'$, which appear in the codiagonal of $\mathbf{C}'$ are positive. To verify these matters we must return to the $\mathcal{G}_\mu$ algorithm, specifically to equations (7.4.5)-(7.4.9). The terms b_{n1}, b'_{n1} are elements of $\mathbf{B} \equiv \mathbf{A}^{-1}$ and $\mathbf{B}' \equiv (\mathbf{A}')^{-1}$ respectively.

Taking inverses of the terms on each side of equation (7.4.5) we find

$$\mathbf{R}^{-1}\mathbf{B}' = \mathbf{B}\mathbf{R}^{-1} \tag{7.5.16}$$

and on equating the $n, 1$ terms we find

$$r_{nn}^{-1} b'_{n1} = b_{n1} r_{11}^{-1}. \tag{7.5.17}$$

The original $\mathbf{A}$ is given by equations (7.5.1)-(7.5.3), so that

$$\mathbf{B} = \mathbf{A}^{-1} = \mathbf{D}\mathbf{E}^{-T}\mathbf{L}\mathbf{E}^{-T}\hat{\mathbf{K}}^{-1}\mathbf{E}^{-1}\mathbf{L}\mathbf{E}^{-1}$$

so that, with $\mathbf{E}^{-1}$ given by equation (2.2.10), it is clear that $b_{n1} > 0$ and thus equation (7.5.17) gives $b'_{n1} > 0$.

We now show that $b'_{n2} > 0$. The matrix $\mathbf{B}'$ is known to be oscillatory; it is thus TN so that the minor $\mathbf{B}'(1, n; 1, 2) \geq 0$; thus

$$\begin{vmatrix} b'_{11} & b'_{12} \\ b'_{n1} & b'_{n2} \end{vmatrix} = b'_{11}b'_{n2} - b'_{n1}b'_{12} \geq 0, \tag{7.5.18}$$

and $b'_{n1} > 0$, $b'_{12} > 0$, $b'_{11} > 0$, imply $b'_{n2} > 0$. We apply a similar argument to show that $b'_{i1} > 0$, $b'_{i2} > 0$:

$$\begin{vmatrix} b'_{i1} & b'_{ii} \\ b'_{n1} & b'_{ni} \end{vmatrix} \geq 0,\ i \geq 2; \quad \begin{vmatrix} b'_{i2} & b'_{ii} \\ b'_{n2} & b'_{ni} \end{vmatrix} \geq 0,\ i \geq 3 \tag{7.5.19}$$

imply $b'_{i1} > 0$, $b'_{i2} > 0$ respectively. We have proved that the procedure will always yield a vector $\mathbf{d}'$ which is strictly positive. Further discussion and results may be found in Gladwell (2002b) [130].

Exercises 7.5

1. Show that there is a 2-parameter system of isospectral beams corresponding to simple scaling, i.e., in which all the masses are scaled by the same factor, the stiffnesses by another, and the lengths by a third one.

2. The argument used in (7.5.17), (7.5.18) is due to Markham (1970) [221]. Show that if $\mathbf{B}$ is O, and an element b_{ij} with $i > j$, i.e., an element in the lower triangle, is zero,then all the elements below and to the left of b_{ij} are also zero. This implies that if $\mathbf{B}$ is O, then it has *staircase structure*, as discussed at the end of Section 7.4.

   Also, if $b_{n1} > 0$ and $b_{1n} > 0$, then $\mathbf{B}$ is a strictly positive matrix.

7.6 Isospectral finite-element models

In Section 2.4 we showed that a finite-element model of a rod in longitudinal vibration had tridiagonal mass and stiffness matrices, the former with positive codiagonal, the latter with negative. The explicit form of the stiffness matrix was given in Ex. 2.4.2. In this section, following Gladwell (1998) [126], Gladwell (1999) [127], we consider how we can find a finite-element system $\mathbf{M}', \mathbf{K}'$ for a rod which is isospectral to a given finite-element system $\mathbf{M}, \mathbf{K}$ for a rod. We first consider a simple way of constructing an isospectral family $\mathbf{M}', \mathbf{K}'$, and then consider a procedure that will yield a large family. See Gladwell (1997) [125] for an earlier attempt to solve this problem.

For simplicity we consider a cantilever rod, i.e., one that is fixed at the left, free at the right. The eigenvalue equation is

$$(\mathbf{K} - \lambda\mathbf{M})\mathbf{y} = \mathbf{0}. \tag{7.6.1}$$

Instead of working with $\mathbf{K}$ and $\mathbf{M}$, we will work with $\tilde{\mathbf{K}} = \mathbf{ZKZ}$ and $\mathbf{M}$; both these are tridiagonal with positive codiagonal, i.e., they are oscillatory (O). We factorise them as

$$\tilde{\mathbf{K}} = \mathbf{A}\mathbf{A}^T, \quad \mathbf{M} = \mathbf{B}\mathbf{B}^T, \tag{7.6.2}$$

where relying on Cryer (1973) [66], we know that $\mathbf{A}, \mathbf{B}$ are lower bidiagonal with positive codiagonals. When reduced to normal form, the equation (7.6.1) is

$$(\tilde{\mathbf{G}} - \lambda\mathbf{I})\mathbf{u} = 0, \tag{7.6.3}$$

where $\tilde{\mathbf{G}} = \mathbf{B}^{-1}\mathbf{K}\mathbf{B}^{-T}$, i.e., $\mathbf{G} = \tilde{\mathbf{B}}^{-1}\tilde{\mathbf{K}}\tilde{\mathbf{B}}^{-T}$ is O:

$$\mathbf{G} = \tilde{\mathbf{B}}^{-1}\mathbf{A}\mathbf{A}^T\tilde{\mathbf{B}}^{-T}. \tag{7.6.4}$$

Thus one way to obtain an isospectral system $\mathbf{M}', \mathbf{K}'$ is to find lower bidiagonal $\mathbf{C}, \mathbf{D}$ with positive codiagonals such that

$$\tilde{\mathbf{K}}' = \mathbf{C}\mathbf{C}^T, \quad \mathbf{M}' = \mathbf{D}\mathbf{D}^T, \tag{7.6.5}$$

and

$$\mathbf{G} = \tilde{\mathbf{B}}^{-1}\mathbf{A}\mathbf{A}^T\tilde{\mathbf{B}}^{-T} = \tilde{\mathbf{D}}^{-1}\mathbf{C}\mathbf{C}^T\tilde{\mathbf{D}}^{-T}. \tag{7.6.6}$$

This holds iff

$$\tilde{\mathbf{B}}^{-1}\mathbf{A} = \tilde{\mathbf{D}}^{-1}\mathbf{C}. \tag{7.6.7}$$

Straightforward algebra shows that this implies

$$c_{ii} = v_i a_{ii}, \; d_{ii} = v_i b_{ii}, \; i = 1, 2, \ldots, n \tag{7.6.8}$$

$$c_{i+1,i} = v_{i+1}a_{i+1,i}, \quad d_{i+1,i} = v_{i+1}b_{i+1,i}, \quad i = 2, 3, \ldots, n-1 \tag{7.6.9}$$

where $(v_i)_1^n$ are arbitrary positive constraints, and

$$a_{11}d_{21} + b_{11}c_{21} = v_2(a_{11}b_{21} + a_{21}b_{11}) = v_2 p. \tag{7.6.10}$$

The general, positive, solution of (7.6.10) is

$$c_{21} = v_2 p \sin^2\theta / b_{11}, \; d_{21} = v_2 p \cos^2\theta / a_{11}, \tag{7.6.11}$$

where $0 < \theta < \pi/2$. This provides an $(n+1)$-parameter family of matrices $\mathbf{M}', \mathbf{K}'$ specified by the $(n+1)$ parameters $(v_i)_1^n$ and θ.

Unless the parameters v_i are chosen properly, the new matrix $\mathbf{K}' = \tilde{\mathbf{C}}\tilde{\mathbf{C}}^T$ will not have the form of a stiffness matrix of a cantilever finite element model of a rod. Such a matrix, with $\mathbf{K}'$ given in Ex. 2.4.2 has the defining property

$$\mathbf{K}'\{1,1,1,\ldots,1\} = \{k_1',0,0,\ldots,0\}. \tag{7.6.12}$$

Equation (7.6.8)-(7.6.11) show that $\mathbf{C}$ has the form

$$\mathbf{C} = \mathbf{N}\mathbf{C}_0,\ \mathbf{N} = diag(v_1, v_2, \ldots, v_n)$$

and $\mathbf{C}_0$ depends only on θ. Thus

$$\mathbf{K}' = \mathbf{N}\tilde{\mathbf{C}}_0\tilde{\mathbf{C}}_0^T\mathbf{N}$$

so that equation (7.6.12) yields

$$\mathbf{N}\tilde{\mathbf{C}}_0\tilde{\mathbf{C}}_0^T\mathbf{N}\{1,1,\ldots,1\} = \{k_1',0,\ldots,0\},$$

i.e.,

$$\tilde{\mathbf{C}}_0\tilde{\mathbf{C}}_0^T\{v_1, v_2, \ldots, v_n\} = \{k_1'/v_1, 0, \ldots, 0\}. \tag{7.6.13}$$

Since $\tilde{\mathbf{C}}_0\tilde{\mathbf{C}}_0^T$ is a non-singular Jacobi matrix, i.e., it is SO, its inverse is positive. Thus, equation (7.6.13) yields positive $(v_i)_1^n$, apart from a single positive factor.

To obtain a wider family we use the general theory of Section 7.4: we form $\mathbf{G}'$ from

$$\mathbf{G} - \mu\mathbf{I} = \mathbf{QR},\ \mathbf{G}' - \mu\mathbf{I} = \mathbf{RQ}, \tag{7.6.14}$$

so that $\mathbf{G}'$ is O. We must show that if $\mathbf{G}$ can be factorised as in (7.6.4), then $\mathbf{G}'$ can be factorised in the form

$$\mathbf{G}' = \tilde{\mathbf{D}}^{-1}\mathbf{C}\mathbf{C}^T\tilde{\mathbf{D}}^{-T}, \tag{7.6.15}$$

where $\mathbf{C}, \mathbf{D}$ are lower bidiagonal with positive codiagonals.

To establish the band forms, we consider how $\mathbf{G}$ was constructed: $\tilde{\mathbf{G}} = \mathbf{B}^{-1}\mathbf{K}\mathbf{B}^{-T}$ or $\mathbf{K} = \mathbf{B}\tilde{\mathbf{G}}\mathbf{B}^T$. This we can write as $\mathbf{H} = \tilde{\mathbf{G}}\mathbf{B}^T$, $\mathbf{K} = \mathbf{BH}$. The equation $\mathbf{BH} = \mathbf{K}$ is

$$\sum_{k=1}^{n} b_{ik}h_{kj} = k_{ij}. \tag{7.6.16}$$

But $\mathbf{K}$ is tridiagonal, so that $k_{ij} = 0$ for $i = 1,2,\ldots,n-2;\ j = i+2,\ldots,n$. The matrix $\mathbf{B}$ is lower bidiagonal, so that (7.6.16) gives

$$b_{i,i-1}h_{i-1,j} + b_{i,i}h_{i,j} = 0,\ i = 1,2,\ldots,n-2;\ j = i+2,\ldots,n.$$

Thus, taking $i = 1$ we find

$$b_{11}h_{1j} = 0 \quad j = 3,4,\ldots,n$$

but taking $i = 2$ in (7.6.16) we have

$$b_{21}h_{1j} + b_{22}h_{2j} = 0 \quad j = 4,5,\ldots,n$$

so that

$$b_{22}h_{2j} = 0 \quad j = 4, 5, \ldots, n$$

and generally

$$h_{ij} = 0 \quad i = 1, 2, \ldots, n-2; \; j = i+2, \ldots, n. \tag{7.6.17}$$

Now consider $\mathbf{H} = \tilde{\mathbf{G}}\mathbf{B}^T$, which is equivalent to

$$h_{ij} = \sum_{k=1}^{n} \tilde{g}_{ik} b_{jk}$$

and when combined with (7.6.17), this gives

$$\tilde{g}_{i,j-1}b_{j,j-1} + \tilde{g}_{ij}b_{jj} = 0 \quad i = 1, 2, \ldots, n-2; \quad j = i+2, \ldots, n.$$

Since $\tilde{g}_{ij} = (-)^{i+j} g_{ij}$, and $\mathbf{G}$ is symmetric, we may write these equations as

$$b_{j,j-1} \begin{bmatrix} g_{j-1,1} \\ g_{j-1,2} \\ \vdots \\ g_{j-1,j-2} \end{bmatrix} = b_{jj} \begin{bmatrix} g_{j1} \\ g_{j2} \\ \vdots \\ g_{j,j-2} \end{bmatrix}, \quad j = 3, 4, \ldots, n. \tag{7.6.18}$$

We will show that these equations mean that the compound matrix $\mathcal{G}_2$ of 2×2 minors of $\mathbf{G}$ has a pattern of zeros like that shown in Figure 7.6.1. Starting from its left hand end, the first $n-3$ terms in the last row of $\mathcal{G}_2$ are

$$G(n-1, n; 1, 2), \; G(n-1, n; 1, 3) \ldots G(n-1, n; 1, n-2).$$

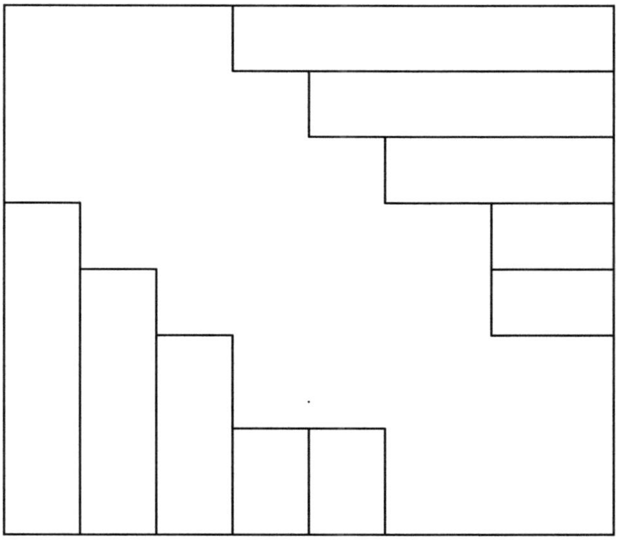

Figure 7.6.1 - The rectangles in the lower left and upper right are zeros.

These are all zero, for, by (7.6.18) with $j = n$,

$$G(n-1, n; 1, k) = \begin{vmatrix} g_{n-1,1} & g_{n-1,k} \\ g_{n,1} & g_{n,k} \end{vmatrix} = 0, \;\; k = 2, \ldots, n-2$$

has its two rows proportional. Now we investigate the first $n-4$ terms in the penultimate row of $\boldsymbol{\mathcal{G}}_2$:

$$G(n-2, n; 1, 2),\; G(n-2, n; 1, 3) \ldots G(n-2, n; 1, n-3).$$

To show that these are all zero we consider the zero determinant

$$\begin{vmatrix} g_{n-2,1} & g_{n-2,1} & g_{n-2,k} \\ g_{n-1,1} & g_{n-1,1} & g_{n-1,k} \\ g_{n1} & g_{n1} & g_{nk} \end{vmatrix} = 0$$

and expand it along its first column to give

$$g_{n-2,1}G(n-1, n; 1, k) - g_{n-1,1}G(n-2, n; 1, k) + g_{n1}G(n-2, n-1; 1, k) = 0. \tag{7.6.19}$$

However, $\mathbf{G}$ given by (7.6.4) is a full matrix with all positive terms so that if any two of the minors in (7.6.19) are zero, then so is the third. But if $k = 2, 3, \ldots, n-3$ then the first is zero, and (7.6.19) with $j = n-1$ shows that the third is zero, and thus the second is also.

Proceeding in this way we find that $G(i, j; 1, k) = 0$ for $3 \leq i < j$, $k = 2, \ldots, i-1$. This provides a non-increasing pattern of zeros for the columns of $\mathcal{G}_2$ in the lower triangle. Now the equation

$$\boldsymbol{\mathcal{R}}_2\boldsymbol{\mathcal{G}}_2 = \boldsymbol{\mathcal{G}}_2'\boldsymbol{\mathcal{R}}_2 \tag{7.6.20}$$

shows that $\mathcal{G}_2'$ has a precisely corresponding pattern, and by tracing the steps in the analysis we can conclude that $\mathbf{G}'$ can be factorised just like $\mathbf{G}$.

We obtain one factorisation

$$\mathbf{G}' = \tilde{\mathbf{D}}_0^{-1}\mathbf{C}_0\mathbf{C}_0^T\tilde{\mathbf{D}}^{-T}, \tag{7.6.21}$$

and then note that equivalently

$$\mathbf{G}' = \tilde{\mathbf{D}}^{-1}\mathbf{C}\mathbf{C}^T\tilde{\mathbf{D}}^{-T}$$

where

$$\mathbf{C} = \mathbf{N}\mathbf{C}_0 \quad \mathbf{D} = \mathbf{N}\mathbf{D}_0$$

and $\mathbf{N}$ is an arbitrary diagonal matrix. Now we choose $\mathbf{N}$, as before, to make $\mathbf{K}' = \tilde{\mathbf{C}}\tilde{\mathbf{C}}^T$ have the form of a stiffness matrix.

Exercises 7.6

1. Use equation (7.6.19) to verify that $\boldsymbol{\mathcal{G}}_2$ and $\boldsymbol{\mathcal{G}}_2'$ have precisely the same staircase patterns, and so show that $\mathbf{G}'$ may be factorised as (7.6.21).

7.7 Isospectral flow, continued

In Section 7.2 we obtained the isospectral flow equation

$$\dot{\mathbf{A}} = \mathbf{AS} - \mathbf{SA}, \tag{7.7.1}$$

which governs the isospectral evolution of a symmetric matrix $\mathbf{A}$; $\mathbf{S}$ is a skew symmetric matrix. In this section we investigate whether the pattern of zero and non-zero elements in $\mathbf{A}$, and the pattern of signs of elements of $\mathbf{A}$, are invariant in this flow. We will restrict our attention to a few types of matrices which appear in vibration problems since the general problem is extremely complicated. Ashlock, Driessel and Hentzel (1997) [13], in a very general discussion of Toda flow, show amongst many results, that staircase patterns are the only patterns that remain invariant under Toda flow. Their paper has a valuable summary of the pertinent literture.

We start with tridiagonal $\mathbf{A}$ and take $\mathbf{S} = \mathbf{A}^{+T} - \mathbf{A}^{+}$, i.e.,

$$\mathbf{A} = \begin{bmatrix} a_1 & -b_1 & & & \\ -b_1 & a_2 & -b_2 & & \\ & \ddots & \ddots & \ddots & \\ & & \ddots & \ddots & -b_{n-1} \\ & & & -b_{n-1} & a_n \end{bmatrix}, \tag{7.7.2}$$

$$\mathbf{S} = \begin{bmatrix} 0 & +b_1 & & & \\ -b_1 & 0 & +b_2 & & \\ & \ddots & \ddots & \ddots & \\ & & \ddots & \ddots & +b_{n-1} \\ & & & -b_{n-1} & 0 \end{bmatrix}.$$

Now $\mathbf{AS} - \mathbf{SA}$ is also tridiagonal, so that $\mathbf{A}$ retains its tridiagonal form, and

$$\dot{a}_i = 2b_{i-1}^2 - 2b_i^2, \quad \dot{b}_i = (a_{i+1} - a_i)b_i, \quad i = 1, 2, \ldots, n \tag{7.7.3}$$

where b_0, b_n are taken to be zero.

We examine the signs of the diagonal and codiagonal elements. The flow is isospectral so that if $\sigma(\mathbf{A}(0)) = (\lambda_i)_1^n$ and all the λ_i are positive, then $\mathbf{A}(t)$, like $\mathbf{A}(0)$ will be positive definite; $a_i > 0$, $i = 1, 2, \ldots, n$. For given i, $b_i(t)$ satisfies $\dot{b}_i(t) = f(t)b_i(t)$, where $f(t) = a_{i+1}(t) - a_i(t)$. This has the solution $b_i(t) = C\exp(F(t))$, where $F(t) = \int_0^t f(t)dt$. Now $f(t)$ is bounded for all t, so that $b_i(t)$ retains the sign of $C = b_i(0)$. Thus $b_i(t)$ is $>, <, = 0$, depending on whether $b_i(0)$ is $>, <, = 0$. We conclude that each codiagonal term retains the sign it had when $t = 0$. In particular, if the signs of the codiagonal terms are all positive, i.e., $\mathbf{A}(0)$ is O, or negative, i.e., $\mathbf{A}(0)$ is SO, then $\mathbf{A}(t)$ is correspondingly O or SO.

Before generalizing this analysis, we introduce some notation. The matrix $\mathbf{S}$ in (7.7.2) is clearly related to $\mathbf{A}$; it may be written as a so-called *Hadamard product*:

$$\begin{bmatrix} 0 & +b_1 & & & \\ -b_1 & 0 & +b_2 & & \\ & \ddots & \ddots & \ddots & \\ & & \ddots & \ddots & +b_{n-1} \\ & & & -b_{n-1} & 0 \end{bmatrix} = \begin{bmatrix} a_1 & -b_1 & & & \\ -b_1 & a_2 & -b_2 & & \\ & \ddots & \ddots & \ddots & \\ & & \ddots & \ddots & -b_{n-1} \\ & & & -b_{n-1} & a_n \end{bmatrix}$$

$$\circ \begin{bmatrix} 0 & -1 & & & \\ +1 & 0 & -1 & & \\ & \ddots & \ddots & \ddots & \\ & & \ddots & \ddots & -1 \\ & & & +1 & 0 \end{bmatrix} \tag{7.7.4}$$

The Hadamard product is quite distinct from the usual matrix product. It is defined only for two matrices $\mathbf{A}, \mathbf{B}$ of the same size, i.e., $\mathbf{A}, \mathbf{B} \in M_{m,n}$, and is given by the pairwise product of corresponding elements. If $\mathbf{C} = \mathbf{A} \circ \mathbf{B}$, then $c_{ij} = a_{ij}b_{ij}$, for $i = 1, 2, \ldots, m$; $j = 1, 2, \ldots, n$. Thus the matrix $\mathbf{S}$ in (7.7.4) may be written $\mathbf{S} = \mathbf{A} \circ \mathbf{Y}$, where

$$\mathbf{Y} = \begin{bmatrix} 0 & -1 & & & \\ +1 & 0 & -1 & & \\ & \ddots & \ddots & \ddots & \\ & & \ddots & \ddots & -1 \\ & & & +1 & 0 \end{bmatrix} \tag{7.7.5}$$

is itself a skew-symmetric matrix. (Clearly, if $\mathbf{A}$ is symmetric and $\mathbf{Y}$ is skew-symmetric, then $\mathbf{A} \circ \mathbf{Y}$ is skew-symmetric.)

This brings us to the next example, in which $\mathbf{A}$ is a periodic Jacobi matrix; now we take

$$\mathbf{A} = \begin{bmatrix} a_1 & -b_1 & & & -b_n \\ -b_1 & a_2 & -b_2 & & \\ & \ddots & \ddots & \ddots & \\ & & \ddots & \ddots & -b_{n-1} \\ & & & -b_{n-1} & a_n \end{bmatrix}, \quad \mathbf{Y} = \begin{bmatrix} 0 & -1 & & & +1 \\ +1 & 0 & -1 & & \\ & \ddots & \ddots & \ddots & \\ & & \ddots & \ddots & -1 \\ -1 & & & +1 & 0 \end{bmatrix}. \tag{7.7.6}$$

It is easy to verify (Ex. 7.7.2) that $\mathbf{A}$ retains its form under the flow (7.7.1) with $\mathbf{S} = \mathbf{A} \circ \mathbf{Y}$, and that all the a_i and b_i retain their signs.

We note in passing that for tridiagonal matrices we have two ways to form an isospectral family: using the operator $\mathcal{G}_\mu$ of Section 7.3, or by using the isospectral flow equation with $\mathbf{S}$ given by (7.7.2). The periodic Jacobi form is not invariant under $\mathcal{G}_\mu$, and it is not clear that there is a factorisation and reversal operation under which it is invariant. The only *algebraic* way to form an isospectral family seems to be to use the spectrum $(\lambda_i)_1^n$ and a second spectrum $(\mu_i)_1^{n-1}$ and reconstruct the matrix as in Section 5.4. For the periodic case, the isospectral flow equation with $\mathbf{S}$ given in (7.7.6), provides a conceptually simpler procedure.

There is a second comment. We showed in Section 7.3 that we can pass from any one Jacobi matrix $\mathbf{J}$ to *any other* isospectral Jacobi matrix $\mathbf{J}'$ in $n-1$ operations $\mathcal{G}_\mu$. It is doubtful that isospectral flow, with $\mathbf{S}$ given by (7.7.2), will lead from one $\mathbf{J}$ to any other isospectral $\mathbf{J}'$ (see Ex. 7.7.7).

We will now show following Gladwell (2002) [129] that this permanence of sign of a tridiagonal matrix under the Toda flow (7.7.1) is a special case of the permanence of the total positivity properties NTN, TP, O, SO under Toda flow. We recall from Section 6.8 that it is the positivity of the corner minors of $\mathbf{A}$ that is crucial in determining whether a TN matrix $\mathbf{A}$ is TP. We first prove a theorem regarding the flow of these corner minors under the Toda flow (7.7.1).

Theorem 7.7.1 *Suppose* $\mathbf{A} \in S_n$ *satisfies (7.7.1), with* $\mathbf{S} = \mathbf{A}^{+T} - \mathbf{A}^+$, $\mathbf{B} = \mathbf{A}^m$, $c_p = B(1,2,\ldots,p;\ n-p+1,\ldots,n)$, *then* $c_p(t)$ *satisfies*

$$\dot{c}_p = (\sum_{j=n-p+1}^{n} a_{jj} - \sum_{j=1}^{p} a_{jj})c_p, \quad p = 1,2,\ldots,n. \tag{7.7.7}$$

Proof. Denote the pth order corner matrix of $\mathbf{B}$ by $\mathbf{B}_p$, and suppose that its columns are $\mathbf{b}_1, \mathbf{b}_2, \ldots, \mathbf{b}_p$. Thus

$$\mathbf{b}_j = [b_{n-p+1,j},\ b_{n-p+2,j}, \ldots, b_{n,n}]^T.$$

Ex. 7.7.3 shows that $\mathbf{B}$ satisfies

$$\dot{\mathbf{B}} = \mathbf{BS} - \mathbf{SB},$$

with $\mathbf{S} = \mathbf{A}^{+T} - \mathbf{A}^+$, so that

$$\dot{b}_{ij} = (a_{ii} - a_{jj})b_{ij} - 2\sum_{k=1}^{j-1} a_{jk}b_{ik} + 2\sum_{k=i+1}^{n} a_{ik}b_{kj},$$

and

$$\dot{\mathbf{b}}_j = a_{jj}\mathbf{b}_j - 2\sum_{k=1}^{j-1} a_{jk}\mathbf{b}_k + \mathbf{C}\mathbf{b}_j, \tag{7.7.8}$$

where $\mathbf{C} \in M_p$ is given by

$$c_{ik} = \begin{cases} a_{ii}, & k = i, \\ 2a_{ik} & k = i+1,\ldots,n \\ 0 & \text{otherwise} \end{cases}$$

for $i, k = n-p+1, \ldots, n$.

Now $c_p = \det(\mathbf{b}_1, \mathbf{b}_2, \ldots, \mathbf{b}_p)$, so that

$$\dot{c}_p = \sum_{j=1}^{p} \det(\mathbf{b}_1, \mathbf{b}_2, \ldots, \ \mathbf{b}_{j-1}, \ \dot{\mathbf{b}}_j, \ \mathbf{b}_{j+1}, \ldots, \mathbf{b}_p). \qquad (7.7.9)$$

Consider the sums obtained by inserting each of the three terms in $\dot{\mathbf{b}}_j$ from (7.7.8) into (7.7.9). The first gives

$$-\sum_{j=1}^{p} a_{jj} c_p.$$

The second gives zero because it is merely a combination of the first $j-1$ columns; the third may be written

$$\textstyle\sum_{j=n-p+1}^{n} a_{jj} c_p. \ \blacksquare$$

We now prove

Theorem 7.7.2 *Let P denote one of the properties TN, NTN, TP, O, SO. If $\mathbf{A}(0) \in S_n$ has property P, then $\mathbf{A}(t)$, given as the solution of (7.7.1) with $\mathbf{S} = \mathbf{A}^{+T} - \mathbf{A}^{+}$ has the same property P.*

Proof. Suppose first that $\mathbf{A}(0)$ is TP. The corner minors c_p of $\mathbf{A}(t)$ are thus positive when $t = 0$; they satisfy

$$\dot{c}_p = f(t) c_p$$

where

$$f(t) = \sum_{j=n-p+1}^{n} a_{jj} - \sum_{j=1}^{p} a_{jj}$$

is bounded: $|f(t)| \le tr(\mathbf{A}(t)) = tr(\mathbf{A}(0))$.

This implies that these corner minors remain positive.

At $t = 0$, *all* the minors of $\mathbf{A}$ are positive. By continuity, therefore, all the minors are positive in some open interval (a, b) around $t = 0$. Suppose if possible that one or more of the minors became zero at $t = b$. $\mathbf{A}(b)$ would be NTN and its corner minors would be positive, so that, by Theorem 6.8.2, it would be TP. This contradiction implies that $\mathbf{A}(t)$ is TP for all t.

Now suppose that $\mathbf{A}(0)$ is TN. By Ando's result, given in Ex. 6.8.3, $\mathbf{A}(0)$ may be approximated arbitrarily closely in the L_1 norm by a TP matrix

$$\mathbf{C}(0, k) = \mathbf{P}(k)\mathbf{A}(0)\mathbf{P}(k)$$

where

$$\mathbf{P}(k) = (p_{ij}), \quad p_{ij} = \exp[-k(i-j)^2].$$

We now suppose $\mathbf{C}(t, k)$ is the solution of

$$\dot{\mathbf{C}}(t, k) = \mathbf{C}(t, k)\mathbf{S}(t, k) - \mathbf{S}(t, k)\mathbf{C}(t, k)$$

where

$$\mathbf{S}(t,k) = \mathbf{C}^{+T}(t,k) - \mathbf{C}^{+}(t,k).$$

By our previous argument, $\mathbf{C}(t,k)$ is TP for all t and all k, and since (Ex. 7.7.3)

$$||\dot{\mathbf{A}}(t) - \dot{\mathbf{C}}(t,k)|| = O(\exp(-k)), \tag{7.7.10}$$

we have

$$\lim_{k\longrightarrow\infty} \mathbf{C}(t,k) = \mathbf{A}(t) : \tag{7.7.11}$$

the minors of $\mathbf{A}(t)$ are the limits, as $k \longrightarrow \infty$, of the (positive) minors of $\mathbf{C}(t,k)$; all the minors of $\mathbf{A}(t)$ are non-negative: $\mathbf{A}(t)$ is NTN.

Finally, suppose $\mathbf{A}(0)$ is O. It is NTN and so, by the previous result, $\mathbf{A}(t)$ is NTN. When $t = 0$, the minors of $(\mathbf{A}(0))^m = \mathbf{B}(0)$ are strictly positive for $m \geq n-1$. The corner minors of $\mathbf{B}(t) = (\mathbf{A}(t))^m$ remain positive. (Ex. 7.7.3) $\mathbf{B}(t)$ is then NTN, with positive corner minors; $\mathbf{B}(t)$ is TP; $\mathbf{A}(t)$ is O.

It now follows trivially that if $\mathbf{A}(0)$ is SO, then so is $\mathbf{A}(t)$. ■

We can immediately apply this result to obtain other isospectral mass reduced stiffness matrices for the discrete beam. Starting from $\mathbf{A}(0)$ in equation (7.5.3), we can form $\mathbf{A}(t)$; $\mathbf{A}(t)$, like $\mathbf{A}(0)$, will be SO. Ex. 7.7.5 shows that the corner minors of $\mathbf{B}(t) = \mathbf{A}^{-1}(t)$ will be strictly positive, and Ex. 7.7.6 shows that the elements in the outer diagonal of $\mathbf{A}(t)$ will be positive. These are the results needed for the reconstruction of $\mathbf{M}', \mathbf{K}', \mathbf{L}'$ from $\mathbf{A}(t)$.

Markham (1970) [221] shows that an oscillatory (or sign-oscillatory) matrix must have staircase form. It may be verified (Ex. 7.7.4) that the isospectral flow with $\mathbf{S} = \mathbf{A}^{+T} - \mathbf{A}^{+}$ preserves such staircase forms. In particular, one may show that the outermost elements of the staircase retain their signs: if they are strictly positive (negative) when $t = 0$, they will remain strictly positive (negative).

Exercises 7.7

1. Write $\mathbf{S} = \mathbf{A}^{+T} - \mathbf{A}^{+}$ as a Hadamard product $\mathbf{S} = \mathbf{A} \circ \mathbf{Y}$.
2. Verify that if $\mathbf{Y}$ is given in (7.7.6), then $\mathbf{A}$ in (7.7.6) retains its form under the flow (7.7.1).
3. Establish the results (7.7.10), (7.7.11).
4. Show that the isospectral flow (7.7.1) with $\mathbf{S} = \mathbf{A}^{+T} - \mathbf{A}^{+}$ preserves staircase forms; these include block banded forms, with no holes.
5. Show that $\mathbf{B} = \mathbf{A}^{-1}$ satisfies the same isospectral flow equation (7.7.1), i.e., $\dot{\mathbf{B}} = \mathbf{BS} - \mathbf{SB}$, and that the corner minors of $\mathbf{B}$ satisfy (7.7.7).
6. Show that if $\mathbf{A}$ has half-bandwidth r, so that $a_{ij} = 0$ if $|i-j| > r$, then the elements in the outdiagonal of $\mathbf{A}$ retain their signs.
7. Find two isospectral matrices $\mathbf{J}, \mathbf{J}'$ with the property that one cannot flow from $\mathbf{J}$ to $\mathbf{J}'$ in a Toda flow with $\mathbf{S}$ given by equation (7.7.2).

Chapter 8

The Discrete Vibrating Beam

A thinking reed - It is not from space that I must seek my dignity, but from the government of my thought. I shall have no more if I possess worlds. By space the universe encompasses and swallows me up like an atom; by thought I comprehend the world.

Pascal's *Pensées*, 348

8.1 Introduction

In this Chapter we shall present in detail the solution of the inverse problem for the discrete spring-mass model of a vibrating beam discussed in Section 2.3. This model is important because it is the simplest model - it is in effect a finite-difference approximation - for a beam with continously distributed mass. See Gladwell (1991) [116] for a qualitative discussion of the customary finite element model of a beam. The inverse problem for a continuous beam will be considered in Chapter 13. The inverse problem for a discrete beam was first considered by Barcilon (1976) [18], Barcilon (1979) [20], Barcilon (1982) [21]. He established that the reconstruction of such a system would require three spectra, corresponding to three different end conditions. The necessary and sufficient conditions for these spectra to correspond to a realizable system, one with positive masses, lengths and stiffnesses, were derived by Gladwell (1984) [104].

Two papers by Sweet (1969) [313], Sweet (1971) [314] consider the discrete model of a beam obtained by using the so-called 'method of straight lines'; he shows that the coefficient matrix obtained in this procedure is (similar to) an oscillatory matrix. See also Gladwell (1991b) [117].

The plan of the Chapter is as follows. In Section 8.2 we show that the (squares of the) natural frequencies of the system are the eigenvalues of an oscillatory matrix. This means that the eigenvalues are distinct and the eigenvectors

$\mathbf{u}_i$ have all the properties derived in Section 6.10. It is found also that not only $\mathbf{u}_i$, but also $\boldsymbol{\theta}_i, \boldsymbol{\tau}_i, \boldsymbol{\phi}_i$, the slopes, moments and shearing forces, have these same properties (Theorem 8.2.2 and Ex. 8.2.1). Theorem 8.2.2 derives an additional result, that the beam always bends away from the axis at a free end. In Section 8.4 the oscillatory properties of the eigenvectors are used in the ordering of the natural frequences of the system corresponding to different end conditions. In Section 8.5 it is shown that while it is possible to take three spectra as the data for the reconstruction, it is better to take one spectrum, that corresponding to a free end, and the end values u_{ni}, θ_{ni} of the normalised eigenvectors, as the basic data. In this way, the conditions on the data may be written as determinantal inequalities. In Section 8.6, a procedure for inversion is presented and it is shown that the conditions (Theorem 8.5.1), which were put forward earlier, are in fact sufficient to ensure that all the physical parameters, masses, lengths and stiffnesses, will be positive. In Section 8.7 a numerical procedure, based on the Block Lanczos algorithm, is described for the actual computation of the physical parameters.

8.2 The eigenanalysis of the cantilever beam

The equations governing the response of the discrete beam were derived in Section 2.3. Equation (2.3.6) shows that vibration with frequency ω is governed by the equation

$$\lambda \mathbf{M}\mathbf{u} = \mathbf{K}\mathbf{u} - \phi_n \mathbf{e}_n - l_n^{-1}\tau_n \mathbf{E}\mathbf{e}_n, \;\; \lambda = \omega^2$$

where $\mathbf{E}$ is given in equation (2.2.10), $\mathbf{e}_n = \{0, 0, \ldots, 1\}$, and ϕ_n and τ_n are the bending moment and shearing force applied at the free end. This means that the free vibrations satisfy

$$\lambda \mathbf{M}\mathbf{u} = \mathbf{K}\mathbf{u}, \tag{8.2.1}$$

which may be reduced to standard form

$$\mathbf{A}\mathbf{v} = \lambda \mathbf{v} \tag{8.2.2}$$

by the substitutions

$$\mathbf{M} = \mathbf{D}^2, \quad \mathbf{v} = \mathbf{D}\mathbf{u}, \quad \mathbf{A} = \mathbf{D}^{-1}\mathbf{K}\mathbf{D}^{-1}. \tag{8.2.3}$$

Theorem 8.2.1 *The matrix* $\mathbf{A}$ *is sign-oscillatory.*

Proof. Equation (2.3.7) shows that

$$\mathbf{K} = \mathbf{E}\mathbf{L}^{-1}\mathbf{E}\hat{\mathbf{K}}\mathbf{E}^T\mathbf{L}^{-1}\mathbf{E}^T$$

where $\mathbf{L}, \hat{\mathbf{K}}$ are diagonal matrices with positive elements.

We recall, from Section 6.7, that a matrix $\mathbf{A}$ is said to be sign-oscillatory (SO) if $\tilde{\mathbf{A}} = \mathbf{ZAZ}$, with $\mathbf{Z} = diag(1, -1, \ldots, (-)^{n-1})$, is oscillatory (O). The matrix

$$\tilde{\mathbf{E}} = \begin{bmatrix} 1 & 1 & & & \\ & 1 & 1 & & \\ & & \ddots & \ddots & \\ & & & 1 & 1 \\ & & & & 1 \end{bmatrix}$$

is NTN (see the beginning of Section 6.6). Also, Ex. 6.7.6 shows that $\tilde{\mathbf{B}} = \tilde{\mathbf{E}}\overline{\mathbf{L}}^{-1}\tilde{\mathbf{E}}$ is NTN, as is its transpose, and hence also $\tilde{\mathbf{K}} = \tilde{\mathbf{B}}\hat{\mathbf{K}}\tilde{\mathbf{B}}^{T}$, and $\tilde{\mathbf{A}} = \mathbf{D}^{-1}\tilde{\mathbf{K}}\overline{\mathbf{D}}^{-1}$. Now, according to Theorem 6.7.3, to show that $\tilde{\mathbf{A}}$ is oscillatory, it is sufficient to show that $\tilde{a}_{i+1,i} > 0, \quad i = 1, 2, \ldots, n-1$. This is easily verified. Thus $\tilde{\mathbf{A}}$ is O, and $\mathbf{A}$ is sign-oscillatory. ∎

Theorem 8.2.1 has important consequences. It means that the eigenvalues $(\lambda_i)_1^n$ are distinct (Corollary to Theorem 6.10.1), that the last element, u_{ni} of each eigenvector $\mathbf{u}_i$ of equation (8.2.1) may be chosen to be (strictly) positive (Corollary to Theorem 6.10.2); note that equation (8.2.3) gives $v_j = d_j u_j$, so that $v_n > 0$ implies $u_n > 0$; and the u_{ji} will satisfy the inequalities (6.10.3). We now prove

Theorem 8.2.2 *The vectors $(\boldsymbol{\theta}_j)_1^n$ are the eigenvectors of a sign-oscillatory matrix.*

Proof. Since $\boldsymbol{\theta} = \mathbf{L}^{-1}\mathbf{E}^T\mathbf{u}$ and thus $\mathbf{u} = \mathbf{E}^{-T}\mathbf{L}\boldsymbol{\theta}$, we have

$$\lambda\mathbf{M}\mathbf{u} = \lambda\mathbf{M}\mathbf{E}^{-T}\mathbf{L}\boldsymbol{\theta} = \mathbf{K}\mathbf{u} = \mathbf{E}\mathbf{L}^{-1}\mathbf{E}\hat{\mathbf{K}}\overline{\mathbf{E}}^{T}\mathbf{L}^{-1}\mathbf{E}^T(\mathbf{E}^{-T}\mathbf{L}\boldsymbol{\theta})$$

so that

$$\lambda(\mathbf{L}\mathbf{E}^{-1}\mathbf{M}\mathbf{E}^{-T}\mathbf{L})\boldsymbol{\theta} = \mathbf{E}\hat{\mathbf{K}}\overline{\mathbf{E}}^{T}\boldsymbol{\theta}$$

or

$$\lambda\mathbf{G}\boldsymbol{\theta} = \mathbf{H}\boldsymbol{\theta}, \quad \mathbf{G}^{-1}\mathbf{H}\boldsymbol{\theta} = \lambda\boldsymbol{\theta}.$$

The matrix $\mathbf{G}$ is O, so that $(\tilde{\mathbf{G}}^{-1})$ is O (Theorem 6.7.5). $\mathbf{H}$ is SO, so that $\tilde{\mathbf{H}}$ is O. Therefore, by Ex. 6.7.7, $(\tilde{\mathbf{G}}^{-1}\tilde{\mathbf{H}})$ is O, and thus $\mathbf{G}^{-1}\mathbf{H}$ is SO. ∎

Theorem 8.2.2 means that the $\boldsymbol{\theta}_i$ must satisfy all the requirements for the eigenvectors of an SO matrix, e.g., $\theta_{n,i} \neq 0$. We now show that, for the particular SO matrix governing the beam, if the $u_{n,i}$ are chosen so that $u_{n,i} > 0$, so that all the minors $u_{n,s}$ of Theorem 6.10.3 are positive, then $\theta_{n,i}$, and hence all the corresponding minors

$$\mathcal{V}_{n,s} = \Theta(n-p+1, n-p+2, \ldots, n; i_1, i_2, \ldots, i_p)$$

will be positive. It is sufficient to prove

Theorem 8.2.3 *Each eigenvector of the cantilever beam satisfies $u_{n,j}\theta_{n,j} > 0$.*

Proof. Choose u_j so that $u_{n,j} > 0$. There is an index r ($1 \le r \le n-1$) such that

i) $u_{i,j} > 0, \quad i = r,\ r+1, \ldots, n,$
ii) $u_{r-1,j} \le 0.$
Note that when $j = 1$, then $r = 1$; we have $u_{0,1} = 0$.
Thus $\theta_{r,j} = (u_{r,j} - u_{r-1,j})/l_r > 0$.
Now, since

$$\boldsymbol{\phi}_j = \lambda_j \mathbf{E}^{-1}\mathbf{M}\mathbf{u}_j$$

then, because of the form of $\mathbf{E}^{-1}$ given in equation (2.2.10), we have

$$\phi_{i,j} > 0 \quad i = r-1, \ldots, n-1.$$

But

$$\boldsymbol{\tau}_j = \mathbf{E}^{-1}\mathbf{L}\boldsymbol{\phi}_j$$

so that, again,

$$\tau_{i,j} > 0 \quad i = r-1, \ldots, n-1.$$

Now consider the equation linking the θ_i and τ_i, namely

$$\theta_{i+1} - \theta_i = k_{i+1}^{-1}\tau_i$$

and sum from r to $n-1$ to obtain

$$\theta_{n,j} - \theta_{r,j} = \sum_{i=r}^{n-1} k_{i+1}^{-1}\tau_{i,j} > 0$$

so that $\theta_{n,j} > 0$. ■

Theorem 8.2.2, while showing that the $\boldsymbol{\theta}_j$ are eigenvectors of a sign-oscillatory matrix, shows that $\mathbf{u}_j$ and $\boldsymbol{\theta}_j$ must both have precisely $j-1$ sign changes. This means that the first mode $\mathbf{u}_1$ will steadily increase, i.e.,

$$0 < u_{1,1} < u_{2,1} < \ldots < u_{n,1},$$

as shown in Figure 8.2.1, while the j-th mode ($j > 1$) will have $j-1$ portions that are convex towards the axis, and one final portion that bends away from the axis, as shown in Figure 8.2.2.

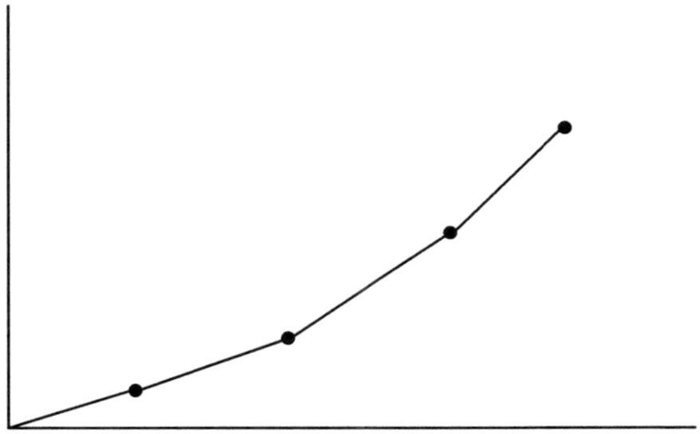

Figure 8.2.1 - The first mode steadily increases

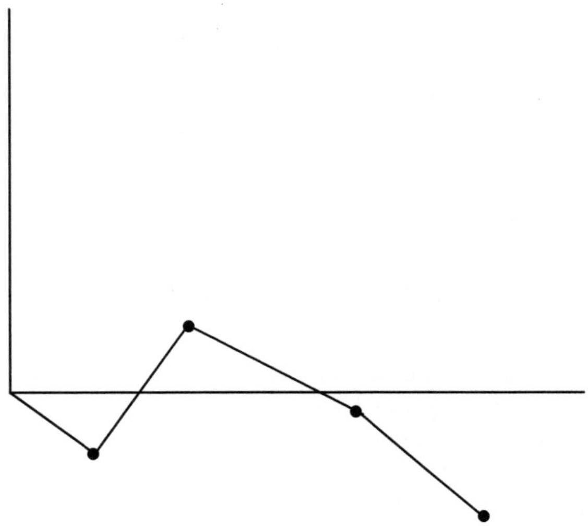

Figure 8.2.2 - The end of the mode bends away from the axis

Exercises 8.2

1. Show that $\boldsymbol{\tau}_j$ and $\boldsymbol{\phi}_j$ are eigenvectors of the equations

$$\begin{aligned}\lambda\hat{\mathbf{K}}^{-1}\boldsymbol{\tau} &= \mathbf{E}^T\mathbf{L}^{-1}\mathbf{E}^T\mathbf{M}^{-1}\mathbf{E}\mathbf{L}^{-1}\mathbf{E}\boldsymbol{\tau}\\ \lambda\mathbf{L}\mathbf{E}^{-T}\hat{\mathbf{K}}^{-1}\mathbf{E}^{-1}\mathbf{L}\boldsymbol{\phi} &= \mathbf{E}^T\mathbf{M}^{-1}\mathbf{E}\boldsymbol{\phi}\end{aligned}$$

and that each is the eigenvector of a sign-oscillatory matrix.

8.3 The forced response of the beam

The equation governing the response to an end shearing force and bending moment is equation (2.3.6), which for vibration of frequency ω becomes

$$\lambda\mathbf{M}\mathbf{u} = \mathbf{K}\mathbf{u} - \phi_n\mathbf{e}_n - l_n^{-1}\tau_n\mathbf{E}\mathbf{e}_n. \tag{8.3.1}$$

Since the eigenvectors $\mathbf{u}_j$ of the clamped-free beam span V_n, and are orthogonal w.r.t. $\mathbf{M}$ and $\mathbf{K}$ we may write

$$\mathbf{u} = \sum_{j=1}^{n}\alpha_j\mathbf{u}_j,$$

and find

$$\alpha_j = (\phi_n u_{n,j} + \tau_n\theta_{n,j})/(\lambda_j - \lambda),$$

where the modes are normalised so that

$$\mathbf{u}_j^T \mathbf{M} \mathbf{u}_k = \delta_{jk}.$$

Thus

$$\mathbf{u} = \sum_{j=1}^{n} \frac{(\phi_n u_{n,j} + \tau_n \theta_{n,j})\mathbf{u}_j}{\lambda_j - \lambda}, \tag{8.3.2}$$

and on multiplying through by $\mathbf{L}^{-1}\mathbf{E}^T$ we find

$$\boldsymbol{\theta} = \sum_{j=1}^{n} \frac{(\phi_n u_{n,j} + \tau_n \theta_{n,j})}{\lambda_j - \lambda} \boldsymbol{\theta}_j. \tag{8.3.3}$$

These two equations completely characterise the forced response of the beam. In the terminology of Bishop and Johnson (1960) [34], equations (8.3.2), (8.3.3) give the *end receptances* for the beam: the displacement (slope) at one coordinate i due to a unit shearing force or bending moment at the end. In particular, for the end displacement and slope we have

$$u_n = \alpha \phi_n + \alpha' \tau_n, \tag{8.3.4}$$

$$\theta_n = \alpha' \phi_n + \alpha'' \tau_n, \tag{8.3.5}$$

where

$$\alpha = \sum_{j=1}^{n} \frac{(u_{n,j})^2}{\lambda_j - \lambda}, \quad \alpha' = \sum_{j=1}^{n} \frac{u_{n,j}\theta_{n,j}}{\lambda_j - \lambda}, \tag{8.3.6}$$

$$\alpha'' = \sum_{j=1}^{n} \frac{(\theta_{n,j})^2}{\lambda_j - \lambda}. \tag{8.3.7}$$

8.4 The spectra of the beam

Now suppose that the left hand end of the beam remains clamped while the conditions at the right hand end are varied. The possible end conditions and eigenvalues, (eigenfrequency)^2, are as follows:

free	$\phi_n = 0 = \tau_n$	$(\lambda_i)_1^n$
sliding	$\theta_n = 0 = \phi_n$	$(\sigma_i)_1^{n-1}$
anti-resonant	$u_n = 0 = \phi_n$ or $\theta_n = 0 = \tau_n$	$(\nu_i)_1^{n-1}$
pinned	$u_n = 0 = \tau_n$	$(\mu_i)_1^{n-1}$
clamped	$u_n = 0 = \theta_n$	$(\gamma_i)_1^{n-2}$

Note that the anti-resonant frequencies are those at which the application of an end bending moment produces no end displacement; we will show that there are $n-1$ such frequencies, and that they are also the frequencies at which the application of an end shearing force produces no end rotation.

We will now relate the various eigenvalues to the receptances derived in Section 8.3. We first state

Theorem 8.4.1 *If* $(p_j)_1^n > 0$ *and* $x_1 < x_2 < \ldots < x_n$*, then the equation*

$$f(x) = \sum_{j=1}^{n} \frac{p_j}{x_j - x} = 0$$

has $n-1$ *real zeros* ξ_j *satisfying*

$$x_j < \xi_j < x_{j+1}.$$

Proof. In each interval (x_j, x_{j+1}), $f(x)$ is strictly increasing from $-\infty$ to $+\infty$, and will cross the x-axis just once. ∎

We now substitute the end conditions in the receptance equations (8.3.6), (8.3.7), starting with the sliding condition;

$$\sum_{j=1}^{n} \frac{(\theta_{n,j})^2}{\lambda_j - \lambda} = 0 \quad \text{has zeros } (\sigma_i)_1^{n-1}. \tag{8.4.1}$$

Making use of Theorem 8.2.3 we may state

$$\sum_{j=1}^{n} \frac{u_{n,j}\theta_{n,j}}{\lambda_j - \lambda} = 0 \quad \text{has zeros } (\nu_i)_1^{n-1}, \tag{8.4.2}$$

and

$$\sum_{j=1}^{n} \frac{(u_{n,j})^2}{\lambda_j - \lambda} = 0 \quad \text{has zeros } (\mu_i)_1^{n-1}. \tag{8.4.3}$$

To find the relative positions of the eigenvalues we need

Theorem 8.4.2 *Suppose* $(p_j)_1^n > 0$, $(q_j)_1^n > 0$, $x_1 < x_2 < \ldots < x_n$,

$$f(x) = \sum_{j=1}^{n} \frac{p_j}{x_j - x}, \quad g(x) = \sum_{j=1}^{n} \frac{q_j}{x_j - x}$$

and that $(\xi_i)_1^{n-1}$, $(\eta_i)_1^{n-1}$ *are the zeros of* $f(x)$, $g(x)$ *respectively. If* $p_j q_i - p_i q_j > 0$ *for* $i > j$ *then* $\xi_i > \eta_i$ *for* $i = 1, 2, \ldots, n-1$.

Proof.

$$p_i g(x) - q_i f(x) = \sum_{j=1}^{n} \frac{p_i q_j - p_j q_i}{x_j - x}.$$

Put $x = \xi_i$, so that $x_i < \xi_i < x_{i+1}$, and divide the sum into two parts, thus

$$p_i g(\xi_i) = \sum_{j=1}^{i-1} \frac{p_j q_i - p_i q_j}{\xi_i - x_j} + \sum_{j=i+1}^{n} \frac{p_i q_j - p_j q_i}{x_j - \xi_i}.$$

Under the stated conditions, each of the numerators and denominators on the right will be positive, so that $g(\xi_i) > 0$, i.e., $g(x)$ has already become positive when $f(x)$ has just become zero, i.e., $\xi_i > \eta_i$. ■

Note that, as in the discussion of positivity in Chapter 6, it is sufficient to have $p_j q_{j+1} - p_{j+1} q_j > 0$ for $j = 1, 2, \ldots, n-1$, for then $p_j q_i - p_i q_j > 0$ for all $i > j$. The converse of Theorem 8.4.2 is *not* true - see Ex. 8.4.1.

We now apply this Theorem, first to σ_i and ν_i. Take $p_j = u_{n,j}\theta_{n,j}$ and $q_j = \theta_{n,j}^2$, then $p_j q_i - p_i q_j = \theta_{n,i}\theta_{n,j}(u_{n,j}\theta_{n,i} - u_{n,i}\theta_{n,j}) = \theta_{n,i}\theta_{n,j}(u_{n,i}u_{n-1,j} - u_{n,j}u_{n-1,i})/\ell_n$. To show that this is positive, we use Theorem 6.10.3 with $p = 2$, $i_1 = j$, $i_2 = i$; it gives

$$\begin{vmatrix} u_{n-1,j} & u_{n-1,i} \\ u_{n,j} & u_{n,i} \end{vmatrix} > 0$$

for $i > j$, and thus $\nu_i > \sigma_i$. We find in an exactly similar way that $\mu_i > \nu_i$. Finally, since the clamped conditions may be obtained by applying the extra constraint $\theta_n = 0$ to the pinned condition, the usual theory of vibration under constraint gives $\gamma_i > \mu_i$.

This gives the following ordering:

$$0 < \lambda_1 < \sigma_1 < \nu_1 < \mu_1 < (\gamma_1, \lambda_2) < \sigma_2 < \nu_2 < \mu_2 < (\gamma_2, \lambda_3) < \ldots (\gamma_{n-2}, \lambda_{n-1})$$

$$< \sigma_{n-1} < \nu_{n-1} < \mu_{n-1} < \lambda_n. \tag{8.4.4}$$

Note that the relative position of γ_j and λ_{j+1} is (so far) indeterminate; in numerical experiments it was always found that $\gamma_j > \lambda_{j+1}$. See Gladwell (1985) [105], Gladwell (1991b) [117].

Exercises 8.4

1. Construct a counterexample to show that the converse of Theorem 8.4.2 is false. Take $n = 3$, $(x_1, x_2, x_3) = (1, 4, 7)$, $(p_1, p_2, p_3) = (4, 1, 4)$, $(q_1, q_2, q_3) = (5, 1, 7)$. Find $\xi_1, \xi_2, \eta_1, \eta_2$ and show that $g(\xi_1) > 0$, $g(\xi_2) > 0$, so that $\xi_1 > \eta_1$, $\xi_2 > \eta_2$, but $p_1 q_2 - p_2 q_1 < 0$, $p_2 q_3 - p_3 q_2 > 0$.

2. Show that if $p_j > 0$, $q_j > 0$, $p_j q_{j+1} - p_{j+1} q_j > 0$ for $j = 1, 2, \ldots, n-1$, then $p_j q_i - p_i q_j > 0$ for all $i > j$. Compare with Theorem 6.8.1.

3. Use equations (8.4.2), (8.4.3) to deduce that

$$\theta_{n,i}^2 = \frac{c_1 \prod_{j=1}^{n-1}(\sigma_j - \lambda_i)}{\prod_{j=1}^{n'}(\lambda_j - \lambda_i)}$$

$$u_{n,i}^2 = \frac{c_2 \prod_{j=1}^{n}(\mu_j - \lambda_i)}{\prod_{j=1}^{n\prime}(\lambda_j - \lambda_i)}$$

where $\prime$ denotes $j \neq i$, and c_1, c_2 are constants.

4. Develop an intuitive argument to show that $\sigma_i < \mu_i$ by considering a clamped-clamped beam made up of two identical cantilevers of length $\ell/2$ welded together at their free ends.

5. The eigenvalues $(\gamma_i)_1^{n-2}$ are the (frequency)^2 values for which the application of a force and moment at the free end produce $u_n = 0 = \theta_n$. Use the equations (8.3.4)-(8.3.7) to show that the γ_i are the roots of

$$\sum_{i,j=1}^{n} \frac{(u_{n,i}\theta_{n,j} - u_{n,j}\theta_{n,i})^2}{(\lambda_i - \lambda)(\lambda_j - \lambda)} = 0.$$

8.5 Conditions on the data for inversion

In the inverse eigenvalue problem for the beam it is required to construct a beam with given eigenvalues. Barcilon showed (for his model) that the beam cannot be uniquely determined from two spectra, and attempted to prove that it could be so determined (apart from a scale factor) from three properly chosen spectra. His procedure (in our notation) was to start from $(\lambda_i, \nu_i, \mu_i)_1^n$ (and note that he had n of each of the ν_i, μ_i, not $n-1$ as in the model of Figure 2.3.1) satisfying

$$\lambda_1 < \nu_1 < \mu_1 < \lambda_2 \ldots \lambda_n < \nu_n < \mu_n$$

and compute the frequencies $(\sigma_i)_1^n$ and $(\gamma_i)_1^{n-1}$ (again note that he had n of the σ_i and $n-1$ of γ_i) using some recurrence relations. For his model it was not possible to prove that the eigenvalues σ_i, γ_i so computed satisfied the complete set of inequalities (similar to (8.4.4)). He had to place subsidiary conditions on $(\lambda_i, \nu_i, \mu_i)_1^n$ in order for the inequalities to be satisfied. His second step was a stripping procedure for computing the parameters l_n, k_n, m_n of the last segment, and for computing the corresponding eigenvalues $(\lambda_i^*, \nu_i^*, \mu_i^*)_1^{n-1}$ of the truncated system obtained by deleting the last segment. The l_n, k_n, m_n were all found to be positive but, even with the extra conditions on the $(\lambda_i, \nu_i, \mu_i)_1^n$, it was not possible to prove that the new (starred) eigenvalues satisfied the necessary orderings, which meant that if the stripping procedure were continued, negative masses, stiffnesses or lengths might be encountered at some stage. He concluded that further conditions must be placed on the data, preferably conditions which could be applied *ab initio*, so eliminating the need for checks at each stage of the stripping procedure. We shall now state such conditions and construct a new stripping procedure.

The spectra, from which will be drawn the data for the inverse problem, may be divided into three parts:

(i) $(\lambda_i)_1^n$; (ii) $(\sigma_i, \nu_i, \mu_i)_1^{n-1}$; (iii) $(\gamma_i)_1^{n-2}$.

Suppose that (i) is given. Each spectrum which is given from (ii) then determines, to within an arbitrary multiplier, the set of coefficients $(\theta_{n,i})^2, (u_{n,i}\theta_{n,i})$ or $(u_{n,i})^2$ respectively, from the eigenvalue equations (8.4.1)-(8.4.3); see Ex. 8.4.3 and an analogous result for $u_{n,i}\theta_{n,i}$. If *any two* of the spectra in (ii) are given, then the two sets of coefficients yield the third set, and hence the third spectrum. (Note that since $u_{n,i}\theta_{n,i} > 0$, there is no ambiguity in taking the square root of $u_{n,i}^2\theta_{n,i}^2$.) However, if two given spectra, say $(\nu_i)_1^{n-1}$ and $(\mu_i)_1^{n-1}$ satisfy the appropriate ordering, $\nu_i < \mu_i$, then the third set $(\sigma_i)_1^{n-1}$ need not satisfy its appropriate ordering, $\sigma_i < \nu_i$. Two counterexamples are provided in Ex. 8.5.1, 8.5.2, and these clearly show that the ordering requirements on the two given spectra e.g., $\nu_i < \mu_i$, are insufficient for the existence of a real model, with positive l_i, k_i, m_i; they do not even ensure the ordering of the remaining spectrum. We now prove the fundamental

Theorem 8.5.1 *A necessary condition for the existence of a real (i.e., positive) model corresponding to data sets $(\lambda_i, u_{n,i}, \theta_{n,i})_1^n$ is that the matrix $\mathbf{P} \in M_{n+1,n}$ given by*

$$\mathbf{P} = \begin{bmatrix} u_{n,1} & u_{n,2} & \cdots & u_{n,n} \\ \theta_{n,1} & \theta_{n,2} & \cdots & \theta_{n,n} \\ \lambda_1 u_{n,1} & \lambda_2 u_{n,2} & \cdots & \lambda_n u_{n,n} \\ \lambda_1 \theta_{n,1} & \lambda_2 \theta_{n,2} & \cdots & \lambda_n \theta_{n,n} \\ \lambda_1^2 u_{n,1} & \lambda_2^2 u_{n,2} & \cdots & \lambda_n^2 u_{n,n} \\ \cdot & \cdot & \cdots & \cdot \end{bmatrix}$$

should have all its minors are positive. Note that the last row of $\mathbf{P}$ *is*

$$\lambda_1^r u_{n,1} \quad \lambda_2^r u_{n,2} \quad \cdots \quad \lambda_n^r u_{n,n}$$

or

$$\lambda_1^r \theta_{n,1} \quad \lambda_2^r \theta_{n,2} \quad \cdots \quad \lambda_n^r \theta_{n,n}$$

according to whether n is even or odd respectively, and $r = [n/2]$.

Proof. Because of the repetitive nature of the rows of $\mathbf{P}$, Theorem 6.8.1 shows that all the minors will be positive iff

$$P(1,2,\ldots,p;\ i,i+1,\ldots,i+p-1) > 0 \quad P(2,3,\ldots,p+1;\ i,i+1,\ldots,i+p-1) > 0 \tag{8.5.2}$$

for $p = 1,2,\ldots,n$ and $i = 1,2,\ldots,n-p+1$.

The proof follows directly from Theorem 6.10.3, for

$$U(n-1,n;\ i,i+1) = \begin{vmatrix} u_{n-1,i} & u_{n-1,i+1} \\ u_{n,i} & u_{n,i+1} \end{vmatrix} > 0.$$

But the recurrence $u_{n-1} = u_n - l_n\theta_n$ yields

$$\begin{aligned} U(n-1,n;\ i,i+1) &= \begin{vmatrix} u_{n,i} - l_n\theta_{n,i} & u_{n,i+1} - l_n\theta_{n,i+1} \\ u_{n,i} & u_{n,i+1} \end{vmatrix} \\ &= l_n \begin{vmatrix} u_{n,i} & u_{n,i+1} \\ \theta_{n,i} & \theta_{n,i+1} \end{vmatrix} = l_n P(1,2;\ i,i+1) > 0 \end{aligned}$$

which we write in abbreviated notation as

$$[u_{n-1}, u_n] = [u_n - l_n\theta_n, u_n] = l_n[u_n, \theta_n] = l_n P(1,2;\ i, i+1).$$

Similarly, the relations between the $u_i, \theta_i, \tau_i, \phi_i$ in Section 2.3 and Theorem 6.10.3 applied to the θ_i (note that Theorem 8.2.2 shows that θ_j is an eigenvector of an SO matrix) give

$$\begin{aligned} 0 &< [\theta_{n-1}, \theta_n] = [\theta_n - k_n^{-1}\tau_{n-1}, \theta_n] = -k_n^{-1}[\tau_{n-1}, \theta_n] \\ &= -k_n^{-1} l_n [\theta_{n-1}, \theta_n] = -k_n^{-1} l_n m_n [\lambda u_n, \theta_n] \\ &= k_n^{-1} l_n m_n [\theta_n, \lambda u_n] = k_n^{-1} l_n m_n P(2,3;\ i, i+1). \end{aligned}$$

Proceeding in this way we may relate the minors occurring in Theorem 6.10.3, for U or Θ, to those appearing in P. Thus

$$\begin{aligned} U(n-2, n-1, n;\ i, i+1, i+2) &= P_{n-2}P(1,2,3;\ i, i+1, i+2) \\ \Theta(n-2, n-1, n;\ i, i+1, i+2) &= Q_{n-2}P(2,3,4;\ i, i+1, i+2) \end{aligned}$$

where

$$P_{n-2} = k_n^{-1} l_n^2 l_{n-1} m_n,\ Q_{n-2} = k_n^{-1} k_{n-1}^{-1} l_n^2 l_{n-1} m_n m_{n-1}$$

and generally

$$\begin{aligned} &U(n-p+1, n-p+2, \ldots, n;\ i, i+1, \ldots, i+p-1) = \\ &\qquad P_{n-p+1}P(1,2,\ldots,p;\ i, i+1, i+p-1) \end{aligned} \tag{8.5.3}$$

$$\begin{aligned} &\Theta(n-p+1, n-p+2, \ldots, n;\ i, i+1, \ldots, i+p-1) = \\ &\qquad Q_{n-p+1}P(2,3,\ldots,p+1;\ i, i+1, \ldots, i+p-1) \end{aligned} \tag{8.5.4}$$

where, as will be important in our discussion later, P_{n-p+1} and Q_{n-p+1} are products of the m_i, l_i, k_i for $i = n-p+2, \ldots, n$. ■

It will be shown below that the condition that **P** is TP is also *sufficient* for the existence of a real model.

Exercises 8.5

1. Construct a counterexample to show that $\lambda_i < \nu_i < \mu_i < \lambda_{i+1}$ does not imply $\lambda_i < \sigma_i < \nu_i < \mu_i < \lambda_{i+1}$. Take $n = 3$, $(\lambda_1, \lambda_2, \lambda_3) = (1, 4, 7)$, $(u_{3,1}, u_{3,2}, u_{3,3}) = (2, 1, 2)$ so that $(\mu_1, \mu_2) = (2, 5)$. Take $\theta_{3,1} = 3/2$, $\theta_{3,2} = 1$ and find $\theta_{3,3}$ so that $\nu_1 < \mu_1$, $\nu_2 < \mu_2$, $\sigma_1 < \mu_1$ but $\sigma_2 > \mu_2$.

2. With the same λ_i and $u_{3,i}$ data, but with $\theta_{3,1} = 1$, find $\theta_{3,3}$ so that $\sigma_1 < \mu_1$, $\sigma_2 < \mu_2$, $\nu_2 < \mu_2$, but $\nu_1 > \mu_1$.

3. Take $n = 3, u_{n,i} = 1, \theta_{n,1} = 1, \lambda_1 = 1$, and find two sets of values of $\theta_{n,2}$ and $\theta_{n,3}$, λ_2, λ_3 so that the positivity conditions of Theorem 8.5.1 are fulfilled and $\gamma_1 < \lambda_2$ in one case, $\gamma_1 > \lambda_2$ in the other. This proves that the relative positions of γ_i and λ_{i+1} are indeterminate.

8.6 Inversion by using orthogonality

In this section we show how the system parameters may be found, at least in theory, from the eigenvalue data, and establish necessary and sufficient conditions on the data for the system parameters to be positive.

Suppose that we are given $(\lambda_i, u_{n,i}, \theta_{n,i})_1^n$ for a cantilever beam, so that $\tau_{n,i} = 0 = \phi_{n,i}$. We will show that we can construct a beam, and that if the data satisfy the condition stated in Theorem 8.5.1, then all the system parameters will be positive.

We start with the system equation

$$\lambda_i \mathbf{M}\mathbf{u}_i = \mathbf{K}\mathbf{u}_i,$$

and, as usual, put $\mathbf{U} = [\mathbf{u}_1, \mathbf{u}_2, \ldots, \mathbf{u}_n]$, $\wedge = diag(\lambda_1, \lambda_2, \ldots, \lambda_n)$. Then

$$\mathbf{MU}\wedge = \mathbf{KU}, \tag{8.6.1}$$

and the orthogonality of the $\mathbf{u}_i$ w.r.t. $\mathbf{K}, \mathbf{M}$ yields

$$\mathbf{U}^T\mathbf{MU} = \mathbf{I},\ \mathbf{U}^T\mathbf{KU} = \wedge. \tag{8.6.2}$$

The first of these equations gives

$$\mathbf{M}^{-1} = \mathbf{UU}^T, \tag{8.6.3}$$

so that

$$\frac{1}{m_j} = \sum_{i=1}^{n} (u_{j,i})^2,\ j = 1, 2, \ldots, n. \tag{8.6.4}$$

Since $(u_{n,i})_1^n$ are known, we have found

$$\frac{1}{m_n} = \sum_{i=1}^{n} (u_{n,i})^2. \tag{8.6.5}$$

The matrix $\mathbf{UU}^T$ is diagonal; its term $j, j-1$ is

$$\sum_{i=1}^{n} u_{j,i} u_{j-1,i} = 0,$$

which, with $u_{j-1} = u_j - l_j\theta_j$ gives

$$\sum_{i=1}^{n} (u_{j,i})^2 - l_j \sum_{i=1}^{n} u_{j,i}\theta_{j,i} = 0,$$

and using (8.6.4) we find

$$\frac{1}{m_j l_j} = \sum_{i=1}^{n} u_{j,i}\theta_{j,i}, \tag{8.6.6}$$

which with $j = n$, yields l_n.

The next step is the determination of k_n. For this we need the explicit expression for $\mathbf{K}$:

$$\mathbf{K} = \mathbf{E}\mathbf{L}^{-1}\mathbf{E}\hat{\mathbf{K}}\mathbf{E}^T\mathbf{L}^{-1}\mathbf{E}^T.$$

This gives

$$\hat{\mathbf{K}}^{-1} = \mathbf{E}^T\mathbf{L}^{-1}\mathbf{E}^T\mathbf{K}^{-1}\mathbf{E}\mathbf{L}^{-1}\mathbf{E}. \tag{8.6.7}$$

Now we use the second of equations (8.6.2) to give

$$\mathbf{K}^{-1} = \mathbf{U}\wedge^{-1}\mathbf{U}^T$$

which, when substituted in (8.6.7), gives

$$\hat{\mathbf{K}}^{-1} = \mathbf{E}^T(\mathbf{L}^{-1}\mathbf{E}^T\mathbf{U})\wedge^{-1}(\mathbf{U}^T\mathbf{E}\mathbf{L}^{-1})\mathbf{E}.$$

But

$$\Theta = [\boldsymbol{\theta}_1, \boldsymbol{\theta}_2, \ldots, \boldsymbol{\theta}_n] = \mathbf{L}^{-1}\mathbf{E}^T\mathbf{U}$$

so that

$$\hat{\mathbf{K}}^{-1} = \mathbf{E}^T\boldsymbol{\Theta}\wedge^{-1}\boldsymbol{\Theta}^T\mathbf{E}$$

which yields

$$\frac{1}{k_j} = \sum_{i=1}^{n} \frac{(\theta_{j,i} - \theta_{j-1,i})^2}{\lambda_i}, \quad j = 1, 2, \ldots, n.$$

But $\theta_{j,i} - \theta_{j-1,i} = k_j^{-1}\tau_{j-1,i}$ so that

$$k_j = \sum_{i=1}^{n} \frac{(\tau_{j-1,i})^2}{\lambda_i}. \tag{8.6.8}$$

Now take $j = n$, then $\tau_{n-1,i} = l_n\phi_{n-1,i} = l_n m_n \lambda_i u_{n,i}$, so that

$$k_n = m_n^2\, l_n^2 \sum_{i=1}^{n} \lambda_i (u_{n,i})^2. \tag{8.6.9}$$

Having found m_n, l_n and k_n we now state the steps in the algorithm to reconstruct the system.

i) set $j = n$.

ii) $\mathrm{u}_{n,i}, \theta_{n,i}, \tau_{n,i} \equiv 0 \equiv \phi_{n,i}$ are known from data.

iii) compute m_j, l_j from equations (8.6.5), (8.6.6).

iv) compute $\begin{array}{rcl} u_{j-1,i} & = & u_{j,i} - l_j\theta_{j,i}, \\ \phi_{j-1,i} & = & \phi_{j,i} + m_j\lambda_i u_{j,i}, \\ \tau_{j-1,i} & = & \tau_{j,i} + l_j\phi_{j-1,i} \end{array}$.

v) compute k_j from (8.6.8).

vi) compute $\theta_{j-1,i} = \theta_{j,i} - \tau_{j-1,i}/k_j$.

vii) set $j = j - 1$. If $j > 1$ go to iii), otherwise stop.

We note that the quantities $(u_{n,i}, \theta_{n,i})_1^n$ will be known only to within arbitrary multiplying factors. If a second, primed, set is related to the first by

$$u'_{j,i} = \alpha u_{j,i}, \; \theta'_{j,i} = \beta\theta_{j,i}, \tag{8.6.10}$$

then the algorithm yields

$$m'_j = m_j/\alpha^2, \; k'_j = k_j/\beta^2, \; l'_j = \alpha l_j/\beta, \tag{8.6.11}$$

or

$$\frac{m'_j l'^2_j}{k'_j} = \frac{m_j l_j^2}{k_j}, \quad j = 1, 2, \ldots, n. \tag{8.6.12}$$

Equations (8.6.11), (8.6.12) define the equivalence class of systems corresponding to the given data. The validity of this inversion procedure is based on

Theorem 8.6.1 *The total positivity of the matrix* $\mathbf{P}$ *of Theorem 8.5.1 is necessary and sufficient for the existence of a real (positive) model having three given spectra, i.e.,* $(\lambda_i)_1^n$ *and two of* $(\sigma_i, \nu_i, \mu_i)_1^{n-1}$.

Proof. The necessity was proved in Theorem 8.5.1. We prove the sufficiency. Consider the equations

$$\mathbf{M}^{-1} = \mathbf{U}\mathbf{U}^T, \quad \mathbf{\Theta} = \mathbf{L}^{-1}\mathbf{E}^T\mathbf{U}$$

and construct the matrix

$$\mathbf{B} \equiv \mathbf{L}^{-1}\mathbf{E}^T\mathbf{M}^{-1} = \mathbf{L}^{-1}\mathbf{E}^T\mathbf{U}\mathbf{U}^T = \mathbf{\Theta}\mathbf{U}^T.$$

Now form the pth compound matrix equation by using the Binet-Cauchy Theorem:

$$\mathcal{B}_p = \mathcal{L}_p^{-1}\mathcal{E}_p^T\mathcal{M}_p^{-1} = \mathbf{\Theta}_p^T\mathcal{U}_p^T.$$

Since $\mathcal{L}_p^{-1}$ and $\mathcal{M}_p^{-1}$ are diagonal matrices, and each principal minor of $\mathcal{E}_p^T$ is unity, the bottom right-hand element of $\mathcal{B}_p$ is

$$b_{NN} = \prod_{k=n-p+1}^{n} (m_k l_k)^{-1} = \sum_{s=1}^{N} \mathcal{V}_{N,s}\mathcal{U}_{N,s}, \tag{8.6.13}$$

where the notation is as in Section 6.2.

We now proceed by induction. Suppose that conditions (8.5.2) are satisfied, and that $l_n, l_{n-1}, \ldots, l_{n-p+2}$ are all positive. Each $\mathcal{U}_{N,S}$ and $\mathcal{V}_{N,S}$ may be expressed, as in equations (8.5.3), (8.5.4), as a product of terms involving m_j, k_j^{-1}, which are all positive, and terms involving $l_n, l_{n-1}, \ldots, l_{n-p+2}$ which are positive by hypothesis. Each such $\mathcal{V}_{N,S}$, $\mathcal{U}_{N,S}$ is thus positive. Therefore, equation (8.6.13) shows that $l_{n-p+1} > 0$. But $l_n > 0$, so that all l_j are positive. ■

8.7 A numerical procedure for the inverse problem

The algorithm described in Section 8.6 has primarily theoretical value. It shows that if the data satsify the conditions in Theorem 8.5.1, then the system parameters constructed by the algorithm will be positive. However, starting as it does from the free end and computing the successive model parameters, the algorithm suffers from the same kind of ill conditioning that was encountered in the inverse problem for the rod in Section 4.3.

To obtain a reliable numerical procedure we use the Block Lanczos algorithm described in Section 5.5. To use this algorithm, we reduce the governing equation (8.2.1) to standard form

$$\mathbf{A}\mathbf{q} = \lambda\mathbf{q}$$

where

$$\mathbf{A} = \mathbf{D}^{-1}\mathbf{K}\mathbf{D}^{-1}, \quad \mathbf{M} = \mathbf{D}^2, \quad \mathbf{q} = \mathbf{D}\mathbf{u}.$$

To apply the Block-Lanczos algorithm to the pentadiagonal matrix $\mathbf{A}$ $(p = 2)$, we use the algorithm starting from the free end (n) rather than the fixed end (1). Thus we need the vectors $\mathbf{x}_1, \mathbf{x}_2$ containing the nth and $(n-1)$st terms of the normalised eigenvectors of $\mathbf{A}$:

$$\begin{aligned} \mathbf{x}_1 &= \{q_{n,1}, q_{n,2}, \ldots, q_{n,n}\} \\ \mathbf{x}_2 &= \{q_{n-1,1}, q_{n-1,2}, \ldots, q_{n-1,n}\}. \end{aligned}$$

Now

$$\begin{aligned} q_{n,i} &= d_n u_{n,i} \\ q_{n-1,i} &= d_{n-1}u_{n-1,i} = d_{n-1}\{u_{n,i} - l_n\theta_{n,i}\}. \end{aligned}$$

Equation (8.6.5) gives m_n, and $d_n = m_n^{\frac{1}{2}}$. Equation (8.6.6) gives l_n and hence $u_{n-1,i}$ and then equation (8.6.4) with $j = n-1$ gives m_{n-1}. Thus the data $(\lambda_i, u_{n,i}, \theta_{n,i})_1^n$ give the vectors $\mathbf{x}_1, \mathbf{x}_2$ which are needed for the Block Lanczos algorithm.

Now suppose that we have computed

$$\mathbf{A} = \mathbf{D}^{-1}\mathbf{K}\mathbf{D}^{-1}$$

from the Block Lanczos algorithm. We must now untangle $\mathbf{A}$ to give $\mathbf{K}$ and $\mathbf{M}$. We do this rather like we did it for the rod, in Section 4.4: we use the static behaviour of the system, as we did in Section 7.5.

First, we apply external static forces f_1, f_2 to masses 1 and 2, and deform the system as shown in Figure 8.7.1.

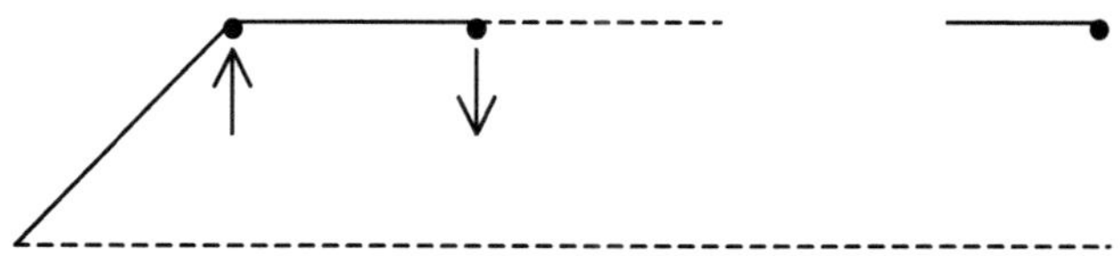

Figure 8.7.1 - Two static forces are needed to deflect all the masses by the same amount

For this configuration $\mathbf{u} = \{1, 1, \ldots, 1\}$, so that $\mathbf{q} = \{d_1, d_2, \ldots, d_n\}$. The static equation is

$$\mathbf{Ku} = \mathbf{f} = \{f_1, -f_2, 0, \ldots, 0\}$$

i.e.,

$$\mathbf{DADu} = \mathbf{f}$$

or

$$\mathbf{Ad} = \{d_1^{-1} f_1, -d_2^{-1} f_2, 0, \ldots, 0\}.$$

Consider this equation. We know $\mathbf{A}$, and we know the last two components d_{n-1}, d_n. But $\mathbf{A}$ is pentadiagonal so that, knowing d_{n-1}, d_n, we can compute $d_{n-2}, \ldots, d_1$, and find $d_1^{-1} f_1, d_2^{-1} f_2$ and hence f_1, f_2.

Having found the masses ($m_j = d_j^2$), we find the lengths. We apply a single force $k_1 l_1^{-1}$ at m_1 and find

$$\mathbf{u}^0 = \{l_1, l_1 + l_2, \ldots, l_1 + l_2 + \cdots l_n\}$$

as shown in Figure 8.7.2.

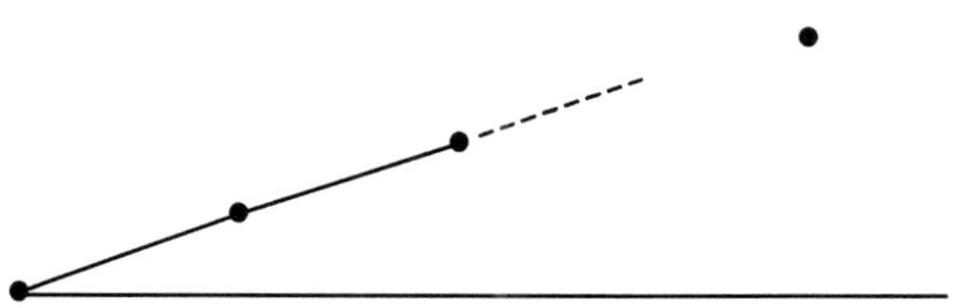

Figure 8.7.2 - One static force will deflect the beam as a straight line

Now the equation

$$\mathbf{Ku}^0 = \{k_1 l_1^{-1}, 0, \ldots, 0\}$$

yields

$$\mathbf{A}\{d_1 l_1, d_2(l_1 + l_2), \ldots, d_n(l_1 + \cdots l_n)\} = \{d_1^{-1} k_1 l_1^{-1}, 0, \ldots, 0\}. \tag{8.7.1}$$

This means that if we invert the equation

$$\mathbf{Ax} = \{1, 0, \ldots, 0\}$$

we will find

$$d_i(l_1 + l_2 + \cdots l_i) = c x_i, \quad c = d_1^{-1} k_1 l_1^{-1}.$$

This yields the l_i, and the theory of Section 8.6 shows that they will all be positive if the data satisfies the conditions of Theorem 8.6.1.

The last step is to find the k_i. Using the form of $\mathbf{E}^{-1}$ in equation (2.2.10), we may write equation (8.7.1) as

$$\mathbf{ADE}^{-T}\mathbf{L}\{1, 1, \ldots, 1\} = \{d_1^{-1}k_1 l_1^{-1}, 0, \ldots, 0\}$$

i.e.,

$$\mathbf{LE}^{-1}\mathbf{DADE}^{-T}\mathbf{L}\{1, 1, \ldots, 1\} = \{k_1, 0, \ldots, 0\}$$

and then as in Section 4.4, we deduce that

$$\mathbf{LE}^{-1}\mathbf{DADE}^{-T}\mathbf{L} = \mathbf{E}\hat{\mathbf{K}}\mathbf{E}^T$$

which gives $\hat{\mathbf{K}}$. The reconstruction is complete.

Chapter 9

Discrete Modes and Nodes

Memory is necessary for all the operations of reason.
Pascal's *Pensées*

9.1 Introduction

The emphasis in all the preceding chapters has been on eigenvalues, and on reconstructing a system from eigenvalue data. In this chapter we turn our attention to eigenvectors. In Sections 9.2, 9.3 we consider the question of constructing a Jacobi matrix that has one or more given eigenvectors, and then go on to constructing a spring mass system from such data. In Section 9.4 we comment on the more difficult problems of constructing a discrete vibrating beam from eigenmode data. Up to this point, all the systems are basically in-line systems, so that the underlying matrices are band matrices, and either oscillatory or sign-oscillatory. In the remaining sections, we widen our study and see what can be said about eigenvectors and their signs, i.e., about modes and nodes, for the equation

$$(\mathbf{K} - \lambda\mathbf{M})\mathbf{u} = \mathbf{0}, \tag{9.1.1}$$

where $\mathbf{K}, \mathbf{M}$ relate to some simple 2-D and 3-D systems, specifically membranes and acoustic cavities. We do not yet have any results about constructing $\mathbf{M}, \mathbf{K}$ from eigenvector data in this case; the properties of the eigenvectors do however provide necessary conditions on the eigenvector data for the masses and stiffnesses of the underlying system to be positive.

Note that in the 2-D and 3-D problems, we will use N to denote the order of the system, and n to label a particular eigenvalue.

9.2 The inverse mode problem for a Jacobi matrix

In this section we consider the problem of constructing a Jacobi matrix that has one or more specified eigenvectors. Following Vijay (1972) [329], Gladwell (1986c) [109] we prove

Theorem 9.2.1 *The vector* $\mathbf{u}$ *is an eigenvector of a Jacobi matrix iff* $S_{\mathbf{u}}^{+} = S_{\mathbf{u}}^{-}$.

Proof. We recall the definitions of $S_{\mathbf{u}}^{+}, S_{\mathbf{u}}^{-}$ from Section 6.9. The necessity, i.e., *only if*, follows from Theorem 6.10.2. To prove sufficiency, i.e., *if*, we need to show first that if $S_{\mathbf{u}}^{+} = S_{\mathbf{u}}^{-}$ then we can find $(a_i)_1^n > 0$, $(b_i)_1^{n-1} > 0$, such that

$$\begin{aligned}
&(a_1 - \lambda)u_1 - b_1 u_2 = 0, \\
&-b_{i-1}u_{i-1} + (a_i - \lambda)u_i - b_i u_{i+1} = 0, \ i = 2, 3, \ldots, n-1 \\
&-b_{n-1}u_{n-1} + (a_n - \lambda)u_n = 0.
\end{aligned} \tag{9.2.1}$$

First, suppose that $(u_i)_1^n \neq 0$, then we may take $(b_i)_1^{n-1} = 1$, $a_i = \lambda + c_i$, $c_i = (u_{i-1} + u_{i+1})/u_i$, $i = 1, 2, \ldots, n$ where $u_0 = 0 = u_{n+1}$. Thus, the matrix

$$\mathbf{C} = \begin{bmatrix} c_1 & -1 & & & \\ -1 & c_2 & -1 & & \\ & \ddots & \ddots & \ddots & \\ & & \ddots & \ddots & -1 \\ & & & -1 & c_n \end{bmatrix}$$

satisfies $\mathbf{Cu} = \mathbf{0}$, and $\mathbf{A} = \lambda \mathbf{I} + \mathbf{C}$. The matrix $\mathbf{C}$, having strictly negative codiagonal, will have distinct eigenvalues $(\kappa_i)_1^n$, one of which will be zero because $\mathbf{C}$ is singular. The matrix $\mathbf{A}$ will have eigenvalues $(\lambda + \kappa_i)_1^n$, so that if λ is chosen so that

$$\lambda \geq \max_{1 \leq i \leq n} (-\kappa_i)$$

then $\mathbf{A}$, having non-negative eigenvalues, will be PSD; $\mathbf{A}$ will be a Jacobi matrix.

What happens when one of the u_i is zero? The condition $S_{\mathbf{u}}^{+} = S_{\mathbf{u}}^{-}$ implies $u_1 \neq 0$, $u_n \neq 0$. Suppose $u_m = 0$ for just one m satisfying $1 < m < n$, then u_{m-1}, u_{m+1} will be non-zero and have opposite signs, so that $u_{m-1}\, u_{m+1} < 0$. The mth line of equation (9.2.1) is

$$b_{m-1}u_{m-1} + b_m u_{m+1} = 0$$

so that a_m, b_{m-1}, b_m may be taken so that

$$a_m = \lambda, \ b_{m-1} = 1, \ b_m = -u_{m-1}/u_{m+1}.$$

The remaining b_i may be chosen so that

$$(b_i)_1^{m-1} = 1, \ (b_i)_m^n = b_m$$

and then

$$a_i = \lambda + c_i, \; c_m = 0, \; c_i = b_i(u_{i-1} + u_{i+1})/u_i, \; i \neq m$$

and again $u_0 = 0 = u_{n+1}$. Now we construct

$$\mathbf{C} = \begin{bmatrix} c_1 & -b_1 & & & \\ -b_1 & c_2 & -b_2 & & \\ & \ddots & \ddots & \ddots & \\ & & \ddots & \ddots & -b_{n-1} \\ & & & -b_{n-1} & c_n \end{bmatrix}$$

which satisfies $\mathbf{Cu} = \mathbf{0}$. Now $\mathbf{A} = \lambda\mathbf{I} + \mathbf{C}$, where λ is chosen as before. This argument may easily be generalised to the case when two or more (non-consecutive) u_i are zero. ■

The next Theorem relates to two given vectors.

Theorem 9.2.2 *Suppose* $\mathbf{u}, \mathbf{v} \in V_n$, *and define* s_i, t_i *as in equation (3.3.6). The necessary and sufficient conditions for* $\mathbf{u}, \mathbf{v}$ *to be eigenvectors of a Jacobi matrix corresponding to two eigenvalues* λ, μ, *unspecified apart from the ordering* $\lambda < \mu$, *are*

(a) $S_{\mathbf{u}}^+ = S_{\mathbf{u}}^-, \; S_{\mathbf{v}}^+ = S_{\mathbf{v}}^-$

(b) $s_n = 0$

(c) either $s_i = 0 = t_i$ or $s_i t_i > 0$ for $i = 1, 2, \ldots, n$.

Proof. The conditions are necessary, for Corollary 6.10.2 yields (a). The orthogonality condition $\mathbf{u}^T\mathbf{v} = 0$ yields (b), while equation (3.3.8) yields (c). Note that a) implies that u_1, v_1 are not zero, so that $s_1 = u_1 v_1 \neq 0$. Hence, $s_1 t_1 > 0$. Also, $s_n = 0$ implies $s_{n-1} = -u_n v_n$; again a) implies that u_n, v_n are not zero so that $s_{n-1} t_{n-1} > 0$. Without loss of generality, we may take $u_1 > 0, \; v_1 > 0$.

The conditions are interesting because they imply that $\mathbf{v}$ has more sign changes than $\mathbf{u}$, i,e., $S_{\mathbf{v}} > S_{\mathbf{u}}$. To see this, we argue as in Theorems 3.3.2, 3.3.3. First, suppose that the first zero of the $\mathbf{u}$-line is $\alpha_1(\lambda) = x$, and of the $\mathbf{v}$-line, $\alpha_1(\mu)$. We prove $\alpha_1(\mu) < \alpha_1(\lambda)$. Suppose if possible that $\alpha_1(\mu) \geq \alpha_1(\lambda) = x$, and that $q < \alpha_1(\lambda) \leq q + 1$ $(1 \leq q < n - 1)$, then all $(u_i)_1^q$ and $(v_i)_1^q$ will be positive, while

$$\begin{aligned} (q + 1 - x)u_q + (x - q)u_{q+1} &= 0 \\ (q + 1 - x)v_q + (x - q)v_{q+1} &\geq 0 \end{aligned}$$

which imply $t_q \leq 0$. On the other hand, $s_q > 0$, which, when used with (3.3.8), provides a contradiction.

Now we show that there is a zero of the **v**-line between any two consecutive nodes of the **u**-line. Let $\alpha, \beta(\alpha < \beta)$ be two neighbouring nodes of the **u**-line and suppose that

$$p - 1 \le \alpha < p, \qquad q < \beta \le q + 1 \quad (p \le q)$$

so that

$$(p - \alpha)u_{p-1} + (\alpha - p + 1)u_p = 0 \tag{9.2.2}$$

$$(q + 1 - \beta)u_q + (\beta - q)u_{q+1} = 0 \tag{9.2.3}$$

and $u_p, u_{p+1}, \ldots, u_q$ have the same sign, say positive. Suppose the **v**-line had no zero in (α, β), and without loss of generality, were positive there. Then $v_p, v_{p+1}, \ldots, v_q$ would be all positive, while

$$(p - \alpha)v_{p-1} + (\alpha - p + 1)v_p \ge 0 \tag{9.2.4}$$

$$(q + 1 - \beta)v_q + (\beta - q)v_{q+1} \ge 0. \tag{9.2.5}$$

On eliminating α between (9.2.2), (9.2.4), and β between (9.2.3), (9.2.5), we find $t_{p-1} \ge 0$, $t_q \le 0$, which, with (c) imply $s_{p-1} \ge 0$, $s_q \le 0$ and therefore $s_q - s_{p-1} \le 0$. But $s_q - s_{p-1} = \sum_{i=p}^{q} u_i v_i > 0$, a contradiction. We can show similarly (Ex. 9.2.1) that the **v**-line has a node to the right of the last node of the **u**-line: the **v**-line has more nodes than the **u**-line.

Now we proceed to the construction. First, suppose that $s_i t_i > 0$ for $i = 1, 2, \ldots, n-1$, then equations (3.3.1), (3.3.6), (3.3.8) show that

$$\begin{aligned} a_i &= \lambda + (\mu - \lambda)\left\{\frac{v_i u_{i+1}}{t_i} + s_{i-1}\frac{(v_{i-1}u_{i+1} - u_{i-1}v_{i+1})}{t_{i-1}t_i}\right\} \\ b_i &= (\mu - \lambda)(s_i/t_i). \end{aligned}$$

Where $i = 2, \ldots, n-1$ in the first formula, $i = 1, \ldots, n-1$ in the second. The two remaining quantities a_1, a_n are given by

$$a_1 = \lambda + (\mu - \lambda)\frac{s_1 u_2}{t_1 u_1}, \quad a_n = \lambda - (\mu - \lambda)\frac{s_{n-1}u_{n-1}}{t_{n-1}u_n}.$$

We may write these equations in the form

$$a_i = \lambda + (\mu - \lambda)c_i, \quad b_i = (\mu - \lambda)d_i.$$

We note that the b_i are positive. Now

$$\mathbf{A} = \lambda\mathbf{I} + (\mu - \lambda)\mathbf{C}$$

where

$$\mathbf{C} = \begin{bmatrix} c_1 & -d_1 & & \\ -d_1 & c_2 & \ddots & \\ & \ddots & \ddots & -d_{n-1} \\ & & -d_{n-1} & c_n \end{bmatrix}.$$

Thus $\mathbf{C}$, having non-zero codiagonal, will have distinct eigenvalues $(\kappa_i)_1^n$. Thus $\mathbf{A}$ will have eigenvalues $\lambda+(\mu-\lambda)\kappa_i$, and $\mathbf{A}$ will be PSD if $\lambda+(\mu-\lambda)\min(\kappa_i) \geq 0$. The slight modifications to the argument which must be made if an s_i is zero, are left to the exercises. ■

Exercises 9.2

1. Show that the conditions (a), (b), (c) of Theorem 9.2.2 imply that the **v**-line will have a node to the right of the last node of the **u**-line.

2. Show that if two consecutive s_i are zero, i.e., $s_{m-1} = 0 = s_m$ $(2 \leq m \leq n-2)$ then $u_m = 0 = v_m$, and deduce that *three* consecutive s_i cannot be zero.

3. Show that if $s_m = 0$ but $s_{m-1} \neq 0$, then b_m may be chosen arbitrarily, e.g., $b_m = \mu - \lambda$. Find a replacement for a_m.

4. Modify the argument to cover the case $s_{m-1} = 0 = s_m$.

9.3 The inverse problem for a single mode of a spring-mass system

We recall from Section 2.2 that the eigenmodes u_j of the system of Figure 2.2.1 are the eigenvectors of the equation

$$\mathbf{E}\hat{\mathbf{K}}\mathbf{E}^T\mathbf{u} = \lambda\mathbf{M}\mathbf{u}. \tag{9.3.1}$$

The matrix $\mathbf{M}^{-1}(\mathbf{E}\hat{\mathbf{K}}\mathbf{E}^T)$ is sign-oscillatory (SO), so the analysis of Section 6.10 applies to the eigenvectors $\mathbf{u}_j$. (Note that $\mathbf{M}^{-1}\mathbf{E}\hat{\mathbf{K}}\mathbf{E}^T$ is not symmetric, but the analysis of SO and O matrices does not depend on symmetry.)

Write $w_i = u_i - u_{i-1}$ so that, with $u_0 = 0$,

$$\mathbf{w} = \mathbf{E}^T\mathbf{u}, \quad \mathbf{u} = \mathbf{E}^{-T}\mathbf{w}.$$

Equation (9.3.1) may be written

$$(\mathbf{E}^T\mathbf{M}^{-1}\mathbf{E})\hat{\mathbf{K}}\mathbf{w} = \lambda\mathbf{w}$$

and again the matrix on the left is SO. This means that the vectors $\mathbf{w}_j$ will have the properties listed in Section 6.10 for the eigenvectors of an SO matrix.

We first prove two theorems regarding the shape of the vector $\mathbf{u}_j$. The first is a simple analogue of the *maximum* principle which appears in elliptic equations.

Theorem 9.3.1 *An eigenmode of (9.3.1) cannot have an interior negative maximum or an interior positive minimum.*

Proof. Suppose $2 \le i \le n-1$. The ith line of (9.3.1) is

$$k_i w_i - k_{i+1} w_{i+1} = \lambda m_i u_i.$$

Suppose u has a relative maximum at u_i. Then $u_i \ge u_{i-1}$, $u_i \ge u_{i+1}$, so that $w_i \ge 0$, $w_{i+1} \le 0$, and hence $u_i \ge 0$. In fact, since w_i, w_{i+1} cannot be simultaneously zero, $u_i > 0$. ■

Theorem 9.3.2 *Two neighbouring u_i can be equal only at a relative maximum or minimum.*

Proof. Suppose $u_i = u_{i-1}$, then $w_i = 0$, so that w_{i-1}, w_{i+1} are non-zero and have opposite signs, i.e.,

$$(u_{i-1} - u_{i-2})(u_{i+1} - u_i) < 0$$

or equivalently

$$(u_i - u_{i-2})(u_i - u_{i+1}) > 0.$$

This implies that $u_i (= u_{i-1})$ is either strictly greater or strictly less than its neighbours u_{i-2} and u_{i+1}: there is a relative maximum or minimum at u_i. ■

The theorems show (Ex. 9.3.1) that $\mathbf{u}_j$ will have $j-1$ portions which bend toward the axis, and a final portion which bends away from the axis, as shown in Figure 9.3.1.

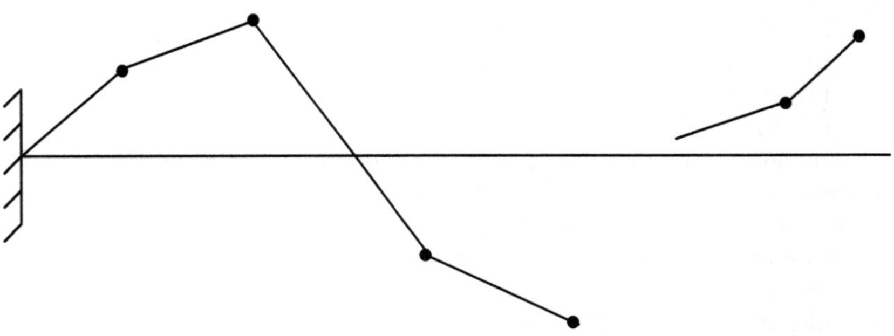

Figure 9.3.1 - The jth mode of a spring-mass system

Theorem 9.3.3 *The necessary and sufficient conditions for $\mathbf{u}$ to be the jth mode of a spring-mass system in the fixed-free configuration are that*

(a) $S^+_{\mathbf{u}} = S^-_{\mathbf{u}} = S^+_{\mathbf{w}} = S^-_{\mathbf{w}} = j - 1$,

(b) $u_1 w_1 > 0$.

Proof. The necessity of these conditions has already been established. To prove sufficiency, we first note that no two of u_i, u_{i-1}, w_i can be simultaneously zero; now we construct a system.

The mode will have a shape like that shown in Figure 9.3.1. Thus u_i will start positive, and will increase $(u_i > 0, w_i > 0)$, until an index r, the first for which

$$u_r > 0, \ w_r \geq 0, \ w_{r+1} < 0.$$

Then u_i will decrease $(u_i > 0, w_i < 0)$ until an index s, the first for which

$$u_s \geq 0, \ u_{s+1} < 0, \ w_s < 0.$$

Now u_i will continue to decrease $(u_i < 0, w_i < 0)$ until an index t, the first for which

$$u_t < 0, \ w_t \leq 0, \ w_{t+1} > 0$$

and then proceed to increase again.

The governing equation (9.3.1) may be written

$$\hat{\mathbf{K}}\mathbf{w} = \lambda \mathbf{E}^{-1}\mathbf{M}\mathbf{u}.$$

Since $\mathbf{E}^{-1}$ is given by equation (2.2.10), we have

$$k_i w_i = \lambda \sum_{k=1}^{n} m_k u_k = \lambda \sigma_i. \tag{9.3.2}$$

This shows that we should take the u_i, and choose m_i so that w_i and σ_i have the same sign.

For the construction, we must choose $(m_i)_1^n > 0$ so that the following conditions hold:

(i) $\sigma_r \geq 0$, with $\sigma_r = 0$ iff $w_r = 0$; then $\sigma_i = \sigma_r + \sum_{k=i}^{r-1} m_k u_k > 0$
(ii) $\sigma_{r+1} < 0$; then $\sigma_i < 0$ for $i = r+1, \ldots, s$
(iii) $\sigma_t \leq 0$, with $\sigma_t = 0$ iff $w_t = 0$; then $\sigma_i < 0$ for $i = s+1, \ldots, t-1$
(iv) $\sigma_{t+1} > 0$,

and so on. Finding $(m_i)_1^n$ with these properties is essentially a linear programming problem. It yields a set of σ_i having the same sign as w_i. If $w_i \neq 0$, then k_i is given by equation (9.3.2), while if a particular w_i is zero, k_i may be given an arbitrary positive value. ■

The question of reconstructing a spring-mass system from modal data was considered by Porter (1970) [267], Porter (1971) [268], but he did not discuss the necessary or sufficient conditions on the modes for the masses and stiffnesses to be positive.

Exercises 9.3

1. Show that the jth mode of a fixed-free spring-mass system will have $j-1$ portions which bend toward the axis, and a final portion which bends away from the axis.

2. Construct a spring-mass system with 7 masses that has third mode $\mathbf{u} = \{1, 2, 1, -1, -2, -1, 1\}$.

9.4 The reconstruction of a spring-mass system from two modes

The construction described in Section 9.3 is far from unique. In this section, following Gladwell (1986c) [109], we shall show that, provided certain conditions are satisfied, there is essentially a unique system for which two given modes are eigenmodes.

We first provide a counterexample to show that even if $\mathbf{u}, \mathbf{v}$ separately satisfy the conditions of Theorem 9.3.3, there may be no system for which they are both eigenmodes, corresponding to two eigenvalues λ, μ, respectively, with $\lambda < \mu$. Write

$$\mathbf{w} = \mathbf{E}^T\mathbf{u}, \quad \mathbf{z} = \mathbf{E}^T\mathbf{v}$$

and suppose

$$\mathbf{u} = \{1, 3, 6\}, \ \mathbf{w} = \{1, 2, 3\}, \ \mathbf{v} = \{1, -1, 4\}, \ \mathbf{z} = \{1, -2, 5\}.$$

The governing equations are

$$\begin{aligned} \lambda m_1 &= k_1 - 2k_2, \ 3\lambda m_2 = 2k_2 - 3k_3, \ 6\lambda m_3 = 3k_3 \\ \mu m_1 &= k_1 + 2k_2, \ -\mu m_2 = -2k_2 - 5k_3, \ 4\mu m_3 = 5k_3 \end{aligned}$$

so that

$$\frac{\mu}{3\lambda} = \frac{5}{6} = \frac{2k_2 + 5k_3}{2k_2 - 3k_3}$$

i.e., $2k_2 = -45k_3$, which is unrealizable.

In order to derive the conditions on the modes, we formalize the elimination procedure used in this counterexample.

The recurrence relations are

$$\lambda m_i u_i = k_i w_i - k_{i+1} w_{i+1}, \ i = 1, 2, \ldots, n-1 \tag{9.4.1}$$

$$\mu m_i v_i = k_i z_i - k_{i+1} z_{i+1}, \ i = 1, 2, \ldots, n-1 \tag{9.4.2}$$

and

$$\lambda m_n u_n = k_n w_n, \ \mu m_n v_n = k_n z_n. \tag{9.4.3}$$

Thus,

$$\frac{\lambda}{\mu} = \frac{w_n}{u_n} \cdot \frac{v_n}{z_n}. \tag{9.4.4}$$

We know that one of the conditions will have to be $S_{\mathbf{u}}^+ = S_{\mathbf{u}}^- = S_{\mathbf{w}}^+ = S_{\mathbf{w}}^-$, and correspondingly $S_{\mathbf{v}}^+ = S_{\mathbf{v}}^- = S_{\mathbf{z}}^+ = S_{\mathbf{z}}^-$. These will entail that u_n, v_n, w_n, z_n will all be non-zero and may be chosen to have the same sign, say positive. The condition $\mu > \lambda$ then demands

$$u_n z_n - v_n w_n > 0. \tag{9.4.5}$$

Eliminating k_i, k_{i+1} in turn from equations (9.4.1), (9.4.2) we find

$$\begin{aligned} m_i(\lambda u_i z_i - \mu v_i w_i) &= k_{i+1}(w_i z_{i+1} - w_{i+1} z_i) \\ m_i(\lambda u_i z_{i+1} - \mu v_i w_{i+1}) &= k_i(w_i z_{i+1} - w_{i+1} z_i) \end{aligned}$$

so that on substituting λ/μ from (9.4.4) we find

$$\lambda m_i p_i = k_{i+1} w_n v_n r_i, \quad \lambda_i m_i q_i = k_i w_n v_n r_i \tag{9.4.6}$$

where

$$\begin{aligned} p_i &= u_i v_n w_n z_i - u_n v_i w_i z_n \\ q_i &= u_i v_n w_n z_{i+1} - u_n v_i w_{i+1} z_n \\ r_i &= w_i z_{i+1} - w_{i+1} z_i. \end{aligned}$$

Thus we may state

Theorem 9.4.1 *The necessary and sufficient conditions for* $\mathbf{u}, \mathbf{v}$ *to be eigenmodes of a (fixed-free) spring-mass system for some eigenvalues* λ, μ $(\lambda < \mu)$, *are*

a) $S_{\mathbf{u}} = S_{\mathbf{w}} < S_{\mathbf{v}} = S_{\mathbf{z}}$

b) $v_n w_n > 0$

c) $u_n z_n - v_n w_n > 0$

d) *for each* i, $1 \le i \le n-1$, *the three quantities* p_i, q_i, r_i *have the same strict sign or are all identically zero; this sign need not be the same for all* i.

Proof. The necessity of the conditions has already been demonstrated. If the conditions hold, and none of the triplets is zero, then equations (9.4.6), for $i = 1, 2, \ldots, n-1$, give the $2(n-1)$ ratios

$$m_1/k_1, \ m_1/k_2; \ m_2/k_2, \ m_2/k_3; \ldots; m_{n-1}/k_{n-1}, \ m_{n-1}/k_n.$$

The final equations (9.4.3), (9.4.4) are left for the ratios m_n/k_n and λ/μ. Thus if we choose say λ and m_n then the system is uniquely determined. If a triplet p_k, q_k, r_k is identically zero then m_k, k_k may be chosen arbitrarily (positive).

We note (Ex. 9.4.1) that the conditions a)-c) preclude the triples p_1, q_1, r_1, or $p_{n-1}, q_{n-1}, r_{n-1}$ from being zero. ∎

In the particular case in which the eigenvalues are consecutive, the conditions may be made sharper, to give

Theorem 9.4.2 *The necessary and sufficient conditions for* $\mathbf{u}, \mathbf{v}$ *to be eigenmodes corresponding to consecutive eigenvalues of the spring-mass system are that*

a) $v_n w_n > 0$

b) $u_n z_n - v_n w_n > 0$

c) $(p_i, q_i, r_i)_1^{n-1} > 0$.

Proof. The necessity of a) and b) follow from (9.4.4) and (9.4.5). The necessity of $(r_i)_1^{n-1} > 0$ is established in Gladwell (1985a) [109]; equation (9.4.6) then shows that $(p_i, q_i)_1^{n-1} > 0$. The sufficiency of the conditions follows as before. ■

Exercises 9.4

1. Show that conditions a)-c) of Theorem 9.4.1 imply $p_1 < 0$. Show also that the assumption $(p_{n-1}, q_{n-1}, r_{n-1}) = 0$ leads to a contradiction.

2. Construct a spring-mass system with first and second modes $\mathbf{u} = \{1, 3, 6, 10, 15\}$, $\mathbf{v} = \{-1, -4, -2, 1, 5\}$.

9.5 The inverse mode problem for the vibrating beam

In this section, we consider the questions of whether and how we may construct a discrete model of a beam, as described in Section 2.3, from a single mode $\mathbf{u}$. As could be expected, this question is considerably more difficult than the corresponding question for a rod. Since the question was definitively answered in Gladwell, Willms, He and Wang (1989) [115], we will merely state the principal results obtained there.

We recall that the eigenvalue problem for the cantilever beam may be obtained from equation (2.3.6):

$$\mathbf{Ku} \equiv \mathbf{EL}^{-1}\mathbf{E}\hat{\mathbf{K}}\mathbf{E}^T\mathbf{L}^{-1}\mathbf{E}^T\mathbf{u} = \lambda\mathbf{Mu}. \tag{9.5.1}$$

The matrix $\mathbf{K}$ is a pentadiagonal SO matrix, so that the eigenvalues are simple, and the eigenvector $\mathbf{u}_j = \mathbf{u}$ has sign count $S_{\mathbf{u}} = j - 1$. As with the rod, we can easily show (Ex. 9.5.1) that

$$\boldsymbol{\theta} = \mathbf{L}^{-1}\mathbf{E}^T\mathbf{u}, \ \boldsymbol{\tau} = \hat{\mathbf{K}}\mathbf{E}^T\boldsymbol{\theta}, \ \boldsymbol{\phi} = \mathbf{L}^{-1}\mathbf{E}\boldsymbol{\tau} \tag{9.5.2}$$

are also eigenvectors of SO matrices, so that $S_{\boldsymbol{\theta}} = S_{\boldsymbol{\tau}} = S_{\boldsymbol{\phi}} = j - 1$ also. We note that although $\boldsymbol{\theta}$ can be formed only when the lengths l_i are known, $\boldsymbol{\theta}$ and the difference $\mathbf{E}^T\mathbf{u}$ will have the same sign count. In considering the construction problem we shall in fact assume that the $(l_i)_1^n$ are given, and seek to construct $(k_i, m_i)_1^n$.

In order to find conditions that must be satisfied by the eigenmodes we need some preliminary results.

Lemma 9.5.1 *If* $\mathbf{u}$ *is not identically zero, and* $S_{\mathbf{u}}^{-} = j-1$, $(j \geq 1)$, *then there is an index* k *and indices* $(q_i)_1^j$ *such that* $1 \leq q_1 < q_2 < \ldots < q_j \leq n$ *and* $(-)^{k+i-1} u_{q_i} > 0$ *for* $i = 1, 2, \ldots, j$. *Conversely, if there exist* k *and* $(q_i)_1^j$ *such that* $(-)^{k+i-1} u_{q_i} > 0$ *for* $i = 1, 2, \ldots, j$, *then* $S_{\mathbf{u}}^{-} \geq j-1$.

Proof. Take q_1 as the index of the first non-zero u_i, and let $(-)^k = sign(u_{q_1})$; then $(-)^k u_{q_1} > 0$. Take q_2 as the index of the first u_i with sign opposite to u_{q_1}, then $(-)^{k+1} u_{q_2} > 0$, and so on. For example, in the sequence 0,1*,0,-4*,2*,0,3,-5*, $S_{\mathbf{u}}^{-} = 3$, so that $j = 4$ and the q_i are the indices of the starred entries; that is, $(q_1, q_2, q_3, q_4) = (2, 4, 5, 8)$. If $S_{\mathbf{u}}^{-} = j-1$, then we can find $(q_i)_1^j$. Conversely, if we can find $(q_i)_1^j$, then $S_{\mathbf{u}}^{-}$ must be at least $j-1$. It may be that $S_{\mathbf{u}}^{-}$ is even larger; in any case $S_{\mathbf{u}}^{-} \geq j-1$. ∎

Lemma 9.5.2 *If* $\mathbf{v} = \mathbf{E}^T \mathbf{u}$, *then* $S_{\mathbf{v}}^{-} \geq S_{\mathbf{u}}^{-}$.

Proof. Note that $v_1 = u_1$, $v_2 = u_2 - u_1, \ldots, v_n = u_n - u_{n-1}$. Suppose that $S_{\mathbf{u}}^{-} = j-1$. Choose k and $(q_i)_1^j$ as in Lemma 9.5.1. Then

$$\begin{aligned} (-)^k v_{q_1} &= (-)^k u_{q_1} > 0 \\ (-)^{k+i-1} v_{q_i} &= (-)^{k+1}(u_{q_i} - u_{q_i-1}) \geq (-)^{k+i-1} u_{q_i} > 0, \; i = 2, \ldots, j \end{aligned}$$

so that, by Lemma 9.5.1, $S_{\mathbf{v}}^{-} \geq j-1$. ∎

Lemma 9.5.3 *If* $\mathbf{v} = \mathbf{E}^T \mathbf{u}$, *then* $S_{\mathbf{v}}^{+} \geq S_{\mathbf{u}}^{+}$. *The proof, following similar lines to that of Lemma 9.5.2, is given in Gladwell, Willms, He and Wang (1989) [115].*

We may now use these Lemmas to prove

Theorem 9.5.1 *If* $l_i > 0$, $i = 1, 2, \ldots, n$, $\mathbf{w} = \mathbf{E}^T \mathbf{L}^{-1} \mathbf{E}^T \mathbf{u}$, *and* $S_{\mathbf{u}} = S_{\mathbf{w}} = j-1$, *then* $S_{\boldsymbol{\theta}} = j-1$. *In addition, if* $m_i > 0$, $k_i > 0$, $i = 1, \ldots, n$, *then* $S_{\boldsymbol{\phi}} = S_{\boldsymbol{\tau}} = j-1$.

Proof. We note that $\mathbf{w}$ has the same sign properties as $\boldsymbol{\tau}$ (see (9.5.2)).
Now $\boldsymbol{\theta} = \mathbf{L}^{-1} \mathbf{E}^T \mathbf{u}$, so that by Lemma 9.5.2, $S_{\boldsymbol{\theta}}^{-} \geq S_{\mathbf{u}}^{-} = j-1$. On the other hand, $\mathbf{w} = \mathbf{E}^T \boldsymbol{\theta}$, so that, by Lemma 9.5.3, $S_{\boldsymbol{\theta}}^{+} \leq S_{\mathbf{w}}^{+} = j-1$. Therefore, $S_{\boldsymbol{\theta}}^{+} \leq j-1 \leq S_{\boldsymbol{\theta}}^{-}$, so that $S_{\boldsymbol{\theta}}^{-} = S_{\boldsymbol{\theta}}^{+} = S_{\boldsymbol{\theta}} = j-1$. This proves the first part. Now consider the converse. Clearly Lemmas 9.5.2, 9.5.3 hold if $\mathbf{E}^T$ is replaced by $\mathbf{E}$ ($\mathbf{E}^T$ is the forward difference operator, $\mathbf{E}$ the backward operator). Since $\boldsymbol{\tau} = \hat{\mathbf{K}} \mathbf{w}$, we have $S_{\boldsymbol{\tau}} = S_{\mathbf{w}}$ if $(k_i)_1^n > 0$. Lemma 9.5.2 applied to $\boldsymbol{\phi} = \mathbf{L}^{-1} \mathbf{E} \boldsymbol{\tau}$ shows that $S_{\boldsymbol{\phi}}^{-} \geq S_{\boldsymbol{\tau}}^{-} = j-1$. Lemma 9.5.3 applied to $\lambda \mathbf{M} \mathbf{u} = \mathbf{E} \boldsymbol{\phi}$ shows that $S_{\boldsymbol{\phi}}^{+} \leq S_{\mathbf{u}}^{+} = j-1$. Therefore, $j-1 \leq S_{\boldsymbol{\phi}}^{-} \leq S_{\boldsymbol{\phi}}^{+} \leq j-1$ so that $S_{\boldsymbol{\phi}} = j-1$. ∎

Suppose that two vectors $\mathbf{u}, \mathbf{w}$ are given. The necessary and sufficient conditions that they should be related in the sense $\mathbf{w} = \mathbf{E}^T \mathbf{L}^{-1} \mathbf{E}^T \mathbf{u}$ for some positive diagonal $\mathbf{L}$ is that the vectors $\boldsymbol{\theta} = \mathbf{E}^{-T} \mathbf{w}$ and $\mathbf{v} = \mathbf{E}^T \mathbf{u}$ should be related by $\mathbf{v} = \mathbf{L} \boldsymbol{\theta}$. This means that $\theta_i = \sum_{k=1}^{i} w_k$ and $v_i = u_i - u_{i-1}$ must be positive, zero or negative in step, that is $\theta_i v_i \geq 0$ with $\theta_i = 0$ iff $v_i = 0$. If $\theta_i \neq 0$ then $l_i = v_i/\theta_i$; if $\theta_i = 0$, then l_i is arbitrary. If $\mathbf{u}, \boldsymbol{\theta}, \mathbf{w}$ are so related, then Theorem 9.5.1 shows that $S_{\mathbf{u}} = S_{\mathbf{w}} = j-1$ implies $S_{\boldsymbol{\theta}} = j-1$.

We now state

Theorem 9.5.2 *Let $\mathbf{u}, \boldsymbol{\theta}, \mathbf{w}$ relate to the jth mode of the cantilever beam. Let $(q_i, r_i, s_i)_1^j$ be the sets of indices for $\mathbf{u}, \boldsymbol{\theta}, \mathbf{w}$ respectively, as in Lemma 9.5.1. Then*

(i) $q_{i-1} < r_i \leq q_i, \quad r_{i-1} < s_i \leq r_i, \quad i = 2, 3, \ldots, j,$

(ii) $s_i \leq q_i, \quad i = 2, 3, \ldots, j$ *and* $s_i \geq q_{i-2} + 2, \quad i = 3, \ldots, j,$

(iii) *if* $u_{q_i - 1} = 0$, *then* $r_i < q_i$; *if* $\theta_{r_i - 1} = 0$, *then* $s_i < r_i$; *in either of these cases, therefore,* $s_i < q_i$,

(iv) *if* $w_{s_i - 1} = 0$, *then* $s_i > q_{i-2} + 2, \quad i = 3, \ldots, j.$

Note: This theorem and Lemmas 9.5.2, 9.5.3 may be considered as codifications and extensions of a discrete form of Rolle's Theorem. They give precision to the intuitively obvious statement, that there must be at least one change of sign in the first differences $\boldsymbol{\theta}, \mathbf{w}$ (that is, the derivatives) between any changes of sign of $\mathbf{u}, \boldsymbol{\theta}$ respectively. The formal proof is given in Gladwell, Willms, He and Wang (1989) [115]. We may now state

Theorem 9.5.3 *Suppose that* $\mathbf{u}$ *and positive* $(l_i)_1^n$ *are given. The necessary and sufficient conditions for them to correspond to the jth mode of a cantilever beam are that*

$$S_{\mathbf{u}} = S_{\mathbf{w}} = j - 1, \text{ where } \mathbf{w} = \mathbf{E}^T \mathbf{L}^{-1} \mathbf{E}^T \mathbf{u}.$$

Proof. The conditions have already been shown to be necessary. We may prove that they are sufficient by actually constructing a set of $(k_i, m_i)_1^n$ which are all positive.

The governing equation (9.5.1) may be written

$$\lambda \mathbf{M} \mathbf{u} = \mathbf{E} \boldsymbol{\phi}, \quad \boldsymbol{\phi} = \mathbf{L}^{-1} \mathbf{E} \hat{\mathbf{K}} \mathbf{w}.$$

We may write this as

$$\hat{\mathbf{K}} \mathbf{w} = \mathbf{E}^{-1} \mathbf{L} \boldsymbol{\phi}, \quad \boldsymbol{\phi} = \lambda \mathbf{E}^{-1} \mathbf{M} \mathbf{u}$$

and because $\mathbf{E}^{-1}$ has the form (2.2.10), we have

$$k_i w_i = \sum_{k=i}^{n} l_k \phi_k = \tau_i, \quad \phi_i = \lambda \sum_{k=i}^{n} m_k u_k \tag{9.5.3}$$

which imply

$$\tau_i = \tau_{i+1} + l_i \phi_i, \quad \phi_i = \phi_{i+1} + \lambda m_i u_i, \quad i = 1, 2, \ldots, n,$$

with $\phi_{n+1} = 0 = \tau_{n+1}$.

We give the construction procedure for the simplest case: $j = 1$. Algorithms and examples relating to the general case may be found in Gladwell, Willms, He and Wang (1989) [115].

When $j = 1$ all the $(u_i, w_i)_1^n$ will be positive. The $(m_i)_1^n$ and λ may be assigned arbitrary positive values; equation (9.5.3b) gives $(\phi_i)_1^n$ which, when substituted in (9.5.3a), yield $(\tau_i)_1^n$. Then $k_i = \tau_i/w_i$, so that the $(k_i)_1^n$ are uniquely determined. ■

Exercises 9.5

1. Show that if $\mathbf{u}$ is the jth eigenvector of (9.5.1), then $\boldsymbol{\theta}, \boldsymbol{\tau}, \boldsymbol{\phi}$ are also jth eigenvectors of SO matrices.

9.6 Courant's nodal line theorem

We now start our discussion of the properties of eigenvectors of a class of systems that includes discrete models of membranes and acoustic cavities. Since the results we obtain are discrete analogues of results relating to continuous systems, we will start by discussing these, principally Courant's Nodal Line Theorem (CNLT), which relates to the Dirichlet eigenfunctions $u(\mathbf{x})$ of elliptic differential equations. It is well-known that such problems have positive eigenvalues with infinity as the only limit point; we label them so that

$$0 < \lambda_1 \le \lambda_2 \le \dots \tag{9.6.1}$$

Now the eigenvalues need not be distinct. If λ_n has multiplicity r we label the eigenvalues so that

$$\lambda_{n-1} < \lambda_n = \lambda_{n+1} = \dots = \lambda_{n+r-1} < \lambda_{n+r}. \tag{9.6.2}$$

CNLT (Courant and Hilbert (1953) [64], Chapter VI, Section 6.) is a theorem of wide applicability with a remarkably simple proof based on the minimax property of the Rayleigh quotient. It relates to the Dirichlet eigenfunctions of elliptic partial differential equations, the simplest and most important of which is the Helmholtz equation

$$\Delta u + \lambda \rho u = 0, \quad \mathbf{x} \in D. \tag{9.6.3}$$

The Dirichlet boundary condition is

$$u(\mathbf{x}) = 0, \quad \mathbf{x} \in \partial D. \tag{9.6.4}$$

Here Δu is the Laplacian, $\rho(\mathbf{x})$ is positive and bounded, and D is a domain in $\mathbb{R}^m$ (m-dimensional Euclidian space). Equations (9.6.3), (9.6.4) govern the spatial eigenmodes of a vibrating membrane with fixed boundary in $\mathbb{R}^2$; and acoustic standing waves in $\mathbb{R}^3$.

The *nodal set* of $u(\mathbf{x})$ is defined as the set of points $\mathbf{x}$ such that $u(\mathbf{x}) = 0$. It is known (Cheng (1976) [53]) that for $D \subset \mathbb{R}^m$, the nodal set of an eigenfunction of (9.6.3), (9.6.4) is locally composed of hypersurfaces of dimension $m-1$. These hypersurfaces cannot end in the interior of D, which implies that they are either closed, or begin and end on the boundary. In particular, therefore, in the plane ($m = 2$), the nodal set of the eigenfunction $u(\mathbf{x})$ of (9.6.3), (9.6.4) is made up of continuous curves, called *nodal lines,* which are either closed, or begin and end on the boundary.

CNLT states that each eigenfunction $u_n(\mathbf{x})$ corresponding to λ_n divides D, by its nodal set, into *at most* n subdomains, called *nodal domains*, or the more informative *sign domains,* in which $u_n(\mathbf{x})$ has one sign. We recall proofs of two versions of CNLT so that we can indicate later how the continuous and discrete results differ from each other. We express the analysis in variational form. Define

$$(u, v)_D = \int_D \boldsymbol{\nabla} u . \boldsymbol{\nabla} v d\mathbf{x}, \quad [u, v]_D = \int_D \rho u v d\mathbf{x}.$$

Here $\boldsymbol{\nabla} = (\frac{\partial}{\partial x_1}, \frac{\partial}{\partial x_2}, \ldots, \frac{\partial}{\partial x_m})$ is the *grad* operator, and

$$\int_D \cdot \, d\mathbf{x} = \int\int\int_D \cdots \int \cdot \, dx_1 dx_2 \ldots dx_m.$$

The fundamental theorem for the Rayleigh quotient

$$\lambda_R = \frac{(u, u)_D}{[u, u]_D}, \tag{9.6.5}$$

is that if u is orthogonal to the first $n-1$ eigenmodes of (9.6.3), (9.6.4), i.e.,

$$[u, u_i]_D = 0, \quad i = 1, 2, \ldots, n-1,$$

then $\lambda_R \geq \lambda_n$, with equality iff $u(\mathbf{x}) = u_n(\mathbf{x})$. We first prove a weak version of CNLT:

Theorem 9.6.1 *Suppose the eigenvalues λ_i of (9.6.3), (9.6.4) are ordered as in (9.6.5), and $u_n(\mathbf{x})$ is an eigenfunction corresponding to λ_n. If λ_n has multiplicity $r \geq 1$, so that (9.6.2) holds, then $u_n(\mathbf{x})$ has at most $n + r - 1$ sign domains.*

Proof. Suppose $u_n(\mathbf{x})$ has p sign domains D_i such that $\bigcup_{i=1}^{p} D_i = D$. Define

$$w_i(\mathbf{x}) = \begin{cases} \beta_i u_n(\mathbf{x}) & \mathbf{x} \in D_i \\ 0 & \text{otherwise} \end{cases}$$

and take

$$v(\mathbf{x}) = \sum_{i=1}^{p} c_i w_i(\mathbf{x}), \quad \sum_{i=1}^{p} c_i^2 = 1. \tag{9.6.6}$$

Since the D_i are disjoint, $(w_i(\mathbf{x}))_1^p$ are orthogonal. Scale the w_i, that is, choose the β_i, so that $[w_i, w_i]_D = 1$, then

$$[v, v]_D = \sum_{i=1}^{p} c_i^2 [w_i, w_i]_D = \sum_{i=1}^{p} c_i^2 = 1.$$

Since $w_i(\mathbf{x})$ satisfies (9.6.3) with $\lambda = \lambda_n$, on D_i, and $w_i(\mathbf{x}) = 0$ on ∂D_i, the divergence theorem gives

$$\begin{aligned}(w_i, w_i)_{D_i} &= \int_D \nabla w_i \cdot \nabla w_i d\mathbf{x} \\ &= \int_D \{div(w_i \nabla w_i) - w_i \triangle w_i\} d\mathbf{x} \\ &= \int_{\partial D_i} w_i \frac{\partial w_i}{\partial n} d\mathbf{x} + \lambda_n \int_{D_i} \rho w_i^2 d\mathbf{x} = \lambda_n.\end{aligned}$$

Thus $(v, v)_D = \sum_{i=1}^{p} c_i^2 (w_i, w_i)_{D_i} = \sum c_i^2 \lambda_n = \lambda_n$, so that $\lambda_R = \lambda_n$. But we may choose $(c_i)_1^p$ so that $[v, u_i]_D = 0, \quad i = 1, 2, \ldots, p-1$, and hence, for that choice, Rayleigh's principle states that $\lambda_R \geq \lambda_p$. Thus $\lambda_p \leq \lambda_n$. Since $\lambda_n < \lambda_{n+r}$, we have $\lambda_p < \lambda_{n+r}$ so that $p < n + r$, $p \leq n + r - 1$. ■

Note that this proof does not require D to be connected. Note also that if λ_n is simple, so that $r = 1$, then the Theorem states that $u_n(\mathbf{x})$ has at most n sign domains. We need to strengthen the result for multiple eigenvalues, reducing the upper bound $n + r - 1$ to n.

To reduce the upper bound in this way we need what is called a *unique continuation theorem.* Loosely speaking, what such a theorem states is that if a solution of (9.6.3) is identically zero in a finite region of D then it is zero throughout D; the only way that it can be *continued* from the zero patch is by taking it identically zero. (Specifically, for those who have a functional analysis background, Jerison and Kenig (1985) [188] proved that if any solution $u \in H_0^1(D)$ of the weak version of (9.6.3) vanishes on a non-empty open subset of a *connected* domain D, then $u \equiv 0$ in D.) Using this result we can prove

Theorem 9.6.2 *Suppose D is connected, the eigenvalues of (9.6.3), (9.6.4) are ordered as in (9.6.5), and $u_n(\mathbf{x})$ is an eigenfunction corresponding to λ_n, then $u_n(\mathbf{x})$ has at most n sign domains.*

Proof. Suppose $u_n(\mathbf{x})$ has $p > n$ sign domains. Define the $w_i(\mathbf{x})$ as before, and define $v(\mathbf{x})$ by (9.6.6) with $c_{n+1} = 0 = \cdots = c_p$, so that $v(\mathbf{x}) \equiv 0$ on $D_{n+1}, \ldots, D_p$. Again we have $\lambda_R = \lambda_n$, and we may choose $(c_i)_1^n$ so that $[v, u_i]_D = 0$, $i = 1, 2, \ldots, n-1$. Thus $v(\mathbf{x})$ is an eigenfunction of (9.6.3), (9.6.4.), but it is identically zero on D_{n+1} and hence, by the unique continuation theorem, it is identically zero on D. This contradiction implies $p \leq n$. ■

We note that the theorem, which is due to Herrmann (1935) [171] and Pleijel (1956) [266], implies that if D is connected, then λ_1 is simple, i.e., $\lambda_1 < \lambda_2$. For any eigenfunction $u_1(\mathbf{x})$ can have at most one sign domain, i.e., it has the same sign throughout D. There cannot be two functions u, v, which are of one sign in a connected domain D and are orthogonal to each other.

Theorem 9.6.3 *Theorem 9.6.2 holds even if D is not connected.*

Proof. Suppose D consists of q connected domains $(D_k)_1^q$. Label the eigenvalues $\lambda_i^{(k)}$ of each D_k increasingly, and suppose the corresponding eigenfunctions are $u_i^{(k)}(\mathbf{x})$. Now assemble the eigenvalue sequences $\{\lambda_i^{(k)}\}$, $k = 1, 2, \ldots, q$; $i = 1, 2, \ldots$ into one non-decreasing sequence $\{\lambda_j\}$ to give the eigenvalues of D. The corresponding eigenfunctions of D are

$$u_j(\mathbf{x}) = \begin{cases} u_i^{(k)}(\mathbf{x}) & \text{on } D_k \\ 0 & \text{elsewhere.} \end{cases}$$

The ordinal number j of a given $\lambda_i^{(k)}$ in this sequence will satisfy $j \geq i$. Theorem 9.6.2 for D_k states that $u_i^{(k)}(\mathbf{x})$ has no more than i sign domains on D_k, so that $u_j(\mathbf{x})$ will have no more than j sign domains on D_k, and it will be zero elsewhere. ∎

9.7 Some properties of FEM eigenvectors

Our aim in the next few sections is to obtain discrete versions of Theorems 9.6.1-9.6.3. In a first step towards achieving this aim, we discuss some properties of eigenvectors of finite element models. We return to the analysis of Section 2.5 and suppose that we are dealing with a FEM model of a membrane with fixed boundary using linear interpolation over *acute* angled triangles, or correspondingly of an acoustic cavity using linear interpolation over tetrahedra with *obtuse* angles between normals to faces. In each of these models, the FEM mesh yields a set of vertices connected by edges to form a graph. There are two kinds of vertices, *boundary vertices*, where $u = 0$ because of the boundary conditions, and the remainder. These *non-boundary vertices* are those that appear in the analysis; they form a graph $\mathcal{G}$ on N vertices $P_i \in \mathcal{V}$ with edge set $\mathcal{E}$. The FEM analysis yields two matrices $\mathbf{K}, \mathbf{M}$ on $\mathcal{G}$ with the properties that if $i \neq j$ then

$$\left.\begin{array}{ll} k_{ij} < 0, \ m_{ij} > 0 & \text{if } (P_i, P_j) \in \mathcal{E} \\ k_{ij} = 0, \ m_{ij} = 0 & \text{otherwise.} \end{array}\right\} \tag{9.7.1}$$

Note that if $(P_i, P_j) \in \mathcal{E}$, we say that P_i, P_j are *adjacent* vertices, and we write $P_i \sim P_j$. The analysis will revolve around *nodal* vertices, i.e., vertices P_i where $u_i = 0$. We first prove

Theorem 9.7.1 *Under conditions (9.7.1), a non-boundary nodal vertex of an eigenvector of (9.1.1) cannot have neighbours that are all of one sign.*

Proof. Suppose P_i, a non-boundary vertex, is nodal, i.e., $u_i = 0$. The ith line of (9.1.1) is

$$\sum (k_{ij} - \lambda m_{ij}) u_j = 0, \tag{9.7.2}$$

where the sum is over those $j(\neq i)$ for which $P_i \sim P_j$; for those j, $k_{ij} - \lambda m_{ij} < 0$. If $u_j \geq 0 (\leq 0)$ for all such j, with at least one inequality strict, then the left-hand side of (9.7.2) would be strictly negative (positive), which is a contradiction. ■

This theorem implies that a non-boundary nodal vertex must either have both positive and negative neighbours, or all nodal neighbours. We may extend this statement to say that a set of nodal vertices of an eigenvector must have positive and negative neighbours: it must separate positive and negative vertex sets. If $\mathcal{G}$ is connected, so that $\mathbf{K}$ and $\mathbf{M}$ are irreducible (see Busacker and Saaty (1965) [46], then we can say more: if an eigenvector $\mathbf{u}$ is non-negative then it must be strictly positive. Such an eigenvector must correspond to the lowest eigenvalue, which must therefore be simple: there cannot be two positive eigenvectors $\mathbf{u}, \mathbf{v}$ which are orthogonal w.r.t. $\mathbf{M}$: $\lambda_1 < \lambda_2$.

There is an important *maximum principle* for the p.d.e. (9.6.3): a solution $u(\mathbf{x})$ cannot have an interior positive minimum or an interior negative maximum (Protter and Weinburger (1984) [271]). To state the discrete version of this principle, we must divide the non-boundary vertices of a FEM mesh into two subsets: vertices adjacent to boundary vertices, that we call *near-boundary vertices*; the remainder, that we term *interior vertices.*

Theorem 9.7.2 *If $\mathcal{G}$ is connected, and (9.7.1) holds, an eigenvector of (9.1.1) cannot have a local positive minimum or a local negative maximum at an interior vertex.*

Proof. By definition, an interior vertex is adjacent only to non-boundary vertices. It is therefore a vertex of an *interior element*, i.e., an element that has no vertices on the boundary. Because of the way in which it is formed, by (2.5.6), the stiffness matrix $\mathbf{K}_e$ of an interior element admits a rigid-body mode, $\{1,1,1\}$ for a triangular mesh, $\{1,1,1,1\}$ for a tetrahedral mesh. If P_i is an interior vertex, all the elements to which P_i belongs are interior elements. This means that after assembling the $\mathbf{K}_e$ to form $\mathbf{K}$ we may deduce that, if P_i is an interior vertex, then $\sum k_{ij} = 0$, where again the sum is taken over all j such that $P_j \sim P_i$. The ith line of (9.1.1) is

$$0 = \sum k_{ij}u_j - \lambda m_{ij}u_j,$$

so that

$$\sum k_{ij}(u_j - u_i) + (\sum k_{ij})u_i = \lambda \sum m_{ij}u_j. \tag{9.7.3}$$

Suppose that there is a local positive minimum at an interior vertex P_i, so that $u_i \geq 0$ and $u_j - u_i \geq 0$ for all j such that $P_j \sim P_i$, and either the first inequality is strict, or the second inequality is strict for at least one j such that $P_j \sim P_i$. (We need the connectedness of $\mathcal{G}$ to be sure that every vertex P_i *does* have a neighbour.) The first sum on the left is non-positive, while the second sum is zero; the sum on the right is non-negative; one of the two sides, left or right, is non-zero This is impossible. ■

This theorem relates to the eigen*vectors* of (9.1.1), but we can immediately reword it to apply to FEM eigen*functions* obtained by linear interpolation from the vertex values. An eigenfunction obtained by linear interpolation can have local maxima and minima only at the vertices of the mesh. We conclude that an eigenfunction cannot have a local positive minimum or a local negative maximum at an interior vertex. Loosely speaking, we may say that a mode may have waves, but not dimples.

One of the mainstays of the theory related to (9.6.3) is the unique continuation theorem. It was this that allowed us to reduce the upper bound on the number of sign domains for eigenfunctions of (9.6.3), (9.6.4), from $n+r-1$ to n. There is no *straightforward* discrete analogue of unique continuation; there is an analogue, as described in Lemma 9.9.2, but it is not straightforward. Figure 9.7.1 shows an example of a FEM eigenmode with zero patches. If the matrices $\mathbf{K}$ and $\mathbf{M}$ are symmetrical about the x-and y-axes, then there will be a mode that is antisymmetrical about both axes, so that the vertex values must have the signs shown. There are four completely zero triangles in the centre, and four other pairs of zero triangles, but the eigenmode is not identically zero.

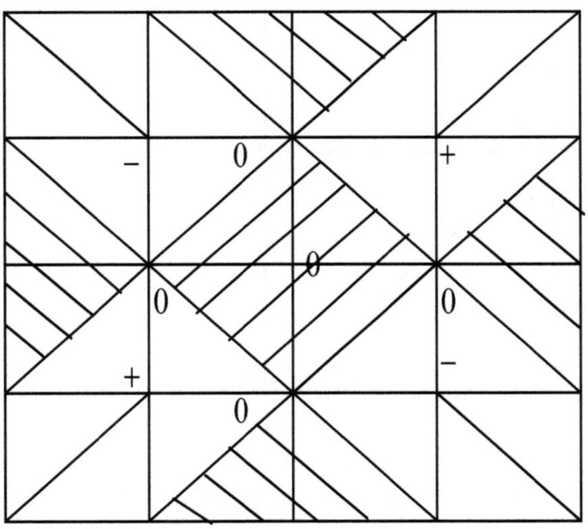

Figure 9.7.1 - An eigenvector can have one or more zero (shaded) polygons

Even though there is no straightforward discrete analogue of unique continuation, we can still obtain discrete analogues of Theorem 9.6.1, 9.6.2. First, we need to find the discrete FEM counterparts of the *sign domains* of the continuous theorems. There are two distinct ways of looking at the piecewise linear function u obtained from an eigenvector of (9.1.1): looking at the values u_i, and particularly at the signs of u_i, at the vertices P_i of $\mathcal{G}$; looking at the subregions with piecewise straight boundaries on which the linearly interpolated $u(\mathbf{x})$ has one sign, either loosely, $u(\mathbf{x}) \geq 0$ (≤ 0) or strictly, $u(\mathbf{x}) > 0$ (< 0).

Consider the first way. The FEM mesh defines a graph $\mathcal{G}$ with N vertices P_i. A FEM vector $\mathbf{u} \in V_N$ associates a value u_i and in particular a sign +, 0, or -, to each vertex P_i of $\mathcal{G}$. We may connect the (strictly) positive vertices by edges of $\mathcal{E}$ to form maximal connected subgraphs of $\mathcal{G}$, called *strong positive sign graphs.* We may do the same with the negative vertices, to form *strong negative sign graphs.* In this way, we can partition the graph $\mathcal{G}$ into disjoint strong positive and strong negative sign graphs, and zero vertices. Figure 9.7.2 shows a graph with 2 strong positive and 2 strong negative sign graphs, each of which has just one vertex. Alternatively, we may partition $\mathcal{G}$ into *weak positive* and *weak negative* sign graphs, by forming maximal connected subgraphs of non-negative, and non-positive vertices, respectively. The graph in Figure 9.7.2 has just one weak positive sign graph, and one weak negative sign graph; these weak sign graphs overlap.

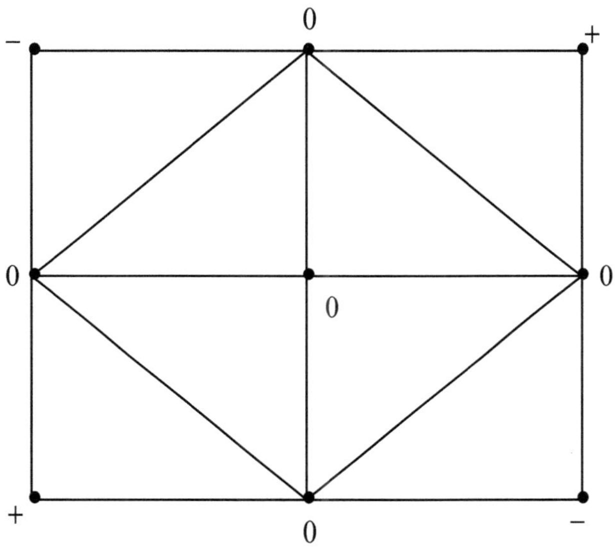

Figure 9.7.2 - The graph has two strong positive and two strong negative sign graphs; it has just one weak positive, and one weak negative sign graph

Two sign graphs S_1, S_2, strong or weak, are said to be *adjacent* if there are vertices $P_1 \in S_1$, $P_2 \in S_2$ such that $P_1 \sim P_2$. We need the following simple but important property:

Lemma 9.7.1 *If two different sign graphs are adjacent, then they have opposite signs.*

Proof. If they had the same sign then one at least would not be maximal. ∎

Note that while two adjacent strong sign graphs are disjoint, two adjacent weak sign graphs may overlap.

Now consider the second way; looking at the signs of the piecewise linear 'eigenfunction' interpolated from the vertex values u_i of an eigenvector $\mathbf{u}$. This 'eigenfunction' is defined on a domain with piecewise straight (in R^2) or piecewise plane (in R^3) boundary, that may be some approximation to an original domain D. We are not concerned with how good the approximation is, nor are we concerned with convergence or taking a 'sufficiently fine' mesh. Thus, we will simply call the FEM domain D, and forget that there might have been some other original domain with perhaps curved boundary. The domain D may be divided, like the graph $\mathcal{G}$, into strong sign subdomains, D_i, on which $u(\mathbf{x})$ has one strict sign, and on the boundaries of which $u(\mathbf{x}) = 0$. Each of these domains will be polygonal in $\mathbb{R}^2$, polyhedral in $\mathbb{R}^3$. In particular, the nodal places of u in R^2 will be piecewise straight lines, either closed or beginning and ending on the boundary, or nodal polygons, as in Figure 9.7.1. In R^3 they will be piecewise plane surfaces which are either closed or begin and end on the boundary, or polyhedra. Instead of using *strong* sign domains, we may use *weak*; they too will have piecewise straight or piecewise plane boundaries. A weak positive and a weak negative sign domain may overlap.

For triangular or tetrahedral meshes corresponding to linear interpolation, there is a clear correspondence between the sign graphs on the one hand and the sign domains on the other. For each strong or weak, positive or negative sign domain there is exactly one strong or weak, positive or negative, sign graph. This means that we can count the number of sign domains by counting the number of sign graphs.

We note however, that the rectangular FEM mesh which is sometimes used in $\mathbb{R}^2$ does not have such simple properties. Inside a rectangle, $u(x, y)$ has a bilinear interpolation

$$u(x, y) = p + qx + ry + sxy.$$

Now all four vertices of the rectangle are neighbours of each other, in the sense that all the off-diagonal entries in the element matrices are non-zero. This is why we show the vertices of the rectangle joined by the diagonals as well as by the sides, as in Figure 9.7.3. (But the intersection of the diagonals is not a vertex of the graph.)

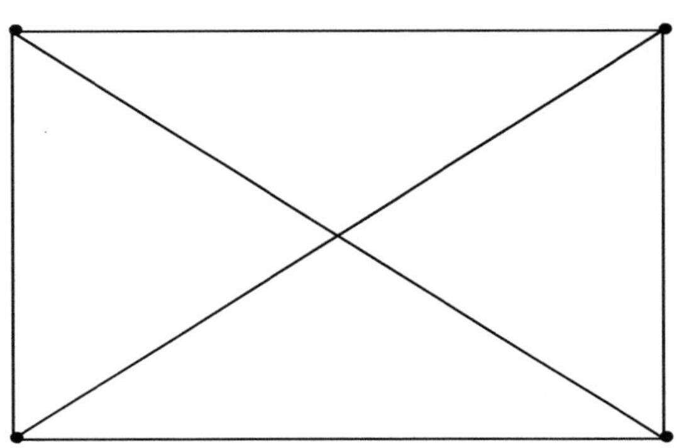

Figure 9.7.3 - A rectangular finite element; each vertex is connected to all the others

It may be shown for this mesh that the element mass matrix is strictly positive, and that the off-diagonal entries of the element stiffness matrix are strictly negative iff the sides a, b of the rectangle satisfy $1/\sqrt{2} < a/b < \sqrt{2}$, i.e., if the rectangle is not too thin. There is a similar result (Ex. 9.7.1) for a rectangular box mesh in $\mathbb{R}^3$. Thus, under these conditions, the matrices $\mathbf{K}, \mathbf{M}$ for the whole mesh will satisfy the inequalities (9.7.1). This means that we can apply the results of the analysis below to the sign *graphs* of a rectangular mesh, but as the example in Figure 9.7.4 shows, we cannot extend them to the sign *domains.* Figure 9.7.4 shows a mesh made up of nine square elements. The vertices A and B are adjacent and have the same sign, so that they belong to the same sign *graph.* However, because nodal lines in an element are now hyperbolic, and not straight, A and B lie in different sign *domains*; there is an intervening negative sign doman between them.

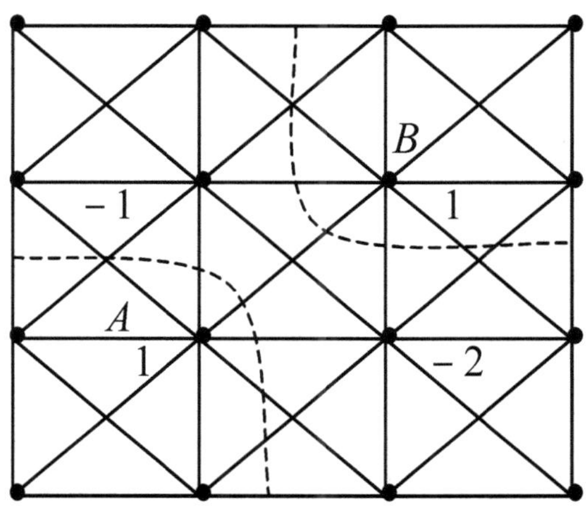

Figure 9.7.4 - Vertices A and B are adjacent, but belong to different sign domains

Exercises 9.7

1. Find the conditions on the ratios of the dimensions of a rectangular box so that the stiffness matrix based on linear interpolation of the assumed modes

$$1, x, y, z, yz, zx, xy, xyz$$

has the sign property (9.7.1).

9.8 Strong sign graphs

The discussion in Section 9.7 should have made it clear that we can study the sign properties of an eigenvector on a graph $\mathcal{G}$ as a problem in its own right, that is, without considering the problem as arising from a FEM model. We will do this and, to simplify the analysis, we will consider the eigenvalue in standard form, namely

$$(\mathbf{A} - \lambda\mathbf{I})\mathbf{u} = 0 \tag{9.8.1}$$

under the following assumption: if $i \neq j$ then

$$a_{ij} = 0 \quad if(P_i, P_j) \notin \mathcal{E}, \; a_{ij} < 0 \quad if(P_i, P_j) \in \mathcal{E}. \tag{9.8.2}$$

We will then show at the end that all the results hold for (9.1.1) under the condition (9.7.1). In this section, we will understand *sign graph* to mean *strong* sign graph. The theorem we are about to prove regarding the number of sign graphs is a discrete analogue of Theorem 9.6.1. In order to prove it, we need to set up a procedure mimicking that used in Theorem 9.6.1, and prove a Lemma, following Davies, Gladwell, Leydold and Stadler (2001) [71].

Suppose $\mathbf{u}$ is an eigenvector of (9.8.1) in the eigenspace of λ_n. Suppose $\mathbf{u}$ has m sign graphs $\mathcal{S}_i$, $i = 1, 2, \ldots, m$. Define m vectors $\mathbf{w}_i$, $i = 1, 2, \ldots, m$, such that

$$\mathbf{w}_i = \begin{cases} \mathbf{u} & \text{on } \mathcal{S}_i \\ \mathbf{0} & \text{otherwise.} \end{cases}$$

Explicitly, let $\mathbf{w}_i = \{w_{i,1}, w_{i,2}, \ldots, w_{i,N}\}$. Then $w_{i,j} = u_j$ if $P_j \in S_j$, and $w_{i,j} = 0$ otherwise. Thus

$$\mathbf{u} = \sum_{i=1}^{m} \mathbf{w}_i.$$

Now form

$$\mathbf{v} = \sum_{i=1}^{m} c_i \mathbf{w}_i. \tag{9.8.3}$$

Using straightforward algebra, we may verify (Ex. 9.8.1) Duval and Reiner's Lemma (Duval and Reiner (1999) [82]).

Lemma 9.8.1

$$\mathbf{v}^T\mathbf{A}\mathbf{v} - \lambda\mathbf{v}^T\mathbf{v} = \sum_{i=1}^{m} c_i^2 \mathbf{w}_i^T(\mathbf{A}\mathbf{u} - \lambda\mathbf{u}) - \frac{1}{2}\sum_{i,j=1}^{m}(c_i - c_j)^2 \mathbf{w}_i^T\mathbf{A}\mathbf{w}_j.$$

This leads to

Theorem 9.8.1 *Any eigenvector corresponding to λ_n has at most $n+r-1$ sign graphs.*

Here the governing equation is (9.8.1), $\mathbf{A}$ satisfies (9.8.2), the $(\lambda_n)_1^N$ are ordered as in (9.6.1), and λ_n has multiplicity r, so that (9.6.2) holds.

Proof. Since none of the $\mathbf{w}_i$ is identically zero and they are disjoint, their linear span has dimension m. It follows that there are real constants $(c_i)_i^m$, not all zero, such that $\mathbf{v}$ is non-zero and is orthogonal to the first $(m-1)$ eigenvectors $(\mathbf{u}_j)_1^{m-1}$ of $\mathbf{A}$, i.e.,

$$\mathbf{v}^T\mathbf{u}_j = 0, \; j = 1, 2, \ldots, m-1.$$

Without loss of generality we can take $\mathbf{v}^T\mathbf{v} = 1$. Therefore, by the minimax theorem (Section 2.10) we have

$$\mathbf{v}^T\mathbf{A}\mathbf{v} \geq \lambda_m. \tag{9.8.4}$$

Now use Lemma 9.8.1 with $\lambda = \lambda_n$, $\mathbf{u} = \mathbf{u}_n$. We find

$$\mathbf{v}^T\mathbf{A}\mathbf{v} - \lambda_n = -\frac{1}{2}\sum_{i,j=1}^{m}(c_i - c_j)^2\mathbf{w}_i^T\mathbf{A}\mathbf{w}_j. \tag{9.8.5}$$

We will show that the sum on the right is non-negative. A term $\mathbf{w}_i^T\mathbf{A}\mathbf{w}_j$ is non-zero only if $\mathbf{w}_i, \mathbf{w}_j$ correspond to adjacent sign graphs; adjacent sign graphs have opposite signs (Lemma 9.7.1); adjacent sign graphs are disjoint. This means that any non-zero product $\mathbf{w}_i^T\mathbf{A}\mathbf{w}_j$ involves only negative, off-diagonal entries in $\mathbf{A}$; therefore

$$\mathbf{w}_i^T\mathbf{A}\mathbf{w}_j = (\pm)(-)(\mp) = +.$$

Therefore, equation (9.8.5) gives

$$\mathbf{v}^T\mathbf{A}\mathbf{v} - \lambda_n \leq 0. \tag{9.8.6}$$

This combined with (9.8.4) states that $\lambda_m \leq \lambda_n$. Since $\lambda_n < \lambda_{n+r}$, we have $\lambda_m < \lambda_{n+r}$, i.e., $m \leq n + r - 1$. ■

Note that we cannot deduce that the inequality in (9.8.6) is strict, because $c_i - c_j$ might be zero for all those pairs i, j for which $\mathbf{w}_i^T\mathbf{A}\mathbf{w}_j$ was (strictly) positive.

As we stated earlier, Theorem 9.8.1 is a discrete counterpart of CNLT in the form of Theorem 9.6.1. Various researchers attempted to reduce the bound $n + r - 1$. Friedman (1993) [96] gave the example of a star on N vertices to show that the bound could not be reduced, as in Theorem 9.6.2, to n. For the star, the second eigenvalue of the so-called Laplacian matrix (Ex. 9.8.2) has multiplicity $N - 2$, and has an eigenvector with $N - 1$ sign graphs. If therefore $N - 1 > 2$, i.e., $N \geq 4$, then a second eigenvector has more than 2 sign graphs. In spite of this counterexample, Duval and Reiner (1999) [82] attempted to reduce the bound to n; the error in their logic is pinpointed in Zhu (2000) [342]; essentially their error lay in thinking that the inequality in (9.8.6) could be made strict. Comments on partly erroneous results put forward by Friedman (1993) [96] and van der Holst (1996) [326] may be found in Davies, Gladwell, Leydold and Stadler (2001) [71].

We note that the distinction between the bounds $n+r-1$ and n appears only when $r > 1$, i.e., λ_n is multiple. Following Gladwell and Zhu (2002) [131] we now show that although it is not possible to reduce the bound $n+r-1$ when λ_n is multiple, it is possible to construct r orthogonal vectors $(\mathbf{u}_j)_n^{n+r-1}$, spanning the eigenspace of λ_n, such that $\mathbf{u}_j$ has at most j sign graphs, $j = n, n+1, \ldots, n+r-1$. In fact, it is possible to go further and construct r linearly independent (but not necessarily orthogonal) vectors spanning the eigenspace of λ_n, such that each of them has at most n sign graphs. We introduce the notation $SG(\mathbf{u})$ for the number of sign graphs of $\mathbf{u}$.

Theorem 9.8.2 *Under the conditions stated in Theorem 9.8.1, if* $\mathbf{u}$ *is an eigenvector corresponding to* λ_n*, and* $SG(\mathbf{u}) = m > n$*, then in the notation of (9.8.3) we may find*

$$\mathbf{v} = \sum_{j=1}^{n} c_j \mathbf{w}_j$$

such that $\mathbf{v}$ *is an eigenvector corresponding to* λ_n*, and* $SG(\mathbf{v}) \le n$*.*

Proof. We can choose c_j, not all zero, such that $\mathbf{v}$ is orthogonal to $(\mathbf{u}_i)_1^{n-1}$. By the minimax theorem $\lambda_R \ge \lambda_n$. By Lemma 9.8.1, $\lambda_R \le \lambda_n$. Thus $\lambda_R = \lambda_n$ and $\mathbf{v}$ is an eigenvector corresponding to λ_n. By its construction, $SG(\mathbf{v}) \le n$. ■

We denote a normalised $\mathbf{v}$ so formed, by $\mathbf{v} = T((\mathbf{w}_j)_1^n, (\mathbf{u}_i)_1^{n-1})$. This $\mathbf{v}$ may not be unique; there is always a non-trivial set $(c_j)_1^n$, but it need not be unique.

Note that in Theorem 9.6.2, for the continuous CNLT, we suppose that the eigenfunction $u_n(\mathbf{x})$ has more than n sign domains, and we construct a purported eigenfunction $v(\mathbf{x})$ orthogonal to $(u_i(\mathbf{x}))_1^{n-1}$, but zero in D_{n+1}; then we use unique continuation of an eigenfunction on a connected domain D to show that $v(\mathbf{x}) \equiv 0$ in D; this contradicted the hypothesis that $v(\mathbf{x})$ was an eigenfunction, i.e., not trivial. In the discrete case we start with an eigenvector $\mathbf{u}_n$ with $SG(\mathbf{u}_n) = m > n$, and construct another $\mathbf{v}$ with $SG(\mathbf{v}) \le n$; the new eigenvector has at least one zero sign graph, but it *is* an eigenvector, and there is no contradiction involved.

We may now prove

Theorem 9.8.3 *Suppose the conditions stated in Theorem 9.8.1 hold. If* λ_n *is an eigenvalue of multiplicity* r*, then we may find* r *orthonormal eigenvectors* $(\mathbf{u}_j)_n^{n+r-1}$ *corresponding to* λ_n*, such that* $SG(\mathbf{u}_j) \le j$*,* $j = n, n+1, \ldots, n+r-1$*.*

Proof. The r-dimensional eigenspace V of λ_n has an orthonormal basis $(\mathbf{v}_j)_n^{n+r-1}$. Theorem 9.8.1 states that $SG(\mathbf{v}_j) \le n+r-1$ for $j = n, n+1, \ldots, n+r-1$. If $SG(\mathbf{v}_n) \le n$, take $\mathbf{u}_n = \mathbf{v}_n$; otherwise $SG(\mathbf{v}_n) > n$. In this case if $(\mathbf{w}_j)_1^m, (m > n)$ are the sign graph vectors of $\mathbf{v}_n$, take $\mathbf{u}_n = T((\mathbf{w}_j)_1^n; (\mathbf{u}_j)_1^{n-1})$, so that $SG(\mathbf{u}_n) \le n$. We now proceed by induction. Suppose we have constructed orthonormal vectors $\mathbf{u}_n, \mathbf{u}_{n+1}, \ldots, \mathbf{u}_{n+s-1} (1 < s < r)$ such that $SG(\mathbf{u}_j) \le j$, for $j = n, n+1, \ldots, n+s-1$. We show how to construct $\mathbf{u}_{n+s}$. First, find a new orthonormal basis $(\mathbf{u}_j)_n^{n+s-1}$, $(\mathbf{x}_j)_{n+s}^{n+r-1}$ for V. If $SG(\mathbf{x}_{n+s}) \le n+s$, then take $\mathbf{u}_{n+s} = \mathbf{x}_{n+s}$; otherwise $SG(\mathbf{x}_{n+s}) > n+s$;

in this case, if $(\mathbf{w}_j)_1^m (m > n+s)$ are the sign graph vectors of $\mathbf{x}_{n+s}$, take $\mathbf{u}_{n+s} = T((\mathbf{w}_j)_1^{n+s}; (\mathbf{u}_j)_1^{n+s-1})$. We may proceed in this way to find $(\mathbf{u}_j)_1^{n+r-1}$ such that $SG(\mathbf{u}_j) \le j$. ■

We now strengthen this result and prove

Theorem 9.8.4 *Suppose the conditions stated in Theorem 9.8.1 hold, and that λ_n is an eigenvalue with multiplicity r, and eigenspace V. There is a basis $(\mathbf{u}_j)_1^{n+r-1}$ for V such that $SG(\mathbf{u}_j) \le n$.*

Proof. We proceed much as in Theorem 9.8.3. We construct $\mathbf{u}_n$ as before, and then use induction: we suppose that we have found a basis $(\mathbf{u}_j)_1^{n+s-1}$, $(\mathbf{x}_j)_{n+s}^{n+r-1}$ for V such that $SG(\mathbf{u}_j) \le n$ for $j = n, n+1, \ldots, n+s-1$, and we show how to construct $\mathbf{u}_{n+s}$. If $SG(\mathbf{x}_{n+s}) \le n$, then $\mathbf{u}_{n+s} = \mathbf{x}_{n+s}$; otherwise $SG(\mathbf{x}_{n+s}) = n+t$, $1 \le t \le r-1$. In this case, let W be the space spanned by the sign graph vectors $(\mathbf{w}_j)_1^{n+t}$ of $\mathbf{x}_{n+s}$: if $\mathbf{w} \in W$, then $\mathbf{w} = \sum_{j=1}^{n+t} c_j \mathbf{w}_j = \mathbf{Wc}$. Let Y be the subspace of W orthogonal to $(\mathbf{u}_j)_1^{n-1}$; Y is not empty because $\mathbf{x}_{n+s} = \sum_{j=1}^{n+t} \mathbf{w}_j \in Y$. If $\mathbf{y} \in Y$, then $\mathbf{y} = \mathbf{Wc}$ and $\mathbf{u}_j^T\mathbf{y} = \mathbf{u}_j^T\mathbf{Wc} = 0$, $j = 1, 2, \ldots, n-1$. Of these $n-1$ constraints on the c_j, $m \le n-1$ are independent; they may be written $\mathbf{Bc} = \mathbf{0}$, where $\mathbf{B} \in M_{m,n+t}$. Then the matrix $\mathbf{B}$ has m linearly independent columns which, by suitably renumbering the $\mathbf{w}_j$, may be taken as the first m. Thus $\mathbf{Bc} = \mathbf{0}$ may be written

$$[\mathbf{B}_1, \mathbf{B}_2]\begin{bmatrix}\mathbf{c}_1\\ \mathbf{c}_2\end{bmatrix} = \mathbf{0}, \tag{9.8.7}$$

where $\mathbf{B}_1 \in M_m$ is non-singular, $\mathbf{B}_2 \in M_{m,n+t-m}$,

$$\mathbf{c}_1 = \{c_1, c_2, \ldots c_m\},\ \mathbf{c}_2 = \{c_{m+1}, \ldots, c_{n+t}\}.$$

The solution space of (9.8.7) is spanned by the $n+t-m$ solutions obtained by taking $c_{2,k}^{(i)} = \delta_{ik}$, $i = m+1, \ldots, n+t$, and then solving for $\mathbf{c}_1^{(i)}$. Each such choice gives a vector $\mathbf{y}_i = \mathbf{Wc}^{(i)}$; these vectors are linearly independent and they span Y; by construction $SG(\mathbf{y}_i) \le m+1 \le n$. At least one of the $\mathbf{y}_i$, say $\mathbf{y}_p$, must be linearly independent of $(\mathbf{u}_j)_n^{n+s-1}$, for $\mathbf{x}_{n+s} \in Y$ is, by construction, linearly independent of $(\mathbf{u}_j)_n^{n+s-1}$. Take $\mathbf{u}_{n+s} = \mathbf{y}_p$, then $SG(\mathbf{u}_{n+s}) \le n$. We may proceed in this way to find $(\mathbf{u}_j)_n^{n+r-1}$ such that $SG(\mathbf{u}_j) \le n$. ■

We conclude this section by discussing some other implications of Lemma 9.8.1.

Suppose that $\mathbf{u}$ is an eigenvector corresponding to a multiple eigenvalue λ_n, so that $\mathbf{Au} = \lambda_n\mathbf{u}$. Suppose that $SG(\mathbf{u}) = m > n$, and $\mathbf{v}$ given by (9.8.3) has been computed so that it is orthogonal to $(\mathbf{u}_j)_1^{n-1}$. Then, as we showed before, $\mathbf{v}$ is also an eigenvector corresponding to λ_n, i.e., $\mathbf{Av} = \lambda_n\mathbf{v}$. Then Lemma 9.8.1 with $\lambda = \lambda_n$ demands

$$\sum_{i,j=1}^{m} (c_i - c_j)^2 \mathbf{w}_i^T \mathbf{A} \mathbf{w}_j = 0. \tag{9.8.8}$$

But, as we showed earlier, $\mathbf{w}_i^T\mathbf{A}\mathbf{w}_j \geq 0$, with strict inequality iff S_i, S_j are adjacent. Equation (9.8.8) implies that if S_i, S_j are adjacent, then $c_i = c_j$. This means that if one sign graph, S_i, is omitted in the construction of $\mathbf{v}$ from the sign graphs of $\mathbf{u}$ (i.e., $c_i = 0$), then any sign graph S_j adjacent to S_i must also be omitted ($c_j = c_i = 0$). On the other hand, if one sign graph S_i is included in $\mathbf{v}$, then any other sign graph S_j adjacent to S_i must be included, and must be included with the same weight as $S_i : c_j = c_i$. This means that in the construction of $\mathbf{v}$ from the sign graphs of $\mathbf{u}$, any connected graph composed of sign graphs of $\mathbf{u}$ must either be included or excluded as a whole. This leads to

Theorem 9.8.5 *Suppose the conditions stated in Theorem 9.8.1 hold. Suppose that* $\mathbf{u}$, *an eigenvector corresponding to* λ_n *has more than* n *sign graphs, so that* $SG(\mathbf{u}) = n + g$, $g \geq 1$. *These sign graphs may be grouped into* $g + s$ *mutually disjoint connected graphs* $(\mathcal{C}_j)_1^{g+s}$, *and* $s \geq 1$.

Proof. If $s < 1$, i.e., $s \leq 0$, then there are at most g connected graphs $\mathcal{C}_j$. If we form a non-trivial eigenvector from the $n+g$, sign graphs of $\mathbf{u}$, by deleting g of them, at least one S_j from each $\mathcal{C}_j$, then *none* of the $\mathcal{C}_j$ will appear; $\mathbf{v}$ will be identically zero. This contradiction implies $s \geq 1$. ■

This theorem has a number of corollaries:

(i) If $\mathbf{u}$ has $m = n+g$ sign graphs, then a connected component $\mathcal{C}_j$ can contain at most n sign graphs. For if one contained $n+1$ sign graphs, then there would be at most $1 + (n + g - n - 1) = g$ connected components. This provides a somewhat restricted counterpart of Theorem 9.6.2.

(ii) If there are n sign graphs in one component $\mathcal{C}_j$, and $n \geq 2$, then $g \geq 2$. For if n sign graphs are in one component $\mathcal{C}_j$, they must constitute an eigenvector; so too will the remaining $n+g-n = g$ sign graphs. If $n \geq 2$, an eigenvector, being orthogonal to $\mathbf{u}_1$, must have at least two sign graphs; $g \geq 2$.

(iii) If $\mathcal{G}$ is connected and $\mathbf{u}_n$ has no zeros then, whether λ_n is simple or multiple, $SG(\mathbf{u}_n) \leq n$. For if there are no zero vertices then all the sign graphs fall into one component.

Exercises 9.8

1. Establish Duval and Reiner's Lemma 9.8.1.

2. Consider the star on N vertices with $a_{11} = N - 1$, $a_{ii} = 1$, $a_{1i} = -1$, $i = 2, \ldots, N$. Show that its eigenvalues are $0, 1, N$.

   Show that the second eigenvalue has multiplicity $N - 2$, and that there is an eigenvector corresponding to λ_2 with $N - 1$ sign graphs.

3. Construct $N - 2$ orthogonal eigenvectors of λ_2 for the star in Ex. 9.8.2 such that $\mathbf{u}_j$ has just j sign graphs, $j = 2, 3, \ldots, N - 1$.

4. For the same star, construct $N-2$ linearly independent eigenvectors $\mathbf{u}_j$ such that each has just 2 sign graphs.

9.9 Weak sign graphs

In order to obtain a proper discrete analogue of Theorem 9.6.2, we must consider weak sign graphs.

Lemma 9.9.1 *Suppose $\mathcal{S}_1, \mathcal{S}_2$ are adjacent weak sign graphs. There is a pair of vertices P_1, P_2 such that $P_1 \in \mathcal{S}_1$, $P_2 \in \mathcal{S}_2 \backslash \mathcal{S}_1$ (i.e., P_2 is in $\mathcal{S}_2$, but not in $\mathcal{S}_1$) and $P_1 \sim P_2$.*

Proof. Without loss of generality, assume $\mathcal{S}_1$ is weak positive and $\mathcal{S}_2$ is weak negative. If $\mathcal{S}_1, \mathcal{S}_2$ are disjoint, then by the definition of adjacency, there exist $P_1 \in \mathcal{S}_1$, $P_2 \in \mathcal{S}_2$ such that $P_1 \sim P_2$; because $\mathcal{S}_1, \mathcal{S}_2$ are disjoint, $P_2 \in \mathcal{S}_2 \backslash \mathcal{S}_1$. Otherwise, $\mathcal{S}_1, \mathcal{S}_2$ have a non-empty intersection $\mathcal{S}_1 \cap \mathcal{S}_2$. $\mathcal{S}_1 \cap \mathcal{S}_2$ is a strict subgraph of $\mathcal{G}$ so that not all vertices $P_1 \in \mathcal{S}_1 \cap \mathcal{S}_2$ can be *interior* vertices in the sense described in Section 9.7. Any boundary vertex P_1 will have the required property: for such a P_1, there will be a vertex P_2 such that $P_2 \sim P_1$, and $u_2 < 0$, i.e., $P_2 \in \mathcal{S}_2 \backslash \mathcal{S}_1$. ■

Now suppose $\mathbf{u}$, an eigenvector corresponding to λ_n, has $m \geq n$ weak sign graphs $\mathcal{S}_i$. We define $\mathbf{w}_i$, $i = 1, 2, \ldots, m$ as before, and we choose c_i, $i = 1, 2, \ldots, m$, not all zero, to make $\mathbf{v}$ given by (9.8.3) orthogonal to $\mathbf{u}_i$, $i = 1, 2, \ldots, m-1$. We prove a continuation result for the coefficients c_i that is a discrete analogue of the unique continuation principle for eigenfunctions.

Lemma 9.9.2 *Suppose $m \geq n$, and two of the weak sign graphs $\mathcal{S}_1$ and $\mathcal{S}_2$ of $\mathbf{u}$ are adjacent, then $c_2 = c_1$.*

Proof. Without loss of generality we may suppose that $\mathcal{S}_1$ is weak positive and $\mathcal{S}_2$ is weak negative. We proceed as in the derivation of equation (9.8.8). The minimax theorem implies $\mathbf{v}^T\mathbf{A}\mathbf{v} \geq \lambda_m$, and Lemma 9.8.1 implies $\mathbf{v}^T\mathbf{A}\mathbf{v} \leq \lambda_n$, and

$$\sum_{i,j=1}^{m} (c_i - c_j)^2 \mathbf{w}_i^T \mathbf{A} \mathbf{w}_j = 0. \tag{9.9.1}$$

Now use Lemma 9.9.1. If $\mathcal{S}_1$ and $\mathcal{S}_2$ are disjoint, then there is a pair P_1, P_2 such that $P_1 \in \mathcal{S}_1$, $P_2 \in \mathcal{S}_2$ and $P_1 \sim P_2$; thus $u_1 > 0$, $u_2 < 0$, $a_{12} < 0$. Thus $\mathbf{w}_1^T\mathbf{A}\mathbf{w}_2 \geq u_1 a_{12} u_2 > 0$, and (9.9.1) implies $c_1 = c_2$.

Otherwise $\mathcal{S}_1, \mathcal{S}_2$ overlap. Since $\mathbf{v}^T\mathbf{A}\mathbf{v} \leq \lambda_n$, $\mathbf{v}$, like $\mathbf{u}$, is in the eigenspace of λ_n, and therefore so is

$$\mathbf{z} = c_1\mathbf{u} - \mathbf{v} = \sum_{j=1}^{m} (c_1 - c_j)\mathbf{w}_j.$$

By definition $w_{j,i} = 0$ unless $P_i \in \mathcal{S}_j$. Choose P_1 and P_2 as in Lemma 9.9.1: $P_1 \in \mathcal{S}_1 \cap \mathcal{S}_2$ implies $u_1 = 0$, i.e., $w_{j,1} = 0$ for all j, so that $z_1 = 0$.

Since $\mathbf{z}$ is in the eigenspace of λ_n, we have

$$\lambda_n \mathbf{z} = \mathbf{A}\mathbf{z} = \sum_{j=1}^{m} (c_1 - c_j)\mathbf{A}\mathbf{w}_j,$$

so that

$$\begin{aligned} \lambda_n z_1 = 0 &= \textstyle\sum_{j=1}^{m} (c_1 - c_j)(\mathbf{A}\mathbf{w}_j)_1, \\ &= \textstyle\sum_{j=1}^{m} (c_1 - c_j) \sum_{i=2}^{N} a_{1i} w_{j,i}, \end{aligned} \tag{9.9.2}$$

where we have used $w_{j,1} = 0$. The term a_{1i}, for $i \geq 2$, is zero unless $P_i \sim P_1$. Since $u_1 = 0$, all such P_i are in $\mathcal{S}_1$ or $\mathcal{S}_2$. The sum in (9.9.2) is therefore over $j = 2$ only:

$$0 = (c_1 - c_2) \sum_{i=2}^{N} a_{1i} w_{2,i}.$$

Since $\mathcal{S}_2$ is weak negative, $a_{1i} w_{2,i} \geq 0$ for $i = 2, \ldots, N$: each term in the sum is non-negative. Since $P_1 \sim P_2$ we have $a_{12} < 0$; since $P_2 \in \mathcal{S}_2 \backslash \mathcal{S}_1$, $w_{2,2} = u_2 < 0$, so that

$$\sum_{i=2}^{N} a_{1i} w_{2,i} \geq a_{12} u_2 > 0,$$

and hence $c_1 = c_2$. ■

We are now in a position to establish

Theorem 9.9.1 *If $\mathcal{G}$ is connected, any eigenvector corresponding to λ_n has at most n weak sign graphs.*

Proof. Suppose, if possible, that $\mathbf{u}$ has m weak sign graphs $\mathcal{S}_i$, $i = 1, 2, \ldots, m$, and $m > n$. At least one of the coefficients c_i, say c_1, is non-zero. Since $n \geq 1$, we have $m \geq 2$. Since $\mathcal{G}$ is connected, $\mathcal{S}_1$ must be adjacent to at least one other weak sign graph, which we label $\mathcal{S}_2$. Lemma 9.9.2 states that $c_2 = c_1$. If $m \geq 3$, one of $\mathcal{S}_1, \mathcal{S}_2$ must be adjacent to one of the remaining sign graphs $\mathcal{S}_i$, $i = 3, \ldots, m$, say $\mathcal{S}_3$, otherwise $\mathcal{G}$ would not be connected. Therefore $c_3 = c_2 = c_1$ by Lemma 9.9.2. In $m - 1$ steps, we conclude that $c_m = c_{m-1} = \cdots = c_2 = c_1$. Hence $\mathbf{v} = c_1\mathbf{u}$. But $\mathbf{v}$ was constructed so that it was orthogonal to $\mathbf{u}_i$ for $i = 1, 2, \ldots, m-1$; if $m > n$, $\mathbf{v}$ is orthogonal to $\mathbf{u}$, contradicting $\mathbf{v} = c_1\mathbf{u}$. Therefore, $m \leq n$. ■

9.10 Generalisation to M, K problems

The proof of Theorem 9.8.1, on strong sign graphs, hinges on two fundamental results: Courant's minimax theorem, and Duval and Reiner's Lemma 9.8.1. Theorem 9.9.1 on weak sign graphs, uses these two, and Lemmas 9.9.1, 9.9.2.

All these intermediate steps may be generalised to give results for the problem (9.1.1), in which $\mathbf{K}$ is PSD, $\mathbf{M}$ is PD, and $\mathbf{K}, \mathbf{M}$ satisfy (9.7.1).

Thus, since $\mathbf{M}$ is PD, the minimax theorem holds for the Rayleigh quotient $\mathbf{v}^T\mathbf{K}\mathbf{v}/\mathbf{v}^T\mathbf{M}\mathbf{v}$. Duval and Reiner's Lemma 9.8.1 may be generalised to read

Lemma 9.10.1

$$\mathbf{v}^T(\mathbf{K}-\lambda\mathbf{M})\mathbf{v} = \sum_{i=1}^{m} c_i^2 \mathbf{w}_i^T(\mathbf{K}-\lambda\mathbf{M})\mathbf{u} - \frac{1}{2}\sum_{i,j=1}^{m}(c_i - c_j)^2 \mathbf{w}_i^T(\mathbf{K}-\lambda\mathbf{M})\mathbf{w}_j.$$

Since $\mathbf{K}$ is PSD and $\mathbf{M}$ is PD, the eigenvalues λ_i are non-negative. This means that when $\mathbf{w}_i, \mathbf{w}_j$ correspond to adjacent sign graphs

$$\mathbf{w}_i^T(\mathbf{K}-\lambda\mathbf{M})\mathbf{w}_j = (\pm)\{(-)-(+)\}(\mp) = +.$$

All the arguments used to establish Theorems 9.8.1, 9.9.1 proceed as before with $\mathbf{A}$ replaced by $\mathbf{K}-\lambda\mathbf{M}$.

Exercises 9.10

1. Establish Lemma 9.10.1.

Chapter 10

Green's Functions and Integral Equations

Mathematicians who are only mathematicians have exact minds, provided all things are explained to them by means of definitions and axioms; otherwise they are inaccurate and unsufferable, for they are only right when the principles are quite clear.

Pascal's *Pensées*

10.1 Introduction

In this and the following two chapters we shall be concerned with the vibration of, and the inverse problems for, three systems with continuously distributed mass: the taut vibrating string, and the rod in longitudinal or torsional vibration. In this section we state the governing differential equation. In Section 10.2 we introduce the Green's function and reformulate the eigenvalue problem giving the natural frequencies as an integral equation. In Section 10.3 we recall the relevant spectral theory for compact self-adjoint operators on a Hilbert space, and in Section 10.4 we apply it to the Green's function integral equation. This chapter thus serves as introductory material for the study of inverse problems in Chapter 11.

The equation governing the free (infinitesinal, undamped) vibration of a taut string having unit tension, mass per unit length $\rho^2(x)$, vibrating with frequency ω is

$$v''(x) + \lambda\rho^2(x)v(x) = 0, \tag{10.1.1}$$

where $\lambda = \omega^2$ and $' \equiv d/dx$. We denote the mass per unit length by $\rho^2(x)$, rather than by $\rho(x)$, to indicate that it is positive, and to avoid continual repetition of $\rho^{1/2}(x)$. The end conditions will be assumed to be

$$v'(0) - hv(0) = 0 = v'(1) + Hv(1), \tag{10.1.2}$$

where $h, H \geq 0$ and h, H are not both zero. This means that the ends $x = 0,\ x = 1$ are attached to fixed supports by the use of springs having stiffnesses h, H respectively. Of course a real (physical) string cannot have a 'free' end in the straightforward sense. However, we can simulate a free end by attaching the end to a device that moves transversely in such a way that the slope of the string at the end remains zero.

The free longitudinal vibrations of a thin straight rod of cross-sectional area $A(x)$, density ρ and Young's modulus E are governed by the equation

$$(A(x)w'(x))' + \lambda A(x)w(x) = 0, \tag{10.1.3}$$

where $\lambda = \rho\omega^2/E$. The end conditions are

$$w'(0) - hw(0) = 0 = w'(1) + Hw(1), \tag{10.1.4}$$

where again $h, H \geq 0$ and h, H are not both zero.

The free torsional vibrations of a thin straight rod of second moment of area $J(x)$, density ρ and shear modulus G are governed by the equation

$$(J(x)\theta'(x))' + \lambda J(x)\theta(x) = 0, \tag{10.1.5}$$

where $\lambda = \rho\omega^2/G$. The end conditions are

$$\theta'(0) - h\theta(0) = 0 = \theta'(1) + H\theta(1). \tag{10.1.6}$$

There is clearly a one-one correspondence $(E, A, \rho, v) \rightarrow (G, J, \rho, \theta)$ between the longitudinal and torsional systems, but we now show that, by means of a transformation of variables, all these systems may be reduced to the same basic equation.

In equation (10.1.3) introduce a new variable ξ, where

$$\xi'(x) = 1/A(x), \quad w(x) = v(\xi). \tag{10.1.7}$$

Then $A(x)w'(x) = A(x)\dot{v}(\xi)\xi'(x) = \dot{v}(\xi)$, where $\cdot \equiv d/d\xi$. Hence $A(Aw')' = \ddot{v}$, and equation (10.1.3) becomes

$$\ddot{v}(\xi) + \lambda\rho^2(\xi)v(\xi) = 0, \tag{10.1.8}$$

with $\rho(\xi) = A(x)$. If

$$\xi(x) = \int_0^x \frac{dt}{A(t)}, \quad 1 = \int_0^1 \frac{dt}{A(t)} \tag{10.1.9}$$

then the end conditions (10.1.4) become

$$\dot{v}(0) - hA(0)v(0) = 0 = \dot{v}(1) + HA(1)v(1). \tag{10.1.10}$$

Since $A(x)$ is positive and bounded, equation (10.1.8) has the same form as (10.1.1), and equation (10.1.10) has the same form as (10.1.2). This means that we may concentrate our attention on equations (10.1.1),(10.1.2).

We showed that equation (10.1.3) could be transformed into (10.1.1) by a simple change of variable. If we assume further smoothness in $A(x)$, that it has a second derivative, then we may transform (10.1.3) into another equation which is often viewed as the standard form, the so-called *Sturm-Liouville* equation.

In equation (10.1.3) put

$$y(x) = f(x)w(x)$$

then

$$\begin{aligned}(Aw')' &= [A(f^{-1}y' - f'f^{-2}y)]' \\ &= Af^{-1}y'' + \{(Af^{-1})' - Af'f^{-2}\}y' - (Af'f^{-2})'y.\end{aligned}$$

Choose the function f to make the terms in y' vanish:

$$(Af^{-1})' - Af'f^{-2} = A'f^{-1} - 2Af'f^{-2}, \text{ i.e., } (Af^{-2})' = 0 \text{ or } f = A^{1/2}.$$

Then

$$(Aw')' + \lambda Aw \equiv fy'' - f''y + \lambda fy = 0$$

or

$$y''(x) + [\lambda - q(x)]y(x) = 0, \tag{10.1.11}$$

where

$$q(x) = f''(x)/f(x). \tag{10.1.12}$$

We note that since (10.1.3) may be transformed into (10.1.1), the latter may be transformed into (10.1.11). In fact if

$$v(x) = y(\xi)/f(\xi), \quad f(\xi) = \rho^{1/2}(x), \ \xi'(x) = f^2(\xi) \tag{10.1.13}$$

then

$$v' = \dot{v}f^2 = f\dot{y} - \dot{f}y, \ v'' = f^2(f\ddot{y} - \ddot{f}y)$$

and

$$v'' + \lambda\rho^2 v \equiv f^2(f\ddot{y} - \ddot{f}y) + \lambda f^4 f^{-1}y = 0$$

so that

$$\ddot{y}(\xi) + [\lambda - q(\xi)]y(\xi) = 0, \tag{10.1.14}$$

where

$$q(\xi) = \ddot{f}(\xi)/f(\xi). \tag{10.1.15}$$

If $\rho(x)$ is continuous in [0,1] then equation (10.1.1) shows that $v(x)$ has a continuous second derivative. If $\rho(x)$ has a simple discontinuity at $x = \xi$ then $v'(x)$ is continuous while $v''(x)$ has a discontinuity at $x = \xi$:

$$v''(x)\Big|_{x=\xi-}^{x=\xi+} = -\lambda v(\xi)\rho^2(x)\Big|_{x=\xi-}^{x=\xi+}. \tag{10.1.16}$$

If $\rho(x)$ therefore is piecewise continuous in (0,1) then $v''(x)$ is piecewise continuous also.

To show that any eigenvalues of (10.1.1), (10.1.2) must be real and positive we may argue as follows: suppose λ, possibly complex, is an eigenvalue, and $v(x)$ a corresponding eigenfunction. Multiply (10.1.1) by $\overline{v(x)}$ and integrate over (0,1):

$$\int_0^1 v''\bar{v}dx + \lambda\int_0^1 \rho^2 v\bar{v}dx = 0.$$

Integrate the first term by parts and use the end conditions (10.1.2):

$$\int_0^1 v'\bar{v}'dx + hv(0)\overline{v(0)} + Hv(1)\overline{v(1)} = \lambda\int_0^1 \rho^2 v\bar{v}dx. \tag{10.1.17}$$

The terms on the left are real; the integral on the right is real and positive; λ is real. The sum on the left can be zero only when $hv(0) = 0 = Hv(1)$. There are two cases to consider i) $h, H > 0$, in this case $v(0) = v'(0) = v(1) = v'(1)$ so that $v(x) \equiv 0$, and there is no eigenfunction $v(x)$. ii) $h = 0 = H$, in this case the supports have no stiffness, and there is an eigenvalue $\lambda = 0$ with eigenfunction $v(x) =$ constant. This is called a *rigid-body mode.* Apart from this case, any eigenvalue is strictly positive.

Any eigenvalues must be *simple*, for if $u(x), v(x)$ were two different eigenfunctions corresponding to the same eigenvalue λ, then

$$u''(x)v(x) - u(x)v''(x) = 0,$$

i.e.,

$$u'(x)v(x) - u(x)v'(x) = \text{Constant}.$$

But at $x = 0$, the end condition (10.1.2) gives

$$u'(0)v(0) - u(0)v'(0) = 0.$$

Thus

$$u'(x)v(x) - u(x)v'(x) = 0,$$

and $u(x), v(x)$ are proportional.

Suppose $v_1(x), v_2(x)$ are eigenfunctions of (10.1.1), (10.1.2) corresponding to different eigenvalues λ_1, λ_2. Then

$$v_1'' + \lambda_1\rho^2 v_1 = 0 = v_2'' + \lambda_2\rho^2 v_2$$

and

$$\int_0^1 (v_1''v_2 - v_2''v_1)dx + (\lambda_1 - \lambda_2)\int_0^1 \rho^2 v_1 v_2 dx = 0.$$

But

$$\int_0^1 (v_1''v_2 - v_2''v_1)dx = [v_1'v_2 - v_2'v_1]_0^1 = 0$$

on account of the end conditions, and hence, since $\lambda_1 - \lambda_2 \neq 0$, v_1 and v_2 are orthogonal in the sense

$$\int_0^1 \rho^2 v_1 v_2 dx = 0.$$

We have shown that *if* equations (10.1.1), (10.1.2) have eigenvalues then they will satisfy

$$0 \leq \lambda_1 < \lambda_2 < \cdots \tag{10.1.18}$$

with equality only as stated above. The corresponding eigenfunctions $v_i(x)$ will be orthogonal; they may be normalised so that

$$\int_0^1 \rho^2 v_i v_j dx = \delta_{ij}. \tag{10.1.19}$$

We have shown that the differential equation we are studying may be presented in three different forms: (10.1.1), (10.1.3) or (10.1.11). For vibration purposes the fundamental equations are the first two: (10.1.1) for the taut string; (10.1.3) for the rod. Equation (10.1.11), called the Sturm-Liouville equation, is introduced as a standard mathematical form because it is easier to analyse, particularly for the asymptotic form of the eigenvalues, and for the inverse problem. Equation (10.1.11) is the one that has been studied by most pure mathematicians, but in our study of vibration problems, we must always remember that it is a secondary equation.

In this chapter, we will study some of the basic properties of the equations, particularly the so-called spectral theory. In Chapter 11 we will study some inverse problems: how to reconstruct the functions $\rho(x)$, $A(x)$ or $q(x)$, appearing respectively in the three forms of the equation.

In the spectral theory there are six main topics:

i) The existence of an infinite sequence of real distinct eigenvalues with only one limit point, $+\infty$. For equations (10.1.1) and (10.1.3) these are all positive apart perhaps for the first, which is zero when $h = 0 = H$.

ii) The completeness of eigenfunctions on [0,1].

iii) The asymptotic form of the eigenvalues and the so-called norming constants.

iv) The interlacing of eigenvalues corresponding to different end constants h, H.

v) The oscillatory properties of eigenfunctions: how many nodes they have.

vi) The interlacing of nodes of neighbouring eigenfunctions.

Each of these topics may be studied in various ways, but there are basically just two avenues of approach: through the study of the differential equation itself; by converting the differential equation to an integral equation and studying that.

Of the six topics, the most difficult is undoubtedly ii), the completeness of the eigenfunctions. In their recent monograph, Levitan and Sargsjan (1991) [212] study completeness by reducing (10.1.11) to an integral equation and then using a variety of approaches to establish completeness. We will approach topics i) and ii) differently, in a way that mimics somewhat the matrix approach to discrete problems, by starting from (10.1.1), converting it to an eigenvalue problem for an integral operator, and establishing the necessary functional analysis. This approach takes more pages than Levitan and Sargsjan's, but we believe it has merit.

For the establishment of the asymptotic form of the eigenvalues we will start from (10.1.11).

Topics v) and vi), nodes and interlacing, were studied by Sturm in his original work. The classical treatment, beautifully presented, may be found in Ince (1927) [185]. Levitan and Sargsjan follow Sturm's approach. We will use the total positivity properties of the integral equation, following the lines of Gantmacher and Krein (1950) [98].

There are two ways to normalise the governing equations and to number the eigenvalues; both have their own advantages and disadvantages, and we shall therefore use both, at different times; we label them V, for vibration, and S, for Sturm-Liouville.

V: *the governing equation is (10.1.1) or (10.1.3), the equation holds for* $x \in [0,1]$*; the end conditions are (10.1.2) or (10.1.4); the eigenvalues are labelled* $(\lambda_i)_1^\infty$*, the eigenfunctions* $(v_i(x))_1^\infty$.

S: *the governing equation is (10.1.11), the equation holds for* $x \in [0,\pi]$*; the end conditions are*

$$y'(0) - hy(0) = 0 = y'(\pi) + Hy(\pi);$$

the eigenvalues are labelled $(\lambda_i)_0^\infty$ *and the eigenfunctions* $(y_i(x))_0^\infty$.

Thus we will use V for the analysis in Sections 10.2-10.8 based on the Green's function approach to equation (10.1.11). We will use S for the study of the asymptotic form of the eigenvalues in Section 10.9, and for the analysis of the inverse problems for the Sturm-Liouville equation (10.1.11) in Chapter 11.

Exercises 10.1

1. Show that the eigenvalues and eigenfunctions of (10.1.1) for $\rho = 1$ and the end conditions (10.1.2) are given by

$$\lambda = \omega_n^2, \quad \omega_n = \alpha_n + \beta_n + (n-1)\pi, \quad n = 1, 2, \ldots$$

where

$$\alpha_n = \arctan(h/\omega_n), \quad \beta_n = \arctan(H/\omega_n)$$

and

$$y_n = \cos(\omega_n x - \alpha_n), \quad n = 1, 2, \ldots$$

Hence, show that ω_n is an increasing function of h and H and that, when h, H are positive, there is just one eigenvalue ω_n in each of the intervals $((n-1)\pi, n\pi), \quad n = 1, 2, \ldots$

2. Consider various special cases of Ex. 10.1.1. Thus,

   a) $h = 0 = H$, then $\omega_n = (n-1)\pi, \quad n = 1, 2, \ldots$ Note: in this case, that was considered earlier, there is a zero eigenvalue with eigenfunction $y_1 = 1$.

b) $h = 0, \quad H = \infty$, then $\omega_n = (n - 1/2)\pi, \quad y_n = \cos\omega_n x$

c) $h = \infty, \quad H = \infty$, then $\omega_n = n\pi, \quad y_n = \sin\omega_n x$

d) h, H finite, then for large n

$$\omega_n = (n-1)\pi + \frac{(h+H)}{(n-1)\pi} + 0\left(\frac{1}{n^3}\right).$$

Note that this expression indicates that it would be an advantage to label the eigenvalues $\lambda_0, \lambda_1, \ldots$ rather than $\lambda_1, \lambda_2, \ldots$

3. Explore how the end conditions change as one equation of (10.1.1), (10.1.3), (10.1.11) is changed to another. Note that the basic equations for vibration purposes are (10.1.1), (10.1.2) and (10.1.3), (10.1.4) in which h, H are non-negative. Note particularly that if (10.1.3) is changed to (10.1.1), i.e., to (10.1.8), the end conditions retain the same form; compare (10.1.10) and (10.1.2). But when (10.1.1) or (10.1.3) is changed to the standard form (10.1.11) the end conditions change: $v'(0) - hv(0) = 0$ becomes $\dot{y}(0) - ky(0) = 0$, and $h > 0$ does not imply $k > 0$.

10.2 Green's functions

The idea of a Green's function is perhaps most easily introduced by considering the static deflection of a string with fixed ends due to a distributed load $f(x)$. The governing equation is

$$-v''(x) = f(x) \tag{10.2.1}$$

and the end conditions are $v(0) = 0 = v(1)$. If instead of a distributed load we consider a single unit concentrated load at $x = s$, then the string will be straight on each side of $x = s$, and have a discontinuity in its slope at $x = s$, as shown in Figure 10.2.1.

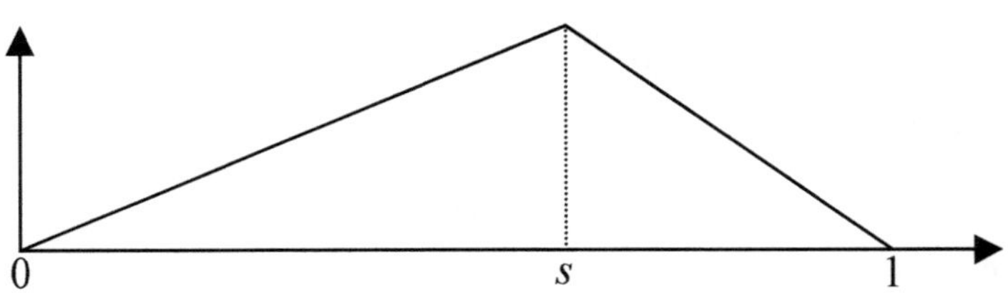

Figure 10.2.1 - The plucked string.

Thus

$$v(x) = \begin{cases} Ax, & 0 \le x \le s, \\ B(1-x), & s < x \le 1. \end{cases} \tag{10.2.2}$$

Equilibrium of the two portions gives

$$\left.\frac{dv}{dx}\right|_{x=s-}^{x=s+} = -1,$$

so that, on using (10.2.2), we find $A+B=1$. Continuity yields $As = B(1-s)$ so that

$$A = (1-s), \quad B = s.$$

We call the resulting deflection $G(x,s)$; thus

$$G(x,s) = \begin{cases} x(1-s), & 0 \le x \le s, \\ s(1-x), & s \le x \le 1. \end{cases} \tag{10.2.3}$$

To obtain the deflection of the string under the action of the distributed load $f(x)$ we combine the actions of the concentrated forces $f(s)ds$ at the locations s; thus

$$v(x) = \int_0^1 G(x,s)f(s)ds. \tag{10.2.4}$$

Clearly, we may generalise this procedure, and define a Green's function for the general end conditions (10.1.2). We introduce two solutions of $v''(x) = 0$: $\phi(x)$ satisfying the condition $\phi'(0) - h\phi(0) = 0$; $\psi(x)$ satisfying $\psi'(1) + H\psi(1) = 0$. Since

$$\phi''(x) = 0 = \psi''(x),$$

we have $\phi(x)\psi''(x) - \phi''(x)\psi(x) = 0$, which on integrating gives $\phi(x)\psi'(x) - \phi'(x)\psi(x) = const.$ We choose this constant as -1, so that

$$\phi(x)\psi'(x) - \phi'(x)\psi(x) = -1, \tag{10.2.5}$$

and define

$$G(x,s) = \begin{cases} \phi(x)\psi(s), & 0 \le x \le s, \\ \phi(s)\psi(x), & s \le x \le 1, \end{cases} \tag{10.2.6}$$

then $G(x,s)$ is continuous at $x = s$, while

$$\left.\frac{\partial G}{\partial x}(x,s)\right|_{x=s-}^{x=s+} = -1.$$

Note that

$$\left.\begin{aligned} \phi(x) &= A(1+hx) \\ \psi(x) &= B(1+H(1-x)) \end{aligned}\right\} \tag{10.2.7}$$

where

$$AB = 1/(h+H+hH),$$

and the conditions $h \ge 0,\ H \ge 0,\ h+H > 0$ ensure that the denominator in (10.2.7) is positive.

We note that the Green's function is symmetric, i.e.,

$$G(x,s) = G(s,x). \tag{10.2.8}$$

and that the functions $\phi(x), \psi(x)$ are positive, $\phi(x)$ increasing while $\psi(x)$ decreasing. In fact, (10.2.5) shows that $\phi(x)/\psi(x)$ is an increasing function of x. There is thus a clear parallel between the Green's function and the Green's *matrix* introduced in Section 10.5.

For our purposes, the most important use of the Green's function is that it reduces the free vibration problem (10.1.1), (10.1.2) to an eigenvalue problem for an integral equation:

$$v(x) = \lambda \int_0^1 \rho^2(s)G(x,s)v(s)ds. \tag{10.2.9}$$

With the changes of variable

$$u(x) = \rho(x)v(x), \quad K(x,s) = \rho(x)\rho(s)G(x,s), \tag{10.2.10}$$

we may transform (10.2.9) into the symmetric equation

$$\int_0^1 K(x,s)u(s)ds = \mu u(x), \tag{10.2.11}$$

in which $K(x,s) = K(s,x)$, and $\mu = 1/\lambda$.

There is a well established body of theory for such integral equations, which we now recall. The theory relates to a compact, self-adjoint linear operator in a separable Hilbert space. In Section 10.3 we summarize the theory regarding the spectrum of such an operator, and in Section 10.4 we apply it to the operator equation (10.2.11).

Exercises 10.2

1. Find the solutions of $(Aw')' = 0$, $\phi(x)$ satisfying (10.1.4a), and $\psi(x)$ satisfying (10.1.4b), and make

$$A(x)\{\phi(x)\psi'(x) - \phi'(x)\psi(x)\} = -1$$

and hence write (10.1.3) as an integral equation

$$w(x) = \lambda \int_0^x A(s)G(x,s)w(s)ds.$$

2. Show that if $v(x)$ satisfies

$$v'' + \lambda\rho^2 v = 0 \quad v(0) = 0 = v'(1)$$

and $\rho(x)$ has a continuous first derivative, then $u = v'$ satisfies $(\rho^{-2}u')' + \lambda u = 0$, $u'(0) = 0 = u(1)$. Hence show that $u(x)$ satisfies

$$u(x) = \lambda \int_0^1 K(x,s)u(s)ds$$

where

$$K(x,s) = \int_{x^+}^{1} \rho^2(t)dt, \quad x^+ = \max(x,s).$$

3. Show that if $w(x)$ satisfies

$$(Aw')' + \lambda Aw = 0, \quad w(0) = 0 = w'(1), \quad A > 1,$$

then $v = Aw'$ satisfies $(Bv')' + \lambda Bv = 0$ where $B = 1/A,\ v'(0) = 0 = v(1)$. Hence, show that

$$v(x) = \lambda \int_0^1 G(x,s)B(s)v(s)ds$$

where

$$G(x,s) = \int_{x^+}^{1} A(t)dt, \quad x^+ = \max(x,s).$$

10.3 Some functional analysis

In the first edition of this book, in order to prove the existence of eigenvalues and eigenfunctions for the integral equation, i.e., operator equation, (10.2.11) we referred the reader to the classical treatment of integral equations in Courant and Hilbert (1953) [64]. Instead, in this edition, we sketch the functional analysis approach to existence by providing the reader with a sign-posted journey through parts of the book *Functional Analysis* by Lebedev, Vorovich and Gladwell (1996) [205]. We refer to definitions and theorems in that book by the abbreviations Def. and Th. respectively.

The journey starts with the definition of a *metric space* X, Def. 2.1.4: a set of elements governed by a *distance metric* $d(x,y)$ satisfying certain distance axioms. After defining an *open ball* or *ϵ-neighbourhood* of a point $x_0 \in X$, Def. 2.2.1, we define an *open set* in X as one in which every point is an interior point. Then, after defining *limit points* Def. 2.2.3, we define a *closed* set as one that contains all its limit points, Def. 2.2.6. We define the *closure* $\bar{S}$ of a set S as the set obtained by adding to S all its limit points, and say, Def. 2.2.7, that S is *dense* in a set T if $\bar{S} \supset T$.

The journey continues through metric spaces, to give the metric space versions of *limit* of a sequence, Def. 2.4.1; *Cauchy sequence*, Def. 2.4.2; and *complete* metric space, Def. 2.5.1: a metric space in which every Cauchy sequence has a limit. The definitions Def. 2.6.1, 2.6.2 and the *completion theorem* Th. 2.6.1 explain how any metric space may be *completed*.

The definition of an *operator* is given in

Definition 10.3.1 *Let X and Y be metric spaces. A correspondence $Ax = y$, $x \in X$, $y \in Y$ is called an **operator** from X into Y, if to each $x \in X$ there corresponds no more than one $y \in Y$. The set of all those $x \in X$ for which there exists a corresponding $y \in Y$ is called the **domain** of A and denoted by*

$D(A)$; the set of all y arising from $x \in X$ is called the ***range*** *of A and denoted by $R(A)$. Thus*

$$R(A) = \{y \in Y; \quad y = Ax, \ x \in X\}.$$

We say that A is an operator ***on*** *$D(A)$* ***into*** *Y, or* ***on*** *$D(A)$* ***onto*** *$R(A)$. We also say that $R(A)$ is the* ***image*** *or* ***map*** *of $D(A)$ under A. The* ***null space*** *of A, denoted by $N(A)$, is the set of all $x \in X$ such that $Ax = 0$.*

A *functional*, Def. 2.7.2 is defined as an operator from X to the real numbers $\mathbb{R}$, or complex numbers $\mathbb{C}$. The definition of a *continuous* operator Def. 2.7.3 is the straightforward analogue of continuity of an ordinary function.

The journey now passes to *linear* spaces (Section 2.8) over $\mathbb{R}$ or $\mathbb{C}$, with the property that if $x, y \in X$ then $\lambda x + \mu y \in X$; when equipped with a norm $||\cdot||$, they become *normed linear spaces*, Def. 2.8.1. After defining a *subspace*, Def. 2.8.4, we define *closed* subspace Def. 2.8.5, *linear dependence* and *independence* Def. 2.8.6; and *dimension*, Def. 2.8.8.

We carry the notion of an operator in a metric space over and define a *linear operator*, Def. 2.9.2, in a normed linear space as one that satisfies $A(\lambda x + \mu y) = \lambda A(x) + \mu A(y)$; define a *continuous* linear operator, and the *norm* of a continuous linear operator from X to Y by (Th. 2.9.1)

$$||A|| = \sup_{x \in D(A)} \frac{||Ax||_Y}{||x||_X}. \tag{10.3.1}$$

A is *continuous*, or *bounded*, iff $||A||$ is finite.

The concepts of metric, $d(x, y)$, and norm, $||x||$, generalise the notions of distance and magnitude in $\mathbb{R}^3$, respectively. We now pass to an *inner product* space X in which an inner product (x, y) is defined for every pair $x, y \in X$. This inner product satisfies the axioms

P1: $(x, x) \geq 0$, and $(x, x) = 0$ iff $x = 0$;

P2: $(x, y) = (\overline{y, x})$;

P3: $(\lambda x + \mu y, z) = \lambda(x, z) + \mu(y, z)$.

Here, $\lambda, \mu \in \mathbb{C}$ and the overbar in P2 denotes complex conjugate. In a *real* inner product space, P2 is replaced by

P2′: $(x, y) = (y, x)$.

In an inner product space we may define a norm by

$$||x|| = (x, x)^{1/2}.$$

That this does in fact provide a norm in the usual sense follows from the *Cauchy-Schwarz* inequality (Th. 2.12.1)

$$|(x, y)| \leq ||x|| \cdot ||y||, \tag{10.3.2}$$

with equality when $x \neq 0,\ y \neq 0$, iff $x = \lambda y$.

For an inner product space X we may define the terms *orthogonal* and *orthonormal*: x and y are *orthogonal* if $(x, y) = 0$; a system $\{g_k\} \subset X$ is *orthonormal* if

$$(g_m, g_n) = \delta_{mn} = \begin{cases} 1 & m = n, \\ 0 & m \neq n. \end{cases} \tag{10.3.3}$$

We may easily extend the concepts of *closed* and *complete* to inner product spaces, and we call a *complete* inner product space a *Hilbert* space H, Def. 2.12.5.

The concept of orthogonality leads to the idea of the *orthogonal decomposition* of a Hilbert space into a closed subspace M and its *orthogonal complement* $N = M^{\perp}$; if $x \in H$, then x may be written

$$x = m + n, \quad m \in M, \quad n \in N. \tag{10.3.4}$$

Clearly, a closed subspace of a Hilbert space is itself a Hilbert space.

This leads to *Riesz's representation theorem* Th. 4.3.3, which states that any continuous (i.e., bounded) linear functional $F(x)$ on H may be expressed as an inner product:

$$F(x) = (x, f) \text{ for every } x \in H, \tag{10.3.5}$$

and $||F|| = ||f||$.

We now define a *separable* Hilbert space H, Def. 4.1.3, one that contains a *countable* (enumerable) *dense subset* $\{f_n\}$. From such a sequence we may, by the usual Gram-Schmidt procedure, construct an orthonormal set $\{g_k\}$ that is dense in H; this will be a *complete* orthonormal system in the sense that if $x \in H$ and $\epsilon > 0$ are given, there is a finite linear combination of the g_k such that

$$||x - \sum_{k=1}^{n} \alpha_k g_k|| \leq \epsilon. \tag{10.3.6}$$

In this case any $x \in X$ has a unique representation

$$x = \sum_{k=1}^{\infty} \alpha_k g_k, \quad \alpha_k = (x, g_k), \tag{10.3.7}$$

and Parseval's equality holds:

$$||x||^2 = \sum_{k=1}^{\infty} |\alpha_k|^2. \tag{10.3.8}$$

It may be argued that almost all existence proofs in Functional Analysis rely on the concept of a *compact* set in a metric space. The concept compact is similar to, but must be sharply distinguished from, the concepts *closed* and *complete*. In brief, $S \subset X$ is *closed* if it contains all its limit points; X is *complete* if every Cauchy sequence in S has a limit point in S. A set $S \subset X$

is *compact* Def. 6.1.1 if every sequence $\{x_n\}$ in S contains a subsequence $\{x_{n_k}\}$ which converges to a point $x \in S$.

The classical *Bolzano-Weierstrass Theorem* (Th. 1.1.2) states that in a finite-dimensional space e.g., $\mathbb{R}^N$, a set S is compact iff it is closed and bounded. This result is false for general metric spaces. To be precise, a compact set $S \subset X$ *is* closed and bounded, but a closed and bounded set is compact *only if* the space X is finite-dimensional.

In order to find a criterion for compactness of a set S in an infinite-dimensional metric space we must generalise the classical *Heine-Borel Theorem*; This uses the concept of an ε-covering.

Definition 10.3.2 *Let X be a metric space, and suppose $S \subset X$. A finite set of N balls $B(x_n, \varepsilon)$ with $x_n \in X$ and $\varepsilon > 0$ is said to be a* ***finite*** *ε-covering of S, if every element of S lies inside one of the balls $B(x_n, \varepsilon)$, i.e.,*

$$S \subset \bigcup_{n=1}^{N} B(x_n, \varepsilon).$$

The set of centers $\{x_n\}$ of a finite ε-covering is called a finite ε-net for S.

Definition 10.3.3 *Let X be a metric space. A set $S \subset X$ is said to be* ***totally bounded*** *if it has a finite ε-covering for every $\varepsilon > 0$.*

Hausdorff's compactness criterion is now

Theorem 10.3.1 *Let X be a complete metric space. A set $S \subset X$ is compact iff it is closed and totally bounded.*

In a compact set the points are, as the word *compact* suggests, close together; the centers x_n form a network, and each point in S is near one of the x_n.

Having the concept of a compact set, we may introduce the idea of a *compact* (linear) *operator.*

Definition 10.3.4 *Let X, Y be metric spaces. A linear operator from X to Y is said to be compact if it maps the unit ball into a compact set in Y.*

Note that the map of the unit ball may not itself *be* a compact set; it is *in* a compact set. We say that it is *precompact*, meaning that it may be made compact by closing it: its closure is compact.

If the range of a linear operator A is finite-dimensional, we say that A is a *finite-dimensional operator.* The Bolzano-Weierstrass Theorem then implies that *a finite-dimensional operator is compact.* We may now use Hausdorff's compactness criterion to obtain a wider class of compact operators.

Theorem 10.3.2 *Let X, Y be metric spaces, and suppose Y is complete. If the sequence of compact linear operators $\{A_n\}$ from X to Y converges uniformly to A, then A is compact.*

Proof. Uniform convergence means $||A-A_n|| \to 0$. Let S be the unit ball in X. Choose $\varepsilon > 0$, and then choose A_n so that $||Ax - A_n x|| < \varepsilon/3$ for all $x \in S$. The operator A_n is compact; therefore the map $A_n(S)$ of A_n is precompact; its closure is compact. Therefore, by Th. 6.2.1, it is totally bounded; there is a finite set $\{x_1, x_2, \ldots, x_m\} \subset S$ such that every point in $A_n(S)$ lies in a ball of radius $\varepsilon/3$ around one of $A_n x_1, A_n x_2, \ldots, A_n x_m$. Choose $x \in S$, then choose i so that

$$||A_n x - A_n x_i|| < \varepsilon/3$$

then

$$||Ax - Ax_i|| \le ||Ax - A_n x + ||A_n x - A_n x_i|| + ||A_n x_i - Ax_i|| \le \varepsilon/3 + \varepsilon/3 + \varepsilon/3 = \varepsilon.$$

This means that the set $A(S)$ is totally bounded and therefore, again by Th. 6.2.1, precompact. (Note that we need Y to be complete.) Thus A is compact. ■

Having introduced one concept, compact, we now introduce another, *self-adjoint*. To do so we suppose from now on that A is a continuous linear operator *on* a Hilbert space H i.e., from H to H; we say $A \in B(H,H)$. If $x, y \in H$, then $G(x) = (Ax, y)$ is a continuous functional on H; therefore, there is an $g \in H$ such that $(Ax, y) = (x, g)$. Clearly, g depends linearly on y, and in fact is the map of y under a new continuous operator A^*, called the *adjoint* of A; thus $g = A^* y$ and

$$(Ax, y) = (x, A^* y). \tag{10.3.9}$$

If $A^* = A$, then A is said to be *self-adjoint*. If A is self-adjoint, the functional

$$F(x) = (Ax, x)$$

is real valued, because

$$F(x) = (Ax, x) = (x, Ax) = \overline{(Ax, x)} = \overline{F(x)}.$$

This functional is extremely important because, if $A \in B(H,H)$ is self-adjoint, then there are two ways to write $||A||$, one from (10.3.1), namely

$$||A|| = \sup ||Ax|| \text{ for } ||x|| = 1 \tag{10.3.10}$$

and another involving $F(x)$, namely

$$||A|| = \sup |F(x)| = \sup |(Ax, x)| \text{ for } ||x|| = 1. \tag{10.3.11}$$

We denote

$$\sup\{F(x)\} = M, \quad \inf\{F(x)\} = m, \text{ for } ||x|| = 1. \tag{10.3.12}$$

Clearly,

$$||A|| = \sup(|M|,\ |m|). \tag{10.3.13}$$

We are now in a position to define an *eigenvalue* of an operator $A \in B(H,H)$.

Definition 10.3.5 *Suppose $A \in B(H,H)$. The scalar μ is called an* ***eigenvalue*** *of A if there is a non-zero $x \in H$ such that $Ax = \mu x$; x is called an* ***eigenvector*** *corresponding to μ.*

Note that we use μ, rather than λ, to denote an eigenvalue, so that we can use $\lambda = 1/\mu$ to denote an eigenvalue of the differential equation (10.1.1). Clearly, any eigenvalue of a self-adjoint operator must be real, for $Ax = \mu x$ implies $(Ax, x) = \mu(x, x)$. See also Ex. 10.3.1.

Theorem 10.3.3 *If $A \in B(H,H)$ is self-adjoint and μ is* ***not*** *an eigenvalue of A, then $R(A - \mu I)$ is dense in H.*

Proof. We need to show that the closure of $R(A - \mu I)$ is H. This is equivalent to saying that if z is orthogonal to all $(A - \mu I)x$, then $z = 0$. If this were so then

$$\begin{aligned} 0 &= (z, (A - \mu I)x) = (z, Ax) - \bar{\mu}(z, x) \\ &= ((A - \bar{\mu} I)z, x) \end{aligned}$$

for all $x \in H$. But, on taking $x = (A - \bar{\mu}I)z$, we find $(A - \bar{\mu}I)z = 0$. If z is not zero, this states that $\bar{\mu}$ is an eigenvalue of A. But A is self-adjoint so that $\bar{\mu}$ is real, i.e., $\bar{\mu} = \mu$; μ is an eigenvalue of A, contrary to hypothesis. ■

We now generalise the concept of an eigenvalue and introduce the concept of the *spectrum* of an operator.

Definition 10.3.6 *Suppose $A \in B(H,H)$. The* ***spectrum*** *of A, denoted by $\sigma(A)$, is the set of all complex numbers μ such that $A - \mu I$ does* ***not*** *have a bounded inverse. The* ***resolvent*** *set $\rho(A)$ is the complement of σ, i.e., $\rho = \mathbb{C} \backslash \sigma$.*

We recall that if $A \in B(H,H)$ then $||Ax|| \le ||A|| \cdot ||x||$; if A is to have a bounded inverse then $||Ax|| \ge k||x||$ for some $k > 0$. We prove

Lemma 10.3.1 *If $A \in B(H,H)$ and $||Ax|| \ge k||x||$ for all $x \in H$ and some $k > 0$, then $R(A)$ is closed.*

Proof. Suppose $\{x_n\} \subset H$ and $Ax_n \to y$. The sequence $\{Ax_n\}$ is a Cauchy sequence, and so therefore is $\{x_n\}$ because $||x_m - x_n|| \le ||Ax_m - Ax_n||/k$. Since H is complete, there is $x \in H$ such that $x_n \to x$. By continuity we have $Ax_n \to Ax$, so that $y = Ax$, i.e., $y \in R(A) : R(A)$ is closed. ■

We may now characterise the resolvent set of a self-adjoint operator.

Theorem 10.3.4 *Suppose $A \in B(H,H)$ is self-adjoint, then $\mu \in \rho(A)$ iff $||(A-\mu I)x|| \ge k||x||$ for all $x \in H$ and some $k > 0$.*

Proof. If $\mu \in \rho(A)$, then $(A - \mu I)$ has a bounded inverse, so that

$$||(A - \mu I)^{-1}|| \cdot ||x||$$

i.e., $||(A - \mu I)x|| \ge ||(A - \mu I)^{-1}||^{-1} \cdot ||x||$.

Conversely, if $||(A-\mu I)x|| \geq k||x||$ for all $x \in H$, then Theorem 10.3.3 states that $R(A-\mu I)$ is dense in H, while Lemma 10.3.1 states that $R(A-\mu I)$ is closed. Thus $R(A-\mu I) = H$, and $||(A-\mu I)x|| \geq k||x||$ states that $(A-\mu I)$ has a bounded inverse, i.e., $\mu \in \rho(A)$. ■

We now show that if $A \in B(H,H)$ self-adjoint then its spectrum is real, non-empty, and lies within the interval $[m, M]$.

Theorem 10.3.5 *If $A \in B(H,H)$ is self-adjoint, then $\sigma(A)$ is a non-empty subset of $[m, M]$, and $m, M \in \sigma(A)$.*

Proof. First we prove that the spectrum is real. For suppose $\mu = \alpha + i\beta$, $\beta \neq 0$, then for all $x \in H$,

$$\begin{aligned} ||(A-\mu I)x||^2 &= (Ax - \alpha x - i\beta x,\ Ax - \alpha x - i\beta x) \\ &= ||(A-\alpha I)x||^2 + \beta^2||x||^2 \\ &\geq \beta^2||x||^2. \end{aligned}$$

Theorem 10.3.4 shows that $\mu \in \rho(A)$. Thus if $\mu \in \sigma(A)$ then μ must be real.

We now show that if $\mu < m$, then $\mu \in \rho(A)$. We have, on the one hand,

$$((A-\mu I)x, x) \leq ||(A-\mu I)x|| \cdot ||x||$$

and, on the other

$$\begin{aligned} ((A-\mu I)x, x) &= (Ax, x) - \mu||x||^2 > m||x||^2 - \mu||x||^2 \\ &> (m-\mu)||x||^2 \end{aligned}$$

so that

$$||(A-\mu I)x|| \geq (m-\mu)||x||$$

so that Theorem 10.3.4 shows that $\mu \in \rho(A)$. We can show similarly that if $\mu > M$, then $\mu \in \rho(A)$. We have thus shown that, *if $\sigma(A)$ exists*, it must lie in $[m, M]$.

We now show that $M \in \sigma(A)$. By the definition of sup, there is a sequence $\{x_n\}$ such that $||x_n|| = 1$, and $(Ax_n, x_n) \to M$. Therefore,

$$\begin{aligned} ||(A-MI)x_n||^2 &= (Ax_n - Mx_n, Ax_n - Mx_n) \\ &= ||Ax_n||^2 - 2M(Ax_n, x_n) + M^2||x_n||^2 \\ &\leq M^2 - 2M(Ax_n, x_n) + M^2 \\ &\leq 2M(M - (Ax_n, x_n)) \to 0. \end{aligned}$$

Thus M, and similarly m, are in $\sigma(A)$. ■

So far, we have shown that a self-adjoint operator $A \in B(H,H)$ *has* a non-empty real spectrum that lies in $[m, M]$. Now we suppose that, in addition to being self-adjoint, A is a compact operator. In that case the spectrum consists entirely of eigenvalues, apart perhaps from zero. This is given in

Theorem 10.3.6 *If $A \in B(H,H)$ is self-adjoint and compact and if $\mu \in \sigma(A)$ and $\mu \neq 0$, then μ is an eigenvalue of A.*

Proof. If $\mu \in \sigma(A)$ then, by definition, $A - \mu I$ does not have a bounded inverse. There is therefore (Ex. 10.3.3) a sequence $\{x_n\}$ such that $||x_n|| = 1$ and $Ax_n - \mu x_n \to 0$ as $n \to \infty$. Since A is compact it maps $\{x_n\}$ into a precompact set. This means that there is a subsequence $\{x_{n_k}\}$ such that $Ax_{n_k} \to y \in H$. We then have

$$x_{n_k} = \mu^{-1}[Ax_{n_k} - (Ax_{n_k} - \mu x_{n_k})] \to \mu^{-1}y$$

and therefore, since A is continuous,

$$\mu y = \lim_{k\to\infty} Ax_{n_k} = Ay.$$

Since $||x_n|| = 1$ and $\mu \neq 0$ we have $||y|| \neq 0$, so that y is an eigenvector corresponding to μ. ■

Since we have already proved that $m, M \in \sigma(A)$, we now know that, provided m, M are not zero, and if A is not zero, one of them at least must be non-zero because of (10.3.13), m and M are eigenvalues of A: a non-zero compact self-adjoint operator has *at least one* real eigenvalue.

Having shown that A has at least one eigenvalue, we now prove

Theorem 10.3.7 *A non-zero compact self-adjoint operator in a Hilbert space H has a finite or infinite sequence of orthonormal eigenvectors $x_1, x_2, \ldots$ corresponding to non-zero eigenvalues $\mu_1, \mu_2, \ldots (|\mu_1| \geq |\mu_2| \geq \ldots)$.*

Proof. By Theorem 10.3.6 there is an eigenvector x_1, with $||x_1|| = 1$, $Ax_1 = \mu_1 x_1$, where

$$\mu_1 = \pm \sup |(Ax, x)|, \quad ||x|| = 1$$

μ_1 is either m or M, and $|\mu_1| = ||A||$.

Rename the Hilbert space H_1, the operator as A_1, let M_1 be the space spanned by x_1, and decompose H_1 into H_2 and M_1 as in equation (10.3.4). The space H_2 is a Hilbert space. If $x \in H_2$, then $A_1 x \in H_2$, for

$$(A_1 x, x_1) = (x, A_1 x_1) = (x, \mu_1 x_1) = \mu_1(x, x_1) = 0.$$

This means that we may define a new operator A_2 in H_2, by

$$A_2 x = A_1 x, \quad x \in H_2.$$

This operator is called the *restriction* of A_1 to H_2; it is clearly a self-adjoint compact linear operator in the Hilbert space H_2. If this operator is not identically zero we may apply Theorem 10.3.6 to it, and find an eigenvector x_2 such that

$$A_2 x_2 = \mu_2 x_2, \quad ||x_2|| = 1.$$

Since $x_2 \in H_2$, we have $(x_2, x_1) = 0$ and, for $||x|| = 1$,

$$|\mu_2| = \sup_{x\in H_2} |(A_1 x, x)| \leq \sup_{x\in H_1} |(A_1 x, x)| = |\mu_1|.$$

We now continue this process; we let M_2 be the space spanned by x_2, decompose H_2 into H_3 and M_2, call A_3 the restriction of A_2 to H_3, and find an eigenvalue μ_3 and eigenvector x_3, and so on.

Generally

$$|\mu_k| = \sup_{x\in H_k} |(Ax,x)| = |(Ax_k,x_k)|, \quad ||x|| = 1 = ||x_k||. \tag{10.3.14}$$

■

Either the process stops after a finite number of steps or it continues indefinitely. In the former case there is an integer n for which the restriction A_{n+1} of A_1 to H_{n+1} is identically zero, i.e.,

$$\sup_{x\in H_{n+1}} |(Ax,x)| = 0, \quad ||x|| = 1. \tag{10.3.15}$$

In this case we obtain a finite sequence of orthonormal eigenvectors $x_1, x_2, \ldots, x_n$. The latter case is the subject of

Theorem 10.3.8 *Suppose $A \in B(H,H)$ is a self-adjoint compact operator. If A has an infinity of eigenvalues, they are enumerable with zero being the only limit point.*

Proof. The procedure described in Theorem 10.3.7 produces a sequence of eigenvalues $\mu_1, \mu_2, \ldots$ such that $|\mu_1| \geq |\mu_2| \geq \cdots$, and corresponding sequence of orthonormal eigenvectors $x_1, x_2, \ldots$ Consider all those eigenvalues satisfying $|\mu| > c$. If there is an infinite sequence $x_1, x_2, \ldots$ corresponding to such eigenvalues, then

$$||Ax_m - Ax_n||^2 = ||\mu_m x_m - \mu_n x_n||^2 = |\mu_m|^2 + |\mu_n|^2 \geq 2c^2. \tag{10.3.16}$$

But since A is compact, the sequence $\{Ax_n\}$ must have a convergent subsequence; this contradicts (10.3.16). Hence there is at most a finite set of eigenvectors corresponding to eigenvectors satisfying $|\mu| > c$. The eigenvalues may be enumerated by placing their absolute values in the intervals $(1,\infty), (1/2,1], (1/3, 1/2], \ldots$; there is a finite number in each of this enumerable set of intervals; the eigenvalues can have zero as their only limit point. ■

Theorem 10.3.9 *Let $A \in B(H,H)$ be a compact self-adjoint operator with eigenvalues μ_i ordered so that $|\mu_1| \geq |\mu_2| \geq \cdots$, and corresponding orthonormal eigenvectors $x_1, x_2, \ldots$ The eigenvectors $\{x_i\}$ are complete in the range of A, i.e., for every $f = Ah$, $h \in H$, the Parseval equality*

$$||f||^2 = \sum_{k=1}^{\infty} |(f,x_k)|^2 \tag{10.3.17}$$

holds.

Proof. First, suppose the process described in Theorem 10.3.7 stops. Take $f = Ah$, and consider

$$g = h - \sum_{k=1}^{n} (h, x_k)x_k. \tag{10.3.18}$$

We have $(g, x_k) = 0, \quad k = 1, 2, \ldots, n$, so that $g \in H_{n+1}$ and hence $x = g/||g||$ satisfies (10.3.15) so that $||Ax|| = 0$, i.e., $Ag = 0$. Thus

$$\begin{aligned} 0 = Ag &= Ah - \textstyle\sum_{k=1}^{n}(h, x_k)Ax_k = Ah - \sum_{k=1}^{n}(Ah, x_k)x_k \\ &= f - \textstyle\sum_{k=1}^{n}(f, x_k)x_k, \end{aligned}$$

so that

$$f = \sum_{k=1}^{n} (f, x_k)x_k.$$

Now consider the case in which the process does not stop. There is an enumerable sequence of eigenvalues $\{\mu_i\}$ with zero as limit point.

Choose $\epsilon > 0$ and then choose N so that if $n > N$, then $|\mu_n|^2 < \epsilon$. Take $n > N$. Suppose $f = Ah$ and consider g given by (10.3.18); $g \in H_n$ so that

$$\frac{||Ag||}{||g||} \leq |\mu_{n+1}|.$$

Thus

$$||Ag|| \leq |\mu_{n+1}| \; ||g|| \leq |\mu_{n+1}| \; ||h||,$$

so that, as before

$$||Ag||^2 = \left|\left| f - \sum_{k=1}^{n} (f, x_k)x_k \right|\right|^2 \leq |\mu_{n+1}|^2 ||h||^2$$

or equivalently

$$0 \leq ||f||^2 - \sum_{k=1}^{n} |(f, x_k)|^2 \leq |\mu_{n+1}| \cdot ||h||^2 \leq \epsilon ||h||^2$$

which implies Parseval's equality

$$\textstyle\sum_{k=1}^{\infty} |(f, x_k)|^2 = ||f||^2. \; \blacksquare$$

We now obtain another result by making a further assumption concerning A; thus we introduce

Definition 10.3.7 *A self-adjoint continuous linear operator A in a Hilbert space H is called **strictly positive** if $(Ax, x) \geq 0$ for all $x \in H$ and $(Ax, x) = 0$ iff $x = 0$.*

For a strictly positive, compact, self-adjoint operator in a Hilbert space the process described in Theorem 10.3.7 can stop only if H itself is finite dimensional. This leads to

Theorem 10.3.10 *Let A be a strictly positive compact self-adjoint operator in an infinite dimensional Hilbert space H. There is an orthonormal system $\{x_n\}$ which is a basis for H, and A has the representation*

$$Ax = \sum_{k=1}^{\infty} \lambda_k (x, x_k) x_k.$$

Proof. Let $y \in H$ and consider

$$y_{n+1} = y - \sum_{k=1}^{n} (y, x_k) x_k,$$

where $\{x_k\}$ is the orthonormal sequence of eigenvectors, as in Theorem 10.3.9. It is easy to show that $\{y_n\}$ is a Cauchy sequence. We wish to prove that its limit is zero. Assume that it is not, i.e., $y_n \to z \neq 0$. Since $y_{n+1} \in H_{n+1}$ we have

$$\frac{(Ay_{n+1}, y_{n+1})}{||y_{n+1}||^2} \leq \mu_{n+1}^2.$$

But $\mu_n \to 0$ as $n \to \infty$ so that passage to the limit gives

$$\frac{(Az, z)}{||z||^2} = 0,$$

which is a contradiction since A is strictly positive. Therefore, $z = 0$ and

$$y = \sum_{k=1}^{\infty} (y, x_k) x_k, \quad y \in H,$$

so that $\{x_k\}$ forms a basis for H, and moreover

$$Ay = \sum_{k=1}^{\infty} (y, x_k) A x_k = \sum_{k=1}^{\infty} \mu_k (y, x_k) x_k. \quad \blacksquare$$

This theorem shows that one can have a strictly positive compact self-adjoint operator only in a *separable* Hilbert space.

Corollary 10.3.1 *Under the condition of Theorem 10.3.10 we can introduce a norm*

$$||x||_A = (Ax, x)^{1/2}$$

and a corresponding inner product

$$(x, y)_A = (Ax, y).$$

The completion of H with respect to this norm is called H_A.

Exercises 10.3

1. Show that eigenvectors x and y, corresponding to two different eigenvectors of a self-adjoint operator A, are orthogonal, i.e., $(x, y) = 0$.

2. Show that the operator A^{-1} is bounded on $R(A)$ iff there is a constant $c > 0$, such that, if $x \in D(A)$, then $||Ax|| \geq c||x||$.

3. Use Ex. 10.3.2 to show that A^{-1} is unbounded iff there is a sequence $\{x_n\}$ such that $||x_n|| = 1$, $||Ax_n|| \to 0$.

4. Show that a compact self-adjoint operator is strictly positive iff its eigenvalues are positive.

10.4 The Green's function integral equation

We must now exhibit the integral operator

$$Au = \int_0^1 K(x,s)u(s)ds \tag{10.4.1}$$

as a stricly positive, self-adjoint, compact operator in a separable Hilbert space. In order to make this identification we need some results about functions.

We start with the space of continuous functions on the closed interval $[0,1]$. We call this $C[0,1]$. The fundamental result about a function $f(x) \in C[0,1]$ is that $f(x)$ is bounded on $[0,1]$, and actually *attains* its upper bound. We may thus form a normed linear space from $C[0,1]$ by using the norm

$$||f||_\infty = \sup_{x\in[0,1]} |f(x)|. \tag{10.4.2}$$

Convergence of a sequence of function $\{f_n(x)\}$ in the norm (10.4.2) is *uniform* convergence. Weierstrass' Theorem on uniform convergence states that a uniformly Cauchy sequence $\{f_n(x)\}$, i.e., a Cauchy sequence in the norm (10.4.2), of uniformly continuous functions on $[0,1]$ converges to a uniformly continuous function. This translates into the statement that $C[0,1]$ under the norm (10.4.2) is *complete.*

We may introduce another norm on $C[0,1]$:

$$||f||_2 = \left\{\int_0^1 (f(x))^2 dx\right\}^{1/2}. \tag{10.4.3}$$

The example in Ex. 10.4.1 shows that $C[0,1]$ is *not* complete under this norm. However, we may use the completion theorem, and complete this space. We may make the space an inner-product space by using the inner product

$$(f,g) = \int_0^1 f(x)g(x)dx. \tag{10.4.4}$$

We call this complete inner-product space, i.e., Hilbert space, $L^2(0,1)$. Here L stands for Lebesgue. Remember that while the elements of $C[0,1]$ are uniformly continuous functions, the elements of $L^2(0,1)$ are equivalence classes of Cauchy sequences of uniformly continuous functions. The space $L^2(0,1)$ is known to be *separable* (Th. 4.1.4).

Now we start to examine the operator A from $L^2(0,1)$ to $\mathrm{L}^2(0,1)$, defined by

$$Au = \int_0^1 K(x,s)u(s)ds$$

where

$$K(x,s) = \rho(x)\rho(s)G(x,s) \tag{10.4.5}$$

and $G(x,s)$ is given in (10.2.6).

The operator A is self-adjoint in $L^2(0,1)$ because $K(x,s)$ is symmetric.

Now we examine the continuity of the operator. Suppose first that $\rho(x) \in C[0,1]$, then $K(x,s) \in C([0,1]\times[0,1])$ so that $K(x,s)$ is bounded on the square, i.e., $K(x,s) \leq M$ and

$$||Au||_\infty = \sup_{x\in[0,1]} |Au| \leq M \sup_{x\in[0,1]} |u| = M||u||_\infty,$$

so that $||A|| \leq M$: A is continuous.

Now examine continuity in $L^2(0,1)$. We have

$$||Au||^2 = \int_0^1 \left\{\int_0^1 K(x,s)u(s)ds\right\}^2 dx.$$

Again, if $K(x,s) \in C([0,1]\times[0,1])$ then $K(x,s) \leq M$ and

$$||Au||^2 \leq M^2 \int_0^1 (u(s))^2 ds \leq M^2||u||^2$$

so that A is continuous. Now suppose that $\rho(x) \in L^2(0,1)$.

Since $K(x,s) = \rho(x)\rho(s)G(x,s)$, and $G(x,s) \in C([0,1]\times[0,1])$ we have $|G(x,s)| \leq M$ and

$$|K(x,s)| \leq \rho(x)\rho(s)M.$$

Thus

$$||Au||^2 \leq M^2 \int_0^1 \rho^2(x) \left\{\int_0^1 \rho(s)u(s)ds\right\}^2 dx.$$

The Schwarz inequality (10.3.2) gives

$$\left\{\int_0^1 \rho(s)u(s)ds\right\}^2 \leq \int_0^1 \rho^2(s)ds \int_0^1 u^2(s)ds,$$

so that

$$||Au||^2 \leq M^2||\rho||^4||u||^2.$$

Thus

$$||A|| \leq M||\rho||^2,$$

and A is continuous.

In order to prove that A is compact we note that if a function $f(x, s)$ is continuous on the unit square, i.e., $f \in C([0,1]\times[0,1])$, then it may be approximated uniformly by a finite sum of the form

$$\sum_{i=1}^{n} \alpha_i(x)\beta_i(s).$$

The Green's function $G(x, s)$ is continuous on the unit square, and is symmetric in x and s. Thus there are functions $\{\alpha_i(x)\}_1^\infty$ such that, given $\varepsilon > 0$, we can find N so that if $n > N$ then

$$\sup |G(x, s) - \sum_{i=1}^{n} \alpha_i(x)\alpha_i(s)| \leq \varepsilon,$$

for $(x, s) \in ([0,1] \times [0,1])$. This means that if

$$K_n(x, s) = \rho(x)\rho(s) \sum_{i=1}^{n} \alpha_i(x)\alpha_i(s),$$

and

$$A_n u = \int_0^1 K_n(x, s)u(s)ds,$$

then A_n is a finite-dimensional operator, and thus compact. If $\rho \in L^2(0,1)$ then A is the limit of a sequence of compact linear operators $\{A_n\}$, and is thus compact by Theorem 10.3.1.

Reader, congratulations if you have read and followed thus far. We have tried to provide a sign-posted journey; clearly, we have not proved every step, but we had no intention of doing that. We could have taken a short cut by merely stating that 'it can be shown that A is compact', but we hope that the route we have taken has been more pleasant and instructive.

What can we conclude from our study? If $\rho(x) \in L^2(0,1)$, the integral equation has a finite or enumerable sequence of positive eigenvalues $\mu_1, \mu_2, \ldots$ satisfying $|\mu_1| > |\mu_2| > \cdots$, and a corresponding set of eigenfunctions $\{u_i\}_0^\infty$ which are orthonormal under the $L^2(0,1)$ norm. However, this result is not as satisfying as we would like, because the eigenfunctions, being in $L^2(0,1)$, are not functions in the ordinary sense, but equivalence classes of Cauchy sequences of functions in $C[0,1]$. Can we say anything more about them?

First, we note that if u satisfies (10.2.11) then v satisfies (10.2.9) where, remember that we now switch $\lambda \to 1/\mu$. Thus, the eigenvalues λ_i of (10.2.9) satisfy $0 < |\lambda_1| \leq |\lambda_2| \leq \ldots$ Actually, we proved earlier that the λ_i are distinct and positive, i.e., they satisfy (10.1.18): $0 < \lambda_1 < \lambda_2 < \ldots$ We have not yet shown that there is an infinity of eigenvalues, nor have we shown, in the Green's function analysis, that they are distinct; we will eventually do this.

We may write (10.2.9) as

$$v(x) = \lambda \int_0^1 \rho(s)G(x,s)u(s)ds. \tag{10.4.6}$$

If $\rho \in L^2(0,1)$ and $u \in L^2(0,1)$ then the integrand in (10.4.6) is integrable in s and uniformly continuous in x, so that the left hand side, $v(x)$, is continuous: $v(x) \in C[0,1]$, and we may properly speak of an eigenfunction. If $\rho \in C[0,1]$ then $v(x)$ actually has a continuous second derivative, and satisfies equation (10.1.1), for on using the form of $G(x,s)$ given in (10.2.6) we see that

$$v(x) = \lambda\psi(x) \int_0^x \rho^2(s)\phi(s)v(s)ds + \lambda\phi(x)\int_x^1 \rho^2(s)\psi(s)v(s)ds \tag{10.4.7}$$

so that

$$\begin{aligned} v(0) &= \lambda\phi(0)\textstyle\int_0^1 \rho^2(s)\psi(s)v(s)ds \\ v(1) &= \lambda\psi(1)\textstyle\int_0^1 \rho^2(s)\phi(s)v(s)ds. \end{aligned}$$

Now, differentiating (10.4.7), which we can do because all the integrands are continuous, we find

$$\begin{aligned} v'(x) &= \lambda\psi'(x)\textstyle\int_0^x \rho^2(s)\phi(s)v(s)ds + \psi(x)\rho^2(x)\phi(x)v(x) \\ &+ \lambda\phi'(x)\textstyle\int_x^1 \rho^2(s)\psi(s)v(s)ds - \phi(x)\rho^2(x)\psi(x)v(x). \end{aligned}$$

Thus

$$\begin{aligned} v'(0) &= \lambda\phi'(0)\textstyle\int_0^1 \rho^2(s)\psi(s)v(s)ds = hv(0) \\ v'(1) &= \lambda\psi'(1)\textstyle\int_0^1 \rho^2(s)\phi(s)v(s)ds = -Hv(1). \end{aligned}$$

Thus $v(x)$ satisfies the stated end conditions. On differentiating a second time, using $\phi''(x) = 0 = \psi''(x)$, we find

$$v''(x) = \lambda(\phi(x)\psi'(x) - \phi'(x)\psi(x))\rho^2(x)v(x)$$

and on account of (10.2.5), this is

$$v''(x) + \lambda\rho^2(x)v(x) = 0.$$

Exercises 10.4

1. Consider the sequence $\{f_n(x)\}$ in $C[0,1]$:

$$f_n(x) = \begin{cases} x^{-\frac{1}{4}} & \frac{1}{n} \le x \le 1 \\ n^{\frac{1}{4}} & 0 \le x \le \frac{1}{n} \end{cases}.$$

Show that $\{f_n(x)\}$ is a Cauchy sequence under the L^2 norm (10.4.3), but $\{f_n(x)\}$ converges to

$$f(x) = x^{-\frac{1}{4}}$$

which is not in $C[0,1]$. Hence $C[0,1]$ is not complete under the L^2 norm.

10.5 Oscillatory properties of Green's functions

In Section 10.4 we showed that when $h \geq 0,\ H \geq 0,\ h+H > 0$, the integral equation (10.2.9) has eigenvalues λ_i satisfying $0 < |\lambda_1| \leq |\lambda_2| \leq \ldots$; if there are an infinity of them, then

$$0 < |\lambda_1| \leq |\lambda_2| \leq \ldots \to \infty.$$

On the other hand, in Section 10.4, we showed that the eigenvalues of the (equivalent) equation (10.1.1) are positive and distinct, i.e.,

$$0 < \lambda_1 < \lambda_2 < \ldots.$$

This means that the Green's function $G(x,s)$ must have some special properties which lead to the eigenvalues being distinct; we now discuss these properties.

We start by defining the interval I, as follows:

$$\begin{aligned} I &= [0,1] \text{ if } h, H \text{ are finite} \\ &= (0,1] \text{ if } h = \infty,\ H \text{ is finite} \\ &= [0,1) \text{ if } h \text{ is finite},\ H = \infty \\ &= (0,1) \text{ if } h = \infty = H. \end{aligned}$$

Note that when $h = \infty$, the end condition $u'(0) - hu(0) = 0$ becomes $u(0) = 0$, i.e., the end $x = 0$ is fixed. This means that I is the set of *movable points* in $[0,1]$. Equations (10.2.6), (10.2.7) show that

$$\begin{aligned} G(x,s) &\geq 0 \text{ for } x, s \in [0,1] \\ &> 0 \text{ for } x, s \in I. \end{aligned}$$

We now introduce the concept of an *oscillatory kernel.*

Definition 10.5.1 *If* $0 < x_1 < x_2 < \cdots < x_n < 1$, *and* $\mathbf{x} = [x_1, x_2, \ldots, x_n]$, *then we say* $\mathbf{x} \in Q$. *If* $x_1, x_n \in I$ *then we say* $\mathbf{x} \in \mathcal{I}$. *A kernel* $K(x,s)$ *on* $[0,1] \times [0,1]$ *is said to be* ***oscillatory*** *if*

i) $K(x,s) > 0$ *for* $x, s \in \mathcal{I}$
ii) $K(\mathbf{x};\mathbf{s}) \geq 0$ *for* $\mathbf{x}, \mathbf{s} \in Q$
iii) $K(\mathbf{x};\mathbf{x}) > 0$ *for* $\mathbf{x} \in Q$

Here

$$K(\mathbf{x};\mathbf{s}) = \begin{vmatrix} K(x_1,s_1) & K(x_1,s_2) & \cdots & K(x_1,s_n) \\ K(x_2,s_1) & K(x_2,s_2) & \cdots & K(x_2,s_n) \\ \cdot & \cdot & \cdots & \cdot \\ K(x_n,s_1) & K(x_n,s_2) & \cdots & K(x_n,s_n) \end{vmatrix}$$

and take note of Ex. 10.5.1 which shows that iii) must necessarily hold for $\mathbf{x} \in \mathcal{I}$.

Theorem 10.5.1 *A kernel* $K(x,s)$ *is oscillatory iff the matrix* $A = (a_{ij}) = (K(x_i,x_j))$ *is an oscillatory matrix for any* $\mathbf{x} \in \mathcal{I}$.

Proof. Suppose the kernel is oscillatory then, in the notation of Section 6.2, if $\alpha = (i_i, i_2, \ldots, i_p)$, $\beta = (j_1, j_2, \ldots, j_p)$ then

$$A(\alpha; \beta) = K(\mathbf{x}^0; \mathbf{s}^0) \geq 0$$

where $x_k^0 = x_{i_k}$, $s_k^0 = x_{j_k}$, $k = 1, 2, \ldots, p$. Thus **A** is TN. Now

$$a_{i,i+1} = K(x_i, x_{i+1}) > 0, \; a_{i+1,i} = K(x_{i+1}, x_i) > 0$$

while

$$\det(\mathbf{A}) = K(\mathbf{x}; \mathbf{x}) > 0.$$

Thus **A** satisfies the three conditions for it to be oscillatory: it is TN, the terms next to the principal diagonal are positive, and it is non-singular. We may reverse this argument to show that if **A** is oscillatory then $K(x, s)$ is an oscillatory kernel. ∎

Note that in addition to being oscillatory, **A** is a strictly positive matrix for $\mathbf{x}, \mathbf{s} \in \mathcal{I}$.

We now show that the Green's function $G(x, s)$ defined in (10.2.6), (10.2.7) is an oscillatory kernel. To do so we recall the definition of a *Green's matrix.*

Definition 10.5.2 *The matrix* $\mathbf{G} = (g_{ij})$ *is called a Green's matrix if*

$$g_{ij} = \begin{cases} a_i b_j, & i \leq j, \\ a_j b_i, & i \geq j, \end{cases}$$

where $(a_i)_1^n, (b_i)_1^n \subset \mathbb{R}$.

Note that **G** is symmetric.

Theorem 10.5.2 *If* $\alpha = (i_i, i_2, \ldots, i_p)$, $\beta = (j_1, j_2, \ldots, j_p)$ *then*

$$G(\alpha; \beta) = a_{k_1} \prod_{r=2}^{p} \begin{vmatrix} a_{k_r} & a_{l_{r-1}} \\ b_{k_r} & b_{l_{r-1}} \end{vmatrix} b_{l_p} \tag{10.5.1}$$

where $k_m = \min(i_m, j_m)$, $l_m = \max(i_m, j_m)$, *provided that* $i_m, j_m < i_{m+1}, j_{m+1}$.

Recall that this means that

$$i_m < i_{m+1}, \quad i_m < j_{m+1}, \quad j_m < i_{m+1}, \quad j_m < j_{m+1}.$$

Proof. If $i_1 < i_2$ but $j_1 \geq i_2$, then the first two rows of the minor are

$$\begin{matrix} g_{i_1,j_1} & g_{i_1,j_2} & \cdots & g_{i_1,j_p} \\ g_{i_2,j_1} & g_{i_2,j_2} & \cdots & g_{i_2,j_p} \end{matrix}$$

but these are

$$\begin{matrix} a_{i_1} b_{j_1}, & a_{i_1} b_{j_2}, & \ldots & a_{i_1} b_{j_p} \\ a_{i_2} b_{j_1}, & a_{i_2} b_{j_2} & \ldots & a_{i_2} b_{j_p} \end{matrix}$$

and are thus proportional, so that the minor is zero. Similarly, if $j_1 < j_2 \le i_1$, the first two columns will be proportional and the minor zero. Thus, we may assume $\max(i_1, j_1) < \min(i_2, j_2)$. Suppose further, for definiteness, that $i_2 \le j_2$ (otherwise the argument proceeds with the first two columns), then the first two row are

$$\begin{array}{llll} a_{k_1}b_{l_1}, & a_{i_1}b_{j_2}, & \ldots & a_{i_1}b_{j_p} \\ a_{j_1}b_{i_2}, & a_{i_2}b_{j_2}, & \ldots & a_{i_2}b_{j_p} \end{array}$$

so that the terms in columns $2, 3, \ldots, p$ are proportional. Multiplying row 2 by a_{i_1}/a_{i_2} and subtracting it from the first, we find the only non-zero term, the first in the first row, to be

$$\begin{aligned} a_{k_1}b_{l_1} - a_{i_1}a_{j_1}b_{i_2}/a_{i_2} &= a_{k_1}b_{l_1} - a_{k_1}a_{l_1}b_{k_2}/a_{k_2} \\ &= \frac{a_{k_1}}{a_{k_2}}\begin{vmatrix} a_{k_2} & a_{l_1} \\ b_{k_2} & b_{l_1} \end{vmatrix} \end{aligned}$$

so that

$$G(\alpha; \beta) = a_{k_1}\begin{vmatrix} a_{k_2} & a_{l_1} \\ b_{k_2} & b_{l_1} \end{vmatrix} \cdot \frac{1}{a_{k_2}} G(\alpha \backslash i_1; \beta \backslash j_1)$$

from which the theorem follows by induction. ■

Theorem 10.5.3 *The Green's matrix* $\mathbf{G}$ *is TN iff all* $(a_i)_1^n, (b_i)_1^n$ *have the same strict sign and*

$$\frac{a_1}{b_1} \le \frac{a_2}{b_2} \le \cdots \le \frac{a_n}{b_n}. \tag{10.5.2}$$

Moreover, $\mathbf{G}$ *will be oscillatory iff* $(a_i)_1^n, (b_i)_1^n$ *have the same strict sign and*

$$\frac{a_1}{b_1} < \frac{a_2}{b_2} < \cdots < \frac{a_n}{b_n}. \tag{10.5.3}$$

Proof. There is no loss in generality in assuming that all $(a_i)_1^n, (b_i)_1^n$ are positive. It was shown in Theorem 10.5.2 that a minor is zero unless

$$i_1, j_1 < i_2, j_2 < \cdots < i_p, j_p.$$

Each of the second order determinants in (10.5.1) is non-negative iff

$$\frac{a_{l_{i-1}}}{b_{l_{i-1}}} \le \frac{a_{k_i}}{b_{k_i}}, \quad i = 1, 2, \ldots, p.$$

This is exactly the condition (10.5.2). $\mathbf{G}$ is TN and $g_{i,i+1} > 0$, $g_{i+1,i} > 0$, so that the only condition to be fulfilled for $\mathbf{G}$ to be oscillatory is that it must be non-singular. Thus each second order determinant in the factorisation of $G(\alpha; \beta)$ must be positive, which is (10.5.3). ■

Corollary 10.5.1 *Let* $\phi(x), \psi(x)$ *be continuous in* $[0, 1]$ *and*

$$K(x, s) = \begin{cases} \phi(x)\psi(s), & 0 \le x \le s \le 1, \\ \phi(s)\psi(x), & 0 \le s \le x \le 1. \end{cases}$$

If $\phi(x)\psi(x) > 0$ in $(0,1)$ and $\phi(x)/\psi(x)$ is an increasing function of x in $(0,1)$ then

$$K(\mathbf{x};\mathbf{s}) \geq 0 \textit{ for } \mathbf{x},\mathbf{s} \in Q.$$

If $\phi(x)\psi(x) > 0$ in I, and $\phi(x)/\psi(x)$ is a strictly increasing function of x in I, then

$$K(\mathbf{x};\mathbf{s}) > 0$$

iff $\mathbf{x},\mathbf{s} \in \mathcal{I}$ and $x_1, s_1 < x_2, s_2 < \cdots < x_n, s_n$.

Theorem 10.5.4 *The Green's function $G(x,s)$ given by (10.2.5), (10.2.6) is oscillatory and a minor $G(\mathbf{x};\mathbf{s}) > 0$ iff $\mathbf{x},\mathbf{s} \in \mathcal{I}$ and $x_1, s_1 < x_2, s_2 < \cdots < x_n, s_n$.*

Proof. Equation (10.2.7) shows that $\phi(x)\psi(x) > 0$ in I. Equation (10.2.5) yields

$$\frac{d}{dx}\left[\frac{\phi(x)}{\psi(x)}\right] = \frac{\phi'(x)\psi(x) - \phi(x)\psi'(x)}{[\psi(x)]^2} = \frac{1}{[\psi(x)]^2} > 0 \text{ in } I$$

so that $\phi(x)/\psi(x)$ is strictly increasing in I, and thus the result follows from the Corollary 10.5.1. ■

In order to ascertain the meaning of the oscillatory character of the Green's function, consider a string under the action of n concentrated forces $(F_i)_1^n$ applied normal to the string at n points $(s_i)_1^n$ in I. The displacement is

$$u(x) = \sum_{i=1}^{n} G(x, s_i)F_i.$$

Thus $G(x,s) > 0$ (condition i) of Definition 10.5.1) means that the displacement due to a single force F occurs 'in the same direction' as the force.

To see the meaning of condition iii) of Definition 10.5.1 we note that the strain energy of the string under the action of the n forces is

$$U = \frac{1}{2}\sum_{i=1}^{n} u(s_i)F_i = \frac{1}{2}\sum_{i=1}^{n}\sum_{j=1}^{n} G(s_i, s_j)F_iF_j,$$

so that condition iii) states that U is positive definite (for forces applied at movable points, i.e., in I).

The essential nature of an oscillatory kernel is evidenced in

Theorem 10.5.5 *Under the action of n forces $(F_i)_1^n$ the displacement $u(x)$ of the string can change its sign no more than $n-1$ times.*

Proof. Suppose that forces $(F_i)_1^n$ are applied at points $(s_i)_1^n$ where $\mathbf{s} \in \mathcal{I}$. If $s_1 > 0$, then

$$u(x) = \phi(x)\sum_{i=1}^{n} F_i\psi(s_i), \quad 0 \leq x \leq s_1,$$

so that $u(x)$ is of one sign in $[0, s_1]$. If

$$\sum_{i=1}^{n} F_i\psi(s_i) = 0$$

then $u(x)$ is identically zero in $[0, s]$. Otherwise, it is of one sign, and can be zero only at $x = 0$, and that only if the string is fixed at $x = 0$, i.e., $h = \infty$.

In the interval $[s_j, s_{j+1}]$, $j = 1, 2, \ldots, n-1$,

$$u(x) = \psi(x)\sum_{i=1}^{j} F_i\phi(s_i) + \phi(x)\sum_{i=j+1}^{n} F_i\psi(s_i).$$

Since $\phi(x), \psi(x)$ are linearly independent, the displacement $u(x)$ is identically zero in $[s_j, s_{j+1}]$ iff

$$\sum_{i=1}^{n} F_i\phi(s_i) = 0 = \sum_{i=j+1}^{n} F_i\psi(s_i).$$

If this is not the case then $u(x)$ can have at most one zero in $[s_j, s_{j+1}]$. For if there were two, say ξ, η such that $s_j \le \xi < \eta \le s_{j+1}$ then $\psi(\xi)\phi(\eta) - \psi(\eta)\phi(\xi) = 0$, contradicting the fact that $\phi(x)/\psi(x)$ is a strictly increasing function.

Finally, if $s_n \le x \le 1$ then

$$u(x) = \psi(x)\sum_{i=1}^{n} F_i\phi(s_i)$$

so that again $u(x)$ has one sign. It is identically zero if

$$\sum_{i=1}^{n} F_i\phi(s_i) = 0,$$

otherwise it can be zero only at $x = 1$, and that only if $H = \infty$. We conclude that $u(x)$ can change sign at most $n - 1$ times, at most once in each of

$$(s_1, s_2],\ [s_2, s_3], \ldots \qquad [s_{n-1}, s_n) \ \blacksquare$$

Exercises 10.5

1. Continuity of the minor in ii) of Definition 10.5.1 shows that it will be non-negative for $(x_i)_1^n$, $(s_i)_1^n$ satisfying $0 \le x_1 < x_2 < \cdots < x_n \le 1$ and $0 \le s_1 < s_2 < \cdots < s_n \le 1$. Use Theorem 6.6.5 to show that iii) necessarily holds for $\mathbf{x} \in \mathcal{I}$.

10.6 Oscillatory systems of functions

In this section we shall derive some basic results that are needed to establish further properties of the eigensolutions.

Let $(\phi_i(x))_1^n$ be a sequence of functions defined on an interval I, ($[0, 1]$, $(0, 1]$, $[0, 1)$, or $(0, 1)$).

Theorem 10.6.1 *The necessary and sufficient condition for the functions $(\phi_i(x))_1^n$ to be linearly dependent is that*

$$\Phi(x_1, x_2, \ldots, x_n; 1, 2, \ldots, n) \equiv \begin{vmatrix} \phi_1(x_1) & \phi_1(x_2) & \ldots & \phi_1(x_n) \\ \phi_2(x_1) & \phi_2(x_2) & \ldots & \phi_2(x_n) \\ \cdot & \cdot & \ldots & \cdot \\ \phi_n(x_1) & \phi_n(x_2) & \ldots & \phi_n(x_n) \end{vmatrix}$$

be zero for any $(x_r)_1^n \in I$.

Proof. The condition is necessary. For if the functions $(\phi_i(x))_1^n$ are linearly dependent then there are constants $(c_i)_1^n$, not all zero, such that

$$\sum_{i=1}^{n} c_i\phi_i(x) = 0 \text{ for } x \in I.$$

This means that for any $(x_r)_1^n \in I$ we have

$$\sum_{i=1}^{n} c_i\phi_i(x_r) = 0, \quad r = 1, 2, \ldots, n. \tag{10.6.1}$$

Since the $(c_i)_1^n$ are not all zero, the determinant of coefficients in (10.6.1) must be zero.

We prove sufficiently by induction. If $n = 1$, then $\Phi = 0$ states that $\phi_1(x_1) = 0$ for any $x_1 \in I$, i.e., $\phi_1(x) \equiv 0$ for $x \in I$.

Suppose therefore that

$$\Phi(x_1, x_2, \ldots, x_n; 1, 2, \ldots, n) = 0 \text{ for all } (x_i)_1^n \in I.$$

We need to prove that the $(\phi_i(x))_1^n$ are linearly dependent. Assume that $(\phi_i(x))_1^{n-1}$ are linearly independent (for if they were dependent then so would the $(\phi_i(x))_1^n$ be), then there are $(x_r)_1^{n-1} \in I$ such that

$$\Phi(x_1, x_2, \ldots, x_{n-1}; 1, 2, \ldots, n-1) \neq 0. \tag{10.6.2}$$

But then, for all $x \in I$

$$\Phi(x_1, x_2, \ldots, x_{n-1}, x; 1, 2, \ldots, n) = 0.$$

Expand this determinant along its last column; the result has the form (10.6.1) in which c_n, being the determinant (10.6.2), is not zero. ∎

Definition 10.6.1 *A sequence of continuous functions $(\phi_i(x))_1^n$ is said to constitute a **Chebyshev sequence** on I if, for any set of real constants $(c_i)_1^n$, not all zero, the function*

$$\phi(x) = \sum_{i=1}^{n} c_i\phi_i(x)$$

does not vanish more than $n - 1$ times on I.

Theorem 10.6.2 *The sequence $(\phi_i(x))_1^n$ is a Chebyshev sequence on I iff*

$$\Phi \equiv \Phi(\mathbf{x};\theta)$$

maintains strictly fixed sign for $\mathbf{x} \in \mathcal{I}$; θ denotes $(1,2,\ldots,n)$.

Proof. If $\Phi = 0$ for some $\mathbf{x} \in \mathcal{I}$ then, and only then, will the equation

$$\sum_{i=1}^{n} c_i\phi_i(x_r) = 0 \quad r = 1,2,\ldots,n$$

has a non-zero solution $(c_i)_1^n$, i.e., the function $\phi(x)$ will have n different zeros. On the other hand, since $\mathcal{I}$ is a connected subset of $\mathbb{R}^N$ and Φ is a continuous function, the fact that $\Phi \neq 0$ in $\mathcal{I}$ means that Φ has strictly fixed sign on $\mathcal{I}$. Without loss of generality we may take $\Phi > 0$. ∎

Definition 10.6.2 *A sequence of continuous $(\phi_i(x))_1^\infty$ will be called a* ***Markov sequence*** *in I if, for each $n = 1,2,\ldots$ the sequence $(\phi_i(x))_1^n$ is a Chebyshev sequence.*

Theorem 10.6.2 shows that $(\phi_i(x))_1^\infty$ is a Markov sequence iff, for $n = 1,2,\ldots$,

$$\Phi(x_1,x_2,\ldots,x_n;1,2,\ldots,n)$$

has the same strict sign for any $\mathbf{x} \in \mathcal{I}$.

We now explore the nature of the zeros of a combination

$$\phi(x) = \sum_{i=1}^{n} c_i\phi_i(x), \qquad \sum_{i=1}^{n} c_i^2 > 0$$

of continuous functions $\phi_i(x)$ in a Chebyshev sequence. By definition, $\phi(x)$ has at most $n-1$ zeros in I. We may divide these zeros into three groups: s *simple nodes* in $(0,1)$, d *double nodes* in $(0,1)$, and p *end-zeros* at 0 or 1 if these are in I. In any two-sided vicinity of a simple node ξ, there are points x_1, x_2 such that $x_1 < \xi < x_2$ and

$$\phi(x_1)\phi(x_2) < 0.$$

In any two-sided vicinity of a double node η, there are points x_1, x_2 such that $x_1 < \eta < x_2$ and

$$\phi(x_1)\phi(x_2) > 0.$$

The statement that $(\phi_i(x))_1^n$ form a Chebyshev sequence means that

$$s + d + p \leq n - 1.$$

We now establish

Theorem 10.6.3 *If the continuous functions* $(\phi_i(x))_1^n$ *form a Chebyshev sequence on* I*, then*

$$s + 2d + p \leq n - 1$$

i.e., in the estimate of the number of zeros, each double node may be counted twice.

Proof. Let $(x_i)_1^m \in I$ satisfy $x_1 < x_2 < \cdots < x_m$. If $\phi(x_k) \neq 0$ for $k = 1, 2, \ldots, m$, then the maximum number of sign changes in the sequence $(\phi(x_k))_1^m$ occurs if, for some integer h (either 0 or 1)

$$(-)^{h+k}\phi(x_k) > 0, \quad k = 1, 2, \ldots, m.$$

If some $\phi(x_k)$ are zero we may assign signs, + or -, to them and obtain different sign change counts for the sequence $(\phi(x_k))_1^m$; the sign change count will be maximum, $m-1$, if for some integer h (either 0 or 1)

$$(-)^{h+k}\phi(x_k) \geq 0, \quad k = 1, 2, \ldots, m.$$

A set of m points with this property is said to have *property Z*.

Consider some examples. Figure 10.6.1 shows $\phi(x)$ with a zero at $x_1 = 0 \in I \equiv [0, 1)$ and two simple nodes in (0,1).

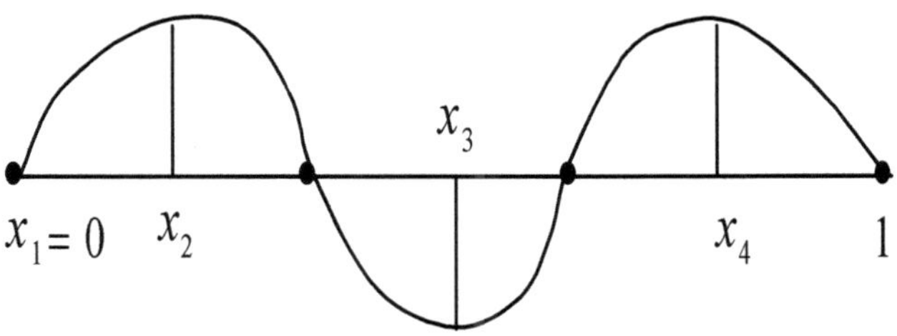

Figure 10.6.1 - $\phi(x)$ has 2 simple nodes.

The points $(x_i)_1^4 \in I$ have property Z. (Note that we are not interested in the value of $\phi(1)$ since 1 is not in I.) Here $s = 2$, $p = 1$ and $m = 4 = s + p + 1$.

Now suppose also that $\phi(x)$ has a double node at x_4, as in Figure 10.6.2, with $I \equiv [0, 1)$.

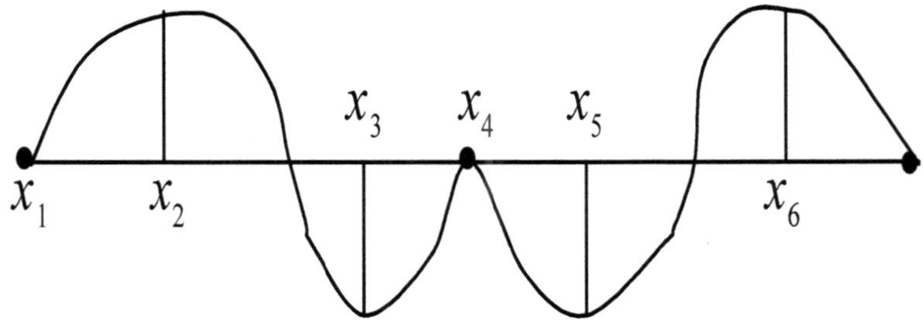

Figure 10.6.2 - $\phi(x)$ has 2 simple nodes and one double node.

The points $(x_i)_1^6 \in I$ have property Z. Here $s = 2$, $d = 1$ and $m = 6 = s + 2d + p + 1$.

In general, if $\phi(x)$ *has* s simple nodes, d double nodes and p end zeros, then we may find

$$m = s + 2d + p + 1$$

points with property Z.

Suppose, if possible, that $s + 2d + p \geq n$, then we may find $n + 1$ points $(x_i)_1^{n+1}$ with property Z, i.e.,

$$(-)^{h+k}\phi(x_k) \geq 0, \quad k = 1, 2, \ldots, n+1. \tag{10.6.3}$$

Since $\phi(x)$ *is a linear combination of* $(\phi_i)_1^n$, the functions $\phi_1, \phi_2, \ldots, \phi_n$, $\phi = \phi_{n+1}$ are linearly dependent. Therefore, by Theorem 10.6.1,

$$\Phi(1, 2, \ldots, n+1; x_1, x_2, \ldots, x_{n+1}) = 0.$$

Expand this zero determinant along its last row; we get

$$\sum_{k=1}^{n+1}(-)^{n+k+1}\phi(x_k)\Phi(1, 2, \ldots, n; x_1, x_2, \ldots, x_{k-1}, x_{k+1}, \ldots, x_n) = 0.$$

Since $(\phi_i(x))_1^n$ form a Chebyshev sequence, the determinants in this equation have the same strict sign, by Theorem 10.6.2. Moreover, by the assumption (10.6.3), the terms $(-)^{n+k+1}\phi(x_k)$ have the same (loose) sign. This means that $\phi(x_k) = 0$ for $k = 1, 2, \ldots, n+1$, but this is impossible: since the $(\phi_i)_1^n$ form a Chebyshev system, $\phi(x)$ has at most $n - 1$ zeros. We conclude that $m \leq n$, i.e., $s + 2d + p \leq n - 1$. ■

We now introduce an extra condition on the function $\{\phi_i(x)\}_1^\infty$, that they are *orthonormal*, and prove the fundamental

Theorem 10.6.4 *If* $\{\phi_i(x)\}_1^\infty$ *is a Markov sequence of continuous functions on* I, *and the* $\phi_i(x)$ *are orthonormal with respect to some inner product, i.e.,* $(\phi_i, \phi_j) = \delta_{ij}$ *then*

1) $\phi_1(x)$ *has no zeros in* I.

2) $\phi_i(x)$ *has* $i - 1$ *simple nodes and no other zeros in* I.

3) $\phi(x) = \sum_{i=j}^k c_i\phi_i(x), \qquad 1 \leq j \leq k, \qquad \sum_{i=j}^k c_i^2 > 0$
has not less than $j - 1$ *simple nodes in* $(0,1)$, *and not more than* $k - 1$ *zeros in* I; *in the notation of Theorem 10.6.3,* $s + 2d + p \leq k - 1$.

Proof. Note that 1) and 2) are particular cases of 3), and all that is left to be proved in 3) is that $\phi(x)$ has not less than $j - 1$ simple nodes.

The functions $(\phi_i)_1^\infty$ form a Markov sequence. This means that if $0 < x_1 < x_2 < \cdots < x_n < 1$, then

$$\Phi(x_1, x_2, \ldots, x_n; 1, 2, \ldots, n)$$

has fixed sign, which we may take to be positive. Let $(\xi_i)_1^s$ be the simple nodes of $\phi(x)$ in (0,1) and define

$$\psi(x) = \Phi(\xi_1, \xi_2, \ldots, \xi_s, x; 1, 2, \ldots, s+1).$$

If $x > \xi_s$ then $\psi(x) > 0$. If $\xi_p < x < \xi_{p+1}$, $p = 1, 2, \ldots, s-1$

$$\psi(x) = (-)^{s-p}\Phi(\xi_1, \xi_2, \ldots, \xi_p, x, \xi_{p+1}, \ldots, \xi_s; 1, 2, \ldots, s+1),$$

while if $x < \xi_1$,

$$\psi(x) = (-)^s\Phi(x, \xi_1, \ldots, \xi_s; 1, 2, \ldots, s+1).$$

Thus $\psi(x)$ changes sign as x passes through each node; $\psi(x)$ has just s zeros, the s simple nodes $(\xi_i)_1^s$. These are the same simples nodes as $\phi(x)$. Therefore,

$$(\psi, \phi) \neq 0.$$

But ψ is a combination of $(\phi_i)_1^{s+1}$ while ϕ is a combination of $(\phi_i)_j^k$; these combinations must overlap, i.e., $s+1 \geq j, \quad s \geq j-1$. ■

Theorem 10.6.5 *Under the conditions of Theorem 10.6.4, the simple nodes of $\phi_i(x)$ and $\phi_{i+1}(x)$ interlace.*

Proof. Any combination

$$\phi(x) = c_i\phi_i(x) + c_{i+1}\phi_{i+1}(x), \quad c_i^2 + c_{i+1}^2 > 0$$

has either $i-1$ or i zeros in (0,1), and all these zeros are simple nodes. ($s \geq i-1$, $s+2d+p \leq i$ imply $d = 0$ and either $s = i-1$, $p = 0$ or 1; or $s = i$, $p = 0$.) Suppose the nodes of $\phi_{i+1}(x)$ are $(\xi_j)_1^i$; write $\xi_0 = 0$, $\xi_{i+1} = 1$, so that

$$0 = \xi_0 < \xi_1 < \cdots \xi_i < \xi_{i+1} = 1$$

and consider

$$\psi(x) = \phi_i(x)/\phi_{i+1}(x).$$

In each of the intervals (ξ_j, ξ_{j+1}), $j = 0, \ldots, i$ the function $\psi(x)$ is continuous, since $\phi_{i+1}(x)$ is non-zero. We now show that $\psi(x)$ is *monotonic* in each of these intervals. Suppose, if possible, that $\psi(x)$ were not monotonic in an interval (ξ_j, ξ_{j+1}). Then there would exist points x_1, x_2, x_3 such that $\xi_j < x_1 < x_2 < x_3 < \xi_{j+1}$ and $\psi(x_1) - \psi(x_2)$, $\psi(x_2) - \psi(x_3)$ have opposite signs. Without loss of generality we may assume $\psi(x_1) < \psi(x_2)$, $\psi(x_3) < \psi(x_2)$. The function $\psi(x)$, being continuous in $[x_1, x_3]$, assumes its maximum value in $[x_1, x_3]$. This maximum must occur at an interior point, x_0, of $[x_1, x_3]$ since $\psi(x_1), \psi(x_3)$ are both less than $\psi(x_2)$. Therefore,

$$\psi(x) - \psi(x_0) \leq 0 \text{ for all } x \in [x_1, x_3]$$

and thus

$$\phi(x) = \phi_{i+1}(x)\{\psi(x) - \psi(x_0)\} = \phi_i(x) - \psi(x_0)\phi_{i+1}(x)$$

retains it sign in the neighbourhood of its zero, x_0. This contradicts the statement that $\phi(x)$ has only simple nodes. Hence $\psi(x)$ is monotonic in each interval (ξ_j, ξ_{j+1}), $j = 0, 1, \ldots, i$.

We now consider the behaviour of $\psi(x)$ near one of the nodes $(\xi_j)_1^i$ of $\phi_{i+1}(x)$. Since $\psi(x)$ is monotonic in each of the intervals (ξ_j, ξ_{j+1}), $j = 0, \ldots, i$, the limits

$$\lim_{x\to\xi_j^-} \psi(x) = L_1, \qquad \lim_{x\to\xi_j^+} \psi(x) = L_2$$

will exist for all $j = 1, 2, \ldots, i$; they may be finite of infinite. If ξ_j is not a node of $\phi_i(x)$ then L_1 and L_2 will be infinite and have opposite signs. We will show that this is the only case that can occur.

Suppose that ξ_j is a node of $\phi_i(x)$, as well as of $\phi_{i+1}(x)$. Then L_1,L_2 may be finite or infinite but will at least have the same sign. Suppose, without loss of generality that $\psi(x)$ is monotonic increasing in (ξ_{j-1}, ξ_j). If $\psi(x)$ is monotonic decreasing in (ξ_j, ξ_{j+1}) there are five possible cases, shown in Figure 10.6.3:

a) $L_1 = \infty$, $L_2 = \infty$

b) $L_1 = \infty$, L_2 finite

c) $L_1 =$ finite, $L_2 = \infty$

d) L_1 finite, $L_2 = L_1$

e) L_1 finite, $L_2 \neq L_1$

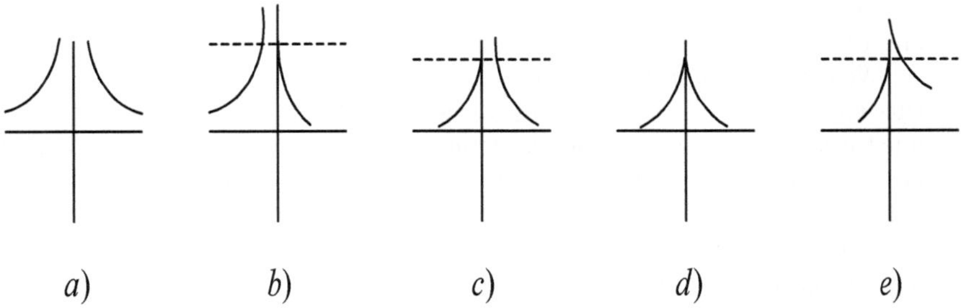

Figure 10.6.3 - $\psi(x)$ is monotonic decreasing in (ξ_j, ξ_{j+1}).

If $\psi(x)$ is monotonic increasing in (ξ_j, ξ_{j+1}) there are just three possible cases shown in Figure 10.6.4:

f) $L_1 = \infty, L_2$ finite

g) L_1 finite, $L_2 = L_1$

h) L_1 finite, $L_2 \neq L_1$

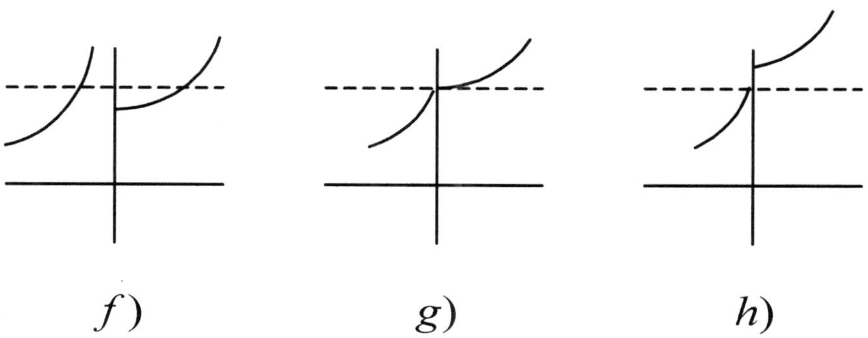

Figure 10.6.4 - $\psi(x)$ is monotonic increasing in (ξ_j, ξ_{j+1}).

In all but cases a), d) there is a line $y = h$, shown, such that $\psi(x)$ crosses this line as x passes through ξ_j. Thus $\psi(x) - h$ changes sign at $x = \xi_j$ and thus

$$\phi(x) = \phi_{i+1}(x)(\psi(x) - h) = \phi_i(x) - h\phi_{i+1}(x)$$

retains its sign as x passes through its zero ξ_j, contradicting the statement that all the zeros of $\phi(x)$ are simple nodes.

Now take case d), suppose $L_1 = L_2 = h$, and consider the function

$$\phi(x, f) = \phi_{i+1}(x)(\psi(x) - f) = \phi_i(x) - f\phi_{i+1}(x)$$

when $f = h$, $\phi(x, h)$ has either $i-1$ or i nodes. Now take $f = h - \varepsilon = h'$, where $\varepsilon > 0$. We may find x_1, x_2 such that $\xi_{j-1} < x_1 < \xi_j < x_2 < \xi_{j+1} \quad \psi(x_1) = \psi(x_2) = h'$.

Since $\phi_{i+1}(x)$ retains its sign and $\psi(x) - h'$ changes its sign as x passes through x_1 and x_2, these points are nodes of $\phi(x, h')$. Thus $\phi(x, f)$ acquires two new nodes as f passes from h to $h - \varepsilon$, but this is impossible since $\phi(x, h)$ and $\phi(x, h')$ both have either $i - 1$ or i nodes.

We conclude that if ξ_j is a node of $\phi_i(x)$ then the only possibility is a). But this means that all the limits L_1, L_2 for $j = 1, 2, \ldots, i$ must be infinite; $\psi(x)$ must assume all values in each interval (ξ_j, ξ_{j+1}), $j = 1, 2, \ldots, i-1$; $\psi(x)$ must have a node in each, and so too must $\phi_i(x)$. But $\phi_i(x)$ has just $i - 1$ nodes, so none of the $(\xi_j)_1^i$ can be nodes of $\phi_i(x)$: case a) cannot occur; $\psi(x)$ must be monotonic increasing in all the intervals (ξ_j, ξ_{j+1}), $j = 0, 1, \ldots, i$, or monotonic decreasing in all of them; the nodes of $\phi_i(x)$ and $\phi_{i+1}(x)$ interlace. ■

10.7 Perron's Theorem and compound kernels

Our aim in this section is to show that the eigenfunction $v_i(x)$ of the integral equation (10.2.9) form a Markov sequence. Following the discussion of total positivity in Chapter 6, we base our analysis on continuous versions of Perron's Theorem and the Cauchy-Binet Theorem. Just as the matrix version of Perron's

Theorem holds for arbitrary positive (square) matrices, not just symmetric ones, so there is a continuous version holding for arbitrary (not necessarily symmetric) positive kernels. However, the proof of the theorem of the arbitrary, non-symmetric, case is beyond the scope of this book. We will therefore state the theorem for the general case but prove it only for the symmetric case, which is in fact all we need. We have

Theorem 10.7.1 *If the continuous kernel $K(x,s)$ satisfies*

$$K(x,s) \geq 0,\ K(x,x) > 0,\ x,s \in (0,1)$$

then the eigenvalue of λ_1 of the integral equation

$$u(x) = \lambda \int_0^1 K(x,s)u(s)ds \tag{10.7.1}$$

which has smallest absolute value is positive and simple; the corresponding eigenfunction $u_1(x)$ has no zero in (0,1).

Proof. In Section 10.3 we showed that a non-zero self-adjoint compact operator A has at least one, non-zero, eigenvalue

$$\mu = \sup_{||x||=1} (Ax, x).$$

When translated into the language of the integral equation (10.7.1), this states that the equation (10.7.1) has an eigenvalue λ_1 satisfying

$$\frac{1}{\lambda_1} = \max\left\{\frac{F(u)}{||u||^2}\right\}, \tag{10.7.2}$$

where

$$F(u) = \int_0^1 \int_0^1 K(x,s)u(x)u(s)dxds,$$

and

$$||u||^2 = \int_0^1 u^2(x)dx.$$

This maximum is actually achieved by $u_1(x)$ which satisfies

$$u_1(x) = \lambda_1 \int_0^1 K(x,s)u_1(s)ds. \tag{10.7.3}$$

Now consider $w_1(x) = |u_1(x)|$. Clearly $||w||^2 = ||u_1||^2$ while $F(w) \geq F(u_1)$, which means that $w_1(x)$ is also an eigenfunction, satisfying (10.7.3), i.e.,

$$w_1(x) = \lambda_1 \int_0^1 K(x,s)w_1(s)ds. \tag{10.7.4}$$

Suppose that $u_1(x)$ had an isolated zero for some $\xi \in (0,1)$. On the basis of $K(\xi,\xi) > 0$ and the continuity of K, we have $K(\xi, s) > 0$, $w_1(s) > 0$ for some interval $(\xi, \xi+\varepsilon)$, $\varepsilon > 0$. Thus, at ξ, the left hand side of (10.7.4) is zero, while the right hand side is positive; this is a contradiction. A zero interval in $u_1(x)$ may be ruled out similarly. This means that any eigenfunction corresponding to λ_1 must have the same sign in (0,1). There cannot be two mutually orthogonal eigenfunctions which maintain fixed sign in (0,1) so that λ_1 must be simple and positive. The proof is thus complete if we can show that if λ is a negative eigenvalue of (10.7.1) then $|\lambda| > \lambda_1$.

Let $v(x)$ be a normalised eigenfunction corresponding to λ, so that

$$v(x) = \lambda \int_0^1 K(x,s)v(s)ds,$$

and therefore

$$|v(x)| \le |\lambda| \int_0^1 K(x,s)|v(s)|ds. \tag{10.7.5}$$

The function $v(x)$, being orthogonal to $w_1(x)$, cannot retain one sign in (0,1) so that there must be strict inequality in (10.7.5). Therefore

$$|v(x)| < |\lambda| \int_0^1 K(x,s)|v(s)|ds$$

and thus

$$(|v|,|v|) < |\lambda| F(|v|).$$

But, by (10.7.2)

$$(|v|,|v|) \ge \lambda_1 F(|v|)$$

so that $|\lambda| > \lambda_1$: λ_1 is the eigenvalue of smallest modulus and is positive and simple. ■

Starting from a kernel $K(x,s)$ on $[0,1]\times[0,1]$ we may use the minors introduced in Section 10.5 to define a *compound kernel* $K(\mathbf{x},\mathbf{s})$ defined on $\bar{Q}\times\bar{Q}$, where $\bar{Q}$ is the n-dimensional simplex

$$0 \le x_1 \le x_2 \le \cdots \le x_n \le 1.$$

If $\mathbf{x}$ is an interior point of $\bar{Q}$ then

$$0 < x_1 < x_2 < \cdots < x_n < 1 \text{ so that } \mathbf{x} \in Q.$$

The place of the Cauchy-Binet Theorem is taken by

Theorem 10.7.2 *If three kernels* $K(x,s), L(x,s), N(x,s)$ *defined on* $[0,1]\times[0,1]$ *are related by*

$$N(x,s) = \int_0^1 K(x,t)L(t,s)dt, \quad x,s \in [0,1]$$

then

$$N(\mathbf{x};\mathbf{s}) = \int_{\bar{Q}} K(\mathbf{x};\mathbf{t})L(\mathbf{t};\mathbf{s})d\mathbf{t}, \quad \mathbf{x},\mathbf{s} \in \bar{Q},$$

where the integration is taken over the simplex $\bar{Q}$.

Proof. The result follows immediately from splitting the integral over the n-dimensional $[0,1] \times [0,1] \times \ldots [0,1]$ into $n!$ integrals over simplices $0 \le x_{i_1} \le x_{i_2} \le \cdots \le x_{i_n} \le 1$. ∎

Theorem 10.7.3 *If* $(\lambda_i)_1^\infty$ *and* $(u_i(x))_1^\infty$ *are eigenvalues and corresponding eigenfunctions of (10.7.1), then*

$$u(\mathbf{x}) = \wedge \int_{\bar{Q}} K(\mathbf{x};\mathbf{s})u(\mathbf{s})d\mathbf{s}, \tag{10.7.6}$$

where

$\wedge = \lambda_{i_1}, \lambda_{i_2} \ldots \lambda_{i_n}, \quad u(\mathbf{s}) = u(\mathbf{s};\alpha), \quad \alpha = (l_1, l_2, \ldots, l_n)$ *and* $1 \le i_1 < i_2 < \cdots i_n$.

Proof. Equation (10.7.1) shows that
$u(\mathbf{x};\alpha) = \lambda_{i_1}, \lambda_{i_2} \ldots \lambda_{i_n} \int_{\bar{Q}} K(\mathbf{x};\mathbf{s})\mathbf{u}(\mathbf{s};\alpha)d\mathbf{s}$. ∎

We may now extend Perron's Theorem to equation (10.7.3).

Theorem 10.7.4 *If the continuous kernel* $K(x,s)$ *satisfies*

$$K(\mathbf{x};\mathbf{s}) \ge 0, \quad K(\mathbf{x};\mathbf{x}) > 0, \quad \mathbf{x},\mathbf{s} \in \mathcal{Q}$$

then the eigenvalue of (10.7.3) which has smallest modulus is positive and simple; the corresponding eigenfunction $u(\mathbf{x})$ *has no zero in* $\mathcal{Q}$.

Proof. The proof in the situation in which $K(\mathbf{x};\mathbf{s})$ is symmetric is the analogue of that in Theorem 10.7.1. ∎

Now we may prove

Theorem 10.7.5 *If the continuous kernel* $K(x,s)$ *satisfies*

$$K(\mathbf{x};\mathbf{s}) \ge 0, \quad K(\mathbf{x};\mathbf{x}) > 0, \quad \mathbf{x},\mathbf{s} \in \mathcal{I}$$

then all the eigenvalues of equation (10.7.1) are positive and simple, i.e., $0 < \lambda_1 < \lambda_2 < \cdots$, *and the corresponding eigenfunctions form a Markov sequence in* I.

Proof. Order the eigenvalues of (10.7.1) so that $|\lambda_1| \le |\lambda_2| \le \cdots$ then the eigenvalue of (10.7.3) that has smallest modulus is $\lambda_1\lambda_2\ldots\lambda_n$. Thus Theorem 10.7.4 states that

a) $\lambda_1\lambda_2\ldots\lambda_n > 0$ b) $\lambda_1\lambda_2\ldots\lambda_n < |\lambda_1\lambda_2\ldots\lambda_{n-1}\lambda_{n+1}|$

for all $n = 2, 3, \ldots$. Thus in turn we have the following: $\lambda_1 > 0$, $\lambda_1 < |\lambda_2|$, $\lambda_1\lambda_2 > 0$ and thus $\lambda_1 < \lambda_2$, $\lambda_1\lambda_2 < |\lambda_1\lambda_3| = \lambda_1|\lambda_3|$, $\lambda_1\lambda_2\lambda_3 > 0$ and thus $\lambda_2 < \lambda_3$, and so on. Theorem 10.7.3 and 10.7.4 shows that the eigenfunction corresponding to the lowest eigenvalue, namely

$$U(\mathbf{x}; \theta) = U(x_1, x_2, \ldots, x_n; \quad 1, 2, \ldots, n),$$

has no zeros in Q, and in fact has fixed sign on $\mathcal{I}$, and this is the necessary and sufficient condition for the sequence $u_i(x)$ to form a Markov sequence on I. ■

Note that we have shown that if $K(x, s)$ is an oscillatory kernel then the corresponding operator A is a strictly positive (compact self-adjoint linear) operator. Thus, Theorem 10.3.9 applies, and the eigenfunctions form a complete orthonormal system in H.

Let us now consider the application of these results to the integral equation governing the vibrations of the string. We recall that we wrote this equation in two ways, namely (10.2.9) and (10.2.11); these are

$$v(x) = \lambda \int_0^1 \rho^2(s) G(x, s) v(s) ds$$

and

$$u(x) = \lambda \int_0^1 K(x, s) u(s) ds.$$

Suppose that $\rho(x)$ is piecewise continuous on [0,1] then, as we showed earlier, $v(x)$, the actual (amplitude of the) string vibration is continuous while $u(x)$ is piecewise continuous.

The Theorems we have proved in this section were phrased in terms of a continuous kernel $K(x, s)$, but clearly this is unnecessarily restrictive. We used the continuity of $K(x, s)$ because we assumed that only $K(x, s) \geq 0$, $K(x, x) > 0$. It was continuity which allowed us to extend $K(x, x) > 0$ to $K(x, s) > 0$ for s near x. If, as for the string, we have $K(x, s) = \rho(x)\rho(s)G(x, s) > 0$ for $x, s \in I$, we do not need to invoke continuity. A similar argument applies to Theorem 10.7.4. We have $K(\mathbf{x}; \mathbf{s}) > 0$ when $\mathbf{x}, \mathbf{s} \in \mathcal{I}$ and $x_1, s_1 < x_2, s_2 < \cdots x_n, s_n$.

We may thus conclude that $U(\mathbf{x}; \theta)$ has fixed sign on $\mathcal{I}$, and hence the corresponding minor $V(\mathbf{x}; \theta)$ formed from the $v_i(x)$ of equation (10.2.9) has fixed sign on $\mathcal{I}$; the $v_i(x)$ form a Markov sequence on I.

Since the $v_i(x)$ form a Markov sequence, they have the properties established in Section 10.6: $v_i(x)$ has exactly $i - 1$ simple nodes in (0,1), and the nodes of $v_i(x)$ and $v_{i+1}(x)$ interlace. Figure 10.7.1 shows typical shapes of a string with end conditions $u(0) = 0 = u'(1)$. Note that we may simulate the 'free' end condition $u'(1) = 0$, by passing the string through a slider at $x = 1$ that keeps the string horizontal there. See Section 2.2. Alternatively, we may simulate a free end by viewing just the left hand half of a symmetrical string stretched between 0 and 2, and considering just the symmetrical modes; these will satisfy $u'(1) = 0$. The modes are qualitatively like the modes $\sin\{(i - 1/2)\pi x\}$ of a uniform string.

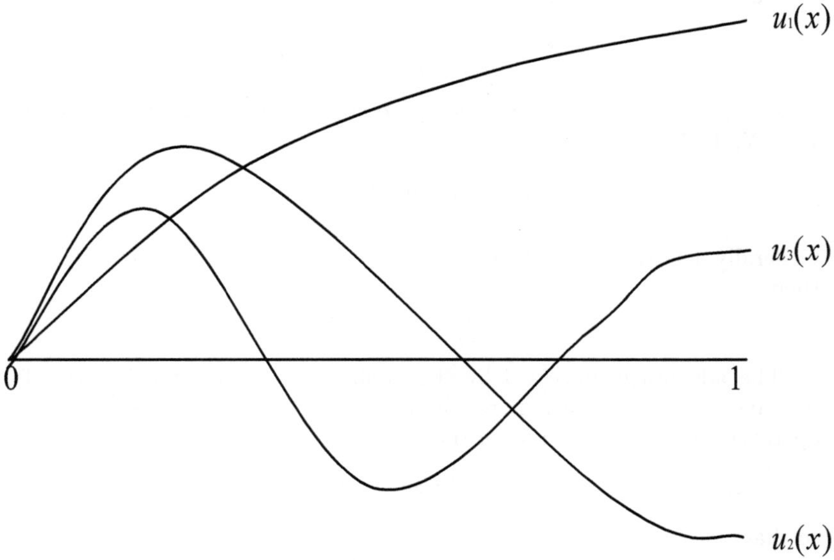

Figure 10.7.1 - Typical modes of a string under the conditions $u(0) = 0 = u'(1)$.

10.8 The interlacing of eigenvalues

In Section 2.9, when discussing vibration under constraint, we used a variational formulation of the matrix eigenvalue problem and, in order to discuss how eigenvalues change under constraint, we used Courant's minimax theorem. This theorem may be extended to a self-adjoint compact operator A in Hilbert space. For simplicity we assume that A is positive definite.

In Section 10.3 we found the greatest eigenvalue of A as

$$\mu_1 = \sup_{x \in H} F(x) = F(x_1)$$

where

$$F(x) = (Ax, x)/||x||^2. \tag{10.8.1}$$

Then we decomposed $H = H_1$ into M_1, the space spanned by x_1, and its orthogonal complement $H_2 : H_1 = M_1 + H_2$, and found

$$\mu_2 = \sup_{x \in H_2} F(x) = F(x_2).$$

Generally,

$$\mu_{n+1} = \sup_{x \in H_{n+1}} F(x) = F(x_{n+1})$$

where M_n is the space spanned by $x_1, x_2, \ldots, x_n$, and $H = M_n + H_{n+1}$. This is the iterative procedure for finding the eigenvalues.

The corresponding minimax procedure is as follows:

$$\mu_1 = \sup_{x \in H} F(x) = F(x_1).$$

Now take $y_1 \in H$, let N_1 be the space spanned by y_1, and decompose H as $H = N_1 + H_2$. Then

$$\mu_2 = \inf_{y_1 \in H} \sup_{x \in H_2} F(x) = F(x_2).$$

Generally, let N_n be the space spanned by $y_1, y_2, \ldots, y_n$, and $H = N_n + H_{n+1}$, then

$$\mu_{n+1} = \inf_{N_n \subset H} \sup_{x \in H_{n+1}} F(x) = F(x_{n+1}).$$

The advantage possessed by the minimax form over the iterative form is seen most clearly when it is required to order the eigenvalues of two different operators A, A'. If it is known that

$$(A'x, x) \geq (Ax, x)$$

so that

$$F'(x) = (A'x, x)/||x||^2 \geq F(x),$$

then

$$\mu'_{n+1} = \inf_{N_n \subset H} \sup_{x \in H_{n+1}} F'(x) \geq \inf_{N_n \subset H} \sup_{x \in H_{n+1}} F(x) = \mu_{n+1} : \tag{10.8.2}$$

the eigenvalues of A' are greater than or equal to those of A; we can compare the eigenvalues because the infs and sups are taken over the same subspaces. By contrast, in the iterative scheme, the subspace H_{n+1} is related to the operator: it is the subspace orthogonal to the space spanned by the previously found eigenvectors $x_1, x_2, \ldots, x_n$.

If in addition

$$(A'x, x) - (Ax, x) = C(x, y)^2, \tag{10.8.3}$$

for some $C > 0$ and $y \in H$, then we can say more. Equation (10.8.3) implies

$$F(x) = F'(x) \text{ if } (x, y) = 0.$$

Thus

$$\mu_n = \inf_{N_{n-1} \subset H} \sup_{x \in H_n} F(x) = \inf_{N'_n \subset H} \sup_{x \in H'_{n+1}} F'(x),$$

where N'_n is the space spanned by the arbitrary $y_1, y_2, \ldots, y_{n-1}$.and y, and $H = N'_n + H'_{n+1}$. But this inf cannot be less than that taken over N_n, so that

$$\mu_n \geq \mu'_{n+1}. \tag{10.8.4}$$

The inequalities (10.8.2), (10.8.4) imply that the eigenvalues of A and A' interlace in the sense

$$\mu'_1 \geq \mu_1 \geq \mu'_2 \geq \mu_2 \geq \cdots$$

We now apply this theory to the eigenvalues of the string under different end conditions. When translated into the language of integral equations, equation (10.8.1) becomes

$$F(u) = \int_0^1 \int_0^1 K(x,s)u(x)u(s)dxds / \int_0^1 u^2(x)dx$$

where

$$K(x,s) = \rho(x)\rho(s)G(x,s),$$

and $G(x,s)$ is given by (10.2.6), $\phi(x), \psi(x)$ by (10.2.7). Since $G(x,s)$ depends on h, H we write it as $G(x,s,h,H)$. Simple algebra shows that

$$G(x,s,h,H') - G(x,s,h,H) = C_1(H-H')(1+hx)(1+hs)$$

and

$$G(x,s,h',H) - G(x,s,h,H) = C_2(h-h')(1+H(1-x))(1+H(1-s))$$

where

$$C_1(h+H'+hH')(h+H+hH) = 1 = C_2(h'+H+h'H)(h+H+hH).$$

This imples that

$$F(u,h,H') - F(u,h,H) = C_1(H-H')(u,w_1)^2/||u||^2$$

and

$$F(u,h',H) - F(u,h,H) = C_2(h-h')(u,w_2)^2/||u||^2$$

where

$$w_1(x) = (1+hx)\rho(x), \quad w_2(x) = (1+H(1-x))\rho(x)$$

and

$$(u,v) = \int_0^1 u(x)v(x)dx.$$

Remembering that μ must be replaced by $1/\lambda$, we may now apply the previous theory as follows:

a) if $H < H'$ then

$$\lambda_n(h,H) \le \lambda_n(h,H') \le \lambda_{n+1}(h,H) \tag{10.8.5}$$

b) if $h < h'$ then

$$\lambda_n(h,H) \le \lambda_n(h',H) \le \lambda_{n+1}(h,H) \tag{10.8.6}$$

if $h < h'$ and $H < H'$ then, by combining a) and b) we find

$$\lambda_n(h,H) \le \lambda_n(h',H) \le \lambda_n(h',H')$$

and

$$\lambda_n(h,H) \le \lambda_n(h',H) \le \lambda_n(h',H') \le \lambda_{n+1}(h',H) \le \lambda_{n+2}(h,H). \qquad (10.8.7)$$

Note that we have used loose inequalities throughout, but in general the inequalities will be strict, as we now show.

We obtained these interlacing results by using the Green's function formulation of the eigenvalue problem. There is another approach using a variational formulation for the original differential equation (10.1.1). The eigenvalue problem (10.1.1), (10.1.2) is equivalent to finding the stationary values of

$$J(v) \equiv \int_0^1 [v'(x)]^2 dx + hv^2(0) + Hv^2(1)$$

subject to

$$(\rho^2 v, v) \equiv \int_0^1 \rho^2(x) v^2(x) dx = 1. \qquad (10.8.8)$$

The following argument may be made rigorous.

We introduce a Lagrange parameter λ and consider

$$G(v) = J(v) - \lambda(\rho^2 v, v).$$

Then

$$\lim_{\varepsilon \to 0} \frac{G(v+\varepsilon\eta) - G(v)}{2\varepsilon} = \int_0^1 v'\eta' dx + hv(0)\eta(0) + Hu(1)\eta(1) - \lambda \int_0^1 \rho^2 v\eta dx.$$

Integrate the first term by parts, rearrange the terms, and equate the whole to zero:

$$-\int_0^1 (v'' + \lambda\rho^2 v)\eta dx - [v'(0) - hv(0)]\eta(0) + [v'(1) + Hv(1)]\eta(1) = 0.$$

This will be zero for all variations $\eta(x)$ only if $v(x)$ satisfies (10.1.1) and (10.1.2).

Suppose that $\{\lambda_n, v_n(x)\}_1^\infty$ are the eigenvalues and eigenfunctions of (10.1.1), (10.1.2), normalised so that

$$(\rho^2 v_m, v_n) = \delta_{mn}.$$

Then

$$\begin{aligned}\int_0^1 v_m'(x)v_n'(x)dx &= [v_m(x)v_n'(x)]_0^1 + \lambda_n\delta_{mn} \\ &= -hv_m(0)v_n(0) - Hv_m(1)v_n(1) + \lambda_n\delta_{mn}. \end{aligned} \qquad (10.8.9)$$

Now consider the variational problem for equation (10.1.1) under the end conditions

$$v'(0) - h'v(0) = 0 = v'(1) + Hv(1), \qquad (10.8.10)$$

where $h' > h$. This is the problem of finding the stationary values of

$$J'(v) = \int_0^1 [v'(x)]^2 dx + h'v^2(0) + Hv^2(1)$$

subject to (10.8.8). Expand $v(x)$ in terms of the eigenfunctions $v_m(x)$;

$$v(x) = \sum_{m=1}^{\infty} c_m v_m(x).$$

Now use the integral (10.8.9) to get

$$J'(v) = \sum_{m=1}^{\infty} \lambda_m c_m^2 + (h' - h)v^2(0). \qquad (10.8.11)$$

The equations giving the values of c_m that make $J'(v)$ stationary are

$$\lambda_m c_m + (h' - h)v_m(0)v(0) - \lambda c_m = 0 \qquad m = 1, 2, \ldots$$

i.e.,

$$c_m = (h' - h)v_m(0)v(0)/(\lambda - \lambda_m)$$

so that the condition

$$v(0) = \sum_{m=1}^{\infty} c_m v_m(0)$$

gives

$$1 = (h' - h) \sum_{m=1}^{\infty} \frac{v_m^2(0)}{\lambda - \lambda_m}. \qquad (10.8.12)$$

This is the analogue of equation (4.3.21), and immediately gives the strict form of (10.8.6):

$$\lambda_n(h, H) < \lambda_n(h', H) < \lambda_{n+1}(h, H). \qquad (10.8.13)$$

The end values, $v_m(0)$, cannot be zero unless $h = \infty$, i.e., the end $x = 0$ is fixed; this case is excluded by $h' > h$. We may employ a similar procedure to get the strict form of (10.8.5).

Exercises 10.8

1. Derive the expression (10.8.11) for the functional $J'(v)$.
2. If $(\lambda'_m)_1^\infty$ are the eigenvalues of (10.1.1) subject to (10.8.10), i.e., $\lambda'_m = \lambda_m(h', H)$, show that

$$1 - (h' - h) \sum_{m=1}^{\infty} \frac{v_m^2(0)}{\lambda - \lambda_m} = \prod_{m=1}^{\infty} \left(\frac{\lambda - \lambda'_m}{\lambda - \lambda_m} \right).$$

and hence deduce that

$$-(h'-h)v_n^2(0) = (\lambda_n - \lambda_n') \prod_{m=1}^{\infty}{}' \left(\frac{\lambda_n - \lambda_m'}{\lambda_n - \lambda_m}\right)$$

where $'$ denotes $m \neq n$.

3. How should the infinite product be interpreted so that, with $h' > h$, the interlacing (10.8.13), i.e., $\lambda_1 < \lambda_1' < \lambda_2 < \ldots$, yields positive values of $v_m^2(0)$. These examples show that knowing $(\lambda_n - \lambda_n')_1^\infty$ we may compute the so-called *norming constants* $(\sigma_n)_1^\infty = (v_n^2(0))_1^\infty$; conversely, knowing $(\lambda_n, \sigma_n)_1^\infty)$, we may compute $(\lambda_n')_1^\infty$. See Elhay, Gladwell, Golub and Ram (1999) [85] for further discussion of eigenvector-eigenvalue relations like (10.8.12).

10.9 Asymptotic behaviour of eigenvalues and eigenfunctions

For the solution of inverse problems in Chapter 11 we shall need to know the *asymptotic behaviour* of the eigenvalues λ_n and eigenfunctions $v_n(x)$, and norming constants for large n. To examine this behaviour it is convenient to suppose that $\rho(x)$ in equation (10.1.1) or $A(x)$ in equation (10.1.3), are sufficiently smooth that the equation (10.1.1) or (10.1.3) may be transformed to the Sturm-Liouville form (10.1.11) with $q(x) \in C[0,\pi]$. We now use the numbering convention **S** described in Section 10.1.

First, we need an existence uniqueness theorem. This is provided by Titchmarsh (1962) [323].

Theorem 10.9.1 *If $q(x) \in C[0,\pi]$ then, for any α there exists a unique solution $y(x,\lambda)$ of (10.1.14) such that $y(0,\lambda) = \sin\alpha, \phi'(0,\lambda) = -\cos\alpha$. For any fixed $x \in [0,\pi]$, (x,λ) is an entire function of λ.*

[Note: Here λ is taken to be complex variable; an entire function of a complex variable λ is one that has no poles in the finite λ-plane.]

On the basis of this theorem we denote the solution of

$$y''(x) + (\lambda - q(x))y(x) = 0 \tag{10.9.1}$$

satisfying the condition

$$\phi(0,\lambda) = 1 \qquad \phi'(0,\lambda) = h \tag{10.9.2}$$

by $\phi(x,\lambda)$. We assume that h is finite and that $q(x) \in C[0,\pi]$.

Write $\lambda = \omega^2$, then (10.9.1) may be written

$$y''(x) + \omega^2 y(x) = q(x)y(x).$$

Treating the right hand side as a forcing function, we may use the so-called *Duhamel solution*

$$\phi(x,\lambda) = A\cos\omega x + B\sin\omega x + \omega^{-1}\int_0^x \sin\omega(x-t)q(t)\phi(t,\lambda)dt, \qquad (10.9.3)$$

where $A = 1, B = h/\omega$. In this equation we can treat ω as a complex variable and can obtain an estimate for $\phi(x,\lambda)$ for large $|\omega|$:

Lemma 10.9.1 *Let $\omega = \sigma + i\tau$. Then there exists $\sigma_0 > 0$ such that for $|\omega| > \sigma_0$*

$$\phi(x,\lambda) = \cos\omega x + O\left(\frac{\exp|\tau|x}{|\omega|}\right) \qquad (10.9.4)$$

uniformly with respect to x in $[0,\pi]$.

Proof. Put $\phi(x,\lambda) = \exp(|\tau|x)f(x)$, then it follows from (10.9.3) that

$$f(x) = (\cos\omega x + h\omega^{-1}\sin\omega x)\exp(-|\tau|x) + \omega^{-1}\int_0^x \sin\omega(x-t)\exp(-|\tau|(x-t))q(t)f(t)dt. \qquad (10.9.5)$$

Let $M = \max_{0\le x\le\pi}|f(x)|$, then equation (10.9.5) gives

$$M \le 1 + \frac{|h|}{|\omega|} + \frac{M}{|\omega|}\int_0^x |q(t)|dt.$$

Thus

$$M \le \left(1 + \frac{|h|}{|\omega|}\right) / \left(1 - \frac{1}{|\omega|}\int_0^x |q(t)|dt\right)$$

provided that the denominator is positive, that is, provided that

$$|\omega| > \int_0^\pi |q(t)|dt.$$

For such ω,

$$|\phi(x,\lambda)| \le M\exp(|\tau|x)$$

so that on substituting this into the integral (10.9.3) we find (10.9.4). ∎

We may now use the estimate (10.9.4) to estimate the eigenvalues of (10.9.1) subject to

$$y'(0) - hy(0) = 0 = y'(\pi) + Hy(\pi); \qquad (10.9.6)$$

we assume that H, like h, is finite. In Section 10.1 we showed that the eigenvalues are real; we may therefore take $\tau = 0$ in (10.9.4) and find

$$\phi(x,\lambda) = \cos\omega x + O(\omega^{-1}).$$

The eigenvalues are the solutions of

$$\phi'(\pi,\lambda) + H\phi(\pi,\lambda) = 0 \qquad (10.9.7)$$

which for large $|\omega|$ becomes

$$-\omega \sin \omega\pi + O(1) = 0 \tag{10.9.8}$$

which clearly has solutions near to integers for large ω. There is in fact only one solution near any large integer n for, on differentiating (10.9.8) with respect to ω, (which is justified because (10.9.8) is actually (10.9.7) which is an analytic function of λ) we find

$$-\omega\pi \cos \omega\pi + O(1)$$

which is not zero near a large integer. We conclude that the eigenvalues which are arranged in the order

$$\lambda_0 < \lambda_1 < \lambda_2 < \dots$$

eventually become positive and near the square of an integer.

To obtain a precise estimate of the eigenvalues we use

Rouché's Theorem *If $f(z)$ and $g(z)$ are analytic within and on a closed contour C and $|g(z)| < |f(z)|$ on C, then $f(z)$ and $f(z)+g(z)$ have the same number of zeros inside C.*

To apply this theorem we take $f(\omega) = -\omega \sin \omega\pi$, $f(\omega) + g(\omega) = \phi'(\pi, \lambda) + H\phi(\pi, \lambda)$, and take C to be a circle with centre O, radius $N + \frac{1}{2}$, in the ω-plane. Then for large enough N, $|g(\omega)| < |f(\omega)|$ on C, so that $f(\omega)$ and $f(\omega) + g(\omega)$ have the same number of zeros inside C.

The eigenvalues λ are real, and both $f(\omega)$ and $f(\omega)+g(\omega)$ are even functions of ω. This means that the zeros, ω, will lie on the real axis, $\omega = \pm\sqrt{\lambda}$ if $\lambda \geq 0$; or on the imaginary axis, $\omega = \pm i\sqrt{|\lambda|}$ if $\lambda < 0$. The number of λ eigenvalues is therefore $\frac{1}{2}*$ (number of zeros of $f(\omega) + g(\omega)$) $= \frac{1}{2}*$ (number of zeros of $f(\omega)$). But the zeros of $f(\omega)$ are $\pm 0, \pm 1, \dots \pm N$; there are $2N + 2$, so that there are $N + 1$ eigenvalues λ inside C. We conclude that

$$\omega_n = n + O(1). \tag{10.9.9}$$

We may now make this estimate more precise by substituting (10.9.9) in (10.9.8). Put $\omega_n = n + \delta_n$, then

$$(n + \delta_n) \sin(\pi\delta_n) + O(1) = 0$$

so that

$$\sin(\pi\delta_n) = O(n^{-1}), \text{ i.e., } \delta_n = O(n^{-1}).$$

This means that, for large n

$$\sqrt{\lambda_n} \equiv \omega_n = n + O(n^{-1}). \tag{10.9.10}$$

We continue to examine this estimate. We can write (10.9.3) and its derivative as

$$\phi(x, \lambda) = \cos \omega x\{1 - \omega^{-1} q_1(x)\} + \omega^{-1} \sin \omega x\{h + q_2(x)\}, \tag{10.9.11}$$

$$\phi'(x,\lambda) = \cos\omega x\{h + q_2(x)\} - \omega\sin\omega x\{1 - \omega^{-1}q_1(x)\}, \tag{10.9.12}$$

where

$$q_1(x) = \int_0^x \sin\omega t q(t)\phi(t,\lambda)dt, \tag{10.9.13}$$

$$q_2(x) = \int_0^x \cos\omega t q(t)\phi(t,\lambda)dt. \tag{10.9.14}$$

Thus

$$q_1(x) = o(1), \qquad q_2(x) = \frac{1}{2}\int_0^x q(t)dt + o(1) \tag{10.9.15}$$

and

$$\phi(x,\lambda) = \cos\omega x + O(\omega^{-1}) \tag{10.9.16}$$

$$\phi'(x,\lambda) = -\omega\sin\omega x + \{h + \frac{1}{2}\int_0^x q(t)dt\}\cos\omega x + o(1) \tag{10.9.17}$$

so that (10.9.7) may be written

$$c\pi\cos\omega\pi - \omega\sin\omega\pi + o(1) = 0 \tag{10.9.18}$$

where

$$c = \frac{1}{\pi}\left(h + H + \frac{1}{2}\int_0^\pi q(t)dt\right). \tag{10.9.19}$$

Equation (10.9.18) gives

$$\tan\omega\pi = c\pi\omega^{-1} + o(\omega^{-1})$$

so that on putting $\omega_n = n + \delta_n$ as before, we find

$$\begin{array}{rcl} \tan\delta_n\pi & = & c\pi n^{-1} + o(n^{-1}) \\ \delta_n & = & cn^{-1} + o(n^{-1}) \\ \sqrt{\lambda_n} = \omega_n & = & n + cn^{-1} + o(n^{-1}) \end{array} \tag{10.9.20}$$

We now consider the asymptotic form of the eigenfunctions. Equations (10.9.11), (10.9.15) give

$$\phi(x,\lambda) = \cos\omega x + h\omega^{-1}\sin\omega x + \frac{1}{2}\omega^{-1}\sin\omega x\int_0^x q(t)dt + o(\omega^{-1}).$$

Substituting for ω_n from (10.9.20), we find

$$\begin{array}{rcl} \phi(x,\lambda_n) & = & \cos nx - cxn^{-1}\sin nx + hn^{-1}\sin nx + \frac{1}{2}n^{-1}\sin nx\int_0^x q(t)dt + o(n^{-1}) \\ & = & \cos nx + n^{-1}\beta(x)\sin nx + o(n^{-1}), \end{array} \tag{10.9.21}$$

where

$$\beta(x) = h - cx + \frac{1}{2}\int_0^x q(t)dt. \tag{10.9.22}$$

To derive the asymptotic expression for the normalised eigenfunctions, we compute the integral

$$\alpha_n^2 = \int_0^\pi \phi^2(x,\lambda_n)dx = \int_0^\pi \{\cos^2 nx + n^{-1}\beta(x)\sin 2nx\}dx + o(n^{-1}).$$

Since $\beta(x)$ is differentiable,

$$\int_0^\pi \beta(x)\sin 2nx dx = O(n^{-1})$$

so that

$$\alpha_n^2 = \frac{\pi}{2} + o(n^{-1}) \tag{10.9.23}$$

and the normalised eigenfunction is

$$y_n(x) = \frac{\phi(x,\lambda_n)}{\alpha_n} = \sqrt{\frac{2}{\pi}}\{\cos nx + n^{-1}\beta(x)\sin nx\} + o(n^{-1}). \tag{10.9.24}$$

So far we have assumed only that $q(x)$ is continuous. If we assume that $q(x)$ has a bounded derivative, then the terms in (10.9.15) are $O(\omega^{-1})$; for example

$$\int_0^x \sin 2\omega t q(t)dt = \left[\frac{-\cos 2\omega t}{2\omega}q(t)\right]_0^x + \frac{1}{2\omega}\int_0^x \cos 2\omega t q'(t)dt = O(\omega^{-1}).$$

In this case the terms $o(1)$, $o(n^{-1})$ in equations (10.9.17)-(10.9.24) may be replaced by $O(n^{-1})$ and $O(n^{-2})$ respectively.

Now consider the case in which $h = \infty$, H is finite. The end condition at $x = 0$ is $y(0) = 0$, and the solution of (10.9.1) satisfying the condition

$$\psi(0,\lambda) = 0, \qquad \psi'(0,\lambda) = 1, \tag{10.9.25}$$

is

$$\psi(x,\lambda) = \omega^{-1}\sin\omega x + \omega^{-1}\int_0^x \sin\omega(x-t)q(t)\psi(t,\lambda)dt, \tag{10.9.26}$$

and we can show as before (see Ex. 10.9.1) that

$$\begin{aligned}\psi(x,\lambda) &= \omega^{-1}\sin\omega x + O(\omega^{-2}) && (10.9.27)\\ \psi'(x,\lambda) &= \cos\omega x + O(\omega^{-1}). && (10.9.28)\end{aligned}$$

This means that the second end condition, (10.9.7), has the form

$$\cos\omega\pi + O(\omega^{-1}) = 0$$

which has solutions near $n + \frac{1}{2}$:

$$\omega_n = n + \frac{1}{2} + \delta_n. \tag{10.9.29}$$

We write $\psi(x,\lambda)$ and $\psi'(x,\lambda)$ as before:

$$\psi(x,\lambda) = \omega^{-1}\sin\omega x\{1+q_2(x)\} - \omega^{-1}\cos\omega x q_1(x), \tag{10.9.30}$$
$$\psi'(x,\lambda) = \cos\omega x\{1+q_2(x)\} + \sin\omega x q_1(x), \tag{10.9.31}$$

where

$$q_1(x) = \int_0^x \sin\omega t q(t)\psi(t,\lambda)dt, \tag{10.9.32}$$

$$q_2(x) = \int_0^x \cos\omega t q(t)\psi(t,\lambda)dt. \tag{10.9.33}$$

Since $\psi(t,\lambda)$ has the form (10.9.27), we have

$$q_1(x) = \frac{1}{2}\omega^{-1}\int_0^x q(t)dt + o(\omega^{-1}) \tag{10.9.34}$$
$$q_2(x) = o(\omega^{-1}) \tag{10.9.35}$$

and

$$\psi'(\pi,\lambda)+H\psi(\pi,\lambda) = \cos\omega\pi+\omega^{-1}\{H+\frac{1}{2}\int_0^\pi q(t)dt\}\sin\omega\pi+o(\omega^{-1}). \tag{10.9.36}$$

Putting $\omega = n+\frac{1}{2}+\delta_n$ we find, as before, that

$$\omega_n = n+\frac{1}{2}+\frac{c}{n+\frac{1}{2}}+o(n^{-1}), \tag{10.9.37}$$

where

$$c = \frac{1}{\pi}\left(H+\frac{1}{2}\int_0^\pi q(t)dt\right). \tag{10.9.38}$$

Similarly, if h is finite and $H=\infty$, then

$$\omega_n = n+\frac{1}{2}+\frac{c}{n+\frac{1}{2}}+o(n^{-1}), \tag{10.9.39}$$

where

$$c = \frac{1}{\pi}\left(h+\frac{1}{2}\int_0^\pi q(t)dt\right). \tag{10.9.40}$$

Finally, consider the case $h=\infty$, $H=\infty$, so that the end conditions are the Dirichlet conditions

$$\psi(0,\lambda) = 0 = \psi(\pi,\lambda).$$

Substituting from (10.9.27) we find that the second condition is

$$\omega^{-1}\sin\omega\pi + O(\omega^{-2}) = 0.$$

For large N, there are as many zeros inside the circle of radius $N+\frac{1}{2}$ as there are zeros of $\omega^{-1}\sin\omega\pi$; there are $2N$ such zeros: $\pm1,\pm2,\ldots,\pm N$. Thus

$$\omega_n = n+1+\delta_n$$

and we find, as before, that

$$\omega_n = n + 1 + \frac{c}{n+1} + o(n^{-1}) \tag{10.9.41}$$

where

$$c = \frac{1}{2\pi}\int_0^\pi q(t)dt. \tag{10.9.42}$$

Again, if $q(x)$ has a bounded derivative, then the terms $o(n^{-1})$ in (10.9.37), (10.9.39), (10.9.41) may be replaced by $o(n^{-2})$. [Note: there are several small errors in Levitan and Sargsjan (1991) [212], as there undoubtedly are in this book; n in their equation (2.19) in Section 1.2.4 should be $n+1$.]

A historical notes is in order. In the many papers on asymptotic estimates, many different assumptions are made regarding the smoothness of $q(x)$: it is continuous; it has a bounded derivative; it has a piecewise continuous derivative; it has a continuous derivative; etc. It is known that if $q(x)$ is continuous it need not have a derivative at all; there is a pathological function that is continuous in $[0, \pi]$ but is differentiable nowhere. However, in the older treatments, e.g., Ince (1927) [185], and some of the Soviet literature, it is assumed implicitly that if $q(x)$ is said to be continuous, then it has a derivative but that this derivative is not necessarily continuous; it is piecewise continuous. Similarly, if $q(x)$ is said to have r continuous derivative then it has a piecewise continuous $(r+1)$th derivative.

One of the most extensive studies of asymptotic estimates of the Sturm-Liouville spectrum was carried out by Hochstadt (1961) [172], who used a variant of the WKB method. He supposes that the mean value of $q(x)$ is zero. Equation (10.9.1) may be reduced to this form by writing it as

$$y''(x) + ({\omega^*}^2 - q^*(x))y(x) = 0 \tag{10.9.43}$$

where

$${\omega^*}^2 = \omega^2 - \bar{q}, \qquad q^*(x) = q(x) - \bar{q}, \qquad \bar{q} = \frac{1}{\pi}\int_0^\pi q(x)dx. \tag{10.9.44}$$

When h, H are finite, and $q(x)$ is twice continuously differentiable, he shows that

$$(\omega_n^2 - \bar{q})^{\frac{1}{2}} = n + b_0 n^{-1} + b_1 n^{-3} + O(n^{-4}), \tag{10.9.45}$$

where

$$b_0 = \frac{h+H}{\pi}, \qquad b_1 = d_1 - d_2, \tag{10.9.46}$$

$$d_1 = \frac{1}{8\pi}\left\{\int_0^\pi [q^*(t)]^2\,dt + q'(\pi) - q'(0) + 4hq^*(0) + 4Hq^*(\pi)\right\}, \tag{10.9.47}$$

$$d_2 = \left(\frac{h+H}{\pi}\right)^2 + \frac{1}{3}\left(\frac{h^3+H^3}{\pi}\right). \tag{10.9.48}$$

Note that $d_1 = 0$ when $q(x) = const$, i.e., $q^*(x) = 0$.

Hochstadt also considered the various special cases in which h or H are infinite. See also Fix (1967) [89], Pőschel and Trubowitz (1987) [269] and Rundell (1997) [294].

Equation (10.9.45) may be written

$$\omega_n = n + a_0 n^{-1} + a_1 n^{-3} + O(n^{-4}) \tag{10.9.49}$$

where

$$a_0 = \frac{1}{\pi}\left(h + H + \frac{1}{2}\int_0^{\pi} q(t)dt\right) = c \tag{10.9.50}$$

$$a_1 = b_1 - \frac{1}{8}\bar{q}^2 - \frac{1}{2}b_0\bar{q}, \tag{10.9.51}$$

where b_1 is given by equation (10.9.46).

Equation (10.9.49) gives

$$\lambda_n = n^2 + 2a_0 + c_0 n^{-2} + O(n^{-3}) \tag{10.9.52}$$

where $c_0 = a_0^2 + 2a_1$.

Equation (10.9.23) gives a first asymptotic estimate of the so-called norming constants

$$\sigma_n = y_n^2(0) = \frac{[\varphi(0,\lambda_n)]^2}{\alpha_n^2} : \quad \sigma_n = \frac{2}{\pi} + O(n^{-2}). \tag{10.9.53}$$

Levitan (1987) [211] shows that if $q(x)$ is twice continuous by differentiable then

$$\sigma_n = \frac{2}{\pi}(1 + e_0 n^{-2} + O(n^{-3}). \tag{10.9.54}$$

Suppose $(\lambda_n)_0^\infty$, $(\mu_n)_0^\infty$ are the eigenvalues of (10.9.1) corresponding to the end conditions

$$y'(0) - h_1 y(0) = 0 = y'(\pi) + Hy(\pi), \tag{10.9.55}$$
$$y'(0) - h_2 y(0) = 0 = y'(\pi) + Hy(\pi), \tag{10.9.56}$$

so that

$$\begin{aligned} \lambda_n^{\frac{1}{2}} &= n + a_0 n^{-1} + a_1 n^{-3} + O(n^{-4}), \\ \mu_n^{\frac{1}{2}} &= n + a_0' n^{-1} + a_1' n^{-3} + O(n^{-4}), \end{aligned}$$

After a long derivation based on Ex. 10.8.2, (with the change of numbering from **V** to **S**) Levitan shows that

$$e_0 = S - \frac{\pi^2}{6}(a_0 - a_0')^2 + a_0 + \frac{a_1' - a_1}{a_0' - a_0}, \tag{10.9.57}$$

where

$$S = \lambda_0 - \mu_0 + \sum_{n=1}^{\infty}[(\lambda_n - \mu_n) - 2(a_0 - a_0')]. \tag{10.9.58}$$

We note that (10.9.52) shows that this series converges.

Note that equation (10.9.54) is important for stating the asymptotic form of σ_n, and not for the actual expression (10.9.54) for e_0, i.e., as a way of finding σ_n; the result in Ex. 10.8.2 (with the change of numbering from **V** to **S**) shows how to find $\sigma_n = v_n^2(0)$ from two spectra. McNabb, Anderssen and Lapwood (1976) [233] discuss the asymptotics of the eigenvalues when there are one or two discontinuities in the potentials.

Exercises 10.9

1. Show that when $h = \infty$,

$$(\omega_n^2 - \bar{q})^{\frac{1}{2}} = n + \frac{1}{2} + b_0(n + \frac{1}{2})^{-1} + b_1 n^{-3} + O(n^{-4})$$

where $b_0 = H/\pi, \quad b_1 = d_1 - d_2,$

$$\begin{aligned} d_1 &= \frac{1}{8\pi}\left\{\int_0^{\pi}[q^*(t)]^2 dt + 4Hq^*(\pi) + q'(0) + q'(\pi)\right\} \\ d_2 &= \left(\frac{H}{\pi}\right)^3 + \frac{1}{3}\frac{H^3}{\pi}. \end{aligned}$$

10.10 Impulse responses

Consider a rod of density ρ, Young's modulus E, cross section $A(x)$ and length 1, free at $x = 0$ and fixed at $x = 1$. Suppose that at time $t' = 0$ the rod is at rest, and is then set in motion by a force $g(t')$ applied at the end $x = 0$. The governing equations are

$$\begin{aligned} \rho A(x)\frac{\partial^2 u}{\partial t'^2} &= \frac{\partial}{\partial x}\left(EA(x)\frac{\partial u}{\partial x}\right) \qquad (10.10.1) \\ EA\left.\frac{\partial u}{\partial x}\right|_{x=0} &= g(t') \\ u(1,t') &= 0, \quad t' > 0 \\ u(x,0) &= 0 \quad = \frac{\partial u}{\partial t'}(x,0), \quad 0 \le x \le 1. \end{aligned}$$

Instead of real time t' we use the scaled time $t = ct'$, $\quad c = \sqrt{E/\rho}$, and put $g(t) = g(t')/E$. We may replace the end force $g(t)$ by a distributed loading $g(x,t)$ over a small interval $(0,\varepsilon)$, so that

$$g(x,t) = \lim_{\varepsilon \to 0}\left(\frac{g(t)}{\varepsilon}\right) = g(t)\delta(x)$$

so that equation (10.10.1) becomes

$$A(x)\frac{\partial^2 u}{\partial t^2} = \frac{\partial}{\partial x}\left(A(x)\frac{\partial u}{\partial x}\right) + g(t)\delta(x). \qquad (10.10.2)$$

Take the Laplace transform of this equation, and put

$$U(x,s)=\int_0^\infty \exp(-st)u(x,t)dt, \quad G(s)=\int_0^\infty \exp(-st)g(t)dt,$$

to obtain

$$s^2A(x)U(x,s)=(A(x)U')'+G(s)\delta(x). \tag{10.10.3}$$

The solution of this equation that satisfies the end condition $U(1,s)=0$ may be written

$$U(x,s)=K(x,s)G(s),$$

so that, by the convolution theorem

$$u(x,t)=\int_0^t k(x,t-\tau)g(\tau)d\tau, \tag{10.10.4}$$

where $k(x,t)$ is the inverse Laplace transform of $K(x,s)$, i.e.,

$$k(x,t)=\frac{1}{2\pi i}\int_\Gamma K(x,s)\exp(st)ds,$$

where Γ is a line $(\gamma-i\infty,\gamma+i\infty)$ lying to the right of the singularities of $K(x,s)$. The function $k(x,t)$ is called the (displacement) impulse response function. Clearly, when $g(\tau)$ is a unit impulse, i.e., $g(\tau)=\delta(\tau)$, then equation (10.10.4) shows that $u(x,t)=k(x,t)$.

If $(\omega_n^2,u_n(x))_0^\infty$ are the (scaled) eigenvalues and normalised eigenfunctions of the free-fixed rod, i.e.,

$$(A(x)u_n'(x))'+\omega_n^2A(x)u_n(x)=0$$

$$u_n'(0)=0=u_n(1)$$

then we may expand $U(x,s)$ in the form

$$U(x,s)=\sum_{n=1}^\infty \alpha_n(s)u_n(x)$$

so that equation (10.10.3) becomes

$$\sum_{n=1}^\infty (s^2+\omega_n^2)A(x)\alpha_n(s)u_n(x)=G(s)\delta(x).$$

Multiplying though by $u_m(x)$ and integrating over (0,1), using orthogonality and the result

$$\int_0^1 u_n(x)\delta(x)dx=u_n(0)$$

we obtain

$$(s^2+\omega_n^2)\alpha_n(s)=G(s)u_n(0)$$

and

$$K(x,s) = \sum_{n=1}^{\infty} \frac{u_n(0)u_n(x)}{s^2+\omega_n^2}$$

for which the inverse is

$$k(x,t) = \begin{cases} \sum \frac{u_n(0)u_n(x)}{\omega_n} \sin \omega_n t, & t > 0 \\ 0 & t \leq 0 \end{cases}. \tag{10.10.5}$$

For a uniform rod

$$u_n(x) = \sqrt{2} \cos\left[\frac{(2n-1)\pi x}{2}\right], \quad \omega_n = \frac{(2n-1)\pi}{2},$$

so that

$$k(x,t) = \frac{4}{\pi} \sum_{n=1}^{\infty} \frac{\cos\left[\frac{(2n-1)\pi x}{2}\right] \sin\left[\frac{(2n-1)\pi t}{2}\right]}{(2n-1)},$$

i.e.,

$$k(x,t) = \frac{1}{2}\left\{S\left[\frac{\pi(x+t)}{2}\right] - S\left[\frac{\pi(x-t)}{2}\right]\right\}, \tag{10.10.6}$$

where

$$S(x) = \frac{4}{\pi} \sum_{n=1}^{\infty} \frac{\sin(2n-1)x}{2n-1}. \tag{10.10.7}$$

Now $S(x)$ is discontinuous at $0,\ \pm\pi,\ \pm 2\pi, \ldots$, and

$$S(x) = sign(x), \quad -\pi < x < \pi. \tag{10.10.8}$$

(Gradshteyn and Ryzhik (1965), 1.4421)

From equation (10.10.8) we may deduce the behaviour of the rod subjected to an impulse at $t = 0$. Thus if $x > t$, then $x + t < 2x < 2$, so that

$$S\left[\frac{\pi(x+t)}{2}\right] = 1, \quad S\left[\frac{\pi(x-t)}{2}\right] = 1,$$

and $k(x,t) = 0$. This may be interpreted as showing that the effect of the impulse moves along the rod with scaled speed 1, i.e., real speed c, and the rod is at rest for $x > t$. Analysis of the partial differential equation (10.10.2) shows that this result is true even when $A(x)$ is not uniform (Courant and Hilbert (1962)). For the uniform rod, behind the initial disturbance, i.e., for $x < t,\ x + t < 2$ we have $k(x,t) = 1/2$. When the disturbance reaches the end $x = 1$ and starts to return we have

$$S\left[\frac{\pi(x+t)}{2}\right] = -1, \quad S\left[\frac{\pi(x-t)}{2}\right] = -1$$

so the step $1/2$ which had stretched from $x = 0$ to $x = 1$ is annihilated starting from $x = 1$. So the process continues indefinitely.

Sometimes it is convenient to use velocity and (scaled) stress as variables, i.e.,

$$v(x,t) = \frac{\partial u}{\partial t}, \quad p(x,t) = A(x)\frac{\partial u}{\partial x}, \tag{10.10.9}$$

then equation (10.10.2) may be written

$$A(x)\frac{\partial v}{\partial t} = \frac{\partial p}{\partial x} + g(t)\delta(x), \quad A(x)\frac{\partial v}{\partial x} = \frac{\partial p}{\partial t}, \tag{10.10.10}$$

and the velocity $v(x,t)$ is given by

$$v(x,t) = \int_0^t \hat{h}(x, t-\tau)g(\tau)d\tau, \tag{10.10.11}$$

where

$$\hat{h}(x,t) = \frac{\partial k}{\partial t}(x,t) \tag{10.10.12}$$

must be interpreted as a generalised function.

Equation (10.10.5) shows that

$$\hat{h}(x,t) = \begin{cases} \sum_{n=1}^{\infty} u_n(0)u_n(x)\cos(\omega_n t), & t \geq 0 \\ 0, & t < 0 \end{cases} \tag{10.10.13}$$

and thus

$$\hat{h}(0,t) = \sum_{n=1}^{\infty} u_n^2(0)\cos(\omega_n t), \quad t \geq 0.$$

For the uniform rod, therefore

$$\hat{h}(0,t) = \begin{cases} 2\sum_{n=1}^{\infty} \cos\left[\frac{(2n-1)\pi t}{2}\right], & t \geq 0 \\ 0, & t < 0 \end{cases}.$$

We note that (Gradshteyn and Ryzhik (1965), 1.4421)

$$\begin{aligned} \int_{-\infty}^t \hat{h}(0,\tau)d\tau &= \int_0^t \hat{h}(0,\tau)d\tau = \frac{4}{\pi}\sum_{n=1}^{\infty}\frac{\sin[(2n-1)\pi t/2]}{2n-1}, \\ &= 1 \qquad (0 < t < 2) \end{aligned}$$

so that for $0 < t < 2$,

$$\hat{h}(0,t) = \delta(t).$$

For larger values of t, $\hat{h}(0,t)$ may be evaluated by using its periodicity, $\hat{h}(0, t+2) = -\hat{h}(0,t)$. For a non-uniform rod it can be shown that

$$\hat{h}(0,t) = \delta(t) + h(t), \tag{10.10.14}$$

where $h(t)$ is continuously differentiable. (See Ex. 10.10.2).

Exercises 10.10

1. Show that $S(x)$ given in (10.10.6) satisfies

$$S(x+\pi) = -S(x), \quad S(x+2\pi) = S(x)$$

and hence show that

$$S(x) = (-)^{|n|}, \quad n\pi < x < (n+1)\pi.$$

2. Show that if the rod is such that its eigenvalues ω_n and eigenfunctions $u_n(x)$ satisfy

$$\omega_n = \frac{(2n-1)\pi}{2}, \quad [u_n(0)]^2 = 2, \quad m = N+1, \ldots$$

then its impulse response may be written in the form (10.10.14), where

$$h(t) = \sum_{m=1}^{N} \left\{ [u_n(0)]^2 \cos(\omega_n t) - 2\cos\frac{(2n-1)\pi t}{2} \right\}.$$

Chapter 11

Inversion of Continuous Second-Order Systems

Certain authors, speaking of their works, say, "My book," "My commentary," "My history," etc. They resemble middle class people who have a house of their own, and always have "My house" on their tongue. They would do better to say, "Our book," "Our commentary," "Our history," etc., because there is in them usually more of other people's than their own.

Pascal's *Pensées*, 43

11.1 A historical review

It was shown in Section 10.1 that the Sturm-Liouville equation can appear in three different forms. The one preferred by pure mathematicians seems to be (10.1.14):

$$y''(x) + [\lambda - q(x)]y(x) = 0. \tag{11.1.1}$$

In vibration problems, the equation

$$u''(x) + \lambda\rho^2(x)u(x) = 0 \tag{11.1.2}$$

appears in the transverse vibrations of a taut string, while

$$(A(x)v'(x))' + \lambda A(x)v(x) = 0 \tag{11.1.3}$$

occurs in the longitudinal or torsional vibrations of a thin straight rod of cross section $A(x)$.

As with all inverse problems (see Parker (1977) [263]), the introduction of Newton (1983) [249], Sabatier (1978) [295], Sabatier (1985) [298], Groetsch (1993) [155], Groetsch (2000) [156] or Kirsch (1996) [193] there are three aspects to the inverse problem:

i) *existence*, i.e., mathematically, is there a function $q(x)$, $\rho(x)$ or $A(x)$, or physically, is there a vibrating system, with the required properties?

ii) *uniqueness*, i.e., is there only one system with these properties?

iii) *construction*, i.e., how can we construct one or more systems from the given data?

These questions, which are closely related, have been gradually elucidated over the past seventy years. In this chapter we will use the numbering convention **S** given in Section 10.1, unless we state otherwise.

Ambarzumian (1929) [3] considered the question of uniqueness in a special case. He considered equation (11.1.1) with the symmetrical end conditions

$$y'(0) = 0 = y'(\pi), \tag{11.1.4}$$

and the equation

$$y''(x) + \lambda y(x) = 0$$

with the same end conditions. He showed that if the two systems have the same spectrum $(\lambda_n)_0^\infty$, where $\lambda_n = n^2$, then $q(x)$ is identically zero. Note that he considered symmetrical end conditions, so that only one spectrum is needed. His proof has a defect in that it relies on a perturbation method which requires $q(x)$ to be small.

The fundamental paper on the inverse problem for the equation (11.1.1) is Borg (1946) [39]. He showed that if $q(x)$ is symmetric, i.e.,

$$q(x) = q(\pi - x), \tag{11.1.5}$$

then the spectrum of equation (11.1.1) corresponding to the end conditions (11.1.4), or to the (Dirichlet) end conditions

$$y(0) = 0 = y(\pi), \tag{11.1.6}$$

determines $q(x)$ uniquely. This validates Ambarzumian's earlier result. (See also Hochstadt and Kim (1970) [174].)

It is important to bear in mind a fundamental feature of equations (11.1.1)-(11.1.3); if the system is symmetrical about the mid-point $x = 1/2$, and the end conditions are symmetrical also then in general one spectrum corresponding to one set of end conditions is sufficient to determine it. If it is not symmetrical then two spectra, corresponding to two different end conditions at one end, are required. In this connection, Gottlieb (1986) [138] constructs some interesting counterexamples. Recall that a uniform string fixed at both ends, i.e., a violin string, has natural frequences ω_i that are all multiples of ω_1; we say that the spectrum (in ω, not λ) is *harmonic*. It is this property that makes the violin a musical instrument: the overtones of a string are all octaves above the fundamental tone. A harmonic spectrum is a special case of a *uniformily spaced spectrum*; here $\omega_{i+1} - \omega_i =$ constant. The uniform string is special in the sense that it has a harmonic spectrum. $\omega_i = (i+1)\pi, i = 0,1,2,\ldots$ for fixed-fixed ends, and a harmonic spectrum $\omega_i = (i+1/2)\pi, i = 0,1,2,\ldots$ for fixed-free ends (see the note on a free end at the beginning of Section 10.1). Gottlieb (1986) [138] constructs piecewise uniform strings with one step that have one harmonic spectrum, either for fixed-fixed or fixed-free ends. (see Section 12.4.) In each case the *other* spectrum is uniformly spaced but not harmonic. His analysis thus highlights the need to consider two spectra to ensure uniqueness.

Borg also considered equation (11.1.1) for two sets of end conditions; one set

$$\cos\alpha y(0) + \sin\alpha y'(0) = 0 = \cos\beta y(\pi) + \sin\beta y'(\pi), \tag{11.1.7}$$

and the other

$$\cos\alpha y(0) + \sin\alpha y'(0) = 0 = \cos\gamma y(\pi) + \sin\gamma y'(\pi), \tag{11.1.8}$$

that differ only at the end $x = \pi$, i.e., $\beta \neq \gamma$. He showed that if $\sin\alpha = 0 = \sin\gamma$, so that (11.1.8) is equivalent to (11.1.6), and $\sin\beta \neq 0$, then two interlacing spectra (as in Section 10.8) determine a unique nonsymmetric function $q(x)$. If $\sin\alpha\sin\beta \neq 0$, then $q(x)$ is uniquely determined by two spectra that are short of the first eigenvalue λ_0 of the first spectrum corresponding to (11.1.7).

Borg's results were extended and simplified by Levinson (1949) [207]. He proved that if the spectra of (11.1.1) for each of the end conditions (11.1.7), (11.1.8) are given, and if $\sin(\gamma-\beta) \neq 0$, that is if (11.1.7), (11.1.8) are not identical, then $q(x)$ is uniquely determined. (Remember that this means that there is not *more* than one $q(x)$, *not* that there is *at least* one $q(x)$.) This result was extended by Hochstadt (1973) [175], Hochstadt (1975a) [177] who considered the extent to which $q(x)$ was determined when some eigenvalues λ_n, μ_n corresponding respectively to the end conditions (11.1.7), (11.1.8), were unknown; see also Hald (1978a) [162], Barcilon (1974c) [16] and further references given there.

For the symmetrical case, Levinson showed that if it is known that (11.1.5) holds almost everywhere in $(0,\pi)$, and if $\alpha+\beta = \pi$, i.e., $h = H$ in (11.1.2), then $q(x)$ is uniquely determined by the spectrum for the end conditions (11.1.7). This result includes Borg's results for (11.1.4) ($h = 0 = H$) and (11.1.6) ($h = \infty = H$) as special cases.

Marchenko (1950) [218], Marchenko (1952) [219], Marchenko (1953) [220] made these results a little sharper. He showed that if $q(x) \in L_1(0,\pi)$ and $\sin(\alpha-\beta) \neq 0$, then the spectra of (11.1.1) corresponding to (11.1.7), (11.1.8) determine $q(x)$ and $\tan\alpha, \tan\beta, \tan\gamma$ uniquely. A full account of the uniqueness theorem may be found in Levinson (1949) [209]. Further results may be found in Hochstadt (1973) [175], Hochstadt (1975b) [178], Hochstadt (1976) [179], Hochstadt (1977) [180], Hochstadt and Lieberman (1978) [181], Sabatier (1979a) [296], Sabatier (1979b) [297], Hald (1984) [165], Seidman (1985) [301], McLaughlin (1986) [228] and Levitan (1987) [211], Kirsch (1996) [193].

These results are all concerned with uniqueness. Basically, they all state that it is not possible to find more than one function $q(x)$ corresponding to two spectra. However, it was shown in Chapter 10 that the eigenvalues of (11.1.1), (11.1.7), and of (11.1.1), (11.1.8) have a number of specific properties, e.g., they interlace, and they have the asymptotic form (10.9.22) if h, H are finite, or one of the others listed in Section 10.9 if h or H is infinite. The question is therefore what are *sufficient* conditions for two sets of numbers $(\lambda_n)_0^\infty$ and $(\mu_n)_0^\infty$ to be the spectra of equation (11.1.1) corresponding to two sets of end conditions, like (11.1.7), (11.1.8). The conditions will, of course, depend on what conditions we demand of $q(x)$.

We note that, when viewed as a purely mathematical problem, this problem is very difficult, as an inspection of the literature will immediately verify. However, the difficulties arise because it is assumed that the data consist of two infinite sequences, either $(\lambda_n, \mu_n)_0^\infty$ or perhaps $(\lambda_n, \sigma_n)_0^\infty$, where $(\sigma_n)_0^\infty$ are the norming constants introduced in Section 10.8. In practical inverse vibration problems, it is not possible to measure more than a (small) finite number of frequencies. In that case we find that the sufficient conditions are that the eigenvalues are positive (they are the squares of the natural frequencies) and interlace as discussed in Section 10.8. We make a few remarks on the mathematical problem for the sake of completeness.

Levitan (1964b) [210] proved the following result. Let $(\lambda_n)_0^\infty, (\mu_n)_0^\infty$ be sets of real numbers satisfying

$$\lambda_0 < \mu_0 < \lambda_1 < \mu_1 < \ldots, \tag{11.1.9}$$

$$\lambda_n^{\frac{1}{2}} = n + a_0 n^{-1} + a_1 n^{-2} + O(n^{-3}), \tag{11.1.10}$$

$$\mu_n^{\frac{1}{2}} = n + a_0' n^{-1} + a_1' n^{-2} + O(n^{-3}), \tag{11.1.11}$$

where $a_0 \neq a_0'$. Then there exists an equation of the form (11.1.1) with a continuous real valued function $q(x)$ and real numbers h, h', H such that $(\lambda_n)_0^\infty$ is the spectrum of (11.1.1) subject to

$$y'(0) - hy(0) = 0 = y'(\pi) + Hy(\pi), \tag{11.1.12}$$

$(\mu_n)_0^\infty$ is the spectrum subject to

$$y'(0) - h'y(0) = 0 = y'(\pi) + Hy(\pi), \tag{11.1.13}$$

and

$$a_0' - a_0 = (h' - h)/\pi. \tag{11.1.14}$$

Note that the asymptotic form (11.1.10) was obtained in Section 10.9 by assuming that $q(x)$ had a bounded derivative, and that the explicit expressions for a_1 given in equation (10.9.51)-(10.9.53), were obtained under the assumption that $q(x)$ was twice continuously differentiable, although actually it would have been sufficient to assume that $q''(x)$ was, say, piecewise continuous, or even just bounded. We showed in Section 10.9 that if it is assumed only that $q(x)$ was continuous, then the asymptotic form of $\lambda_n^{\frac{1}{2}}$ was (10.9.20). Thus we note that the sufficient conditions (11.1.10), (11.1.11) are stronger than the necessary condition (10.9.20). As Levitan (1964b) [210] shows, there is a similar mismatch between necessary and sufficient conditions if it is required that $q(x)$ have, say,

r continuous derivatives. See Levitan and Sargsjan (1991) [212] and references given there.

Levitan's result is a refinement of results contained in the final section of Gel'fand and Levitan (1951) [100]. In that paper the data for the inverse problem are $(\lambda_n)_0^\infty$ and the norming constants $(\sigma_n)_0^\infty$. They showed that if $\lambda_n^{\frac{1}{2}}$ had the asymptotic form (11.1.10), and σ_n had the asymptotic form (10.9.54), then there exists a continuous function $q(x)$ for which $(\lambda_n, \sigma_n)_0^\infty$ are the spectral constants corresponding to (11.1.12). Note that there are various special cases of these results corresponding to h or H being infinite.

After these few remarks on sufficient conditions, we come to the third question: How does one construct $q(x)$ from the given data? Here the fundamental paper is Gel'fand and Levitan (1951) [100]. They show that $q(x)$ and the constants h, H are uniquely determined from $(\lambda_n, \sigma_n)_0^\infty$. They develop a procedure for reconstructing $q(x)$ based on an earlier paper by Marchenko (1950) [218]; we describe a modified form of this procedure in this chapter.

Three papers by Krein (1951a) [200], Krein (1951b) [201], Krein (1952) [202], considered the question of uniqueness, existence and reconstruction for the taut string (equation (11.1.2)). He used his theory of the extension of positive definite functions. His results were generally stated without proof, and his methods have been used only by a few later authors; see Gopinath and Sondhi (1971) [137], and Landau (1983) [204].

Gopinath and Sondhi (1970) [136], in considering the determination of the shape of the human vocal tract from accoustical measurements, encountered Webster's horn equation (10.10.10), and devised two methods for its inversion. The first is in the spirit of Gel'fand and Levitan, and can be replaced by the analysis of Section 11.6. The second is set in the time domain and relies on the impulse response described in Section 10.10. This formulation was improved and extended by Gopinath and Sondhi (1971) [137] and is described in Section 11.11. A recent review of the vocal tract inverse problem was given by Sondhi (1984) [309]. The interconnections between all the various procedures for the inversion of second-order problems have been analysed by Burridge (1980) [45]; he pays particular attention to the case in which the cross-sectional area function $A(x)$ is discontinuous. See also Hald (1984) [165].

One of the early strands of research into actually constructing the potential in a Sturm-Liouville problem, used some finite difference/finite element approximation to the governing equation. One of the difficulties that had to be faced was that the eigenvalues derived from a discrete approximation diverge, with increasing mode number, from those predicted by the differential equation. Paine and his colleagues made a detailed study of this problem. See Paine and de Hoog (1980) [258], Paine, de Hoog and Andersen (1981) [259], Paine (1982) [256], Paine (1984) [257], Andrew and Paine (1985) [10], Andrew and Paine (1986) [11]. A study of inverse problems for Sturm-Liouville systems modelled as discrete (Jacobi) systems was carried out by Andersson (1970) [5]; he did not have the results on inverse problems for Jacobi matrices at his disposal. See also Barcilon (1974a) [14]. Hald (1972) [159] made a detaled study of the problem, later Hald (1977) [161] paid particular attention to the Sturm-Liouville problem with symmetric potential. Hald (1978b) [163] discussed the discrete system obtained by applying the Rayleigh-Ritz procedure to the continuous problem, and considered the limiting case in which the number of terms in the Fourier series expansions of $q(x)$ tends to infinity. A modern version of such an approximation procedure may be found in Section 11.9. Barcilon (1983) [22] attempted to derive the (continuous) density of the string from the known solution to the inverse problem for the discrete system, but his procedure does not lend itself to computation. A straightforward and comparitively simple solution of the inverse Sturm-Liouville system for a rod, using a piecewise uniform model, is given in Section 12.1.

11.2 Transformation operators

The fundamental step in the elucidation of all three aspects, uniqueness, existence and reconstruction, is the introduction of the *Gel'fand-Levitan-Marchenko transformation operator.* This operator relates the solution of one Sturm-Liouville equation to another.

Consider two equations, a base equation

$$\phi''(x) + (\lambda - p(x))\phi(x) = 0, \quad x \geq 0, \tag{11.2.1}$$

subject to the single boundary condition

$$\phi'(0) - h\phi(0) = 0, \tag{11.2.2}$$

and another equation

$$\psi''(x) + (\lambda - q(x))\psi(x) = 0, \quad x \geq 0, \tag{11.2.3}$$

subject to another boundary condition

$$\psi'(0) - h'\psi(0) = 0. \tag{11.2.4}$$

We seek an operator of the form

$$\psi(x) = \phi(x) + \int_0^x K(x,y)\phi(y)dy, \tag{11.2.5}$$

that transforms a solution of (11.2.1), (11.2.2) into a solution of (11.2.3), (11.2.4).

Differentiation of (11.2.5) gives

$$\psi'(x) = \phi'(x) + K(x,x)\phi(x) + \int_0^x K_x(x,y)\phi(y)dy$$

where

$$K_x(x,y) = \frac{\partial K}{\partial x}(x,y).$$

A second differentiation gives

$$\psi''(x) = \phi''(x) + \frac{dK(x,x)}{dx}\phi(x) + K(x,x)\phi'(x) + K_x(x,x)\phi(x) + \int_0^x K_{xx}(x,y)\phi(y)dy.$$

This is the first term in (11.2.3). The last term is

$$q(x)\psi(x) = q(x)\phi(x) + \int_0^x q(x)K(x,y)\phi(y)dy.$$

This leaves the second term:

$$\begin{aligned} \lambda\psi(x) &= \lambda\phi(x) + \lambda\textstyle\int_0^x K(x,y)\phi(y)dy \\ &= \lambda\phi(x) + \textstyle\int_0^x K(x,y)\{p(y)\phi(y) - \phi''(y)\}dy. \end{aligned}$$

We evaluate the last integral in this expression by parts twice:

$$\int_0^x K(x,y)\phi''(y)dy = [K(x,y)\phi'(y) - K_y(x,y)\phi(y)]_0^x + \int_0^x K_{yy}(x,y)\phi(y)dy.$$

Collect the terms in these equations to write (11.2.3); the result is as follows:

$$\begin{aligned} &\phi''(x) + (\lambda - q(x))\phi(x) + \frac{dK(x,x)}{dx}\phi(x) + K(x,x)\phi'(x) \\ &+ K_x(x,x)\phi(x) - [K(x,y)\phi'(y) - K_y(x,y)\phi(y)]_{y=0}^{y=x} \\ &+ \int_0^x \{K_{xx}(x,y) - K_{yy}(x,y) + (p(y) - q(x))K(x,y)\}\phi(y)dy. \end{aligned}$$

Now use the facts that $\phi(x)$ satisfies (11.2.1), and that

$$K_x(x,x) + K_y(x,x) = \frac{dK(x,x)}{dx}$$

to obtain

$$\left\{\frac{2dK(x,x)}{dx} + p(x) - q(x)\right\}\phi(x) + K(x,0)\phi'(0) - K_y(x,0)\phi(0) +$$

$$\int_0^x \{K_{xx}(x,y) - K_{yy}(x,y) + (p(y) - q(x))K(x,y)\}\phi(y)dy$$

This equation is satisfied identically by taking

$$K_{xx}(x,y) - K_{yy}(x,y) + (p(y) - q(x))K(x,y) = 0, \qquad 0 \le y \le x \le \pi, \quad (11.2.6)$$

$$\frac{dK}{dx}(x,x) = \frac{1}{2}(q(x) - p(x)), \qquad 0 \le x \le \pi, \quad (11.2.7)$$

$$K(x,0)\phi'(0) - K_y(x,0)\phi(0) = 0, \qquad 0 \le x \le \pi. \quad (11.2.8)$$

Now we examine the boundary conditions at $x = 0$. Clearly

$$\psi(0) = \phi(0), \quad \psi'(0) = \phi'(0) + K(0,0)\phi(0). \tag{11.2.9}$$

If h, h' are finite then $\phi(0) \neq 0$, $\phi'(0) = h\phi(0)$ imply

$$K_y(x,0) - hK(x,0) = 0, \qquad 0 \le x \le \pi, \tag{11.2.10}$$

and

$$\psi'(0) = (h + K(0,0))\psi(0),$$

so that

$$K(0,0) = h' - h, \tag{11.2.11}$$

and hence, with (11,2,7),

$$K(x,x) = h' - h + \frac{1}{2}\int_0^x (q(y) - p(y))dy. \tag{11.2.12}$$

Note that if $h = \infty$, so that $\phi(0) = 0$, then equation (11.2.8) implies

$$K(x,0) = 0, \qquad 0 \le x \le \pi, \tag{11.2.13}$$

and $\psi(x)$ satisfies $\psi(0) = 0$, i.e., $h' = \infty$. In that case equation (11.2.12) is replaced by

$$K(x,x) = \frac{1}{2}\int_0^x (q(y) - p(y))dy. \tag{11.2.14}$$

With this kernel $K(x,y)$, the equation (11.2.5) transforms a solution of (11.2.1), (11.2.2) into a solution of (11.2.3), (11.2.4).

11.3 The hyperbolic equation for $K(x,y)$

The kernel $u(x,y) = K(x,y)$ satisfies the hyperbolic equation (11.2.6), i.e.,

$$u_{xx} - u_{yy} + (p(y) - q(x))u = 0, \qquad 0 \le y \le x \le \pi \tag{11.3.1}$$

in the upper triangle OIC shown in Figure 11.3.1. The characteristics for this equation are the lines $x \pm y = const.$ The kernel has the value (11.2.12), i.e.,

$$u(x,x) = h' - h + \frac{1}{2}\int_0^x (q(t) - p(t))dt, \tag{11.3.2}$$

on the characteristic $x = y$, and satisfies the condition (11.2.10), i.e.,

$$u_y(x,0) - hu(x,0) = 0, \qquad 0 \le x \le \pi \tag{11.3.3}$$

on the x-axis.

First we discuss how $u(x,y)$ may be continued to the lower triangle OID so that the boundary condition (11.3.3) is satisfied. There are three cases:
i) $\mathbf{h = 0}$. Now $u_y(x,0) = 0$ so that we continue $u(x,y)$ to the lower triangle as an *even* function of y, i.e.,

$$u(x,-y) = u(x,y);$$

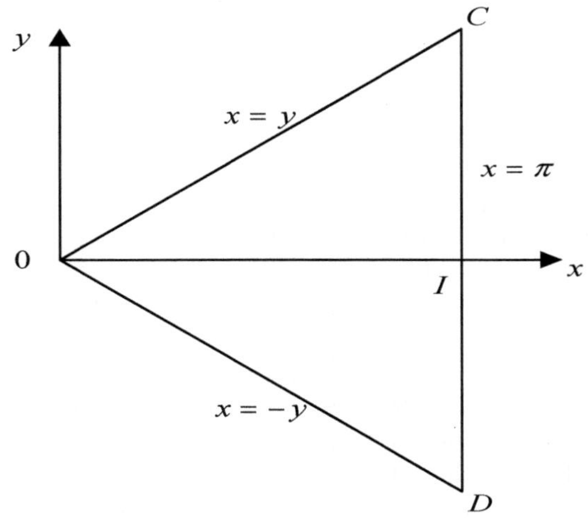

Figure 11.3.1 - $0 \leq y \leq x \leq \pi$ *in the upper triangle* OIC.

then

$$u(x,-x) = h' + \frac{1}{2}\int_0^x (q(t) - p(t))dt.$$

ii) h $= \infty$. Now $u(x,0) = 0$ so that we continue $u(x,y)$ as an *odd* function of y, i.e.,

$$u(x,-y) = -u(x,y)$$

so that, according to (11.2.14),

$$-u(x,-x) = u(x,x) = \frac{1}{2}\int_0^x (q(t) - p(t))dt.$$

iii) h is finite and not zero. Define

$$u(x,y) = \exp(-hy)K(x,y) \tag{11.3.4}$$

then

$$u_y(x,0) = K_y(x,0) - hK(x,0) = 0.$$

This means that we should continue $u(x,y)$ as an even function of y. The values of $u(x,y)$ on the characteristics are

$$u(x,x) = u(x,-x) = \exp(-hx)K(x,x)$$

where $K(x,x)$ is given by (11.2.12). Since $K(x,y)$ satisfies (11.3.1), $u(x,y)$ satisfies

$$u_{xx} - u_{yy} - 2h\ sign(y)u_y + (p(y) - q(x) - h^2)u = 0 \tag{11.3.5}$$

throughout the triangle OCD, i.e., $0 \le |y| \le x \le \pi$. Here

$$sign(y) = \begin{cases} +1 & y > 0 \\ -1 & y < 0 \end{cases}$$

and $p(-y) = p(y)$.

The equations (11.3.1), (11.3.5) are hyperbolic partial differential equations. The existence and uniqueness properties of such equations are the subject matter of treatises on p.d.e.'s. In keeping with the philosophy of this book, we shall not assume that the reader is acquainted with these properties, and will derive them *ab initio*.

There are two fundamental questions regarding the p.d.e.'s (11.3.1) and (11.3.5): what boundary data lead to a unique solution? How can we find this unique solution from the boundary data? It transpires that there are two kinds of suitable boundary data, giving rise to two problems:

The *Goursat problem* in which u is given on the characteristics $x = \pm y$.

The *Cauchy problem* in which u and u_x are given on the side CD, i.e., on $x = \pi$, $-\pi \le y \le \pi$.

In both these cases we can reduce the solution of the p.d.e. to the solution of a *Volterra integral equation*, and we can show that this equation has a unique solution.

The Goursat problem

We first consider cases i) and ii); the governing equation is equation (11.3.1).

Stokes' theorem in 2-D is

$$\int_S\int \left(\frac{\partial w_2}{\partial x} - \frac{\partial w_1}{\partial y}\right) dxdy = \int_\Gamma w_1 dx + w_2 dy, \tag{11.3.6}$$

where Γ is the boundary of the region S, traversed counter-clockwise.

Apply this theorem to the rectangle $OBPA$ in Figure 11.3.2, with $w_1 = u_y$, $w_2 = u_x$. Then

$$\frac{\partial w_2}{\partial x} - \frac{\partial w_1}{\partial y} = u_{xx} - y_{yy} = f(x,y)u,$$

where

$$f(x,y) = q(x) - p(y).$$

The L.H.S. of equation (11.3.6) is thus

$$\int_S\int f(x,y)u(x,y)dxdy, \tag{11.3.7}$$

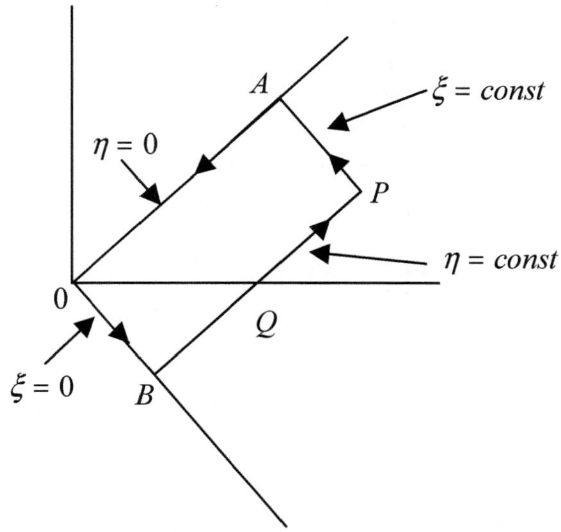

Figure 11.3.2 - The rectangle OBPA.

where S is the rectangle $OBPA$. The R.H.S. of equation (11.3.6) is made up of four line integrals, along $OB + BP + PA + AO$. To evaluate these integrals, it is convenient to introduce the so-called *characteristic coordinates*

$$\xi = \frac{1}{2}(x + y), \qquad \eta = \frac{1}{2}(x - y). \tag{11.3.8}$$

Equivalently,

$$x = \xi + \eta, \qquad y = \xi - \eta. \tag{11.3.9}$$

The partial derivates in these coordinates are

$$\frac{\partial}{\partial \xi} = \frac{\partial}{\partial x}\frac{\partial x}{\partial \xi} + \frac{\partial}{\partial y}\frac{\partial y}{\partial \xi} = \frac{\partial}{\partial x} + \frac{\partial}{\partial y},$$

$$\frac{\partial}{\partial \eta} = \frac{\partial}{\partial x}\frac{\partial x}{\partial \eta} + \frac{\partial}{\partial y}\frac{\partial y}{\partial \eta} = \frac{\partial}{\partial x} - \frac{\partial}{\partial y}.$$

Consider the integral along OB

$$I_1 = \int_{OB} w_1 dx + w_2 dy.$$

On OB, $x = \eta, \quad y = -\eta$, so that $dx = d\eta, \quad dy = -d\eta$, and $w_1 dx + w_2 dy = (u_y - u_x)d\eta = -u_\eta d\eta$, so that

$$I_1 = \int_{OB} u_\eta d\eta = -[u]_O^B = -u(B) + u(O).$$

On BP, $\eta = const$, $dx = dy = d\xi$, so that

$$I_2 = \int_{BP} u_\xi d\xi = [u]_B^P = u(P) - u(B).$$

Similarly

$$I_3 = u(P) - u(A), \quad I_4 = u(O) - u(A).$$

Thus the R.H.S. of (11.3.6) is

$$2u(P) - 2u(A) - 2u(B) + 2u(O)$$

since A has coordinates $((x + y)/2, (x + y)/2)$ and B has coordinates $(x - y)/2, (y - x)/2$, we find

$$\begin{aligned} u(x, y) &= u((x+y)/2, (x+y)/2)) + u((x-y)/2, (y-x)/2) \\ &\quad -u(0,0) + \frac{1}{2}\int_S\int f(x', y')u(x', y')dx'dy'. \end{aligned} \tag{11.3.10}$$

This equation expresses $u(x, y)$ as a sum of two parts: the first, comprising the first three terms, is made up of data on the characteristics; the second is an integral over the rectangle $OBPA$.

In the characteristic coordinates, equation (11.3.10) is

$$u(\xi + \eta, \xi - \eta) = u(\xi, \xi) + u(\eta, -\eta) - u(0, 0)$$

$$+\frac{1}{2}\int_0^\eta \left\{\int_0^\xi f(\sigma + \tau, \sigma - \tau)u(\sigma + \tau, \sigma - \tau)d\sigma\right\} d\tau, \tag{11.3.11}$$

which has the form of a Volterra integral equation. We note that when $h = 0$, $u(\eta, -\eta) = u(\eta, \eta)$; when $h = \infty$, $u(\eta, -\eta) = -u(\eta, \eta)$ and $u(0, 0) = 0$.

Equation (11.3.11) has a unique solution for given data on the characteristics. For if there were two solutions, then their difference, $u = u_1 - u_2$, would satisfy

$$u(\xi, \eta) = \frac{1}{2}\int_0^\eta \left\{\int_0^\xi f(\sigma + \tau, \sigma - \tau)u(\sigma + \tau, \sigma - \tau)d\sigma\right\} d\tau. \tag{11.3.12}$$

The classical way to show that this equation has only the trivial solution is as follows. The function f is bounded: $|f| \leq 2M$. This means that $v = |u|$ satisfies

$$v(\xi, \eta) \leq MV(\xi, \eta)$$

where

$$V(\xi, \eta) = \int_0^\eta \int_0^\xi v(\sigma + \tau, \sigma - \tau)d\sigma d\tau.$$

Suppose $0 \leq \xi \leq k$, and $0 \leq \eta \leq k$, then

$$v(\xi, \eta) \leq MV(k, k)$$

so that

$$V(k,k) \le Mk^2V(k,k).$$

If $v(\xi,\eta)$ is not identically zero in $[0,k]\times[0,k]$, then this inequality is clearly impossible if $Mk^2 < 1$. Choose k_0 so that $Mk_0^2 < 1$, then $u(\xi,\eta) \equiv 0$ in $[0,k_0]\times[0,k_0]$. Now suppose $(\xi,\eta) \in [0,\sqrt{2}k_0]\times[0,\sqrt{2}k_0]$, then for (ξ,η) outside $[0,k_0]\times[0,k_0]$ we have

$$v(\xi,\eta) \le MV(\sqrt{2}k_0,\sqrt{2}k_0)$$

so that

$$V(\sqrt{2}k_0,\sqrt{2}k_0) \le MV(\sqrt{2}k_0,\sqrt{2}k_0)(2k_0^2 - k_0^2)$$

which, with $Mk_0^2 < 1$, provides a contradiction. By continuing this argument inductively we deduce that $v(\xi,\eta) = 0$, i.e., $u(\xi,\eta) = 0$.

The extension of this argument to case iii), when h is finite but not zero, is a little complicated but not essentially difficult. Proceeding exactly as before we find the equation corresponding to (11.3.10) to be

$$u(x,y) = u((x+y)/2,(x+y)/2) + u((x-y)/2,(y-x)/2) -$$

$$-u(0,0) + \frac{1}{2}\int_S\int f(x',y')u(x',y')dx'dy' - h\int_S\int sign(y')u_y dx'dy' \quad (11.3.13)$$

where

$$f(x,y) = q(x) - p(y) + h^2, \quad (11.3.14)$$

and $p(-y) = p(y)$. Again we can use Stokes' theorem to write the last term as a sum of line integrals; see Ex. 11.3.1.

The resulting equation is again a Volterra integral equation with a unique solution.

The Cauchy problem

We proceed as in the Goursat problem, but now apply Stokes' theorem to the triangle ABP in Figure 11.3.3.

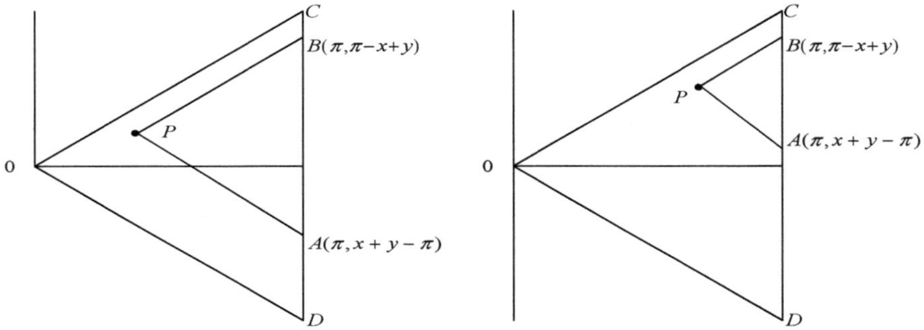

Figure 11.3.3 - The triangle ABP when a) $x+y<\pi$ and b) $x+y>\pi$.

In cases i) and ii) the R.H.S. of equation (11.3.6) is (11.3.7) while the L.H.S. is the sum of line integrals along $AB + BP + PA$. Now

$$
\begin{aligned}
I_1 &= \int_{AB} w_1 dx + w_2 dy = \int_{AB} u_x dy \\
I_2 &= \int_{BP} w_1 dx + w_2 dy = \int_{BP} u_\xi d\xi = [u]_B^P = u(P) - u(B) \\
I_3 &= \int_{PA} w_1 dx + w_2 dy = \int_{PA} u_\eta d\eta = -[u]_P^A = u(P) - u(A).
\end{aligned}
$$

This yields the Volterra integral equation

$$
2u(x,y) = u(\pi, x+y-\pi) + u(\pi, \pi - x + y)
$$

$$
-\int_{x+y-\pi}^{\pi-x+y} u_x(\pi,t)dt + \int_S\int f(x',y')u(x',y')dx'dy' \tag{11.3.15}
$$

where now S is the triangle ABP. Again, in case iii) there is an extra term

$$
-h\int_S\int sign(y)u_y dx'dy' \tag{11.3.16}
$$

to be added to the R.H.S. of (11.3.15). This is given in Ex. 11.3.3.

In all cases $u(x,y)$ is given as the solution of a Volterra integral equation; the solution is uniquely determined by the values of u and u_x on the line CD, i.e., $x = \pi,\ -\pi \le y \le \pi$.

We now show how the uniqueness of solution of the hyperbolic equation for $K(x,y)$ may be used to show the uniqueness of an inverse problem for the Sturm-Liouville equation.

Exercises 11.3

1. Show that the integral in (11.3.13) may be written

$$
\begin{aligned}
I &= \int_S\int sign(y)u_y dxdy = -2\int_0^{x-y} u(s,0)ds \\
&\quad -\int_\eta^\xi u(\sigma+\eta,\sigma-\eta)d\sigma + \int_0^\eta u(\sigma+\eta,\sigma-\eta)d\sigma \\
&\quad +\int_0^\eta u(\xi+\tau,\xi-\tau)d\tau + \int_0^\xi u(\sigma,\sigma)d\sigma + \int_0^\eta u(\tau,-\tau)d\tau.
\end{aligned}
$$

2. Show that the integral in (11.3.16) may be written

$$
\begin{aligned}
I &= \int_S\int sign(y)u_y dxdy = \int_\xi^{\pi-\eta} u(\sigma+\eta,\sigma-\eta)d\sigma \\
&\quad -\int_\eta^\xi u(\xi+\tau,\xi-\tau)d\tau + \int_\xi^{\pi-\xi} u(\xi+\tau,\xi-\tau)d\tau - 2\int_{x+y}^{\pi} u(s,0)ds
\end{aligned}
$$

when $y > 0,\ x + y < \pi$ and

$$I = \int_S \int sign(y) u_y dxdy = \int_{\xi}^{\pi-\eta} u(\sigma+\eta, \sigma-\eta) d\sigma - \int_{\eta}^{\pi-\xi} u(\xi+\tau, \xi-\tau) d\tau$$

when $y > 0,\ x + y \geq \pi$.

11.4 Uniqueness of solution of an inverse problem

With the uniqueness results of Section 11.3 we are now in a position to show that the potential $p(x)$ in (11.2.1) is uniquely determined by two spectra corresponding to two different conditions at one end of $(0, \pi)$.

Theorem 11.4.1 *Suppose that there were two potentials* $p(x),\ q(x) \in C[0, \pi]$ *with the following properties:*

i) $y'' + (\lambda - p)y = 0, \quad y'(0) - h_1 y(0) = 0 = y'(\pi) + H_1 y(\pi)$
has spectrum $(\lambda_n)_0^\infty$*;*

ii) $y'' + (\lambda - p)y = 0, \quad y'(0) - h_1 y(0) = 0 = y'(\pi) + H_1' y(\pi)$
has spectrum $(\mu_n)_0^\infty$*;*

iii) $y'' + (\lambda - q)y = 0, \quad y'(0) - h_2 y(0) = 0 = y'(\pi) + H_2 y(\pi)$
has spectrum $(\lambda_n)_0^\infty$*;*

iv) $y'' + (\lambda - q)y = 0, \quad y'(0) - h_2 y(0) = 0 = y'(\pi) + H_2' y(\pi)$
has spectrum $(\mu_n)_0^\infty$*.*

If $H_2 \neq H_2'$*, then* $p(x) = q(x),\ h_1 = h_2,\ H_1 = H_2,\ H_1' = H_2'$*.*

Proof. First we use the known asymptotic forms for the eigenvalues. Equation (10.9.20) states that

$$\sqrt{\lambda_n} = n + cn^{-1} + o(n^{-1})$$

where

$$c = \frac{1}{\pi}(h + H + \frac{1}{2} \int_0^\pi q(x) dx).$$

Thus, since i) and iii) have the same spectrum

$$h_1 + H_1 + \frac{1}{2} \int_0^\pi p(x) dx = h_2 + H_2 + \frac{1}{2} \int_0^\pi q(x) dx \tag{11.4.1}$$

and because ii) and iv) have the same spectrum

$$h_1 + H_1' + \frac{1}{2} \int_0^\pi p(x) dx = h_2 + H_2' + \frac{1}{2} \int_0^\pi q(x) dx. \tag{11.4.2}$$

Now we transform a solution $\phi_n(x)$ of i) corresponding to the eigenvalue λ_n, into a solution $\psi_n(x)$ of iii) with the same eigenvalue:

$$\psi_n(x) = \phi_n(x) + \int_0^x K(x, y) \phi_n(y) dy,$$

and, according to (11.2.11), (11.2.12), we have

$$K(0,0) = h_2 - h_1, \tag{11.4.3}$$

$$K(\pi,\pi) = h_2 - h_1 + \frac{1}{2}\int_0^{\pi} (q(x) - p(x))dx. \tag{11.4.4}$$

We examine the boundary condition at $x = \pi$:

$$\begin{aligned} \psi_n(\pi) &= \phi_n(\pi) + \textstyle\int_0^\pi K(\pi,y)\phi_n(y)dy, \\ \psi_n'(\pi) &= \phi_n'(\pi) + K(\pi,\pi)\phi_n(\pi) + \textstyle\int_0^\pi K_x(\pi,y)\phi_n(y)dy, \end{aligned}$$

so that

$$\begin{aligned} \psi_n'(\pi) + H_2\psi_n(\pi) &= \phi_n'(\pi) + H_1\phi_n(\pi) + \{K(\pi,\pi) + H_2 - H_1\}\phi_n(\pi) \\ &\quad + \int_0^{\pi} \{K_x(\pi,y) + H_2K(\pi,y)\}\phi_n(y)dy. \end{aligned}$$

Now $\phi_n'(\pi) + H_1\phi_n(\pi) = 0$, and equations (11.4.1), (11.4.4) show that

$$K(\pi,\pi) + H_2 - H_1 = 0. \tag{11.4.5}$$

Thus

$$\int_0^{\pi} \{K_x(\pi,y) + H_2K(\pi,y)\}\phi_n(y)dy = 0. \tag{11.4.6}$$

But the $\{\phi_n\}$ form a complete orthogonal set on $(0,\pi)$, so that

$$K_x(\pi,y) + H_2K(\pi,y) = 0. \tag{11.4.7}$$

Applying the same argument to ii) and iv), we find

$$K_x(\pi,y) + H_2'K(\pi,y) = 0 \tag{11.4.8}$$

and, since $H_2 \neq H_2'$ by hypothesis,

$$K_x(\pi,y) = 0 = K(\pi,y). \tag{11.4.9}$$

This holds for $0 \le y \le \pi$, and therefore also for $-\pi \le y \le \pi$. But in Section 11.3 we showed that if $K(x,y)$ satisfies this condition, then $K(x,y) \equiv 0$ in $0 \le |y| \le x \le \pi$. Now equation (11.2.7) implies $p(x) = q(x)$, (11.4.3) implies $h_1 = h_2$, (11.4.5) implies $H_1 = H_2$, (11.4.2) implies $H_1' = H_2'$. ■

In this proof we have assumed that $h_1, h_2, H_1, H_2, H_1', H_2'$ are all finite, but the argument may easily be adapted to the situation in which some of these are infinite.

As we noted in the historical review, if it is known that $q(x)$ is symmetric about $\frac{\pi}{2}$, i.e., $q(x) = q(\pi - x)$, then $q(x)$ is uniquely determined from one spectrum corresponding to symmetrical end conditions, i.e., $h = H$. For since the governing equation (11.1.1) and the end conditions

$$y'(0) - hy(0) = 0 = y'(\pi) + hy(\pi) \tag{11.4.10}$$

are invariant under the transformation $x \to \pi - x$, the solutions of (11.1.1) satisfying (11.4.10) must satisfy $y(x) = \pm y(-x)$. Since the lowest eigenfunction $y_0(x)$ can have no zero in $(0, \pi)$, the even eigenfunctions must satisfy $y'(\frac{\pi}{2}) = 0$, while the odd ones must satisfy $y(\frac{\pi}{2}) = 0$. This means that the given spectrum $(\lambda_n)_0^\infty$ must split into two: $(\lambda_{2n})_0^\infty$ corresponding to

$$y'(0) - hy(0) = 0 = y'(\frac{\pi}{2})$$

and $(\lambda_{2n+1})_0^\infty$ corresponding to

$$y'(0) - hy(0) = 0 = y(\frac{\pi}{2}).$$

We thus have two spectra which will uniquely determine $q(x)$ on $[0, \frac{\pi}{2}]$; the symmetry then gives $q(x)$ on $[\frac{\pi}{2}, \pi]$. For other uniqueness theorems, see McLaughin (1986) [228], McLaughlin and Rundell (1987) [230].

The literature on uniqueness and existence of solutions of inverse problems for the various forms of equations (11.1.1)-(11.1.3) is so vast that one can only make some pointers to the literature. Hald (1984) [165] is useful for a review of the early research. Other studies of problems with discontinuous $q(x)$, in (11.1.1) or $A(x)$, in (11.1.3) include Willis (1985) [334], Kobayashi (1988) [197], Andersson (1988a) [6], (1988b) [7], Coleman and McLaughlin (1993a) [62], (1993b) [63].

11.5 The Gel'fand-Levitan integral equation

The transformation operator introduced in equation (11.2.5) transforms a solution $\phi(x)$ of equation (11.2.1) subject to the single end condition (11.2.2) into a solution $\psi(x)$ of a new equation (11.2.3) subject to the single end condition (11.2.4). But, as in Section 11.4, we require more of the transformation: that it produce a complete orthonormal set (c.o.s.) of eigenfunctions for the new equation (11.2.3) subject to two end conditions, at 0 and π.

Denote the unique solution of (11.2.1), (11.2.2) satisfying $\phi(0) = \xi$ by $\phi(x, \lambda, \xi)$. The eigenfunction $\phi_n(x)$ of (11.2.1) subject to the end conditions

$$\phi'(0) - h\phi(0) = 0 = \phi'(\pi) + H\phi(\pi) \tag{11.5.1}$$

is therefore

$$\phi_n(x) = \phi(x, \lambda_n, \phi_n(0)). \tag{11.5.2}$$

We are going to construct new orthonormal eigenfunctions $\psi(x)$ of (11.2.3) subject to end conditions

$$\psi'(0) - h'\psi(0) = 0 = \psi'(\pi) + H'\psi(\pi), \tag{11.5.3}$$

from equation (11.2.5). We denote the new eigenvalues by $(\mu_n)_0^\infty$ and write

$$\chi_n(x) = \phi(x, \mu_n, \psi_n(0)), \tag{11.5.4}$$

$$\psi_n(x) = \chi_n(x) + \int_0^x K(x,t)\chi_n(t)dt. \tag{11.5.5}$$

Note that $\chi_n(x)$ is the solution of equation (11.2.1), (11.2.2) for $\lambda = \mu_n$, while $\psi_n(x)$ is to be the nth orthonormal eigenfunctions of equation (11.2.3) subject to (11.5.3).

The eigenfunctions $\{\phi_n\}_0^\infty$ of the base problem do form a c.o.s. on $(0, \pi)$. This means that if $g \in L^2(0, \pi)$, then

$$||g||^2 \equiv \int_0^\pi [g(x)]^2 dx = \sum_{n=0}^\infty a_n^2, \tag{11.5.6}$$

where

$$a_n = (g, \phi_n) = \int_0^\pi g(x)\phi_n(x)dx. \tag{11.5.7}$$

This implies also that if $g, h \in L^2(0, \pi)$, then

$$(g, h) = \int_0^\pi g(x)h(x)dx = \sum_{n=0}^\infty a_n b_n,$$

where $b_n = (h, \phi_n)$.

The eigenfunctions $\{\psi_n\}_0^\infty$ are to form a c.o.s. on $(0, \pi)$, so that

$$||g||^2 = \sum_{n=0}^\infty a_n'^2, \tag{11.5.8}$$

where $a_n' = (g, \psi_n)$. Equation (11.5.5) shows that

$$\psi_n = \chi_n + K\chi_n, \tag{11.5.9}$$

where K is the operator defined by

$$Ku = \int_0^x K(x,t)u(t)dt. \tag{11.5.10}$$

Now

$$(Ku, v) = \int_0^\pi \left\{ \int_0^t K(t,x)u(x)dx \right\} v(t)dt$$

and on interchanging the order of integration we see that

$$(Ku, v) = \int_0^\pi \left\{ \int_x^\pi K(t,x)v(t)dt \right\} u(x)dx.$$

The adjoint operator K^* is defined by

$$(Ku, v) = (u, K^*v),$$

so that

$$K^*v = \int_x^\pi K(t,x)v(t)dt. \tag{11.5.11}$$

Return to equation (11.5.9); we can write

$$a'_n = (g, \chi_n) + (g, K\chi_n)$$

so that the equation

$$0 = \sum_{n=0}^{\infty} a_n'^2 - \sum_{n=0}^{\infty} a_n^2$$

can be written

$$0 = \sum_{n=0}^{\infty} \{(g,\chi_n)^2 + 2(g,\chi_n)(g,K\chi_n) + (g,K\chi_n)^2 - (g,\phi_n)^2\} \tag{11.5.12}$$

Put $K^*g = G$ then, since $G \in L^2(0,\pi)$, we have

$$(g, G) - \sum_{n=0}^{\infty} (g,\phi_n)(G,\phi_n) = 0,$$

and this is equivalent to

$$(Kg, g) - \sum_{n=0}^{\infty} (g,\phi_n)(g,K\phi_n) = 0. \tag{11.5.13}$$

Similarly

$$(G,G) - \sum_{n=0}^{\infty} (G,\phi_n)^2 = 0$$

is equivalent to

$$(KK^*g, g) - \sum_{n=0}^{\infty} (g, K\phi_n)^2 = 0. \tag{11.5.14}$$

Now form the combined equation (11.5.12) + 2*(11.5.13) + (11.5.14) and group the terms to get

$$0 = S_1 + S_2 + S_3 + S_4, \tag{11.5.15}$$

where

$$\begin{aligned} S_1 &= \textstyle\sum_{n=0}^{\infty}\{(g,\chi_n)^2 - (g,\phi_n)^2\}, \\ S_2 &= 2\textstyle\sum_{n=0}^{\infty}\{(g,\chi_n)(g,K\chi_n) - (g,\phi_n)(g,K\phi_n)\}, \\ S_3 &= \textstyle\sum_{n=0}^{\infty}\{(g,K\chi_n)^2 - (g,K\phi_n)^2\}, \\ S_4 &= 2(g,Kg) + (g,KK^*g). \end{aligned}$$

In order to represent these products of integrals as multiple integrals we use the simple identity

$$\int_0^{\pi} g(x)dx \int_0^{\pi} h(y)dy = \int_0^{\pi}\int_0^{x} g(x)h(y)dydx + \int_0^{\pi}\int_0^{y} g(x)h(y)dxdy$$

obtained by dividing the square $(0,\pi)\times(0,\pi)$ into two triangles. This yields

$$\begin{aligned}
S_1 &= 2\int_0^\pi\int_0^x g(x)g(y)F(x,y)dydx,\\
S_2 &= 2\int_0^\pi\int_0^x g(x)g(y)\int_0^x K(x,t)F(t,y)dtdydx\\
&\quad +2\int_0^\pi\int_0^y g(x)g(y)\int_0^x K(x,t)F(t,y)dtdxdy,\\
S_3 &= 2\int_0^\pi\int_0^y g(x)g(y)\int_0^x K(x,t)\left[\int_0^y K(y,s)F(s,t)ds\right]dtdxdy,\\
S_4 &= 2\int_0^\pi\int_0^x g(x)g(y)K(x,y)dydx + 2\int_0^\pi\int_0^y\int_0^x K(x,t)K(y,t)dtdxdy
\end{aligned}$$

where

$$F(x,y)=\sum_{n=0}^{\infty}\{\chi_n(x)\chi_n(y)-\phi_n(x)\phi_n(y)\}, \tag{11.5.16}$$

so that equation (11.5.15) gives

$$\int_0^\pi\int_0^x g(x)g(y)\left\{J(x,y)+\int_0^y J(x,t)K(y,t)dt\right\}dydx=0, \tag{11.5.17}$$

where

$$J(x,y)=K(x,y)+\int_0^x K(x,t)F(t,y)dt+F(x,y). \tag{11.5.18}$$

Since $g(x)$ is an arbitrary function in $L^2(0,\pi)$, equation (11.5.17) implies

$$J(x,y)+\int_0^y J(x,t)K(y,t)dt=0, \qquad 0\le y\le x\le\pi. \tag{11.5.19}$$

For fixed x, this is a homogeneous Volterra integral equation for $J(x,y)$, and we may argue exactly as in Section 11.3 that its only solution is $J(x,y)=0$, for $0\le y\le x$. Thus

$$K(x,y)+\int_0^x K(x,t)F(t,y)dt+F(x,y)=0, \qquad 0\le y\le x\le\pi. \tag{11.5.20}$$

This is the *Gel'fand-Levitan integral equation* for $K(x,y)$. Note that for fixed x, (11.5.19) is a *Volterra* equation for $J(x,y)$; on the other hand, for fixed x, (11.5.20) is a *Fredholm* equation for $K(x,y)$.

There is one matter in this analysis that needs to be examined: the convergence of the series in equation (11.5.16). There are two ways to approach this question: examine the asymptotic form of the terms in the series and find the conditions under which the series is convergent; make an assumption that will obviate the question by turning the infinite series into a finite one. We shall follow the latter course.

We started this section by taking a base problem consisting of equation (11.2.1) and end conditions (11.5.1); the c.o.s. of eigenfunctions of this problem is $\{\phi_n\}_0^\infty$. We then used the operator K to construct a new c.o.s. of eigenfunctions $\{\psi_n\}_0^\infty$ for a new problem. The orthonormal eigenfunctions ψ_n were constructed from the solutions $\chi_n = \phi(x, \mu_n, \psi_n(0))$ of the base equation (11.2.1) with $\lambda = \mu_n$, and with initial conditions $\chi_n(0) = \psi_n(0)$, $\chi_n'(0) = h\psi_n(0)$. Now we introduce the *Truncation Assumption*

$$\mu_n = \lambda_n, \qquad \psi_n(0) = \phi_n(0) \text{ for } n = N+1, \ldots$$

This means that, for $n = N+1, \ldots$

$$\chi_n(x) = \phi(x, \mu_n, \psi_n(0)) = \phi(x, \lambda_n, \phi_n(0)) = \phi_n(x)$$

so that

$$F(x,y) = \sum_{n=0}^{N} \{\chi_n(x)\chi_n(y) - \phi_n(x)\phi_n(y)\}. \tag{11.5.21}$$

We now prove

Theorem 11.5.1 *Let $F(x,y)$ be given by (11.5.21), and suppose that $K(x,y)$ is continuous in y, $0 \le y \le x \le \pi$, for each fixed x, $0 \le x \le \pi$. Then there exists at most one solution of equation (11.5.20).*

Proof. We need to show that the homogeneous Fredholm integral equation

$$f(y) + \int_0^x F(t,y)f(t)dt = 0, \qquad 0 \le y \le x, \tag{11.5.22}$$

has only the zero solution. Multiply (11.5.22) by $f(y)$ and integrate from 0 to x to obtain

$$\int_0^x [f(y)]^2 dy + \int_0^x \int_0^x F(t,y)f(t)f(y)dtdy = 0. \tag{11.5.23}$$

The function

$$g(y) = \begin{cases} f(y) & , \quad 0 \le y \le x, \\ 0 & , \quad x < y \le \pi, \end{cases}$$

is in $L^2(0,\pi)$, so that

$$\int_0^x [f(y)]^2 dy = \int_0^\pi [g(y)]^2 dy = \sum_{m=0}^{\infty} a_m^2,$$

where

$$a_m = \int_0^\pi g(y)\phi_m(y)dy = \int_0^x f(y)\phi_m(y)dy.$$

On the other hand

$$\int_0^x \int_0^x F(t,y)f(t)f(y)dtdy = \sum_{m=0}^{N} (b_m^2 - a_m^2) = \sum_{m=0}^{\infty} (b_m^2 - a_m^2),$$

where

$$b_m = \int_0^x f(y)\chi_m(y)dy = \int_0^\pi g(y)\chi_m(y)dy,$$

and we have used $\chi_m(y) = \phi_m(y)$ for $m > N$ to give $b_m = a_m$ for $m > N$. Equation (11.5.23) now gives

$$\sum_{n=0}^{\infty} b_m^2 = 0,$$

that is

$$b_m = 0, \qquad m = 0, 1, \ldots$$

We must show that this implies $g(y) = 0$. This is equivalent to showing that $b_m = 0$, $m = 1, 2, \ldots$ implies $a_m = 0$, $m = 0, 1, 2, \ldots$ Now

$$b_m = (g, \chi_m) = \sum_{n=0}^{\infty} c_{mn} a_n \qquad m = 0, 1, 2, \ldots \tag{11.5.24}$$

where $c_{mn} = (\chi_m, \phi_n)$. If $m > N$, then $\chi_m = \phi_m$ and $c_{mn} = \delta_{mn}$, so that $b_m = 0$ implies $a_m = 0$. Thus the sum in (11.5.24) is over $n = 0, 1, \ldots, N$, and we have the $N + 1$ equations

$$0 = \sum_{n=0}^{N} c_{mn} a_n, \qquad m = 0, 1, \ldots, N.$$

If there is a pair m', n' such that $\chi_{m'} = \phi_{n'}$ then equation (11.5.24) with $m = m'$ gives $a_{n'} = 0$ so that, when $m \neq m'$, the term with $n = n'$ may be omitted. This means that we need consider only those m, n for which $\chi_m \neq \phi_n$. Renumber these $0, 1, \ldots, N'$. We need to show that these $N' + 1$ equations have only the trivial solution, i.e., their determinant of coefficients is not zero.

The equations

$$\chi_m'' + (\mu_m - p)\chi_m = 0 = \phi_n'' + (\lambda_n - p)\phi_n$$

yield

$$(\lambda_n - \mu_m)\chi_m\phi_n = \chi_m''\phi_n - \chi_m\phi_n''$$

so that

$$(\lambda_n - \mu_m)c_{mn} = [\chi_m'\phi_n - \chi_m\phi_n']_0^\pi.$$

Since χ_m and ϕ_n satisfy the same condition at $x = 0$, we have

$$(\lambda_n - \mu_m)c_{mn} = d_m e_n \tag{11.5.25}$$

where

$$d_m = \chi_m'(\pi) + H\chi_m(\pi), \qquad e_n = \phi_n(\pi).$$

Both d_m, e_n are non-zero, and thus the determinant of coefficients is

$$\det(C) = \prod_{n=1}^{N'} d_n e_n \det(1/(\lambda_n - \mu_m))$$

and it may easily be shown (Ex. 11.5.1) that this is non-zero when, as we know, $\lambda_n - \mu_m \neq 0$ for all m, n. ■

We have proved that, under the Truncation Assumption (TA), the Gel'fand-Levitan integral equation has at most one solution. In fact it is a *degenerate* integral equation with a solution of the form

$$K(x,y) = \sum_{m=0}^{N} \{F_m(x)\chi_m(y) - G_m(x)\phi_m(y)\}. \tag{11.5.26}$$

On substituting (11.5.26) into (11.5.20) and equating multiples of $\chi_m(y), \phi_m(y)$ to zero we find

$$F_m(x) + \sum_{n=0}^{N} \{b_{mn}(x)F_n(x) - c_{mn}(x)G_n(x)\} + \chi_m(x) = 0 \tag{11.5.27}$$

$$G_m(x) + \sum_{n=0}^{N} \{c_{nm}(x)F_n(x) - d_{mn}(x)G_n(x)\} + \phi_m(x) = 0 \tag{11.5.28}$$

for $m = 0, 1, \ldots, N$, where

$$\begin{aligned} b_{mn}(x) &= b_{nm}(x) = \textstyle\int_0^x \chi_m(t)\chi_n(t)dt \\ c_{mn}(x) &= \textstyle\int_0^x \chi_m(t)\phi_n(t)dt \\ d_{mn}(x) &= d_{nm}(x) = \textstyle\int_0^x \phi_m(t)\phi_n(t)dt. \end{aligned}$$

We may verify (Ex. 11.5.2) that these equations do have a unique solution, as stated by Theorem 11.5.1.

When we first introduced the transformation operator K in Section 11.2, we showed that $K(x, y)$ must satisfy the hyperbolic differential equation (11.2.6). In this section we showed that $K(x, y)$ must satisfy the integral equation (11.5.20). In order to relate these two equations we note (Ex. 11.5.3) that $F(x, y)$ given by (11.5.21) satisfies the hyperbolic equation

$$F_{xx}(x,y) - F_{yy}(x,y) + (p(y) - p(x))F(x,y) = 0. \tag{11.5.29}$$

It is not difficult to show (Ex. 11.5.4) that if K satisfies (11.5.20) then it satisfies the differential equation (11.2.6), where $q(x)$ is given by (11.2.7)

Exercises 11.5

1. Show that if $\mu_m \neq \lambda_n$ for all $m, n = 0, 1, \ldots, N$, then the matrix $C = (c_{mn}) = (1/(\mu_m - \lambda_n))$ is non-singular.

2. Show that the equations (11.5.27), (11.5.28) have a unique solution. Hint: consider the homogeneous equations obtained by omitting $\chi_m(x), \phi_m(x)$; multiply the first by $F_m(x)$, the second by $G_m(x)$ and add the equations for $m = 0, 1, \ldots, N$.

3. Show that if $F(x,y)$ is given by (11.5.21) then it satisfies equation (11.5.29).

4. Show that if $K(x,y)$ satisfies equation (11.5.20) then

$$L(x,y) + \int_0^x L(x,t)F(t,y)dt + [K(x,0)F_x(0,y) - K_y(x,0)F(0,y)] = 0$$

where

$$L(x,y) \equiv K_{xx}(x,y) - K_{yy}(x,y) + (p(y) - q(x))K(x,y)$$

and $q(x)$ is related to $p(x)$ by equation (11.2.7). Show that the term in square brackets is zero, and hence, by Theorem 11.5.1, $L(x,y) = 0$ for $0 \leq y \leq x \leq \pi$; this is equation (11.2.6).

5. Show that the solutions of equations (11.5.26), (11.5.27) may be written

$$\begin{aligned} F_n(x) &= -\{\chi_n(x) + \textstyle\int_0^x K(x,y)\chi_n(y)dy\} = -\psi_n(x) \\ G_n(x) &= -\{\phi_n(x) + \textstyle\int_0^x K(x,y)\phi_n(y)dy\}. \end{aligned}$$

11.6 Reconstruction of the Sturm-Liouville system

First, we recapitulate what we have achieved in this chapter so far. We have shown that by starting with one $S-L$ system

$$y''(x) + (\lambda - p(x))y(x) = 0, \tag{11.6.1}$$

$$y'(0) - hy(0) = 0 = y'(\pi) + Hy(\pi), \tag{11.6.2}$$

with eigenvalues $(\lambda_n)_0^\infty$ and c.o.s. of eigenfunctions $(\phi_n)_0^\infty$ we may, by introducing the operator K, form a new $S-L$ system

$$y''(x) + (\lambda - q(x))y(x) = 0, \tag{11.6.3}$$

$$y'(0) - h'y(0) = 0 = y'(\pi) + H'y(\pi), \tag{11.6.4}$$

with eigenvalues $(\mu_n)_0^\infty$ and c.o.s. of eigenfunctions $(\psi_n)_0^\infty$ given by equation (11.2.5). In order to find the new system we need the $(\mu_n)_0^\infty$ and the end values $(\psi_n(0))_0^\infty$ of the eigenfunctions $\psi_n(x)$ which are yet to be found.

We can find these, as in Section 10.8, from two spectra of equation (11.6.3), $(\mu_n)_0^\infty$ corresponding to the end conditions (11.6.4), and $(\nu_n)_0^\infty$ corresponding to

$$y'(0) - h_1'y(0) = 0 = y'(\pi) + H'y(\pi).$$

Changing equation (10.8.12) to the numbering system S we find that $(\nu_n)_0^\infty$ are the roots of

$$1 = (h_1' - h')\sum_{n=0}^{\infty} \frac{\psi_n^2(0)}{\lambda - \mu_n}. \tag{11.6.5}$$

The Truncation Assumption allows us to write this

$$1 = (h_1' - h')\left\{\sum_{n=0}^{N}\frac{\psi_n^2(0)}{\lambda - \mu_n} + \sum_{n=N+1}^{\infty}\frac{\phi_n^2(0)}{\lambda - \lambda_n}\right\}. \tag{11.6.6}$$

This gives $N+1$ equations

$$1 = (h_1' - h')\left\{\sum_{n=0}^{N}\frac{\psi_n^2(0)}{\nu_m - \mu_n} + \sum_{n=N+1}^{\infty}\frac{\phi_n^2(0)}{\nu_m - \lambda_n}\right\}, \qquad m = 0, 1, \ldots, N, \tag{11.6.7}$$

for the $N+1$ quantities $\{\psi_n(0)\}_0^N$. As in Ex. 11.5.1, the determinant of coefficients is not zero. To check that the $\psi_n^2(0)$ are indeed positive, we write (11.6.6) as

$$1 - (h_1' - h')\left\{\sum_{n=0}^{N}\frac{\psi_n^2(0)}{\lambda - \mu_n} - f(\lambda)\right\} = \prod_{m=0}^{N}\left(\frac{\lambda - \nu_m}{\lambda - \mu_m}\right) g(\lambda) \tag{11.6.8}$$

where for definiteness we take $h_1' > h'$. The functions $f(\lambda), g(\lambda)$ are positive for $0 < \lambda < \nu_m$, and the μ_n, ν_n interlace according to

$$\mu_0 < \nu_0 < \mu_1 < \cdots \tag{11.6.9}$$

Multiplying (11.6.8) throughout by $(\lambda - \mu_n)$ and then putting $\lambda = \mu_n$ we find

$$(h_1' - h')\psi_n^2(0) = (\nu_n - \mu_n)\prod_{m=0}^{N}{}'\left(\frac{\mu_n - \nu_m}{\mu_n - \mu_m}\right) g(\mu_n) \tag{11.6.10}$$

so that the interlacing gives $\psi_n^2(0) > 0$.

Taking an arbitrary value of h_1' has the disadvantage that the $\psi_n^2(0)$ depend on h_1' and h'. If $h_1' = \infty$, then equation (11.6.6) takes the simpler form

$$\sum_{n=0}^{N}\frac{\psi_n^2(0)}{\lambda - \mu_n} + \sum_{n=N+1}^{\infty}\frac{\phi_n^2(0)}{\lambda - \lambda_n} = 0. \tag{11.6.11}$$

This yields

$$\sum_{n=0}^{N}\frac{\psi_n^2(0)}{\nu_m - \mu_n} = f(\nu_m), \qquad m = 0, 1, \ldots, N \tag{11.6.12}$$

for $\psi_n^2(0)$, $n = 0, 1, \ldots N$.

We are now ready to proceed to the reconstruction. We need to find $q(x), h', H'$ such that the first $N+1$ eigenvalues of (11.6.3), (11.6.4) are the specified $(\mu_n)_0^N$, and the first $N+1$ end values of the normalised eigenfunctions are $(\psi_n(0))_0^N$. We take the following steps:

Step 1: Choose a base system (11.6.1), (11.6.2), and find $\{\phi_n(x)\}_0^N$, $\{\chi_n(x)\}_0^N$ given by (11.5.2), (11.5.4) respectively. Under the Truncation Assumption, the values of $\mu_{N+1}, \mu_{N+2}, \ldots$, which are not part of the data, are taken to be $\lambda_{N+1}, \lambda_{N+2}, \ldots$ respectively. We must therefore choose the base system so that $\mu_N < \lambda_{N+1}$. The simplest choice for the base system would be to take $p(x) = 0$, and h, H each to be 0 or ∞. If for example $h = 0$, $H = \infty$ then (11.6.1), (11.6.2) reduce to

$$y'' + \lambda y = 0 \tag{11.6.13}$$

$$y'(0) = 0 = y(\pi) \tag{11.6.14}$$

and

$$\phi_n(x) = \sqrt{\frac{2}{\pi}} \cos\left(n + \frac{1}{2}\right) x, \qquad \lambda_n = \left(n + \frac{1}{2}\right)^2 \tag{11.6.15}$$

$$\chi_n = \psi_n(0) \cos \omega_n x, \qquad \mu_n = \omega_n^2. \tag{11.6.16}$$

This choice for a base system would therefore be appropriate provided that $\mu_N < \left(N + \frac{3}{2}\right)^2$, i.e., $\omega_N < \left(N + \frac{3}{2}\right)$. Since the μ_n are to be the eigenvalues of some $S - L$ system, they must have the asymptotic form given in Section 10.9. Depending on the end conditions, they must therefore have the form (10.9.20), (10.9.39) or (10.9.41); in any of these cases, $\omega_N < \left(N + \frac{3}{2}\right)$ for large enough N. If of course ω_n had the form (10.9.41), then it would be more appropriate to take $h = \infty$, $H = \infty$, so that $h' = \infty$.

Step 2: Form $F(x, y)$ given by (11.5.21) and solve equations (11.5.27), (11.5.28) for $\{F_n(x), G_n(x)\}_0^N$.

Step 3: Form $K(x, y)$ from equation (11.5.26).

Step 4: Form $q(x)$ from equation (11.2.7).

Step 5: Find h' from equation (11.2.11).

Step 6: Find H'.

For this final step we proceed as follows. Since $d_{mn}(\pi) = \delta_{mn}$, equation (11.5.28) gives

$$\sum_{n=0}^{N} c_{nm} F_n(\pi) + \phi_m(\pi) = 0$$

where, as in (11.5.23), $c_{nm} = c_{nm}(\pi)$. Differentiating (11.5.28) w.r.t. x and putting $x = \pi$, we find

$$\sum_{n=0}^{N} c_{nm} F_n'(\pi) + \phi_m(\pi) K(\pi, \pi) + \phi_m'(\pi) = 0.$$

Now use Ex. 11.5.5, which shows that $F_n(\pi) = -\psi_n(\pi)$, and the fact that $\phi'_m(\pi) + H\phi_m(\pi) = 0$, to give

$$\sum_{n=0}^{N} c_{nm}\{\psi'_m(\pi) + (H - K(\pi,\pi))\psi_n(\pi)\} = 0.$$

But $\det(C) \neq 0$, so that

$$\psi'_m(\pi) + (H - K(\pi,\pi))\psi_n(\pi) = 0. \tag{11.6.17}$$

This means

$$H' = H - K(\pi,\pi). \tag{11.6.18}$$

Apart from the introduction of the Truncation Assumption, the analysis described in this chapter so far is the classical Gel'fand-Levitan inversion of the Sturm-Liouville equation. While the method has great theoretical value, it is impractical; the stumbling block is Step 2, the solution of the equations for $F_n(x), G_n(x)$, and the subsequent steps 3,4 which give $q(x)$ by differentiating $K(x,y)$.

In Section 11.9 we describe other methods that use the partial differential equation satisfied by $K(x,x)$.

Exercises 11.6

1. Show that equation (11.2.12), (11.6.18) imply

$$c' \equiv \frac{1}{\pi}\left(h' + H' + \frac{1}{2}\int_0^\pi q(t)dt\right) = c \equiv \frac{1}{\pi}\left(h + H + \frac{1}{2}\int_0^\pi p(t)dt\right).$$

Since we took $\mu_n = \lambda_n$ for $n = N+1, \ldots$, this equation must hold; see the asymptotic form (10.9.22).

11.7 An inverse problem for the vibrating rod

The inversion procedure that we have described so far has been for the Sturm-Liouville equation (11.2.1). As we have already pointed out, this is not the basic equation for vibrating systems. In this section, at the risk of repetition, we show how the ideas behind the $S-L$ inversion may be adapted to the rod equation (11.1.3).

We start with a base problem

$$(A(x)u'(x))' + \lambda A(x)u(x) = 0, \qquad 0 \leq x \leq \pi \tag{11.7.1}$$

write $A(x) = a^2(x)$, and scale the independent variable x so that $0 \leq x \leq \pi$. The eigenfunctions $u_n(x)$ of (11.7.1) subject to some end conditions yet to be described, are orthonormal with weight function $a^2(x)$, i.e.,

$$\int_0^\pi a^2(x)u_m(x)u_n(x)dx = \delta_{mn}$$

so that the functions

$$\phi_n(x) = a(x)u_n(x), \qquad n = 0, 1, \ldots \tag{11.7.2}$$

form a c.o.s. Provided that $a(x) \in C^2(0, \pi)$, $\phi(x) = a(x)u(x)$ satisfies

$$\phi'' + (\lambda - p)\phi = 0, \tag{11.7.3}$$

where

$$p(x) = a''(x)/a(x). \tag{11.7.4}$$

Suppose that the end condition at $x = 0$ is

$$\phi'(0) - h\phi(0) = 0, \tag{11.7.5}$$

then the corresponding end condition for $u(x)$ is

$$a(0)u'(0) + (a'(0) - ha(0))u(0) = 0.$$

Without loss of generality we may choose the base system so that

$$a(0) = 1, \qquad a'(0) - ha(0) = 0. \tag{11.7.6}$$

This means $a(x)$ is the solution of (11.7.3) for $\lambda = 0$, that satisfies the end condition (11.7.5) and $\phi(0) = 1$, and the base rod is free at $x = 0$.

The rod that is to be constructed is governed by

$$(B(x)v'(x))' + \lambda B(x)v(x) = 0, \qquad 0 \le x \le \pi. \tag{11.7.7}$$

Write $B(x) = b^2(x)$, $\psi(x) = b(x)v(x)$, then

$$\psi'' + (\lambda - q)\psi = 0, \tag{11.7.8}$$

where

$$q(x) = b''(x)/b(x). \tag{11.7.9}$$

We now use the operator K to link ψ to ϕ:

$$\psi(x) = \phi(x) + \int_0^x K(x,t)\phi(t)dt, \tag{11.7.10}$$

i.e.,

$$b(x)v(x) = a(x)u(x) + \int_0^x K(x,t)a(t)u(t)dt. \tag{11.7.11}$$

As we know from Section 11.2, this operator transforms the solution of (11.7.3) satisfying $\phi(0) = 1$, and (11.7.5), into a solution of (11.7.8) satisfying $\psi(0) = 1$ and

$$\psi'(0) - (h + K(0,0))\psi(0) = 0. \tag{11.7.12}$$

This last condition is equivalent to

$$b(0)v'(0) + \{b'(0) - (h + K(0,0)b(0)\}v(0) = 0. \tag{11.7.13}$$

If we choose the new system so that

$$b(0) = 1, \qquad b'(0) - (h + K(0,0)b(0) = 0, \tag{11.7.14}$$

then $v'(0) = 0$: the new rod is free at $x = 0$. In this case $b(x)$ is the solution of (11.7.8) for $\lambda = 0$ satisfying the conditions (11.7.14). Thus $b(x)$ is related to $a(x)$ by equation (11.7.10):

$$b(x) = a(x) + \int_0^x K(x,t)a(t)dt. \tag{11.7.15}$$

In particular, if we choose $h = 0$, $a(x) = 1$, then

$$b(x) = 1 + \int_0^x K(x,t)dt. \tag{11.7.16}$$

The remainder of the analysis is as before: $K(x,y)$ satisfies

$$K(x,y) + \int_0^x K(x,t)F(t,y)dt + F(x,y) = 0, \qquad 0 \le y \le x \tag{11.7.17}$$

where

$$F(x,y) = a(x)a(y)\sum_{n=0}^{N}(w_n(x)w_n(y) - u_n(x)u_n(y)). \tag{11.7.18}$$

Here $\chi_n(x) = a(x)w_n(x)$ is the solution of (11.7.3) with $\lambda = \mu_n$ satisfying $\chi_n(0) = \psi_n(0)$, $\chi'_n(0) = h\chi_n(0)$. This means that $w_n(0) = v_n(0)$, $w'_n(0) = 0$. Here $(\mu_n)_0^N$ are the eigenvalues of the new system and $v_n(0)$ is the end value of the corresponding normalised eigenfunction $v_n(x)$, and $u_n(x)$ is the normalised eigenfunction of (11.7.1).

Again we must choose the base system so that $\mu_N < \lambda_{N+1}$. If we make the choice $a(x) = 1, h = 0, H = \infty$ then this means $\sqrt{\mu_N} = \omega_N < (N + \frac{3}{2})$.

The solution of equation (11.7.17) has the form

$$K(x,y) = a(x)a(y)\sum_{n=0}^{N}\{F_n(x)w_n(y) - G_n(x)u_n(y)\}$$

where $F_n(x), G_n(x)$ satisfy

$$F_m(x) + \sum_{n=0}^{N}\{b_{mn}(x)F_n(x) - c_{mn}(x)G_n(x)\} + w_m(x) = 0, \tag{11.7.19}$$

$$G_m(x) + \sum_{n=0}^{N}\{c_{nm}(x)F_n(x) - d_{mn}(x)G_n(x)\} + u_m(x) = 0, \tag{11.7.20}$$

and

$$\begin{aligned} b_{mn}(x) &= \int_0^x a^2(t)w_m(t)w_n(t)dt, \quad c_{mn}(x) = \int_0^x a^2(t)w_m(t)u_n(t)dt \\ d_{mn}(x) &= \int_0^x a^2(t)u_m(t)u_n(t)dt. \end{aligned}$$

We note that $d_{mn}(\pi) = \delta_{mn}$.

It is important to note that the $b(x)$ generated by the construction procedure will always be positive. We show this by supposing that $b(x_0) = 0$ for some $x_0 \in [0, \pi]$ and arriving at a contradiction. By analogy with Ex. 11.5.5, we have

$$a(x)F_n(x) = -\{\chi_n(x) + \int_0^x K(x,y)\chi_n(y)\} \qquad (11.7.21)$$

$$a(x)G_n(x) = -\{\phi_n(x) + \int_0^x K(x,y)\phi_n(y)\} \qquad (11.7.22)$$

where $\chi_n(x) = \phi(x, \mu_n, c_1)$, $\phi_n(x) = \phi(x, \lambda_n, c_2)$, and $\phi(x, \lambda, c)$ denotes the solution of (11.7.3) for $\phi(0) = c$, satisfying (11.7.5). But if χ_n, ϕ_n are solutions of (11.7.3), then $a(x)F_n(x), a(x)G_n(x)$ given by (11.7.21), (11.7.22), are solutions of (11.7.8). Thus

$$\begin{aligned} a(x)F_n(x) &= -b(x)v(x, \mu_n, c_1) \\ a(x)G_n(x) &= -b(x)v(x, \lambda_n, c_2) \end{aligned}$$

where $v(x, \lambda, c)$ denotes the solution of (11.7.7) satisfying $v(0) = c$, $v'(0) = 0$. Note that $v(x, \mu_n, c_1)$ is an unnormalised eigenfunction of the new system. This means that if $b(x_0) = 0$, then $F_n(x_0) = 0 = G_n(x_0)$ for $n = 0, 1, \ldots, N$, and hence, from equations (11.7.19), (11.7.20), $w_n(x_0) = 0 = u_n(x_0)$, for $n = 0, 1, \ldots, N$. But $u_n(x)$ is the nth eigenfunction of the base system, and when $n = 0$, $u_0(x)$ has no zero in $[0, \pi]$, except possibly at $x = \pi$ when $H = \infty$. Therefore, the only possibility is $x_0 = \pi$, $H = \infty$, and then $w_n(\pi) = 0$, $n = 0, 1, \ldots, N$ also. This means that $(\mu_n)_0^N$ are eigenvalues of the base problem and $\mu_n = \lambda_n$, $n = 0, 1, \ldots, N$. This contradiction implies $b(x_0) \neq 0$. Since $b(0) = 1$ and $b(x)$ is continuous, we must have $b(x) > 0$ for $x \in [0, \pi]$.

To conclude this section we return to the end conditions. The base problem, in the $S - L$ form (11.7.3), has end conditions

$$\phi'(0) - h\phi(0) = 0 = \phi'(\pi) + H\phi(\pi).$$

In terms of u, these are

$$u'(0) = 0 = a(\pi)u'(\pi) + (a'(\pi) + Ha(\pi))u(\pi)$$

where we have taken $a'(0) = ha(0)$. The end condition for the $S - L$ form of the new problem are

$$\psi'(0) - h'\psi(0) = 0 = \psi'(\pi) + H'\psi(\pi).$$

In terms of v, these are

$$v'(0) = 0 = b(\pi)v'(\pi) + (b'(\pi) + (H - K(\pi,\pi)b(\pi))v(\pi),$$

where we have used (11.6.18) to give $H' = H - K(\pi, \pi)$.

We note that while the choices (11.7.6), (11.7.14) for $a(x), b(x)$, make the analysis straightforward, it is not necessary to make these choices. Again, if we know the eigenvalues of the new rod for the end $x = 0$ free and fixed, then we can find $(v_n(0))_0^N$ as in Section 11.6. For examples of reconstruction, see Gladwell and Dods (1987) [111]. See also Andersson (1988a) [6], (1988b) [7] for a detailed study of the inverse problem for equation (11.7.1), see Knobel and Lowe (1993) [195]. For the case in which $A(x)$ is rough, see Coleman (1989) [61], Coleman and McLaughlin (1993a) [62], (1993b) [63].

11.8 An inverse problem for the taut string

In Section 10.1, we showed how the three forms of the Sturm-Liouville equation, (10.1.1), (10.1.3) and (10.1.11) were related. In approaching the taut string, it is somewhat easier to start from (10.1.3), the rod, rather than from the standard form (10.1.11). We recall part of the analysis in Section 10.1, and make a few changes in the way we normalise variables.

Suppose $u(x)$ satisfies equation (11.7.1), i.e.,

$$(A(x)u'(x))' + \lambda A(x)u(x) = 0 \tag{11.8.1}$$

and the end conditions

$$u'(0) - hu(0) = 0 = u'(\pi) + Hu(\pi).$$

Scale $A(x)$ so that

$$\int_0^\pi \frac{dt}{A(t)} = \pi$$

and introduce a new variable ξ by the equation

$$\xi = \int_0^x \frac{dt}{A(t)}$$

so that $0 \le \xi \le \pi$, and $\xi'(x) = 1/A(x)$. Put

$$u(x) = y(\xi), \quad A(x) = \rho(\xi)$$

then $A(x)u'(x) = \dot{y}(\xi)$, and equation (11.8.1) becomes

$$\ddot{y}(\xi) + \lambda\rho^2(\xi)y(\xi) = 0.$$

The end conditions become

$$\dot{y}(0) - hA(0)y(0) = 0 = \dot{y}(\pi) + HA(\pi)y(\pi).$$

We note that the new spring constants are scaled versions $hA(0), HA(\pi)$ of the old, but that the end conditions

$$u'(0) = 0 = u(\pi), \quad u(0) = 0 = u(\pi) \tag{11.8.2}$$

remain invariant:

$$\dot{y}(0) = 0 = y(\pi), \quad y(0) = 0 = y(\pi). \tag{11.8.3}$$

This means that, under either of these two sets of end conditions, the string has the same eigenvalues as the rod, and in particular the asymptotic forms of the eigenvalues are the same.

If therefore we are given two sequences of eigenvalues $(\mu_n)_0^\infty, (\nu_n)_0^\infty$ which purport to be the eigenvalues of a taut string under the two sets of end conditions (11.8.2), we must first scale them, (effectively to find the length, L, of the string to which they correspond) so that they correspond to a string of length π. They will then have the asymptotic forms

$$\mu_n = [(n + \frac{1}{2})\pi]^2 + \alpha_n, \quad \nu_n = [(n+1)\pi]^2 + \beta_n.$$

Given $(\mu_n)_0^\infty, (\nu_n)_0^\infty$, we then find the end values $y_n(0)$ of the normalised eigenfunctions from the fact that

$$\sum_{n=0}^{\infty} \frac{y_n^2(0)}{\lambda - \mu_n} = 0$$

has roots $(\nu_n)_0^\infty$. We use this in truncated form, as in Section 11.6 to find $(y_n(0))_0^N$. We note that

$$\int_0^\pi \rho^2(\xi) y_n^2(\xi) d\xi = \int_0^\pi A(x) w_n^2(x) dx = 1$$

and $y_n(0) = w_n(0)$. This means that we have the data needed to find the new rod $B(x)$ as in (11.7.7).

We now reverse the analysis given at the beginning of this section. Thus we scale $B(x)$ so that

$$\int_0^\pi \frac{dt}{B(t)} = \pi$$

and then *introduce* a new variable ξ by

$$\xi = \int_0^\pi \frac{dt}{B(t)}, \quad 0 \le \xi \le \pi.$$

The new mass density of the string is

$$\rho(\xi) = B(x).$$

11.9 Some non-classical methods

In this section we explain in general terms the theory behind some recent approximate methods for inverting the Sturm-Liouville equation. The theory is largely due to Rundell and Sacks (1992a) [292], Rundell and Sacks (1992b) [293]; see also Lowe, Pilant and Rundell (1992) [216] and Rundell (1997) [294]. The distinguishing feature of the methods is that they rely on the hyperbolic equation satisfied by $K(x, y)$, rather than on the Gel'fand Levitan integral equation.

We start by recalling analysis from Section 11.2 onwards. The base problem is

$$y'' + (\lambda - p)y = 0, \qquad 0 \le x \le \pi, \tag{11.9.1}$$

$$y'(0) - hy(0) = 0 = y'(\pi) + Hy(\pi). \tag{11.9.2}$$

The eigenfunctions $(\phi_n)_0^\infty$ of this problem form a c.o.s. This means that if

$$f(x) = \sum_{n=0}^{N} a_n \phi_n(x)$$

then

$$a_n = (f, \phi_n).$$

Suppose that $(\mu_n)_0^\infty$ is a new spectrum and, in the notation of (11.5.4),

$$\chi_n(x) = \phi(x, \mu_n, 1).$$

If we expand $f(x)$ in terms of these functions:

$$f(x) = \sum_{n=0}^{N} b_n \chi_n(x),$$

then we can find the b_n from

$$a_m = (f, \phi_m) = \sum_{n=0}^{N} b_n(\chi_n, \phi_m) = \sum_{n=0}^{N} c_{nm} b_n, \qquad m = 0, 1, \ldots, N. \tag{11.9.3}$$

We showed in equation (11.5.25) that

$$(\lambda_m - \mu_n)c_{nm} = (\chi_n'(\pi) + H\chi_n(\pi))\phi_m(\pi).$$

The end value, $\phi_m(\pi)$, is not zero; we are assuming that H is finite. As in Section 11.5, if $\chi_{n'}'(\pi) + H\chi_{n'}(\pi) = 0$ for some n', then $\mu_{n'}$ is an eigenvalue of the base system with $\chi_{n'}(x)$ being a (not necessarily normalised) eigenfunction; i.e., $\chi_{n'}(x) = c\phi_{m'}(x)$. In that case, equation (11.9.3) with $m = m'$ yields $a_{m'} = b_{n'}$ and there are just $N-1$ equations for the remaining b_n. In any case, we can solve equation (11.9.3) for $b_0, b_1, \ldots, b_N$. This is the first result we will use.

In the classical method we suppose first that we have two spectra of a $S-L$ equation with (unknown) potential $q(x)$ corresponding to two sets of end conditions

$$\begin{aligned}(\mu_n)_0^\infty &\text{ for } y'(0)-h'y(0)=0=y'(\pi)+H'y(\pi)\\ (\nu_n)_0^\infty &\text{ for } y'(0)-h_1'y(0)=0=y'(\pi)+H'y(\pi).\end{aligned} \tag{11.9.4}$$

Note that the conditions at $x=\pi$ are the same, while those at $x=0$ are different. We then used equation (11.6.6), or preferably (11.6.11), to find the end values $(\psi_n(0))_0^\infty$ corresponding to (11.9.4). (Of course we introduced the Truncation Assumption, so that we had to find only $(\psi_n(0))_0^N$.) We then introduced the operator K and found $K(x,y)$ so that the normalised eigenfunctions $(\psi_n(x))_0^\infty$ corresponding to (11.9.4) were given by

$$\begin{aligned}\chi_n(x) &= \phi(x,\mu_n,\psi_n(0))\\ \psi_n(x) &= \chi_n(x)+\int_0^x K(x,y)\chi_n(y)dy.\end{aligned}$$

The particular $K(x,y)$ that transforms $(\chi_n(x))_0^\infty$ into a c.o.s. $(\psi_n(x))_0^\infty$ is the solution of the Gel'fand-Levitan integral equation (11.5.20). The potential $q(x)$ is given by

$$q(x)=p(x)+\frac{2dK(x,x)}{dx}$$

and the values of h',H' are given by (11.2.11) and (11.6.18):

$$h'=h+K(0,0),\quad H'=H-K(\pi,\pi).$$

Rundell and Sacks proceed differently. They suppose that we are given two spectra $(\mu_n)_0^N,(\nu_n)_0^N$ corresponding to a $S-L$ equation

$$y''+(\lambda-q)y=0 \tag{11.9.5}$$

under two sets of end conditions that differ at $x=\pi$ (not at 0 as in the classical approach):

$$(\mu_n)_0^N \text{ for } y'(0)-h'y(0)=0=y'(\pi)+H_1'y(\pi), \tag{11.9.6}$$

$$(\nu_n)_0^N \text{ for } y'(0)-h'y(0)=0=y'(\pi)+H_2'y(\pi). \tag{11.9.7}$$

Now we find $K(x,y)$ so that

$$\begin{aligned}\chi_n(x) &= \phi(x,\mu_n,1),\\ \psi_n(x) &= \chi_n(x)+\int_0^x K(x,y)\chi_n(y)dy\end{aligned}$$

give (unnormalised) eigenfunctions of (11.9.5) corresponding to the end conditions (11.9.6), while

$$\begin{aligned}\theta_n(x) &= \phi(x,\nu_n,1),\\ \tau_n(x) &= \theta_n(x)+\int_0^x K(x,y)\theta_n(y)dy\end{aligned}$$

gives (unnormalised) eigenfunctions of (11.9.5) corresponding to the end conditions (11.9.7). Note that $K(x,y)$ will be given by the theory of Section 11.2, not by that of Section 11.5. This means that $K(x,y)$ will satisfy the hyperbolic equation (11.2.6) and the boundary condition (11.2.10).

The basic theory of Section 11.2 states that the transformation operator transforms a solution of the base equation (11.9.1) satisfying (11.9.2a) into a solution of (11.9.5) and (11.9.6a). This means that $(\psi_n(x))_0^N$ will be eigenfunctions of (11.9.5) corresponding to the end conditions (11.9.6) if

$$\psi_n'(\pi) + H_1'\psi_n(\pi) = 0, \qquad n = 0, 1, \ldots, N. \tag{11.9.8}$$

Similarly $(\tau_n(x))_0^N$ will be eigenfunctions of (11.9.5) corresponding to the end conditions (11.9.7) if

$$\tau_n'(\pi) + H_2'\tau_n(\pi) = 0, \qquad n = 0, 1, \ldots, N. \tag{11.9.9}$$

The problem is therefore to find a solution of the hyperbolic equation (11.2.6) that satisfies equations (11.9.8), (11.9.9). Before considering how to do this we make some preliminary simplifications.

If the given sequences $(\mu_n)_0^\infty, (\nu_n)_0^\infty$ are indeed the spectra of some $S-L$ equation (11.9.5) corresponding to (11.9.6), (11.9.7) respectively, then they must have one of the asymptotic forms listed in Section 10.9: (10.9.47) if h' is finite; Exercise 10.9.1 if $h' = \infty$. Assume for the sake of argument that $h' = \infty$ then, by examining the sequences we can recover $\bar{q}$ from either of the two equations

$$\lim_{n\to\infty}\left\{(\mu_n - \bar{q})^{\frac{1}{2}} - (n + \frac{1}{2})\right\} = 0 = \lim_{n\to\infty}\left\{(\nu_n - \bar{q})^{\frac{1}{2}} - (n + \frac{1}{2})\right\}. \tag{11.9.10}$$

With this $\bar{q}$, form $q^*(x) = q(x) - \bar{q}$ as in (10.9.44). The starred system has eigenvalues $\mu_n^* = \mu_n - \bar{q}$, $\nu_n^* = \nu_n - \bar{q}$ corresponding to (11.9.6), (11.9.7) respectively. We note that even if the equation (11.9.5) was derived from a physical system with positive eigenvalues, and the limits show that μ_n, ν_n will eventually exceed $\bar{q}$, there is no guarantee that *all* the starred quantitites μ_n^*, ν_n^* will be positive.

We now consider the reduced, i.e., starred, system and drop the asterisks.

We showed in Section 11.2 that if $h' = \infty$, then we must take a base system with $h = \infty$. Then $K(x,0) = 0$, $0 \le x \le \pi$ so that $K(x,y)$ is continued as an odd function of y into the lower triangle in Figure 11.3.1. For simplicity we take $p(x) = 0$, so that equation (11.2.14) gives $K(\pi,\pi) = 0$. We choose $H = 0$ so that the base system is

$$\begin{aligned} \phi'' + \lambda\phi &= 0, \qquad 0 \le x \le \pi, \\ \phi(0) &= 0 = \phi'(\pi), \end{aligned} \tag{11.9.11}$$

with eigenvalues $\lambda_n = \left(n + \frac{1}{2}\right)^2, n = 0, 1, \ldots$, and eigenfunctions

$$\phi_n(x) = \sqrt{\frac{2}{\pi}}\sin\left(n + \frac{1}{2}\right)x.$$

We define $\phi(x,\lambda)$ as the solution of (11.9.11) satisfying

$$\phi(0,\lambda)=0, \qquad \phi'(0,\lambda)=1.$$

This means that if $\omega=|\lambda|^{\frac{1}{2}}$, then

$$\phi(x,\lambda)=\frac{\sin\omega x}{\omega},\ \text{if}\ \lambda>0;\quad \frac{\sinh\omega x}{\omega},\ \text{if}\ \lambda<0.$$

We now define

$$\begin{aligned}\chi_n(x) &= \phi(x,\mu_n), \qquad n=0,1,\ldots,N,\\ \theta_n(x) &= \phi(x,\nu_n), \qquad n=0,1,\ldots,N.\end{aligned}$$

and construct $\psi_n(x),\tau_n(x)$, the eigenfunctions of (11.6.3) corresponding to (11.9.6), (11.9.7) respectively, by using the transformation operator $K(x,y)$:

$$\begin{aligned}\psi_n(x) &= \chi_n(x)+\int_0^x K(x,t)\chi_n(t)dt \\ \tau_n(x) &= \theta_n(x)+\int_0^x K(x,t)\theta_n(t)dt.\end{aligned} \tag{11.9.12}$$

Now consider the equation (11.9.8). Equation (11.9.12) gives

$$\psi_n(\pi)=\chi_n(\pi)+\int_0^\pi K(\pi,y)\chi_n(y)dy$$

and

$$\psi_n'(\pi)=\chi_n'(\pi)+\int_0^\pi K_x(\pi,y)\chi_n(y)dy+K(\pi,\pi)\chi_n(\pi).$$

But

$$0=\psi_n'(\pi)+H_1'\psi_n(\pi)=\chi_n'(\pi)+H_1'\chi_n(\pi)+\int_0^\pi\{K_x(\pi,y)+H_1'K(\pi,y)\}\chi_n(y)dy. \tag{11.9.13}$$

This equation gives the inner products of the function $f_1(y)=K_x(\pi,y)+H_1'K(\pi,y)$ with respect to the $\chi_n(y)$. But knowing these we can use the analysis leading to (11.9.3) to find the inner products with respect to $(\sin nt)_1^{N+1}$, because $\sin nt$ is the solution of the base problem under the Dirichlet conditions $\phi(0)=0=\phi(\pi)$. Proceeding in exactly the same way for the second spectrum $(\nu_n)_0^N$, we can find the inner products of the function $f_2(y)=K_x(\pi,y)+H_2'K(\pi,y)$ with respect to $(\theta_n(y))_0^N$, and hence to $(\sin nt)_1^{N+1}$. By taking multiples of $f_1(y)$ and $f_2(y)$ we find

$$K(\pi,y)=\sum_{n=1}^{N+1}a_n\sin ny, \qquad K_x(\pi,y)=\sum_{n=1}^{N+1}b_n\sin ny. \tag{11.9.14}$$

Note that these expansions give $K(\pi,0)=0=K(\pi,\pi)$ and $K_x(\pi,0)=0$, as required (recall that $K(x,0)\equiv 0$), but they make $K_x(\pi,\pi)=0$ which is an unnecessary restriction. We recall that when $h = \infty$, $K(x,y)$ is an odd function of y, and the expansions (11.9.14) *are* odd in y.

Now return first to equation (11.2.7) which states

$$q(x) = \frac{2dK(x,x)}{dx}, \tag{11.9.15}$$

and then to equation (11.3.15) which expresses $K(x,y)$ in terms of K and K_x on the line $x = \pi$, and an integral over the triangle ABP of Figure 11.3.3. When $x = y$, the triangle is as shown in Figure 11.9.1, so that equation (11.3.15) gives

$$\begin{aligned} 2K(x,x) &= K(\pi, 2x-\pi) + K(\pi,\pi) - \int_{2x-\pi}^{\pi} K_x(\pi,t)dt \\ &\quad + \int_x^{\pi} \left\{ \int_{2x-s}^{s} q(s)K(s,t)dt \right\} ds. \end{aligned}$$

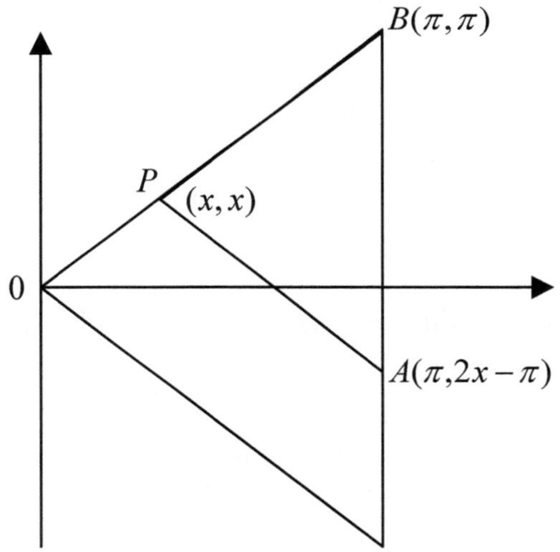

Figure 11.9.1 - The triangle ABP when $x = y$.

We have no space to refer to the many other numerical methods, see for example Brown, Samko, Knowles and Marletta (2003) [41]. On differentiating w.r.t. x we find

$$\frac{2dK(x,x)}{dx} = 2K_y(\pi, 2x-\pi) + 2K_x(\pi, 2x-\pi) - 2\int_x^{\pi} q(s)K(s, 2x-s)ds.$$

This equation provides the basis for an iterative solution to the problem. Putting

$$G(x) = 2K_y(\pi, 2x-\pi) + 2K_x(\pi, 2x-\pi),$$

we use the equation in the form

$$q_{m+1}(x) = G(x) - 2\int_x^\pi q_m(s)K(s, 2x - s)ds \tag{11.9.16}$$

to obtain a new value of $q(x)$ from an existing one. Treating the R.H.S. of (11.9.16) as the result of operating on q_m by an operator T, we have

$$q_{m+1} = Tq_m.$$

The potential $q(x)$ is thus sought as a fixed point of the mapping Tq.

The actual numerical implementation is not our primary concern; for that see, say, Rundell (1997) [294]. In principle we can proceed as follows:

Step 1: Start from some initial approximation, for example $q_0(x) = G(x)$. Put $m = 0$.

Step 2: Solve the Cauchy problem

$$K_{xx} - K_{yy} - q_m K = 0$$

with K, K_x given by (11.9.14) on $x = \pi$. This can be done using standard numerical procedures.

Step 3: Form $q_{m+1}(x)$ from equation (11.9.16). Put $m = m + 1$ and return to step 2 until convergence is achieved.

11.10 Some other uniqueness theorems

The fundamental uniqueness theorem in Section 11.4 showed that the potential $q(x)$ and the end constants h, H were uniquely determined by two spectra. The crucial step in the analysis was that the completeness of the eigenfunctions meant that equation (11.4.6) implied $K(\pi, y) = 0 = K_x(\pi, y)$ for $0 \le y \le \pi$. But this is the Cauchy data for the hyperbolic equation (11.2.6); since the data is zero, $K(x, y) = 0$ for $0 \le y \le x \le \pi$, and $p(x) = q(x)$, $h_1 = h_2$ and $H_1 = H_2$.

The fundamental uniqueness theorem uses two spectra corresponding to two different end constants at one end. We now show that if just one spectrum is known, then there are various other sets of auxiliary data which will lead to a unique system. As in Section 11.4, we phrase the uniqueness theorem in terms of general end conditions, i.e., end constants that are neither zero nor infinite. The special cases in which one or both are zero or infinite may be covered by straightforward modifications of the argument.

Theorem 11.10.1 *Suppose that there were two potentials $p(x), q(x) \in C[0, \pi]$ with the following properties:*

i) $y'' + (\lambda - p)y = 0, \quad y'(0) - h_1 y(0) = 0 = y'(\pi) + H_1 y(\pi)$ *has the spectrum* $(\lambda_n)_0^\infty$ *and eigenfunctions* $(\phi_n(y))_0^\infty$*;*

ii) $y'' + (\lambda - q)y = 0, \quad y'(0) - h_2 y(0) = 0 = y'(\pi) + H_2 y(\pi)$ *has the same spectrum* $(\lambda_n)_0^\infty$ *and eigenfunctions* $(\psi_n(y))_0^\infty$*;*
and ***one*** *of the following properties holds:*

iii) $\frac{\psi_n(\pi)}{\psi_n(0)} = \frac{\phi_n(\pi)}{\phi_n(0)} \qquad n = 0, 1, 2, \ldots$

iv) $\frac{\psi_n'(\pi)}{\psi_n'(0)} = \frac{\phi_n'(\pi)}{\phi_n'(0)} \qquad n = 0, 1, 2, \ldots$

v) $\frac{\psi_n^2(\tau)}{\int_0^\pi \psi_n^2(x)dx} = \frac{\phi_n^2(\tau)}{\int_0^\pi \phi_n^2(x)dx} \qquad \tau = 0 \text{ or } \pi, \quad n = 0, 1, 2, \ldots$

vi) $\frac{\psi_n'^2(\tau)}{\int_0^\pi \psi_n^2(x)dx} = \frac{\phi_n'^2(\tau)}{\int_0^\pi \phi_n^2(x)dx} \qquad \tau = 0 \text{ or } \pi, \quad n = 0, 1, 2, \ldots$

vii) $p(x) = p(\pi - x), \quad q(x) = q(\pi - x), \quad h_1 = H_1, \quad h_2 = H_2$
then $p(x) = q(x), \quad h_1 = h_2, \quad H_1 = H_2.$

Proof. $\psi_n(x)$ is related to $\phi_n(x)$ by

$$\psi_n(x) = \phi_n(x) + \int_0^x K(x,y)\phi_n(y)dy \tag{11.10.1}$$

so that

$$\psi_n'(x) = \phi_n'(x) + K(x,x)\phi_n(x) + \int_0^x K_x(x,y)\phi_n(y)dy \tag{11.10.2}$$

$$\begin{aligned}
\psi_n(0) &= \phi_n(0) \\
\psi_n'(0) - h_2\psi_n(0) &= \phi_n'(0) - h_1\phi_n(0) + \{K(0,0) + h_1 - h_2\}\phi_n(0) \\
\psi_n'(\pi) + H_2\psi_n(\pi) &= \phi_n'(\pi) + H_1\phi_n(\pi) + \{K(\pi,\pi) + H_2 - H_1\}\phi_n(\pi) \\
&\quad + \int_0^\pi \{K_x(\pi,y) + H_2 K(\pi,y)\}\phi_n(y)dy.
\end{aligned} \tag{11.10.3}$$

Since **i)** and **ii)** have the same spectrum,

$$h_1 + H_1 + \frac{1}{2}\int_0^\pi p(x)dx = h_2 + H_2 + \frac{1}{2}\int_0^\pi q(x)dx,$$

but

$$K(x,x) = h_2 - h_1 + \frac{1}{2}\int_0^x \{q(x) - p(x)\}dx$$

so that $K(0,0) + h_1 - h_2 = 0 = K(\pi,\pi) + H_2 - H_1$.

Since $\psi_n'(\pi) + H_2\psi_n(\pi) = 0 = \phi_n'(\pi) + H_1\phi_n(\pi)$, equation (11.10.3) implies $K_x(\pi,y) + H_2 K(\pi,y) = 0$ as in (11.4.7).

Now bring in the extra information:

iii) Since $\psi_n(0) = \phi_n(0)$, we have $\psi_n(\pi) = \phi_n(\pi)$ and thus, from (11.10.1),

$$\int_0^\pi K(\pi,y)\phi_n(y)dy = 0 \qquad n = 0, 1, 2, \ldots$$

and $K(\pi, y) = 0$, so that $K_x(\pi, y) = 0$, and the conclusion follows as before.

iv) Expressing $\psi_n'(\pi)$ and $\psi_n'(0)$ in terms of ϕ_n, we find, after some manipulations, that

$$(h_1 H_2 - h_2 H_1)\phi_n(0)\phi_n(\pi) = \phi_n'(0)\int_0^\pi K_x(\pi, y)\phi_n(y)dy.$$

Let $n \to \infty$, then the Riemann-Lebesque Lemma states that

$$\lim_{n\to\infty}\int_0^\pi K_x(\pi, y)\phi_n(y)dy = 0.$$

Thus $h_1 H_2 - h_2 H_1 = 0$, and $K_x(\pi, y) = 0$, and we proceed as before.

v) We need to get an expression for $\int_0^\pi \phi_n^2(x)dx = \rho_n$.

We show that two eigenfunctions satisfying i) are orthogonal, i.e., $\int_0^\pi \phi_m(x)\phi_n(x)dx = 0$ by taking the two equations

$$\phi_n'' + (\lambda_n - p)\phi_n = 0 = \phi_m'' + (\lambda_m - p)\phi_m,$$

multiplying the first by ϕ_m, the second by ϕ_n, subtracting the resulting equations and integrating over $(0, \pi)$. To find ρ_n, we need to take the equation $\phi'' + (\lambda - p)\phi = 0$ for λ_n and for another λ infinitesimally close to it. We proceed as follows.

Let $\phi = \phi(x, \lambda, c)$ be the solution of

$$\phi'' + (\lambda - p)\phi = 0, \quad \phi'(0) - h_1\phi(0) = 0, \quad \phi(0) = c. \tag{11.10.4}$$

Then, on letting $\bullet = \partial/\partial\lambda$, we find

$$\dot{\phi}'' + (\lambda - p)\dot{\phi} + \phi = 0, \quad \dot{\phi}'(0) - h_1\dot{\phi}(0) = 0, \quad \dot{\phi}(0) = 0. \tag{11.10.5}$$

Multiplying (11.10.4a) by $\dot{\phi}$, (11.10.5a) by ϕ, subtracting and integrating over $(0, \pi)$, and putting $\lambda = \lambda_n$, so that $\phi = \phi_n$, we find

$$[\dot{\phi}_n\phi_n' - \dot{\phi}_n'\phi_n]_0^\pi = \int_0^\pi \phi_n^2 dx.$$

At the lower limit, the L.H.S. is zero, at the upper limit it is

$$-(\dot{\phi}_n'(\pi) + H_1\dot{\phi}_n(\pi))\phi_n(\pi) = \int_0^\pi \phi_n^2(x)dx. \tag{11.10.6}$$

We may carry out the same calculation for $\psi_n(x)$, and find

$$-(\dot{\psi}_n'(\pi) + H_2\dot{\psi}_n(\pi))\psi_n(\pi) = \int_0^\pi \psi_n^2(x)dx. \tag{11.10.7}$$

Since $K(x, y)$ is independent of λ, equations (11.10.1), (11.10.2) when differentiated w.r.t. λ, give

$$\dot{\psi}_n'(\pi) + H_2\dot{\psi}_n(\pi) = \dot{\phi}_n'(\pi) + H_1\dot{\phi}_n(\pi). \tag{11.10.8}$$

Thus **v)** with (11.10.6), (11.10.7) yield

$$\frac{\psi_n^2(\tau)}{\phi_n^2(\tau)} = \frac{\psi_n(\pi)}{\phi_n(\pi)} \qquad n = 0, 1, \ldots$$

If $\tau = 0$, then $\psi_n(0) = \phi_n(0)$ yields $\psi_n(\pi) = \phi_n(\pi)$; if $\tau = \pi$, then again $\psi_n(\pi) = \phi_n(\pi)$. But if $\psi_n(\pi) = \phi_n(\pi)$, then equation (11.10.1) shows that

$$\int_0^{\pi} K(\pi, y)\phi_n(y)dy = 0,$$

so that $K(\pi, y) = 0$, and we proceed as before.

vi) On using (11.10.6)-(11.10.8) we find

$$\frac{\psi_n'^2(\tau)}{\phi_n'^2(\tau)} = \frac{\psi_n(\pi)}{\phi_n(\pi)}.$$

If $\tau = 0$, then the L.H.S. is h_2^2/h_1^2. Thus

$$\psi_n(\pi) = \phi_n(\pi) + \int_0^{\pi} K(\pi, y)\phi_n(y)dy = \left(h_2^2/h_1^2\right)\phi_n(\pi).$$

Again, $K(\pi, y) = 0$. If $\tau = \pi$, then $\psi_n(\pi) = \left(H_1^2/H_2^2\right)\phi_n(\pi)$ and again $K(\pi, y) = 0$.

vii) The potential and the end conditions are invariant under the transformation $x \to \pi - x$. Thus, all the eigenfunctions must be either symmetric or antisymmetric about $x = \frac{\pi}{2}$. More precisely $\phi_{2n}(x), \psi_{2n}(x)$ are symmetric while $\phi_{2n+1}(x), \psi_{2n+1}(x)$ are antisymmetric. Thus

$$\frac{\psi_n(\pi)}{\phi_n(0)} = \pm 1 = \frac{\phi_n(\pi)}{\phi_n(0)}$$

so that this is a special case of **iii)**. ■

Corresponding to each of the sets of auxiliary data in **iii)-vii)** we may devise a way to estimate $K(\pi, y)$ and $K_x(\pi, y)$ for $0 \le y \le \pi$. We may then proceed as in Section 11.9 to construct the potential.

Hochstadt and Lierberman (1978) [181] considered the problem of determining $q(x)$ in $[0, \frac{\pi}{2}]$ from knowledge of $q(x)$ in $[\frac{\pi}{2}, \pi]$ and one spectrum, say that for the Dirichlet end conditions $y(0) = 0 = y(\pi)$. The non-classical method described in Section 11.9 lends itself well to this problem.

Suppose the Dirichlet spectrum is $(\mu_n)_0^{\infty}$; it must have the asymptotic form (10.9.41), i.e.,

$$\sqrt{\mu_n} = \omega_n = n + 1 + \frac{c}{n+1} + o(n^{-1}),$$

where

$$c = \frac{1}{2\pi}\int_0^{\pi} q(x)dx$$

is known.

Without loss of generality we take $p(x) = 0$ in the base problem. Let $\chi_n(x)$ be the solution of

$$\chi_n'' + \mu_n \chi_n = 0 \qquad \chi_n(0) = 0, \quad \chi_n'(0) = 1$$

and let

$$\psi_n(x) = \chi_n(x) + \int_0^x K(x,y)\chi_n(y)dy.$$

The equation $\psi_n(\pi) = 0$ is

$$\chi_n(\pi) + \int_0^\pi K(\pi,y)\chi_n(y)dy = 0$$

which yields $K(\pi, y),\ 0 \le y \le \pi$. The kernel K satisfies $K(x,0) = 0$, and

$$K(x,x) = \frac{1}{2}\int_0^x q(x)dx = \frac{1}{2}\int_0^\pi q(x)dx - \frac{1}{2}\int_x^\pi q(x)dx.$$

Since c is known, and $q(x)$ is known for $x \ge \frac{\pi}{2}$, so is $K(x,x)$.

We need to recall the arguments we used in Section 11.3. We considered the Goursat problem in which $u(x,y)$ is known on the two characteristics $x = \pm y$, for $0 \le x \le \pi$, and we showed that $u(x,y)$ is uniquely determined. Under Dirichlet conditions the kernel $K(x,y) = u(x,y)$ is an odd function of y, so that $K(x,0) = 0$. The uniqueness result is therefore that $K(x,y)$ is determined in the region $0 \le y \le x \le \pi$ if it is known on the two parts $y = 0,\ 0 \le x \le \pi$; and $x = y$ for $0 \le x \le \pi$, of the boundary.

But we can argue just as in Section 11.3 that if $K(x,y)$ is known, as indicated by the asterisks on the two parts $x = \pi,\ 0 \le y \le \pi$ and $\frac{\pi}{2} \le x = y \le \pi$ of the boundary of the shaded region in Figure 11.10.1a, then it is known in that region. That means that we can find $K(x,y)$ on the third part of the boundary: $y = \pi - x,\ \frac{\pi}{2} \le x \le \pi$. Now consider the new shaded region in Figure 11.10.1b. The kernel K is known on the two parts $y = 0,\ y = \pi - x$ for $\frac{\pi}{2} \le x \le \pi$, of the boundary, again indicated by asterisks; therefore it is known throughout that shaded region, and therefore $K(\frac{\pi}{2}, y)$ and $K_x(\frac{\pi}{2}, y)$ are known for $0 \le y \le \frac{\pi}{2}$. Finally, we consider a Cauchy problem for the shaded region in Figure 11.10.1c; K and K_x are known on $x = \frac{\pi}{2}$, so that K is known throughout. Thus $K(x,x)$ is known for $0 \le x \le \frac{\pi}{2}$, and

$$q(x) = \frac{2dK(x,x)}{dx}.$$

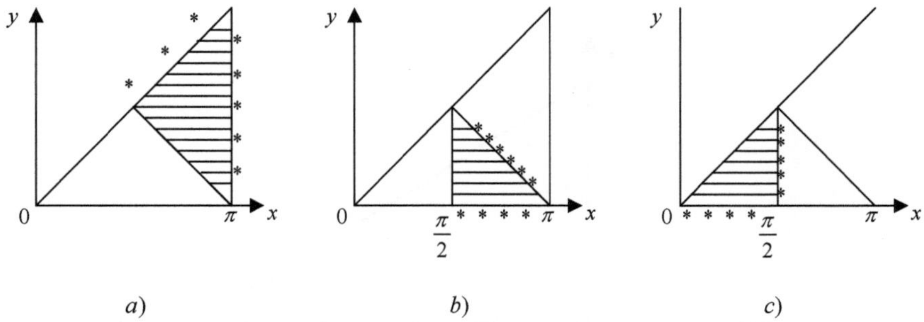

Figure 11.10.1 - Boundary value problems on 3 triangles.

11.11 Reconstruction from the impulse response

In this section we describe analysis, derived by Gopinath and Sondhi, by which $A(x)$ can be reconstructed from the impulse response $\hat{h}(0,t)$ of Section 10.10. See Gopinath and Sondhi (1970) [136], (1971) [137], Sondhi and Gopinath (1971) [308] and Sondhi (1984) [309]. Suppose a unit impulse is applied to the free end, $x = 0$, at time $t = 0$. It is intuitively clear that the response $\hat{h}(0,t)$ at the end of the rod at time t is independent of the shape of the rod for $x > \ell$, where $\ell = t/2$. This is because any effect on $\hat{h}(0,t)$ due to the shape for $x > t/\ell$ would not be felt until *after* time t, the time taken for a disturbance moving with (scaled) speed 1 to reach $x = \ell$ and return. Sondhi and Gopinath demonstrate the converse, namely that knowledge of $\hat{h}(t)$ for $0 \le t \le 2$ is sufficient (and necessary) for the determination of $A(x)$ for $0 \le x \le 1$.

The solution is based upon the following observation. Suppose the rod is at rest at time $t = t_o$, i.e., $v(x,t_o) = 0 = p(x,t_o)$, for $0 \le x \le 1$, and a force is applied at the free end $x = 0$. At time $t = t_o + a$ the rod will still be at rest for $x \ge a$, because the scaled wave speed is 1. Integrating the first of equations (10.10.10), we obtain

$$A(x)[v(x,t)]_{t_o}^{t_o+a} = A(x)v(x,t_o+a) = \int_{t_o}^{t_o+a} \frac{\partial p}{\partial x} dt$$

and on a second integration, w.r.t. x, we find

$$\int_0^a A(x)v(x,t_o+a)dx = \int_{t_o}^{t_o+a} p(0,t)dt. \qquad (11.11.1)$$

If now, for every a, we could find a force $p(0,t)$ such that $v(x,t_o+a) = 1$ for $0 \le x \le a$ then, for that case, equation (11.11.1) would give

$$\int_0^a A(x)dx = \int_{t_o}^{t_o+a} p(0,t)dt. \qquad (11.11.2)$$

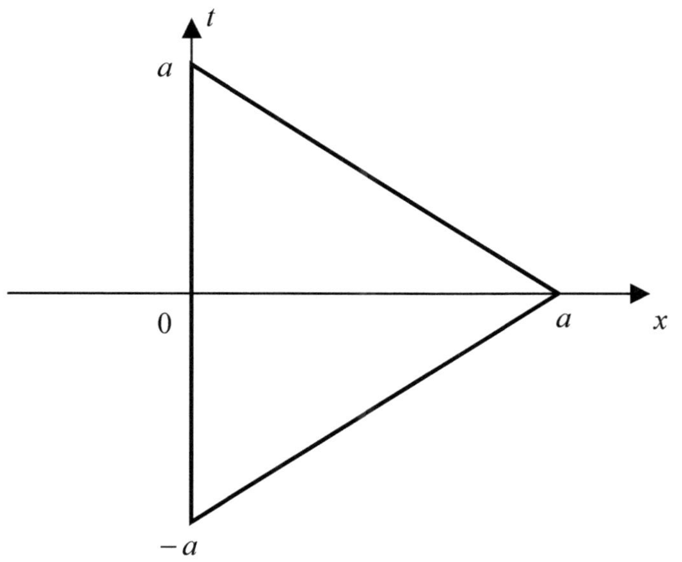

Figure 11.11.1 - The region in the x,t plane.

Thus the integral of $A(x)$, and hence $A(a)$, would be determined as a function of a. We now show that such a force exists, and can be determined from a knowledge of $\hat{h}(0,t)$.

If $v(x,t), p(x,t)$ satisfy equation (10.10.10), then so do $v(,-t), -p(x,-t)$ and, by superposition

$$\begin{aligned} V(x,t) &= v(x,t)+v(x,-t), \\ P(x,t) &= p(x,t)-p(x,-t). \end{aligned}$$

Trivially, $V(x,t)=2, P(x,t)=0$ is such a solution. The analysis of the Cauchy Problem in Section 11.3 states that this is the unique solution in the triangular region in Figure 11.11.1 which satisfies the conditions $V(0,t)=2, P(0,t)=0$, for $-a \leq t \leq a$. Thus if $p(0,t)$ is such that $V(0,t)=2, P(0,t)=0$ for $-a \leq t \leq a$, then everywhere in the triangle, $V(x,t)=2, P(x,t)=0$. In particular, when $t=0$, $V(x,0)=2v(x,0)=2$ implies $v(x,0)=1$; this gives the $v(x,t)$ required in equation (11.11.2) if t_0 is taken to be $-a$. To find the required pressure $p(0,t)$, we note that since the rod is rest at $t=-a$, equation (10.10.11) gives

$$v(0,t)=\int_{-a}^{t}\hat{h}(0,t-\tau)p(0,t)d\tau$$

so that if $v(0,t)+v(0,-t)=2$ then

$$\int_{-a}^{t}\hat{h}(0,t-\tau)p(0,\tau)d\tau+\int_{-a}^{-t}\hat{h}(0,t-\tau)p(0,\tau)d\tau=2.$$

The solution of this equation depends on a; we therefore write

$$p(0,\tau) := f(a,\tau).$$

Now using the fact that $p(0,\tau)$ is even in τ, and that equation (10.10.14) yields

$$\hat{h}(t) = \delta(t) + h(t)$$

we find

$$f(a,t) + \frac{1}{2}\int_{-a}^{a} h(|t-\tau|)f(a,\tau)d\tau = 1. \tag{11.11.3}$$

Once $f(a,\tau)$ is known, equation (11.11.2) gives

$$\int_0^a A(x)dx = \int_0^a f(a,\tau)dt.$$

Equation (11.11.3) may be written in operator form

$$(I + H_a)f(a,t) = 1.$$

Sondhi and Gopinath show that if $\hat{h}(t)$ is the impulse response of an actual rod, then the operator $I + H_a$ will be positive definite, so that equation (11.11.3) will have a unique solution. They show moreover that the corresponding $A(a)$ will be positive, provided that it is continuous. In addition, they show that if $I + H_a$ is positive definite, then there is a rod (i.e., an $A(x)$) which has this impulse response.

We now apply the procedure to the problem of Section 11.6, i.e., the reconstruction of a rod which has, from some index on, the same eigenvalues and end values of nomalised eigenfunctions, as the uniform rod, i.e.,

$$\omega_i = \omega_i^0, \quad u_i(0) = u_i^0(0), \quad i = n+1, n+2, \ldots$$

where

$$\omega_i^0 = \frac{(2i+1)\pi}{2}, \quad [u_i^0(0)]^2 = 2.$$

In this case $h(t)$ is given by Ex. 10.10.2, and the kernel $h(|t-\tau|)$ is degenerate. Since $f(a,t)$ is even in t, we may write equation (11.11.3) as

$$f(a,t) + \frac{1}{2}\int_0^a \{h(|t-\tau|) + h(|t+\tau|)\}f(a,\tau)d\tau = 1, \quad 0 \le t \le a. \tag{11.11.4}$$

Now the kernel is

$$\begin{aligned} H(t,\tau) &= \tfrac{1}{2}\{h(|t-\tau|) + h(|t+\tau|)\} \\ &= \textstyle\sum_{i=0}^{n}\{[u_i(0)]^2 \cos\omega_i t \cos\omega_i\tau - [u_i^0(0)]^2 \cos\omega_i^0 t \cos\omega_i^0\tau\}. \end{aligned}$$

Since the kernel is degenerate, the solution may be found by a straightforward matrix inversion. Thus equation (11.11.4) gives

$$f(a,t) = 1 + \sum_{i=0}^{n}\{a_i(a)[u_i(0)]^2 \cos\omega_i t - b_i(a)[u_i^0(0)]^2 \cos\omega_i^0 t\},$$

and on substituting this into equation (11.11.4) we find

$$
\begin{aligned}
a_i + \int_0^a \cos \omega_i \tau d\tau + \sum_{j=0}^{n} (b_{ij} a_j - c_{ij} b_j) &= 0, \\
b_i + \int_0^a \cos \omega_i^0 \tau d\tau + \sum_{j=0}^{n} (c_{ji} a_j - d_{ij} b_j) &= 0,
\end{aligned}
$$

where

$$
\begin{aligned}
b_{ij} &= [u_j(0)]^2 \textstyle\int_0^a \cos \omega_i \tau \cos \omega_j \tau d\tau, \\
c_{ij} &= [u_j^0(0)]^2 \textstyle\int_0^a \cos \omega_i \tau \cos \omega_j^0 \tau d\tau, \\
d_{ij} &= [u_j^0(0)]^2 \textstyle\int_0^a \cos \omega_i^0 \tau \cos \omega_j^0 \tau d\tau.
\end{aligned}
$$

Once $f(a,t)$ has been found, $A(x)$ may be computed from equation (11.11.2). This completes the inversion.

This analysis has intimate connections to the whole area of inverse scattering; see for example Burridge (1980) [45],Bube and Burridge (1983) [43], Landau (1983) [204], Bruckstein and Kailath (1987) [42], Chadan and Sabatier (1989) [52]. Further references may be found in Gladwell (1993) [120].

Chapter 12

A Miscellany of Inverse Problems

Symmetry is what we see at a glance; based on the fact that there is no reason for any difference, and based also on the face of man; whence it happens that symmetry is only wanted in breadth, not in height or depth.
Pascal's *Pensées*, 28

12.1 Constructing a piecewise uniform rod from two spectra

All the uniqueness proofs and construction algorithms described in Chapter 11 relate to the construction of a continuous system (i.e., continuous $q(x), A(x)$ or $\rho(x)$). The basic data is two infinite sequences which may be two spectra corresponding to different end conditions, or one spectrum and some auxiliary data, as in Theorem 11.10.1. If two finite data sets are given then they are either complemented by using the Truncation Assumption, as in Section 11.5, or the system is approximated numerically, as in Section 11.9. In this section we show how a *piecewise uniform* rod may be constructed so that it has precisely two given finite spectra; we do not use the Truncation Assumption. Andersson (1990) [8] was the first to provide a constructive algorithm; we follow the analysis given in Gladwell (1991c) [118], which places Andersson's algorithm in the context of inversion algorithms in seismology and transmission line theory, see Bube and Burridge (1983) [43], Bruckstein and Kailath (1987) [42] and Gladwell (1993) [120].

Andersson considered a vibrating rod, i.e., equation (11.1.3), viz.

$$(A(x)v'(x))' + \omega^2 A(x)v(x) = 0 \tag{12.1.1}$$

subject to the end conditions

$$\text{i)} \qquad v'(0) = 0 = v'(L); \text{ ii)} \qquad v(0) = 0 = v'(L); \tag{12.1.2}$$

these correspond to free-free and fixed-free ends respectively. He showed that if there were given $n+1$ frequencies $(\omega_k)_0^n$ satisfying

$$0 = \omega_0 < \omega_1 < \ldots \omega_n = \pi n/2L, \tag{12.1.3}$$

and such that the even ω_j were eigenvalues for equation (12.1.1) for i), the odd for ii), then there exists a unique rod with piecewise constant $A(x)$, such that

$$A(x) = A_j, \qquad (j-1)\triangle \leq x \leq j\triangle \qquad j = 1, 2, \ldots, n, \tag{12.1.4}$$

where $\triangle = L/n$, $\quad A_1 = 1$, as shown in Figure 12.1.1.

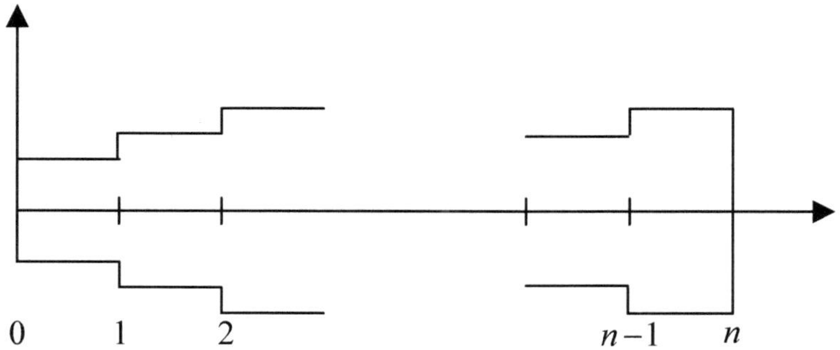

Figure 12.1.1 - A stepped rod with n segments.

In seismology and transmission line theory, a medium with parameters that are constant over equal intervals of depth $\triangle$, such as (12.1.4), is called a *Goupillard medium*. In transmission line theory, as in most inverse scattering problems, the data do not relate to eigenvalues; there are no eigenvalues, or so-called bound states. Instead the data refer to the response to an input. One way of expressing the data uses the reflected wave $U(t)$ at equal intervals $2\triangle$, due to an incoming wave $D(t)$ also sampled at intervals $2\triangle$. One of the fundamental questions is to ask whether a given reflected wave and incoming wave actually correspond to a Goupillard medium. This is the question: 'Are the data realisable?' The realisability criterion can be phrased by introducing the Z-transforms, $U(z)$ and $D(z)$, or $U(t)$ and $D(t)$, and defining the left-reflection function

$$R(z) = \frac{U(z)}{D(z)}$$

and then putting

$$f_1(z) = z^{-1}R(z).$$

The realizability criterion is

$$M(f_1) \equiv \sup_{|z|\leq 1} |f_1(z)| \leq 1. \tag{12.1.5}$$

Schur (1917) [300] constructed an algorithm to test whether a function $f_1(z)$ satisfies (12.1.5), that is, is bounded by 1 on the unit disc. The algorithm is based on the fundamental see Gilbarg and Trudinger (1977) [102] *Maximum Modulus Principle*:

The maximum modulus of a function $f(z)$ (of the complex variable $z = x+iy$) which is regular (holomorphic) in a closed region, always lies on the boundary of that region.

Note that $f(z)$ is said to have a *maximum modulus* at z_0 if $|f(z_0)| \geq |f(z)|$ for all z in some neighbourhood $|z - z_0| \leq \rho$ of z_0. An important corollary of the principle is that *if $f(z)$ has a maximum modulus at an interior point z_0 of a region in which it is regular, then $f(z) = f(z_0)$ throughout the region.*

Schur's algorithm is based on the fact that if $|\gamma| < 1$, then

$$w = \frac{z - \gamma}{1 - \bar{\gamma} z} \tag{12.1.6}$$

maps $|z| \leq 1$ onto $|w| \leq 1$, and $|z| = 1$ onto $|w| = 1$. His algorithm is based on the recurrence

$$f_j(z) = \frac{1}{z} \bullet \frac{f_{j-1}(z) - \gamma_j}{1 - \bar{\gamma}_j f_{j-1}(z)}, \quad j = 2, 3, \ldots \tag{12.1.7}$$

where $\gamma_j = f_{j-1}(0)$. Suppose $M(f_{j-1}) \leq 1$. There are two possibilities: either $|\gamma_j| = 1$, in which case the condition $M(f_{j-1}) \leq 1$ and the maximum modulus principle forces $f_{j-1}(z) = \gamma_j$, so that the sequence terminates at $f_{j-1}(z)$; or $|\gamma_j| < 1$, in which case $M(f_j) \leq 1$. Thus the condition (12.1.5) used with the recurrence (12.1.7) leads to a finite or infinite sequence $\gamma_2, \gamma_3, \ldots$ with the property $|\gamma_j| \leq 1$, where the inequality is strict except possibly for the last one. We note in particular that if $M(f_1) = 1$, then $M(f_j) = 1$ for all j, and if the sequence terminates at $j = n + 1$, it will do so with $|f_n| = 1$.

Now we formulate the vibration problem so that we obtain a recurrence of the form (12.1.6). First we replace equation (12.1.1) by two coupled first-order equations, namely

$$v'(x) = i\omega p(x)/A(x), \quad p'(x) = i\omega A(x) v(x).$$

Note that $i\omega p(x) = A(x)v'(x)$, so that it is $v(x)$ and $p(x)$ that are continuous at a point at which $A(x)$ is discontinuous. Put $\eta(x) = \{A(x)\}^{\frac{1}{2}}$ and define *down* and *up* quantities

$$D = \frac{1}{2}(\eta v + \eta^{-1} p), \quad U = \frac{1}{2}(\eta v - \eta^{-1} p). \tag{12.1.8}$$

These satisfy the equations

$$D' = i\omega D + \eta' \eta^{-1} U, \quad U' = -i\omega U + \eta' \eta^{-1} D,$$

so that if $A(x) =$ constant, then $\eta' = 0$ and

$$D' = i\omega D, \quad U' = -i\omega U$$

which have the solutions

$$D = D_0 \exp(i\omega x), \quad U = U_0 \exp(-i\omega x). \tag{12.1.9}$$

Suppose $A(x)$ has the form (12.1.4). Define the quantities

$$D_j = D(j\triangle+), \quad U_j = U(j\triangle+), \quad D_j^* = D(j\triangle-), \quad U_j^* = U(j\triangle-) \tag{12.1.10}$$

where + or - indicates a value just to the right or left of $j\triangle$, respectively. Equations (12.1.9) show that

$$D_j^* = \exp(i\omega\triangle)D_{j-1}, \quad U_j^* = \exp(-i\omega\triangle)U_{j-1}.$$

Put $\exp(i\omega\triangle) = z^{\frac{1}{2}}$, then

$$\begin{bmatrix} D_j^* \\ U_j^* \end{bmatrix} = \begin{bmatrix} z^{\frac{1}{2}} & 0 \\ 0 & z^{-\frac{1}{2}} \end{bmatrix} \begin{bmatrix} D_{j-1} \\ U_{j-1} \end{bmatrix}. \tag{12.1.11}$$

Let

$$\mathbf{H}_j = \frac{1}{2} \begin{bmatrix} \eta_j & \eta_j^{-1} \\ \eta_j & -\eta_j^{-1} \end{bmatrix}$$

then equation (12.1.8) and the continuity of v and p across a discontinuity of $A(x)$ give

$$\begin{bmatrix} D_{j-1} \\ U_{j-1} \end{bmatrix} = \mathbf{H}_j \begin{bmatrix} v_{j-1} \\ p_{j-1} \end{bmatrix}, \quad \begin{bmatrix} D_{j-1}^* \\ U_{j-1}^* \end{bmatrix} = \mathbf{H}_{j-1} \begin{bmatrix} v_{j-1} \\ p_{j-1} \end{bmatrix}$$

so that

$$\begin{bmatrix} D_{j-1} \\ U_{j-1} \end{bmatrix} = \mathbf{H}_j \mathbf{H}_{j-1}^{-1} \begin{bmatrix} D_{j-1}^* \\ U_{j-1}^* \end{bmatrix}. \tag{12.1.12}$$

The matrix $\mathbf{\Theta}_j = \mathbf{H}_j \mathbf{H}_{j-1}^{-1}$ may be written

$$\mathbf{\Theta}_j = \frac{1}{\sigma_j} \begin{bmatrix} 1 & -\gamma_j \\ -\gamma_j & 1 \end{bmatrix} \tag{12.1.13}$$

where

$$\sigma_j = (1 - \gamma_j^2)^{\frac{1}{2}}, \quad \gamma_j = (A_{j-1} - A_j)/(A_{j-1} + A_j). \tag{12.1.14}$$

We can combine equations (12.1.11), (12.1.12) to obtain

$$\begin{bmatrix} D_j^* \\ U_j^* \end{bmatrix} = \begin{bmatrix} z^{\frac{1}{2}} & 0 \\ 0 & z^{-\frac{1}{2}} \end{bmatrix} \mathbf{\Theta}_j \begin{bmatrix} D_{j-1}^* \\ U_{j-1}^* \end{bmatrix}. \tag{12.1.15}$$

Put $U_j^*/D_j^* = f_j(z)$, then equation (12.1.15) gives

$$f_j(z) = \frac{1}{z} \bullet \frac{f_{j-1}(z) - \gamma_j}{1 - \gamma_j f_{j-1}(z)}, \tag{12.1.16}$$

which, since γ_j is real $(\gamma_j = \bar{\gamma}_j)$, is precisely Schur's recurrence (12.1.7).

Before considering the inverse problem of reconstructing the cross-sections A_j from the spectra, we consider the simpler problem of computing the spectra from the cross-sections.

Suppose we are given $(A_j)_1^n$, with $A_1 = 1$, and we wish to find the eigenvalues corresponding to the end conditions i) and ii). Suppose the rod is vibrating with frequency ω and the condition $v'(L) = 0$ is satisfied, then without loss of generality we can take $v(L) = 1$; then $p_n = 0$, $v_n = 1$, so that $D_n^* = \eta_n/2 = U_n^*$ and $f_n(z) = 1$. The values of $v(0) = v_0$, $p(0) = p_0$ are related to D_0, U_0 by

$$D_0 = \frac{1}{2}\{\eta_0 v_0 + \eta_0^{-1} p_0\}, \quad U_0 = \frac{1}{2}\{\eta_0 v_0 - \eta_0^{-1} p_0\}$$

so that

$$\begin{aligned} \eta_0^2 \frac{v_0}{p_0} &= \frac{D_0 + U_0}{D_0 - U_0} = \frac{z^{-\frac{1}{2}} D_1^* + z^{\frac{1}{2}} U_1^*}{z^{-\frac{1}{2}} D_1^* - z^{\frac{1}{2}} U_1^*} \\ &= \frac{1 + g(z)}{1 - g(z)} \end{aligned} \tag{12.1.17}$$

where

$$g(z) = z f_1(z). \tag{12.1.18}$$

In the forward problem, we are given $f_n(z) = 1$ and we are given the $(\gamma_j)_2^n$ with $|\gamma_j| < 1$. We may thus compute $f_{n-1}(z), f_{n-2}(z), \ldots, f_1$ using the recurrence (12.1.16) in its reverse form:

$$f_{j-1}(z) = \frac{z f_j(z) + \gamma_j}{1 + \gamma_j z f_j(z)}, \quad j = n, n-1, \ldots, 2. \tag{12.1.19}$$

The mapping of $z f_j(z)$ onto $f_{j-1}(z)$ has the form (12.1.6). Thus the region $|z f_j(z)| \le 1$ is mapped onto $|f_{j-1}(z)| \le 1$, and $|z f_j(z)| = 1$ is mapped onto $|f_{j-1}(z)| = 1$. But $f_n(z) = 1$, so that each $(f_j(z))_1^n$ has $|f_j(z)| = 1$ when $|z| = 1$, i.e., when ω is real. Thus the function $w = g(z)$ maps $|z| \le 1$ onto $|w| \le 1$, and $|z| = 1$ onto $|w| = 1$. When $g(z)$ is expressed in terms of z it has the form

$$g(z) = z P_{n-1}(z)/Q_{n-1}(z), \tag{12.1.20}$$

where $P_{n-1}(z), Q_{n-1}(z)$ are polynomials of degree $n-1$. Thus $g(z)$ maps the circle $|z| = 1$ into itself n times.

Equation (12.1.19) shows that *if* $f_j(z^{-1}) = 1/f_j(z)$, *then* $f_{j-1}(z^{-1}) = 1/f_{j-1}(z)$. But $f_n(z^{-1}) = 1 = 1/f_n(z)$, so that indeed

$$f_j(z^{-1}) = 1/f_j(z), \quad j = 1, 2, \ldots, n, \tag{12.1.21}$$

and hence

$$g(z^{-1}) = 1/g(z). \tag{12.1.22}$$

The mapping of $|z| = 1$ into itself caused by $g(z)$ produces two sets of n points on $|z| = 1$ of significance, namely

$$\begin{aligned} \mathcal{A} &= \{z; |z| = 1 \text{ and } g(z) = 1\} \\ \mathcal{B} &= \{z; |z| = 1 \text{ and } g(z) = -1\} \end{aligned} .$$

The points in $\mathcal{A}$ correspond to values of z for which, according to (12.1.17), $p_0 = 0$; the z values give values of ω which are eigenvalues of i). Similarly the z values on $\mathcal{B}$ give $v_0 = 0$, so that ω corresponds to an eigenvalue of ii). The known interlacing of these two sets of eigenvalues means that the points of $\mathcal{A}$ and $\mathcal{B}$ will interlace on the circle $|z| = 1$ Equation (12.1.22) shows that if z is a member of either set, then $z^{-1} = \bar{z}$ is a member of the same set. Figure 12.1.2 shows the arrangement of the two sets when $n = 2$ and $n = 3$. Since $f_n(1) = 1$, the recurrence (12.1.19) shows that $f_j(1) = 1$ for $j = n, n-1, \ldots, 1$. Thus $g(1) = 1$: 1 is in $\mathcal{A}$. On the other hand $f_j(-1) = (-1)^{n-j}$, so that $g(-1) = (-1)^n$: -1 is in $\mathcal{A}$ if n is even, in $\mathcal{B}$ if n is odd. It may easily be verified that there are $n+1$ values of z in $\mathcal{A} \cup B$ which satisfy

$$0 \leq \arg(z) \leq \pi. \tag{12.1.23}$$

If these points are $z_k = \exp(i\theta_k)$, where $0 = \theta_0 < \theta_1 < \ldots \theta_n = \pi$, then $\omega_k = \theta_k/(2\triangle) = n\theta_k/(2L)$, so that $0 = \omega_0 < \omega_1 < \ldots < \omega_n = n\pi/(2L)$. These points z, other than $z = \pm 1$, yield $n-1$ points $\bar{z} = z^{-1}$ on the lower half of the circle. Thus the system also has the eigenvalues

$$\omega_{n+j} = \pi/\triangle - \omega_{n-j}, \quad j = 0, 1, \ldots, n. \tag{12.1.24}$$

Since $z = \exp(2i\omega\triangle)$ is a periodic function of ω with period $\pi/\triangle$, each value of z gives rise to an infinite sequence of eigenvalues with equal spacing $\pi/\triangle$, and each z^{-1} gives another such sequence. Thus the system not only has the eigenvalues $(\omega_j)_0^{2n}$, but also

$$\omega_{mn+j} = \frac{m\pi}{\triangle} + \omega_j, \quad j = 0, 1, \ldots, n \text{ if } m \text{ is even} \tag{12.1.25}$$

$$\omega_{mn+j} = \frac{m\pi}{\triangle} + \omega_{n-j}, \quad j = 0, 1, \ldots, n \text{ if } m \text{ is odd.} \tag{12.1.26}$$

Now consider the inverse problem, that of determining the γ_j from the spectrum. We are given $n+1$ eigenvalues ω_j satisfying (12.1.3). We must use them to construct $g(z)$ and hence $f_1(z)$, and then find the γ_j which will lead eventually to $f_n(z) = 1$.

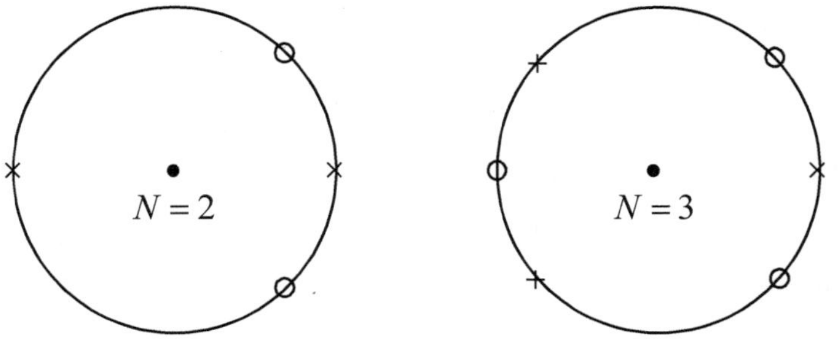

Figure 12.1.2 - The members of $A(\times)$ and $B(\bigcirc)$ interlace on the circle

First consider the case in which n is even: $n = 2m$. Of the $n + 1 = 2m + 1$ eigenvalues, $m + 1$ are even, corresponding to i), m are odd, corresponding to ii). The set $\mathcal{A}$ consists of $2m$ points: $z_0 = 1, z_{2m} = -1$, and the $m - 1$ pairs $z_{2j}, z_{2j}^{-1}, j = 1, 2, \ldots, m - 1$. The $2m$ odd z's in $\mathcal{B}$ occur in m pairs $z_{2j-1}, z_{2j-1}^{-1}, j = 1, 2, \ldots, m$. Thus equation (12.1.17) gives

$$\frac{\eta_0^2 v_0}{p_0} = \frac{1 + g(z)}{1 - g(z)} = \frac{\mathcal{C}\prod_{j=1}^{m}(z - z_{2j-1})(z - z_{2j-1}^{-1})}{(z^2 - 1)\prod_{j=1}^{m-1}(z - z_{2j})(z - z_{2j}^{-1})}, \tag{12.1.27}$$

so that $g(z) = 1$ when z is a root of the denominator, and $g(z) = -1$ when z is a root of the numerator. The constant $\mathcal{C}$ must be chosen so that $g(0) = 0$, i.e., $\mathcal{C} = -1$; the numerator of $g(z)$ will thus have no constant term, while the highest powers of z^{2m}, in the denominator will cancel, so that $g(z)$ will have the form (12.1.20). Denote the right hand side of equation (12.1.27) by $f(z)$, so that

$$\frac{1 + g(z)}{1 - g(z)} = f(z) = \zeta. \tag{12.1.28}$$

The function f will map the open, connected region $\mathcal{D} = \{z : |z| < 1\}$ into an open, connected region in the ζ-plane. When $|z| = 1$ we can easily verify that $f(z)$ given by (12.1.27) satisfies$\overline{f(z)} = -f(z)$, so that $\bar{\zeta} = -\zeta : \zeta$ lies on the imaginary axis. The function f maps $z = 0$ onto $\zeta = 1$ so that we may conclude that f maps $|z| \leq 1$ into the right hand half plane i.e., if $|z| \leq 1$, then $\mathcal{R}\{f(z)\} \geq 0$; if $|z| = 1$, Then $\mathcal{R}\{f(z)\} = 0$. Since the given eigenvalues ω_j, corresponding to i) and ii) interlace, the members of $\mathcal{A}$ and $\mathcal{B}$ interlace, then as we proceed counterclockwise around $|z| = 1$ starting at $z = 1$, the points of $\mathcal{A}$ and $\mathcal{B}$ are mapped successively onto the point at infinity and the origin in the ζ-plane. Equation (12.1.28) implies

$$g(z) = \frac{f(z) - 1}{f(z) + 1} = \frac{\zeta - 1}{\zeta + 1}.$$

But if $\mathcal{R}(\zeta\} = \xi \geq 0$, then $|\zeta - 1| \leq |\xi + 1|$, so that $|g(z)| \leq 1$. We conclude that $g(z)$, and hence by the *Schwarz lemma*, $f_1(z)$, is bounded by 1 on the unit disc.

Now apply Schur's algorithm to produce a sequence $(f_j(z))_1^n$. The form of $g(z)$ given by (12.1.27) leads to a form

$$f_1(z) = p_{n-1}(z)/Q_{n-1}(z) \tag{12.1.29}$$

with real coefficients. Therefore, all γ_j will be real. Equation (12.1.27) shows that $g(z)$ has the properties

$$g(z^{-1}) = 1/g(z), \quad g(1) = 1.$$

Therefore $f_1(z^{-1}) = 1/f_1(z)$, and $f_1(1) = 1$. Equation (12.1.16) now shows that

$$f_j(z^{-1}) = 1/f_j(z), \quad f_j(1) = 1, \quad j = 1, 2, \ldots, n \tag{12.1.30}$$

because the statement is true for $j = 1$. Thus $f_j(z)$ will have the form

$$f_j(z) = P_{n-j}(z)/Q_{n-j}(z), \quad j = 1, 2, \ldots, n, \tag{12.1.31}$$

so that the sequence will terminate with $f_n(z) = 1$ as required, and the γ_j will satisfy

$$-1 < \gamma_j < 1, \quad j = 2, 3, \ldots, n; \gamma_{n+1} = 1. \tag{12.1.32}$$

Since $A_1 = 1$ by assumption, these γ_j lead to a unique set of finite, positive $(A_j)_1^n$ as required. We stress that the single condition (12.1.5) ensures the existence of the γ_j satisfying (12.1.32).

For computational purposes, Schur's algorithm leads to a recurrence relation for the coefficients in the polynomials $P_{n-j}(z)$ and $Q_{n-j}(z)$. Let

$$P_{n-j}(z) = \sum_{k=0}^{n-j} a_{n-j,k} z^k, \quad Q_{n-j}(z) = \sum_{k=0}^{n-j} b_{n-j,k} z^k.$$

Equation (12.1.27) yields the values of $a_{n-j,k}$ and $b_{n-j,k} (k = 0, 1, \ldots, n-j)$ from data. Equation (12.1.31) states that

$$a_{n-j,k} = b_{n-j,n-j-k} \quad k = 0, 1, \ldots, n-j$$

so that the sequences $\{a_{n-j,k}\}_0^{n-j}$ and $\{b_{n-j,k}\}_0^{n-j}$ consist of the same numbers, in opposite orders.

The recurrence (12.1.16) yields

$$\begin{aligned} \gamma_j &= a_{n-j+1,0}/b_{n-j+1,0} \\ a_{n-j,k} &= a_{n-j+1,k+1} - \gamma_j b_{n-j+1,k+1}, \quad k = 0, 1, \ldots, n-j \; . \\ b_{n-j,k} &= b_{n-j+1,k} - \gamma_j a_{n-j+1,k}, \quad k = 0, 1, \ldots, n-j \end{aligned}$$

In its simplest terms, the algorithm has three steps; we have adapted the procedure of Kailath and Lev-Ari (1987) [189]:

I. Take the coefficients of $P_{n-1}(z)$ from equation (12.1.27) and construct $G_0 \in M_{2,n}$:

$$\mathbf{G}_0 = \begin{bmatrix} a_0 & a_1, & \ldots & a_{n-1} \\ a_{n-1} & a_{n-2} & \ldots & a_0 \end{bmatrix}, \quad a_k = a_{n-1,k}.$$

II. Compute $\gamma_n = a_0/a_{n-1}$ and construct

$$\mathbf{G}'_1 = \begin{bmatrix} 1 & -\gamma_2 \\ -\gamma_2 & 1 \end{bmatrix} \mathbf{G}_0 = \begin{bmatrix} 0 & a'_0 & a'_1 & \dots & a'_{n-3} & a'_{n-2} \\ a'_{n-2} & a'_{n-3} & \cdot & \dots & a'_0 & 0 \end{bmatrix}.$$

III. Shift the top row of the matrix formed in II to the left and delete the last column to form $G_1 \in M_{2,n-1}$:

$$\mathbf{G}_1 = \begin{bmatrix} a'_0 & a'_1 & \dots & a'_{n-2} \\ a'_{n-2} & a'_{n-3} & \dots & a'_0 \end{bmatrix}$$

and go to step I.

Bruckstein and Kailath (1987) [42] showed that Schur's algorithm is computationally stable and efficient.

Note that by making minor changes in the analysis (See Ex. 12.1.2) we can construct a Goupillard model of a rod from n interlacing eigenvalues $0 < \omega_1 < \omega_2 < \ldots < \omega_n = n\pi/(2L)$ corresponding to the end conditions

$$\text{i)}\;\; v'(0) = 0 = v(L),\ \text{odd}\ \omega_j \qquad \text{ii)}\;\; v(0) = 0 = v(L),\ \text{even}\ \omega_j. \qquad (12.1.33)$$

However, it is *not* possible to use the essentially algebraic method described here to construct the A_i from the eigenvalues $0 < \omega_1 < \omega_2 < \ldots < \omega_n$ corresponding to the general end condition

$$v'(0) = 0 = v'(L) + Hv(L); \quad v(0) = 0 = v'(L) + Hv(L).$$

This is because ω will appear in the analysis as itself, and not just in the form $\exp(2i\omega\triangle)$.

It is possible to modify the analysis (see Ex. 12.1.2) so that it can be applied to a piecewise uniform string, governed by equation (10.1.1), but now the model will consist of a string with density $\rho^2(x)$ satisfying $\rho(x) = \eta_j^2$, $x_{j-1} < x < x_j$, where $\eta_j^2 * (x_j - x_{j-1}) = \text{constant}$, $j = 1, 2, \ldots, n$.

Exercises 12.1

1. Make the necessary modifications to the analysis of this section so that it applies to the case of n odd. Take $n = 2m - 1$. Show that $\mathcal{A}$ consists of $z_0 = 1$ and $m - 1$ pairs $z_{2j}, z_{2j}^{-1}, j = 1, 2, \ldots, m-1$, and $\mathcal{B}$ consists of $z_{2m-1} = -1$ and $m-1$ pairs $z_{2j-1}, z_{2j-1}^{-1}, j = 1, 2, \ldots, m-1$. Hence show that
$$\frac{\eta_0^2 v_0}{p_0} = \frac{1+g(z)}{1-g(z)} = C\frac{z+1}{z-1} \bullet \prod_{j=1}^{m-1} \frac{(z - z_{2j-1})(z - z_{2j-1}^{-1})}{(z - z_{2j})(z - z_{2j}^{-1})},$$
where again $g(0) = 0$ implies $C = -1$.
2. Make the necessary changes so that it applies to (12.1.33).

3. For the string governed by (10.1.1) with $\rho(x) = \eta^2(x)$, the appropriate up and down quantities are given by (12.1.8), where now

$$v'(x) = i\omega p(x), \quad p'(x) = i\omega\eta^4(x)v(x).$$

Note that it is v and v', i.e., p that are continuous at discontinuities of $\rho(x)$. Show that

$$D' = i\omega\eta^2 D + \eta'\eta^{-1}U, \quad U' = -\omega\eta^2 U + \eta'\eta^{-1}D$$

and that when η =const,

$$D = D_0 \exp(i\omega\eta^2 x) \qquad U = U_0 \exp(-i\omega\eta^2 x).$$

This means that we must choose intervals of uniformity so that $\eta_j^2 * (x_j - x_{j-1})$ =const. Now when the γ_j have been found, one must also find the points x_j of discontinuity.

12.2 Isospectral rods and the Darboux transformation

We denote the spectrum of the rod governed by the equation

$$(Av')' + \lambda Av = 0, \tag{12.2.1}$$

and the end conditions

$$A(0)v'(0) - kv(0) = 0 = A(\pi)v'(\pi) + Kv(\pi), \tag{12.2.2}$$

by $\sigma(A, k, K)$. If two such rods have the same spectrum i.e.,

$$\sigma(A_1, k_1, K_1) = \sigma(A_2, k_2, K_2), \tag{12.2.3}$$

we say that they are *isospectral.*

The simplest, almost trivial, pair of isospectral rods is obtained by physically turning the rod and restraints around so that

$$A_2(x) = A_1(\pi - x), \quad k_2 = K_1, \quad K_2 = k_1.$$

This will have no effect on the spectrum, so that

$$\sigma(A(x), k, K) = \sigma(A(\pi - x), K, k). \tag{12.2.4}$$

To avoid complications we shall henceforth assume that $A(x) = a^2(x)$ is a positive, twice continuously differentiable function of x. This is unnecessarily restrictive, but at this time we are not interested in discussing the finer points of analysis. We leave it to the reader to see what regularity conditions are sufficient for various points of the analysis.

To obtain the next simplest pair we note that if v satisfies (12.2.1), then $w = Av'$ satisfies

$$(A^{-1}w')' + \lambda A^{-1}w = 0,$$

which is precisely (12.2.1) with A replaced by A^{-1}. Now consider the end conditions. We have

$$w = Av', \qquad w' = -\lambda Av.$$

Thus if the original rod is a cantilever, with $v(0) = 0 = v'(\pi)$, then the new rod satisfies $w'(0) = 0 = w(\pi)$, so that it is a reversed cantilever. The cantilever cannot have a zero eigenvalue so that we conclude

$$\sigma(A, \infty, 0) = \sigma(A^{-1}, 0, \infty)$$

and using (12.2.4) we deduce also that

$$\sigma(A(x), \infty, 0) = \sigma(A^{-1}(\pi - x), \infty, 0).$$

This is a result that has been known for many years, see Eisner (1967) [83], Benade (1976) [26], and was recently pointed out again by Ram and Elhay (1995) [285]; they examined many other interesting dualities.

If the original rod is free, so that $v'(0) = 0 = v'(\pi)$, then $w(0) = 0 = w(\pi)$, so that the new rod is supported. But the free rod has a zero eigenvalue with eigenfunction $v = 1$, for which $w = 0$. Thus the zero eigenvalue will not appear in the spectrum for the supported rod. We conclude that

$$\sigma'(A, 0, 0) = \sigma(A^{-1}, \infty, \infty)$$

where $'$ indicates that the zero eigenvalue has been omitted.

To conduct a more systematic search for isospectral pairs, we reduce (12.2.1) to standard Sturm-Liouville form, as in Section 10.1. Write

$$A = a^2, \quad y = av, \tag{12.2.5}$$

then

$$Av' = a^2v' = ay' - a'y, \tag{12.2.6}$$

so that (12.2.1) reduces to the Sturm-Liouville form

$$y'' + (\lambda - p)y = 0, \tag{12.2.7}$$

where

$$a'' - pa = 0. \tag{12.2.8}$$

For given A or a, there is a unique p, but for given p there are many a. This allows us to obtain further isospectral sets. Although rather obvious, and observed already in Bernoulli and Euler, the indeterminacy introduced by the Liouville transformation in the inverse eigenvalue problem seems to have been systematically studied first by Hochstadt (1975a) [177]. He proved that classical uniqueness theorems for Sturm-Liouville problems hold, *modulo* a Liouville transformation: if a_0 is one a corresponding to a given p, then variation of parameters gives the general solution

$$a(x) = a_0(x)\left\{d_0 + d_1 \int_0^x \frac{ds}{a_0^2(s)}\right\}, \quad d_0, d_1 \text{ constant}.$$

The normalization condition $a(0) = 1$, gives $d_0 = 1$, so that

$$a(x) = a_0(x)\left\{1 + d_1 \int_0^x \frac{ds}{a_0^2(s)}\right\}. \tag{12.2.9}$$

The constant d_1 must be chosen so that $A > 0$ for $0 \leq x \leq \pi$; this happens iff

$$1 + d_1\rho > 0, \text{ where } \rho = \int_0^\pi \frac{ds}{a_0^2(s)}. \tag{12.2.10}$$

If v_0, v are solutions of (12.2.1) corresponding to the same y, then

$$a_0v_0 = y = av.$$

A simple calculation shows that if v_0 satisfies the conditions

$$A_0(0)v_0'(0) - k_0v_0(0) = 0 = A_0(\pi)v_0'(\pi) + K_0v_0(\pi) \tag{12.2.11}$$

then v satisfies (12.2.2) with

$$k = k_0 - d_1 \quad K = K_0(1 + d_1\rho) + d_1a_0(\pi) \tag{12.2.12}$$

where ρ is given by (12.2.10). Thus, provided that d_1 satisfies

$$-j_0 < d_1 < k_0, \quad j_0 = \frac{K_0}{K_o\rho + a_0(\pi)} \tag{12.2.13}$$

we have a one-parameter family of rods with positive spring constraints:

$$\sigma(A, k, K) = \sigma(A_0, k_0, K_0).$$

In particular, if $k_0 = \infty = K_0$, then $k = \infty = K$, and

$$\sigma(A, \infty, \infty) = \sigma(A_0, \infty, \infty),$$

provided only that d_1 satisfies (12.2.10).

In a series of papers, Isaacson and Trubowitz (1983) [186], Isaacson, McKean and Trubowitz (1984) [187], Dahlberg and Trubowitz (1984) [68], Trubowitz and his co-workers have given a complete characterisation of the isospectral potentials $p(x)$ for the Sturm-Liouville problem (12.2.7) with different sets of boundary conditions. Coleman and McLaughlin (1993a) [62], Coleman and McLaughlin (1993b) [63] extended this analysis to equation (12.2.1) with Dirichlet boundary conditions. In this section we have a more modest aim: to show how to obtain families of rods isospectral to a given one, following Gladwell and Morassi (1995) [122].

The analysis is based on the fundamental result that if A and B are two linear operators then $AB + \mu$ and $BA + \mu$ have the same eigenvalues except perhaps for μ. For if $AB + \mu$ has eigenvalue λ then there is a $u \neq 0$ such that $(AB + \mu)u = \lambda u$. Thus $ABu = (\lambda - \mu)u$, so that $\lambda \neq \mu$ implies $Bu \neq 0$. Now $B(ABu) = BA(Bu) = (\lambda - \mu)Bu$, i.e., $(BA + \mu)Bu = \lambda(Bu)$. Since $Bu \neq 0$, λ is an eigenvalue of $BA + \mu$.

To apply this to our situation we factorise the operator

$$D^2 - p + \mu = (D + \alpha)(D - \alpha) = D^2 - \alpha' - \alpha^2.$$

Thus $p = \alpha' + \alpha^2 + \mu$. Put $\alpha = g'/g$, so that $p = (g''/g) + \mu$. This means that g satisfies

$$g'' + (\mu - p)g = 0. \tag{12.2.14}$$

Now y satisfies

$$y'' + (\lambda - p)y = 0, \tag{12.2.15}$$

then

$$0 = (D^2 - p + \lambda)y = \{(D + \alpha)(D - \alpha) + \lambda - \mu\}y = 0$$

so that $z = (D - \alpha)y$ satisfies

$$\{(D - \alpha)(D + \alpha) + \lambda - \mu\}z = 0$$

i.e., $(D^2 + \alpha' - \alpha^2 + \lambda - \mu)z = 0$. Write this as

$$z'' + (\lambda - q)z = 0 \tag{12.2.16}$$

where

$$q = -\alpha' + \alpha^2 + \mu = p - 2\alpha' = p - 2(\ell n g)''. \tag{12.2.17}$$

We can interpret this analysis, called the Darboux Lemma or the Darboux Transformation, after Darboux (1882) [69], Darboux (1915) [70], in various ways. We can say that, starting from one system with potential p and solution y, we can find another system with potential q and solution

$$z = (D - \alpha)y = y' - \frac{g'y}{g} = \frac{[g, y]}{g}, \tag{12.2.18}$$

where the *bracket* is defined by

$$[g, y] := gy' - g'y. \tag{12.2.19}$$

Alternately we can say that, given two solutions, y of (12.2.15), and g of (12.2.14), we can form a solution z of (12.2.16) given by (12.2.18), where q is related to p by (12.2.17).

Note that $\lambda \neq \mu$. It may be shown (Ex. 12.2.1) that, when $\lambda = \mu$, the general solution of (12.2.16) is

$$z = \frac{1}{g}\left(1 + d\int_0^x g^2(s)ds\right), \quad d = \text{constant}. \tag{12.2.20}$$

Suppose that we have a rod $A(x)$ with spectrum $\{\lambda_n\}_0^\infty$ corresponding to end conditions (12.2.2). Transforming to Sturm-Liouville form, we have a set of eigenfunctions y_n satisfying

$$y_n'' + (\lambda_n - p)y_n = 0, \tag{12.2.21}$$

where p is given by (12.2.8), and the end conditions

$$y_n'(0) - hy_n(0) = 0 = y_n'(\pi) + Hy_n(\pi), \tag{12.2.22}$$

where

$$h = k + a'(0)/a(0) \quad H = K - a'(\pi)/a(\pi). \tag{12.2.23}$$

In particular the zeroth eigenfunctions y_0 will satisfy

$$y_0'' + (\lambda_0 - p)y_0 = 0. \tag{12.2.24}$$

Taking $\mu = \lambda_0$, $g = y_0$ we deduce that

$$z_n = \frac{1}{y_0}[y_0, y_n] \tag{12.2.25}$$

is a solution of

$$z_n'' + (\lambda_n - q)z_n = 0 \tag{12.2.26}$$

where

$$q = p - 2(\ell n y_0)''. \tag{12.2.27}$$

We can use this result only if y_0 is positive in $0 \leq x \leq \pi$. This will be the case if k, K are finite. Since y_0, y_n satisfying the same conditions (12.2.22), z_n will satisfy

$$z_n(0) = 0 = z_n(\pi). \tag{12.2.28}$$

This means that the eigenfunction of the new Sturm-Liouville system will satisfy Dirichlet end conditions. We must now find a function $b(x)$, or in fact a family of such $b(x)$ corresponding to q.

The original S-L system was (12.2.7). As we showed earlier, there is a family of rods with cross sections $A(x) = a^2(x)$, associated with this p. If $a_0(x)$ is one such, then each member of the family may be written

$$a(x) = a_0(x)\left\{1 + d_1\int_0^x \frac{ds}{a_0^2(s)}\right\}. \tag{12.2.29}$$

We note that if d_1 satisfies (12.2.10), then $a(x)$ will be positive throughout $[0, \pi]$; otherwise $a(x)$ will change sign once in $[0, \pi]$. All the $a(x)$ will satisfy (12.2.8). On replacing λ by 0 in the preceding analysis, we find that

$$b = \frac{1}{y_0}[y_0, a] \tag{12.2.30}$$

satisfies $b'' - qb = 0$. For this b to correspond to a proper rod, it must have one sign throughout $[0, \pi]$. First, we show that $b(x)$ can have at most one zero in any interval in which $a(x)$, given by (12.2.29) is of one sign. For suppose $b(x)$ had two such zeros, $x_1, x_2 (x_1 < x_2)$ in such an interval, then by Rolle's theorem, $[y_0, a]'$ must be zero at an intermediate point. But

$$[y_0, a]' = (y_0 a' - y_0' a)' = y_0 a'' - y_0'' a = -\lambda_0 y_0 a \neq 0,$$

which is a contradiction.

There are two cases:

i) a, given by (12.2.29) is positive throughout $[0, \pi]$. Now $a > 0$, $1 + d_1 \rho > 0$ (see 12.2.10)). Now b can have at most one zero in $[0, \pi]$, and so it will have no zero if it has the same sign at 0 and π. A simple calculation shows that

$$b(0) = -ka(0), \qquad b(\pi) = Ka(\pi). \tag{12.2.31}$$

Since k, K are related to k_0, K_0 by (12.2.12), $b(x)$ will have one sign throughout if

$$d_1 > k_0 \text{ or } -\frac{1}{\rho} < d_1 < -j_0, \tag{12.2.32}$$

where j_0 is given by (12.2.13).

ii) $a(x)$, given by (12.2.29) has one zero in $[0, \pi]$. Now $a(\xi) = 0$ for some $\xi \in [0, \pi]$, and $d_1 \leq -1/\rho$. Since $b(\xi) = d_1/a_0(\xi) < 0$, $b(\xi)$ will have the same sign throughout iff $b(0) < 0, b(\pi) < 0$, i.e., if $d_1 < -j_0$. But since $d_1 \leq -1/\rho$, this is satisfied automatically. We conclude that (12.2.29), (12.2.30) provide a proper rod with fixed end conditions if $d_1 > k_0$ or $d_1 < -j_0$. Note that in both cases the intermediate system specified by $a(x), k, K$ will not be proper because the inequalities (12.2.13) will not be satisfied.

Note that the restriction $\lambda \neq \mu$ in the original analysis relating to $AB + \mu$ and $BA + \mu$, means that the new rod $b(x)$, with fixed ends, will not have the eigenvalue λ_0, so that

$$\sigma'(A_0, k_0, K_0) = \sigma(B, \infty, \infty), \tag{12.2.33}$$

where the prime indicates that λ_0 has been deleted.

If the original rod is free ($k_0 = 0 = K_0$), then $\lambda_0 = 0$ and $y_0 = g$. Now equation (12.2.20) states that the general solution of $b'' - qb = 0$ is

$$b = \frac{1}{g}\left(1 + d\int_0^x g^2(s)ds\right). \tag{12.2.34}$$

This will be positive in $[0, \pi]$ provided $1 + d\int_0^\pi g^2(s)ds > 0$. Again

$$\sigma'(A_0, 0, 0) = \sigma(B, \infty, \infty). \tag{12.2.35}$$

We now show that λ_n, and $w_n = z_n/b$ given by (12.2.25), are in fact the $(n-1)$th eigenvalue and eigenfunction of the B rod. First, we show that there is a zero of y_n between two zeros of z_n. If x_1, x_2 are two consecutive zeros of z_n, then

$$0 = [y_0, y_n]|_{x_1}^{x_2} = \int_{x_1}^{x_2} (y_0 y_n'' - y_0'' y_n)ds = (\lambda_0 - \lambda_n)\int_{x_1}^{x_2} y_0 y_n ds.$$

But y_0 has constant sign throughout $[0, \pi]$, so that y_n must change sign, and have a zero, between x_1 and x_2. Now we show that there is a zero of z_n between consecutive zeros of y_n. This follows from (12.2.25), namely

$$z_n = y_n' - \frac{y_0'}{y_0} y_n$$

when $y_n = 0, z_n = y_n'$. But y_n' has opposite signs at successive zeros of y_n. Thus z_n changes sign, and therefore has a zero, between zeros of y_n. We conclude that the zeros of y_n and z_n interlace. But y_n has n zeros in $(0, \pi)$ while $z_n(0) = 0 = z_n(\pi)$. Therefore z_n has $(n-1)$ zeros in $(0, \pi)$; it is the $(n-1)$th eigenfunction. We may thus rewrite (12.2.33) as

$$\lambda_n(A_0, k_0, K_0) = \lambda_{n-1}(B, \infty, \infty). \tag{12.2.36}$$

The foregoing analysis breaks down where y_0 has a zero at an end, as it does when one or other end of the original rod is fixed. For such cases, and to eliminate the $'$ in (12.2.33), we must modify the analysis of reversing the order of the factors in the differential equation twice. Crum (1955) [65] has a different approach to finding pairs of solutions to the Sturm-Liouville equation.

Exercises 12.2

1. Show that the general solution of equation (12.2.16) is given by equation (12.2.20).

2. Equation (12.2.36) states that the $(n-1)$th eigenvalue of one rod is equal to the nth eigenvalue of another. This means that the $(n-1)$th eigenvalue of (12.2.26) is equal to the nth of (12.2.21). Examine the asymptotic forms of the two spectra as given by equations (10.9.19), (10.9.20) to show that they are consistent with this statement.

12.3 The double Darboux transformation

Suppose we have a rod $A_0(x)$ with spectrum $\{\lambda_n\}_0^\infty$ corresponding to end conditions (12.2.2). Transforming to S-L form, we have a set of eigenfunctions y_n satisfying

$$y_n'' + (\lambda_n - p)y_n = 0$$

and some end conditions

$$y_n'(0) - hy_n(0) = 0 = y_n'(\pi) + Hy_n(\pi),$$

as before. We now choose a particular eigenvalue and eigenfunction λ_m, y_m; m does not need to be zero. Thus y_m satisfies $y_m'' + (\lambda_m - p)y_m = 0$. Applying the Darboux lemma, we find a non-trivial solution

$$z_n = \frac{1}{y_m}[y_m, y_n], \quad n \neq m \tag{12.3.1}$$

of

$$z_n'' + (\lambda_n - q)z_n = 0 \tag{12.3.2}$$

where

$$q = p - 2(\ell n y_m)''. \tag{12.3.3}$$

On the other hand, the second part of the Darboux lemma, equation (12.2.20), states that the general solution of the equation

$$z_m'' + (\lambda_m - q)z_m = 0 \tag{12.3.4}$$

is

$$z_m = \frac{1}{y_m}\left(1 + d\int_0^x y_m^2(s)ds\right). \tag{12.3.5}$$

We now apply the Darboux lemma to equations (12.3.2), (12.3.4), and deduce that if $n \neq m$, then

$$w_n = \frac{1}{z_m}[z_m, z_n] \tag{12.3.6}$$

is a non-trivial solution of

$$w_n'' + (\lambda_n - r)w_n = 0 \tag{12.3.7}$$

where

$$\begin{aligned} r &= q - 2(\ell n z_m)'' \\ &= p - 2(\ell n(y_m z_m))''. \end{aligned} \tag{12.3.8}$$

We now examine w_n and r. First, we note that equation (12.3.5) gives

$$y_m z_m = 1 + d\int_0^x y_m^2(s)ds. \tag{12.3.9}$$

If y_m has been normalised so that $\int_0^\pi y_m^2(s)ds = 1$, then $y_m z_m$ will be positive, and so r will be continuous, if $d > -1$. We now evaluate w_n: it is

$$w_n = \frac{1}{z_m}(z_m z_n' - z_m' z_n) = z_n' - \frac{z_m'}{z_m} z_n.$$

But equation (12.3.1) shows that

$$z_n' = \frac{y_m y_n'' - y_m'' y_n}{y_n} - \frac{y_m'}{y_m} z_n = (\lambda_m - \lambda_n) y_n - \frac{y_m'}{y_m} z_n$$

so that

$$w_n = (\lambda_m - \lambda_n) y_n - \frac{(y_m z_m)}{(y_m z_m)} z_n.$$

But since $z_n(0) = 0$,

$$\begin{aligned} z_n &= \frac{y_m y_n' - y_m' y_n}{y_m} = \frac{1}{y_m}\int_0^x (y_m y_n'' - y_m'' y_n) ds \\ &= \frac{(\lambda_m - \lambda_n)}{y_m}\int_0^x y_m y_n ds. \end{aligned}$$

This means that w_n has a factor $(\lambda_m - \lambda_n)$, so that is we define

$$w_n^0 = \frac{w_n}{\lambda_m - \lambda_n}$$

and use (12.3.9) to give $y_m z_m$, we find

$$w_n^0 = y_n - \frac{d y_m \int_0^x y_m(s) y_n(s) ds}{1 + d\int_0^x y_m^2(s) ds}. \tag{12.3.10}$$

We see that this is a non-trivial solution of (12.3.7) even when n in that equation is equal to m. It may also be shown (Ex. 12.3.1) that w_n^0 is normalised so that $\int_0^\pi [w_n^0(s)]^2 ds = 1$.

Now we must find the corresponding rods. We started with a rod with $a(x)$ satisfying

$$a'' - pa = 0. \tag{12.3.11}$$

Applying the Darboux lemma to this equation and $y_m'' + (\lambda_m - p) y_m = 0$ we find that

$$b = \frac{1}{y_m}[y_m, a] \tag{12.3.12}$$

satisfies

$$b'' - qb = 0. \tag{12.3.13}$$

Now apply the Darboux lemma to this equation and (12.3.5), and we find that

$$c = \frac{1}{z_m}[z_m, b] \tag{12.3.14}$$

satisfies

$$c'' - rc = 0. \tag{12.3.15}$$

We can find c as we found w_n^0:

$$c(x) = a(x) - \frac{dy_m(x)[y_m, a]}{\lambda_m\{1 + d\int_0^x y_m^2(s)ds\}}. \tag{12.3.16}$$

Note that just as $a(x)$ is one solution of (12.3.11), and $b(x)$ is one solution of (12.3.13), so $c(x)$ is one solution of (12.3.15); other solutions may be found as in Section 12.2, see equation (12.2.9).

We now consider whether $c(x)$ is of one sign in $[0, \pi]$. Suppose the end conditions for the original rod were (12.2.2), i.e.,

$$A(0)v'(0) - k_1 v(0) = 0 = A(\pi)v'(\pi) + K_1 v(\pi).$$

Equations (12.2.5), (12.2.6) show that these transform to

$$[a, y](0) - k_1 y(0)/a(0) = [a, y](\pi) + Ky(\pi)/a(\pi)$$

so that the end values of $c(x)$ given by (12.3.16) satisfy

$$\frac{c(0)}{a(0)} = 1 + \frac{dk_1 y_m^2(0)}{\lambda_m a^2(0)} = \beta_0 \tag{12.3.17}$$

$$\frac{c(\pi)}{a(\pi)} = 1 - \frac{dK_1 y_m^2(\pi)}{\lambda_m(1+d)a^2(\pi)} = \beta_1. \tag{12.3.18}$$

Note that unless the original rod is fixed ($y_m(0) = 0$), or free ($k = 0$) at the left hand end, the new $c(x)$ will not be normalised so that $c(0) = 1$. We now show that if $d > -1$ then β_0 and β_1 are both positive. Let v_m be the mth mode of the original rod; then $(Av_m')' + \lambda_m A v_m = 0$ so that

$$\begin{aligned}\lambda_m &= \lambda_m \int_0^\pi A v_m^2 dx = -\int_0^\pi v_m (Av_m')' dx \\ &= [-v_m(Av_m')]_0^\pi + \int_0^\pi A v_m'^2 ds \\ &= k_1 v_m^2(0) + K_1 v_m^2(\pi) + \int_0^\pi A v_m'^2 dx\end{aligned}$$

so that $\lambda_m > k_1 v_m^2(0) + K_1 v_m^2(\pi)$ and hence

$$\begin{aligned}\beta_0 &> \frac{(1+d)k_1 v_m^2(0) + K_1 v_m^2(\pi)}{\lambda_m} > 0 \\ \beta_1 &> \frac{(1+d)k_1 v_m^2(0) + K_1 v_m^2(\pi)}{\lambda_m(1+d)} > 0.\end{aligned}$$

These inequalities hold provided that $k_m m_m^2(0)$ and $K_1 v_m^2(\pi)$ are not both zero, i.e., provided that at most one end of the rod is free (k_1 or K_1 is zero) or fixed ($v_m(0)$ or $v_m\,(\pi)$ is zero).

We now have a one-parameter family of rods $c(x) = c(x, d)$ defined for $x \in [0, \pi]$, $\quad d > -1$; each member of the family is positive at $x = 0$ and $x = \pi$ and, when $d = 0$, $c(x, 0) = a(x)$ is positive for $x \in [0, \pi]$. To show that $c(x, d)$ must be positive for all $x \in [0, \pi]$, $\quad d > -1$, we use the following *deformation lemma.*

Lemma 12.3.1 *Let h_t, $0 \leq t \leq 1$, be a family of real valued functions on $a \leq x \leq b$, which is jointly continuously differentiable in t and x. Suppose that for every t, h_t has a finite number of zeros in $[a, b]$, all of which are simple, and has boundary values with signs that are independent of t. Then h_0 and h_1 have the same number of zeros in $[a, b]$.*

This is a slightly extended version of Lemma 3 in Pőschel and Trubowitz (1987) [269] (p. 41); they simply supposed that h_t *has boundary values that are independent of* t, but it may easily by seen that their proof holds if only the *signs* of these boundary values are independent of t.

It may easily be seen that $c(x, d)$ can have only a finite number of zeros, and that these must be simple (Ex. 12.3.2), so that the lemma implies that $c(x, d)$, like $c(x, 0) = a(x)$, must have no zeros, and thus be positive, for $x \in [0, \pi]$, and $d > -1$. We may use the deformation lemma to show that $c(x)$ is positive for the limiting cases in which each end of the rod is either free or supported.

We now examine the end conditions for the new rod. The eigenfunctions of the new rod are $u_n = w_n^0/c$.

A tedious, but straightforward calculation shows that the new rod has end conditions

$$C(0)u'(0) - k_2u(0) = 0 = C(\pi)u'(\pi) + K_2u(\pi)$$

where $C(x) = c^2(x)$, and

$$k_2 = \beta_0 k_1, \qquad K_2 = \beta_1 K_1.$$

Thus

$$\sigma(A, k_1, K_1) = \sigma(C, k_2, K_2)$$

and in particular

$$\sigma(A, 0, 0) = \sigma(C, 0, 0)$$

and

$$\sigma(A, \infty, \infty) = \sigma(C, \infty, \infty).$$

It must be remembered, of course, that the particular C that is formed from a given A depends on the end conditions corresponding to the original rod, and the value of m that is chosen in the Darboux transformation.

Exercises 12.3

1. Show that w_n^0 given by (12.3.10) is normalised so that

$$\int_0^\pi [w_n^0(s)]^2 = 1.$$

2. Show that the zeros of $c(x, d)$ given by (12.3.16) are simple.

12.4 Gottlieb's research

H.P.W. Gottlieb has been carrying out research into various vibrating systems - rods, strings, beams, membranes, plates, etc. - amongst other matters, since 1984. In this section we briefly describe some of these researches, those related to strings, rods and beams.

We start by considering one of his early papers, Gottlieb (1986) [138], which builds on earlier papers by Levinson (1976) [208] and Sakata and Sakata (1980) [299]. We made a comment about Gottlieb (1986) [138] in Section 11.1; Gottlieb's work was motivated by the fact, central to the analysis of Chapter 11, that two spectra, corresponding to two different conditions at one end of the string, are needed to determine the string density uniquely.

Consider the string shown in Figure 12.4.1, with a density $\rho^2(x)$ that has one step, at $x = 0$.

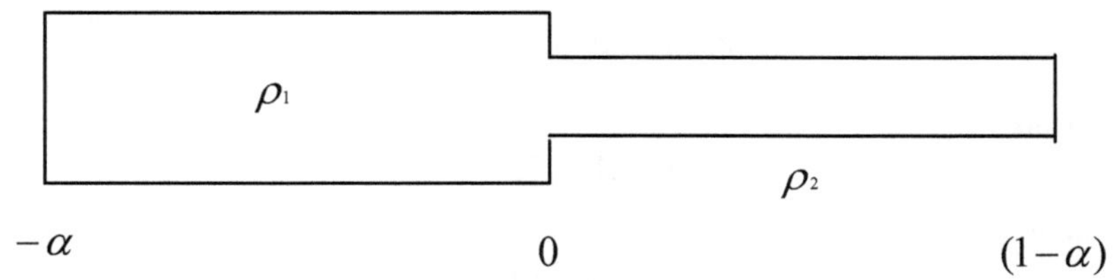

Figure 12.4.1 - A stepped string.

For fixed ends, at $-\alpha$ and $(1-\alpha)$, the end and continuity conditions are

$$u(-\alpha) = 0 = u(1-\alpha), \quad [u]_0 = 0 = [u']_0.$$

Thus

$$u(x) = \begin{cases} u_1(x) & \text{for} \quad x \in [\alpha, 0] \\ u_2(x) & \text{for} \quad x \in [0, 1-\alpha] \end{cases}$$

where

$$u_1'' + \rho_1^2\omega^2 u_1 = 0 = u_2'' + \rho_2^2\omega^2 u_2.$$

Thus

$$\begin{aligned} u_1(x) &= A\sin\{\rho_1\omega(x+\alpha)\} \\ u_2(x) &= B\sin\{\rho_2\omega(1-\alpha-x)\} \end{aligned}$$

so that the continuity conditions at $x = 0$ give

$$\begin{aligned} A\sin(\rho_1\omega\alpha) &= B\sin\{\rho_2\omega(1-\alpha)\} \\ \rho_1 A\cos(\rho_1\omega\alpha) &= -\rho_2 B\cos\{\rho_2\omega(1-\alpha)\}. \end{aligned}$$

This gives the frequency equation

$$\rho_2\sin(\rho_1\omega\alpha)\cos\{\rho_2\omega(1-\alpha)\} + \rho_1\sin\{\rho_2\omega(1-\alpha)\}\cos(\rho_1\omega\alpha) = 0. \quad (12.4.1)$$

This is the frequency equation for the general case of a string with one step, as shown in Figure 12.4.1. Gottlieb examines the special case in which

$$\rho_1\alpha = \rho_2(1-\alpha). \tag{12.4.2}$$

Now (12.4.1) reduces to

$$(\rho_1+\rho_2)\sin(2\rho_1\omega\alpha) = 0$$

which has the spectrum

$$\omega_n = n\pi/(2\rho_1\alpha), \quad n = 1, 2, \ldots \tag{12.4.3}$$

The spectrum is harmonic: $\omega_n = n * \omega_1$. To compare this spectrum with that of a uniform string of uniform density ρ^2, fixed at $x = -\alpha, 1-\alpha$, we note that the governing equations are

$$u'' + \rho^2\omega^2 u = 0, \quad u(-\alpha) = 0 = u(1-\alpha)$$

so that

$$u = A\sin\{\rho\omega(x+\alpha)\},$$

where

$$\sin(\rho\omega) = 0.$$

Now

$$\omega_n = n\pi/\rho, \quad n = 1, 2, \ldots \tag{12.4.4}$$

If $\rho = 2\rho_1\alpha$, then the two spectra, (12.4.3) and (12.4.4) are identical.

To distinguish between the two strings, we must examine their spectra for fixed-free ends. Now (Ex. 12.4.1) the frequency equation for the stepped string is

$$\cos(2\rho_1\omega\alpha) = (\rho_2-\rho_1)/(\rho_2+\rho_1), \tag{12.4.5}$$

so that the spectrum is uniformly spaced, but not harmonic. The frequency equation for the uniform string is $\cos\rho\omega = 0$, with harmonic spectrum $\omega_n = (n-\frac{1}{2})\pi/\rho$, $n = 1, 2, \ldots$

Gottlieb (1986) [138] considers other strings, and extends his analysis to multi-segment strings, some with harmonic spectra, in Gottlieb (1987a) [139].

For the special case (12.4.2), the discontinuous string in Figure 12.4.1 is isospectral to the uniform string, for fixed-fixed ends. In Gottlieb (1988a) [141] and the somewhat simpler paper Gottlieb (2002) [149], Gottlieb examines continuous isospectral strings, as we now describe. Start with a string governed by

$$\nu''(\xi) + \lambda\phi(\xi)\nu(\xi) = 0, \tag{12.4.6}$$

with fixed ends at 0,1, so that

$$\nu(0) = 0 = \nu(1). \tag{12.4.7}$$

We seek transformations to a new coordinate x and new displacement u, that preserve the structural form of the governing equation, and the fixed end conditions. Let

$$x = x(\xi), \quad v(\xi) = \gamma(x)u(x), \tag{12.4.8}$$

where $\gamma(x)$ is some positive non-singular function of x. We wish to find $x(\xi)$ and $\gamma(x)$, so that the new displacement u satisfies

$$\ddot{u}(x) + \lambda f(x)u(x) = 0 \tag{12.4.9}$$

where $f(x)$ is some new density, dual to the density function $\phi(\xi)$, and $\cdot = d/dx$. Now

$$v' = \frac{dv}{d\xi} = \frac{dv}{dx} \cdot \frac{dx}{d\xi} = x'\dot{v} = x'(\gamma\dot{u} + \dot{\gamma}u)$$

and

$$\begin{aligned} v'' &= x''(\gamma\dot{u} + \dot{\gamma}u) + (x')^2(\gamma\ddot{u} + 2\dot{\gamma}\dot{u} + \ddot{\gamma}u) \\ &= \gamma(x')^2\ddot{u} + (x''\gamma + 2(x')^2\dot{\gamma})\dot{u} + (x''\dot{\gamma} + (x')^2\ddot{\gamma})u. \end{aligned} \tag{12.4.10}$$

To maintain the form of the equations, one must have

$$x''\dot{\gamma} + (x')^2\ddot{\gamma} = 0 \tag{12.4.11}$$

$$x''\gamma + 2(x')^2\dot{\gamma} = 0. \tag{12.4.12}$$

Equation (12.4.11) may be written $(x'\dot{\gamma})' = 0$; and since $x'\dot{\gamma} = \gamma'$, we have $\gamma'' = 0$ and

$$\gamma(x(\xi)) = -a\xi + b. \tag{12.4.13}$$

Equation (12.4.12) implies

$$x''\gamma^2 + 2(x')^2\gamma\dot{\gamma} = 0$$

i.e., $(x'\gamma^2)' = 0$, so that $x' = c/\gamma^2$, i.e.,

$$\frac{dx}{d\xi} = \frac{c}{(-a\xi + b)^2}, \quad x(\xi) = \frac{-c}{a(-a\xi + b)} + d.$$

The requirements that $x(0) = 0$ and $x(1) = 1$ give

$$c = abd, \quad ad = a - b$$

so that on taking $ad = -1$ we find

$$x = \frac{\xi}{1 + a(1 - \xi)}, \quad \xi = \frac{(1 + a)x}{1 + ax}, \quad \gamma = \frac{1 + a}{1 + ax}; \tag{12.4.14}$$

clearly, we must take $a > -1$. Now (12.4.10) gives

$$f(x) = \phi(\xi(x))/x'^2 = \dot{\xi}^2\phi(\xi(x))$$

and since $\dot{\xi} = (1+a)/(1+ax)^2$ we have

$$f(x) = \frac{(1+a)^2}{(1+ax)^4}\phi\left(\frac{(1+a)x}{1+ax}\right). \tag{12.4.15}$$

The relation between the solutions $u(x)$ and $v(\xi)$ is

$$u(x) = \frac{(1+ax)}{1+a}v\left(\frac{(1+a)x}{1+ax}\right). \tag{12.4.16}$$

We stress that the system (12.4.9) with fixed end conditions is isospectral to (12.4.6), (12.4.7), for all values of $a > -1$.

The transformation from one coordinate ξ to another, x, has a group structure. First we note that if $\xi = \frac{(1+a)x}{1+ax}$, then $x = \frac{\xi}{1+a(1-\xi)} = \frac{(1+a')\xi}{1+a'\xi}$ where $a' = -a/(1+a)$: thus, if a characterises $\xi \to x$ then a' characterises $x \to \xi$. Note that if $a' = -a/(1+a)$ then $a = -a'/(1+a')$, and that $a > -1$ implies $a' > -1$ and *vice versa*. This shows that each transformation has an inverse. Now consider a product of transforms. Suppose

$$x_1 = \frac{(1+a_1)x_2}{1+a_1x_2}, \quad x_2 = \frac{(1+a_2)x_3}{(1+a_2x_3)}$$

then

$$x_1 = \frac{(1+a_1+a_2+a_1a_2)x_3}{1+(a_1+a_2+a_1a_2)x_3} = \frac{(1+a_{1,2})x_3}{1+a_{1,2}x_3}$$

where

$$a_{1,2} = a_1 + a_2 + a_1a_2 = a_2 + a_1 + a_2a_1. \tag{12.4.17}$$

We note that

$$(1+a_{1,2}) = (1+a_1)(1+a_2), \tag{12.4.18}$$

so that $a_1 > -1, a_2 > -1$ implies $a_{1,2} > -1$: the product of two transformations is a transformation, and (12.4.17) shows that the product is commutative. There is an identity transformation, $a = 0$, and the associative property holds (Ex. 12.4.2): the transformations form a group, a one-parameter Lie group.

Now consider the density functions. When f and ϕ are linked by (12.4.15) then we say that f is the *dual* of ϕ with respect to a. Since $1+ax = 1/(1+a'\xi)$ and $1+a = 1/(1+a')$, we can rewrite (12.4.15) as

$$\phi(\xi) = \frac{(1+a')^2}{(1+a'\xi)^4}f\left(\frac{(1+a')\xi}{1+a'\xi}\right).$$

This shows that ϕ is the dual of f with respect to a'. We may express this symbolically as

$$f = D(\phi, a) \to \phi = D(f, a')$$

and now we may verify (Ex. 12.4.3) that if

$$f_2 = D(f_1, a_1), \quad f_3 = D(f_2, a_2) \text{ then } f_3 = D(f_1, a_{1,2}). \tag{12.4.20}$$

This means that the dual w.r.t. a_2 of the dual of f_1 w.r.t. a_1 is just another dual of f_1, with respect to $a_{1,2}$.

Gottlieb (1986) [138] provides some examples. For the simplest, we start with $\phi(\xi) = 1$, then

$$f(x) = (1+a)^2/(1+ax)^4. \tag{12.4.21}$$

Both these systems have spectrum $\lambda_n = \omega_n^2$, where

$$\omega_n = n\pi, \quad n = 1, 2, \ldots$$

for fixed-fixed ends. The eigenfunctions are

$$v_n(\xi) = \sin(n\pi\xi), \quad u(x) = \frac{(1+ax)}{1+a}\sin\left\{\frac{n\pi(1+a)x}{1+ax}\right\}$$

and we note that while the nodes of the former are the equidistant points

$$\xi_m = m/n, \quad m = 1, 2, \ldots n-1,$$

the nodes of the latter are

$$x_m = m/\{n + a(n-m)\}, \quad m = 1, 2, \ldots, n-1.$$

Gottlieb calls (12.4.21) the Borg density because it was discussed by Borg (1946) [39]. Another example is given in Ex. 12.4.4. Gottlieb (1987b) [140] studied isospectral beams. In the notation of Section 13.7, his analysis is as follows. Start with the governing equation (13.1.4):

$$\frac{d^2}{dx^2}\left(r(x)\frac{d^2u}{dx^2}\right) = \lambda a(x)u(x) \tag{12.4.22}$$

and introduce a new variable $s = s(x)$ so that (12.4.22) reduces to the standard form

$$\frac{d^4v(s)}{ds^4} + \frac{d}{ds}\left(A(s)\frac{dv}{ds}\right) + B(s)v(s) = \lambda v(s). \tag{12.4.23}$$

As in Section 13.7 we write

$$b(s) = \left(\frac{a(x)}{r(x)}\right)^{\frac{1}{4}}, \quad c^2(s) = (a(x)r^3(x))^{\frac{1}{4}}$$

so that

$$r(x) = c^2(s)b^{-1}(x), \quad a(x) = c^2(s)b^3(s)$$

where

$$\frac{ds}{dx} = b(s). \tag{12.4.24}$$

In terms of s, equation (12.4.22) is

$$(b(c^2(bu')')')' = \lambda b^2c^2u \tag{12.4.25}$$

where $' = d/ds$. Put

$$u(x) = v(s)/\{b(s)c(s)\}$$

then

$$bu' = (v' - \theta v)/c \tag{12.4.26}$$

$$c^2(bu')' = c\{v'' - (\theta + \gamma)v' - (\theta' - \theta\gamma)v\} \tag{12.4.27}$$

$$(c^2(bu')')' = c\{v''' - \theta v'' - (2\theta' + \phi)v' + (-\theta'' + \theta\phi)v\} \tag{12.4.28}$$

$$(b(c^2(bu')')')' = bc\{v'^v + (Av')' + Bv\} \tag{12.4.29}$$

where

$$\frac{b'}{b} = \beta, \quad \frac{c'}{c} = \gamma, \quad \theta = \beta + \gamma, \quad \phi = \gamma' + \gamma^2$$

$$A = -3\theta' - \theta^2 - \phi, \quad B = (-\theta'' + \theta\phi)' + (-\theta'' + \theta\phi)\theta. \tag{12.4.30}$$

This means that the transformed system (12.4.23) will correspond to a uniform beam if $A = 0 = B$. We will of course have to check which, if any, of the end conditions are preserved. The only end condition that is preserved in all cases is the clamped condition:

$$u = 0 = \frac{du}{dx} \Longrightarrow v = 0 = \frac{dv}{ds}.$$

Equations (12.4.30) shows that one solution of $A = 0 = B$ is given by $\theta = 0 = \phi$. Since (12.4.26), (12.4.28) show that, when $\theta = 0 = \phi$,

$$bu' = v'/c, \quad (c^2(bu')')' = cv''',$$

any such solution will preserve a sliding-end condition. We explore this solution: $\theta = (bc)'/bc$ so that $\theta = 0$ implies $bc =$ constant; $\phi = c''/c$, so that $\phi = 0$ implies $c(s) = ps + q$, where p, q are constants. The coordinate transformation becomes

$$\frac{ds}{dx} = (ps + q)^{-1}$$

so that

$$\frac{(ps+q)^2}{2p} = x + d.$$

We choose the constants so that $x = 0, 1$ correspond respectively to $s = 0, 1$:

$$1 + ks = (1 + Kx)^{\frac{1}{2}}, \quad 1 + k = (1 + K)^{\frac{1}{2}}.$$

The original beam is given by

$$r(x) = r_0(1 + Kx)^{\frac{3}{2}}, \quad a(x) = a_0(1 + Kx)^{-\frac{1}{2}}.$$

As we noted earlier, this beam will have the same spectrum as a uniform beam for clamped-clamped and clamped-sliding end conditions.

Another solution is given by $b(s) =$ constant. Now $\theta = \gamma$ so that $A = 0$ implies $2\gamma' + \gamma^2 = 0$, and thus $\gamma'' + \gamma\gamma' = 0$ and $-\theta'' + \theta\phi = -\gamma'' + \gamma(\gamma' + \gamma^2) = 0$ so that $B = 0$. Now $2\gamma' + \gamma^2 = 2c''/c - c'^2/c^2 = 0$ and $c = (ps+q)^2$. Since $x = s$ we have

$$r(x) = r_0(px+q)^4 \quad a(x) = a_0(px+q)^4.$$

Other examples are given in the exercises. Gottlieb (1987b) [140] studies many other cases in detail; see also Abrate (1995) [1].

Gottlieb has studied isospectral membranes and plates in Gottlieb (1988) [142], Gottlieb (1991) [144], Gottlieb (1992b) [146], Gottlieb (1993) [147], Gottlieb (2000) [148], Gottlieb (2004a) [150]. In a recent paper, Gottlieb showed that the only mappings of the mapping that transform the membrane equation onto another membrane equation are conferral mappings.

Exercises 12.4

1. Set up the frequency equation for the stepped string in Figure 12.4.1 for fixed-free end conditions and obtain (12.4.5); find ω_n.

2. Show that the product of transformations defined by (12.4.14) is associative, i.e., $a_{(1,2),3} = a_{1,(2,3)}$.

3. Verify (12.4.17).

4. Show that if $\phi(\xi) = (1+b\xi)^n$, then

$$f(x) = (1+a)^2(1+cx)^n/(1+ax)^{n+4}$$

where $c = a + b + ab$.

5. Show that in the special case $n = -2, c = 0$, the dual string is just the original string turned around, i.e., $f(x) = \{1 + b(1-x)\}^{-2}$.

6. The composition law for the transformation group is (12.4.17). Show that if $\alpha = \ell n(1+a)$ then the composition law becomes additive: $\alpha_{1,2} = \alpha_1 + \alpha_2$.

7. Another possible solution of (12.4.30) is given by $A = 0, \theta'' = \theta\phi$. Explore this solution.

8. Explore solutions of $A = 0 = B$ by seeking $b = b_0 S^\mu, c = c_0 S^\nu$ where $S = ps + q$, and μ, ν are to be determined.

12.5 Explicit formulae for potentials

We discussed at length in Chapters 10, 11 what spectral data are necessary to determine the *potential* in Sturm-Liouville equation. By *potential* we mean either $q(x)$ in (10.1.14), $\rho(x)$ (or $\rho^2(x)$) in (10.1.8) or $A(x)$ in (10.1.3). In general, as we have found, there is no explicit formula for a potential; rather, it is found after a long process involving integral and/or differential equations. In this section we describe some explicit formulae that have been found for various particular cases. We will give few derivations since these are generally very lengthy; instead we will make references to the original papers.

We start with Gel'fand and Levitan (1953) [101]. They considered equation (10.1.14) under the free-free end condition $y'(0) = 0 = y'(1)$, with $q(x) \in C^1(0,1)$ and

$$\int_0^1 q(x)dx = 0. \tag{12.5.1}$$

They showed that if $(\lambda_n)_1^\infty$ denote the eigenvalues of (10.1.14), and $(\mu_n)_1^\infty$ are the corresponding eigenvalues of the same equation with $q(x) \equiv 0$, i.e.,

$$-y''(x) = \lambda y(x), \quad y'(0) = 0 = y'(1),$$

then

$$\sum_{n=1}^{\infty}(\lambda_n - \mu_n) = \frac{1}{4}(q(0) + q(1)). \tag{12.5.2}$$

Halberg and Kramer (1960) [158] extended this result to the end conditions

$$y'(0) - hy(0) = 0 = y'(1) + Hy(1). \tag{12.5.3}$$

If h, H are finite, then (12.5.2) holds; if h is finite and $H = \infty$, then

$$\sum_{n=1}^{\infty}(\lambda_n - \mu_n) = \frac{1}{4}(q(0) - q(1)); \tag{12.5.4}$$

if $h = \infty$, H is finite, then

$$\sum_{n=1}^{\infty}(\lambda_n - \mu_n) = \frac{1}{4}(q(1) - q(0)); \tag{12.5.5}$$

and if $h = \infty = H$ (the ends are fixed) then

$$\sum_{n=1}^{\infty}(\lambda_n - \mu_n) = -\frac{1}{4}(q(0) + q(1)). \tag{12.5.6}$$

Barcilon (1974d) [17] gives an alternative derivation of these results. Note that all these formulae give a sum or difference of values of q at the end points as a function of the effect of q on the eigenvalues.

Barcilon (1983) [22] examined the string equation (10.1.8) and considered the eigenfunctions for a *part* of the string; we change his formulation somewhat. Barcilon first proves

Theorem 12.5.1 *Let* $(\lambda_n)_1^\infty$ *be the spectrum of*

$$u'' + \lambda\rho u = 0, \tag{12.5.7}$$

$$u(0) = 0 = u(1), \tag{12.5.8}$$

and $(\mu_n)_1^\infty$ *the spectrum of*

$$u'' + \lambda\rho u = 0,$$

$$u(0) = 0 = u'(1). \tag{12.5.9}$$

If the function $\rho(x)$ *is continuous and bounded away from zero, then*

$$\rho(1) = \frac{1}{\mu_1}\prod_{n=1}^{\infty}\frac{\lambda_n^2}{\mu_{n+1}\mu_n}. \tag{12.5.10}$$

We may make a (rather weak) check on this by considering the case $\rho \equiv 1$ (Ex. 12.5.1).

Now we consider the string with ends at 0 and x, and find the spectra by simply scaling the coordinate so that the string occupies $(0, 1)$. This gives the

Corollary 12.5.1 *If* $(\lambda_n(x))_1^\infty$ *and* $\{\mu_n(x)\}_1^\infty$ *are the spectra of (12.5.7) for the end conditions* $u(0) = 0 = u(x)$, $u(0) = 0 = u'(x)$ *respectively, then*

$$\rho(x) = \frac{1}{x^2\mu_1(x)}\prod_{n=1}^{\infty}\frac{[\lambda_n(x)]^2}{\mu_{n+1}(x)\mu_n(x)}. \tag{12.5.11}$$

Barcilon's formula (12.5.11) involves *two* spectra, for two different end conditions at x. Pranger (1989) [270] expresses $\rho(x)$ in terms of just *one* spectrum $\{\lambda_n(x)\}$ for (12.5.7) subject to $u(0) = 0 = u(x)$. He shows that if

$$s(x) = \sum_{n=1}^{\infty}\{\lambda_n(x)\}^{-1}, \tag{12.5.12}$$

if $\rho(x)$ is positive, has a continuous first derivative, and has a second derivative in L^2, then $\rho(x)$ is given by the remarkable explicit formula

$$\rho(x) = \left(\frac{d^2}{dx^2} + \frac{2}{x}\frac{d}{dx}\right)s(x). \tag{12.5.13}$$

Gottlieb (1992a) [145] considers some examples and counter-examples of this formula. First, if $\rho(x) \equiv 1$, then $\lambda_n(x) = (n\pi/x)^2$ so that

$$s(x) = \frac{x^2}{\pi^2}\sum_{n=1}^{\infty}\frac{1}{n^2}$$

and substitution into (12.5.13) recovers $\rho(x) = 1$. See also Ex. 12.5.2. Prager considers some other explicit formulae.

Gottlieb (1992a) [145] also considers some cases of discontinuous $\rho(x)$ to show that, as Pranger himself thought, his formula holds under wider conditions than he assumed.

Exercises 12.5

1. Take $\rho \equiv 1$ in (12.5.7) and find the eigenvalues λ_n and μ_n for (12.5.7) subject to (12.5.8) and (12.5.9) respectively. Check that equation (12.5.10) gives $\rho(1) = 1$. [Use the identity

$$\prod_{n=1}^{\infty}\left(1 - \frac{1}{4n^2}\right) = \frac{2}{\pi},$$

given in 0.2622 in Gradshteyn and Ryzhik (1965) [152]]

2. The string with density given by (12.4.21), i.e.,

$$\rho(y) = (1+a)^2/(1+ay)^4 \qquad 0 \leq y \leq 1$$

is isospectral to the uniform string. Use the scaling $xy = \xi$ to find $\lambda_n(x)$ and hence recover $\rho(x)$ from equation (12.5.13).

12.6 The research of Y.M. Ram et al.

For the whole of his scientific career, Ram's research has been related, more or less, to some aspect of inverse problems, interpreted in a loose sense. Since much of this work does not fit neatly into just one category, we have chosen to devote this section to it. It is impossible to do justice to it in so short a space and therefore we limit our treatment to the questions that he and his colleagues asked and the methods they used to answer them. We limit our attention to those papers related to undamped vibrating systems.

One of the earliest papers is Ram, Braun and Blech (1988) [272]. This paper is in the tradition of modal analysis, see, for example, Berman (1984) [27]. They ask the following question: for a system with *unknown* mass and stiffness matrices $\mathbf{M}$ and $\mathbf{K}$, but with its first n eigenmodes and eigenvalues known from modal testing, how can one find approximations to the first n eigenmodes and eigenvalues of a modified system $\mathbf{M} + \delta\mathbf{M}, \mathbf{K} + \delta\mathbf{K}$? They show that upper bounds to the eigenvalues are given by the eigenvalue problem

$$(\boldsymbol{\Lambda} + \boldsymbol{\Phi}^T\delta\mathbf{K}\boldsymbol{\Phi})\mathbf{x} = \tau(\mathbf{I} + \boldsymbol{\Phi}^T\mathbf{M}\boldsymbol{\Phi})\mathbf{x}. \qquad (12.6.1)$$

They illustrate their analysis by an example of a vibrating beam. The bounds obtained in this paper were upper bounds to the eigenvalues of the modified structure because they were found as stationary values of the Rayleigh quotient in a constrained subspace. Ram and Braun (1990a) [273] obtain both upper and lower bounds by judicious use of the independent definition of eigenvalues discussed in Section 2.10, and the fact that decreasing (increasing) the stiffness, i.e., the strain energy, of a structure, decreases (increases) the eigenvalues. They show moreover that both the upper and lower bounds that they obtain are optimal. This paper gives the clearest introduction to the search for upper and lower bounds. In Ram and Braun (1990b) [274] they extend their results to obtain upper and lower bounds to n eigenvalues, not necessarily the lowest n, by using Lehman's optimal interval (see Parlett (1980) [264], pp. 198-202). In Ram, Blech and Braun (1990) [275], this analysis is placed in an abstract matrix setting and related to previous matrix analytical results. Braun and Ram (1991) [40] give various examples of the application of this analysis. Ram and Blech (1991) [277] prove a nice result regarding the addition of an *oscillator* of stiffness k and mass m to a vibrating system with a single spatial direction of motion: *the eigenvalues of the original system that are less than k/m increase, while those greater than k/m decrease.* They introduce their analysis with the example given in Ex. 12.6.1.

Ram and Braun (1991) [278] apply the analysis derived in Ram, Braun and Blech (1988) [272] to the inverse problem: find $\delta\mathbf{M}, \delta\mathbf{K}$ to give a specified spectrum. They apply their result to some simple examples. See also Ram and Braun (1993) [280] for more examples.

Ram and Caldwell (1992) [279] consider the reconstruction of a spring mass system with a single direction of motion in which the masses are not simply in-line, as in Jacobi systems, but form a multiply connected system. The data consist of various spectra obtained by anchoring one or more of the masses to the ground. The solution obtained is not unique; note that the graph of the system is not a tree, as in the system considered by Duarte (1989) [81], and described in Section 5.7.

We have already referred to Ram (1993) [276] in Section 4.5. He applies the result in Ram and Blech (1991) [277]to the situation when an *oscillator* of stiffness k and mass m is attached to the free end of an in-line mass-spring system. Ram and Gladwell (1994) [289] consider a finite element model of a vibrating string, for which both the stiffness and the mass matrices are tridiagonal, and show that both these matrices may be constructed from a single eigenvalue and two eigenvectors. Since this method is extremely sensitive to error, they also use overdetermined data. Ram (1994a) [281] discusses the reconstruction of the mass-spring model of a beam in transverse vibration described in Section 2.3 from three eigenvectors, one eigenvalue and the total mass and length of the beam. Unfortunately, no criteria are given for deciding whether the mode/eigenvalue data will lead to a realistic model; see Gladwell, Willms, He, and Wang (1989) [115] for a discussion of this matter. In Ram (1994b) [282] he returns to the idea introduced in Ram and Blech (1991) [277] to enlarge a *spectral gap*: to modify a system so that the modified eigenvalues $\lambda_i^*, \lambda_{i+1}^*$ satisfy $\lambda_i^* < \lambda_i, \lambda_{i+1}^* > \lambda_{i+1}$. He shows that this may be accomplished by judiciously adding *two* oscillators, k_1, m_1 and k_2, m_2 with $k_1/m_1 > \lambda_{i+1}$ and $k_2/m_2 < \lambda_i$. Certain specific conditions, which he states, must be satisfied.

In Ram (1994c) [283] he considers the continuous model for an axially vibrating rod, in the form

$$(ru')' + \lambda a u = 0, \tag{12.6.2}$$

and shows that r and a may be reconstructed from two eigenvalues, the corresponding eigenmodes and the total mass of the rod. He states the conditions on the given modes that ensure that $r(x), a(x)$ will be positive, and gives a number of examples.

Ram and Elhay (1995a) [285] attempt a difficult problem, the reconstruction of the mass and stiffness matrices, $\mathbf{M}, \mathbf{K}$ of a system from modal and spectral data, on the assumption that $\mathbf{M}, \mathbf{K}$ are symmetric band matrices. This construction procedure still leaves many important questions open for further research.

Ram and Elhay (1996) [284] consider the theory of dynamic absorbers, and their use in dynamic modification problems.

In the important paper, Sivan and Ram (1997) [306], the authors confront the realisation that the mass and stiffness matrices for a given kind of system will have a specific form. They consider the forms associated with a general mass-spring system as in Ram and Caldwell (1992) [279]; the mass matrix is diagonal, while the stiffness matrix has negative (or non-positive) off-diagonal elements and is diagonally dominant.

They start with raw spectral data $\Lambda^* = \text{diag}(\lambda_1^*, \lambda_2^*, \ldots, \lambda_n^*)$ and modal data $\mathbf{\Phi}^*$. In theory, the mass and stiffness matrices $\mathbf{M}, \mathbf{K}$ should satisfy

$$\mathbf{\Phi}^T\mathbf{M}\mathbf{\Phi} = \mathbf{I}, \quad \mathbf{\Phi}^T\mathbf{K}\mathbf{\Phi} = \mathbf{\Lambda}$$

so that, in theory

$$\mathbf{M} = \mathbf{\Phi}^{-T}\mathbf{\Phi}^{-1}, \quad \mathbf{K} = \mathbf{\Phi}^{-T}\mathbf{\Lambda}\mathbf{\Phi}^{-1}.$$

But in general, the given matrices Φ^*, Λ^* will not yield a diagonal $\mathbf{M}$, or $\mathbf{K}$ with the required positivity properties. They therefore pose the problem of finding Φ, Λ, near, in some sense, to $\mathbf{\Phi}^*, \mathbf{\Lambda}^*$, such that $\mathbf{M}, \mathbf{K}$, computed from (12.6.3) do have the correct form. They divide the problem into two parts:

Problem 12.6.1 *Given $\mathbf{\Phi}^*$, determine $\mathbf{\Phi}$ such that $\mathbf{M} = \mathbf{\Phi}^{-T}\mathbf{\Phi}^{-1}$ is a positive diagonal matrix, and minimises $||\mathbf{\Phi}^* - \mathbf{\Phi}||$.*

Problem 12.6.2 *Given $\mathbf{\Lambda}^*$ and $\mathbf{\Phi}$, determine $\mathbf{\Lambda}$ which minimises $||\Lambda^* - \Lambda||$, such that $\mathbf{K} = \mathbf{\Phi}^{-T}\mathbf{\Lambda}\mathbf{\Phi}^{-1}$ has the required form.*

They give algorithms for solving both these problems. This paper provides a promising starting point for realistic construction procedures.

Ram and Elhay (1998) [287] is related to isospectral Jacobi systems, and contains a novel fixed-point approach to constructing a particular Jacobi matrix.

Sivan and Ram (1999) [307] return to the analysis of Ram and Braun (1991) [278]. They start from the equations $\mathbf{K}\mathbf{\Phi} = \mathbf{M}\mathbf{\Phi}\mathbf{\Lambda}$ and partition $\mathbf{\Phi}, \mathbf{\Lambda} \in M_n$ into

$$\mathbf{\Phi} = [\mathbf{\Phi}_1|\mathbf{\Phi}_2], \quad \Lambda = \begin{bmatrix} \mathbf{\Lambda}_1 & \\ & \mathbf{\Lambda}_2 \end{bmatrix}$$

where $\mathbf{\Phi}_1 \in M_{n,m}, \mathbf{\Lambda}_1 \in M_m$ are given. They pose the problem of finding $\tilde{\mathbf{\Phi}}_1 = \mathbf{\Phi}_1\mathbf{W}$, and $\tilde{\mathbf{\Lambda}}_1 = diag(\tilde{\lambda}_i)$ and mass and stiffness matrices $\tilde{\mathbf{M}} = \mathbf{M} + \hat{\mathbf{M}}$, $\tilde{\mathbf{K}} = \mathbf{K} + \hat{\mathbf{K}}$ to minimise the norm of

$$R = (\tilde{\mathbf{M}})^{-1/2}(\tilde{\mathbf{K}}\tilde{\mathbf{\Phi}}_1 - \tilde{\mathbf{M}}\tilde{\mathbf{\Phi}}_1\tilde{\mathbf{\Lambda}}_1)$$

and apply their analysis to some simple spring mass problems. The greatest problem to be overcome is that of ensuring that the matrices $\tilde{\mathbf{M}}, \tilde{\mathbf{K}}$ have prescribed forms.

Burak and Ram (2001) [44] treat the eigenvalue equation $(\mathbf{K} - \lambda\mathbf{M})\mathbf{u} = \mathbf{0}$ by writing both $\mathbf{K}$ and $\mathbf{M}$ as sums

$$\mathbf{K} = \sum \alpha_i\mathbf{K}_i, \quad \mathbf{M} = \sum \beta_i\mathbf{M}_i$$

where $\mathbf{K}_i, \mathbf{M}_i$ are matrices with fixed elements that reflect the *connectivity*, the *graph*, of the system, and α_i, β_i are unknown parameters. When the system is an in-line mass-spring system, the parameters may be obtained from two modes and an eigenvalue, as in Ram and Gladwell (1994) [289].

Ram and Elishakoff (2004) [288] return to the problem of reconstructing a rod cross-sectional area from a mode, for both discrete and continuous models. For the continuous model, the governing equation is (10.1.3):

$$(A(x)u'(x))' + \lambda A(x)u(x) = 0.$$

They concentrate on the problem of finding $A(x)$ when $u(x)$ is a *polynomial*, and discuss particular low order polynomials for the fundamental and the first few overtones of a free-free rod.

In conclusion, we note that the research conducted by Ram and his colleagues demonstrates the complexity of inverse problems: the data must be available from testing or elsewhere, the construction algorithms must be robust, and the model that is constructed must be realistic - it should satisfy all the necessary *positivity* and *connectivity* constraints.

Ram and his colleagues have made important advances in many diverse aspects of these matters; in spite of this there is still ample opportunity for more research in fulfilling all the requirements of a satisfactory solution to the many inverse problems that arise in vibration theory.

Exercises 12.6

1. Consider a uniform taut spring of unit length, fixed at $x = 0$, free at $x = 1$ (the end $x = 1$ is attached to a massless ring that slides at right angles to the string). Find its eigenvalues. Now replace the slider by an oscillator of mass m and stiffness k. Show that the eigenvalues $\lambda_i < k/m$ increase, while those with $\lambda_i > k/m$ decrease.

Chapter 13

The Euler-Bernoulli Beam

There is enough light for those who only desire to see, and enough obscurity for those who have a contrary disposition.
Pascal's *Pensées*, 430

13.1 Introduction

The free undamped infinitesimal vibrations, of frequency ω, of a thin straight beam of length L are governed by the Euler-Bernoulli equation

$$\frac{d^2}{dx^2}\left(EI(x)\frac{d^2u(x)}{dx^2}\right) = A(x)\rho\omega^2 u(x), \quad 0 \le x \le L. \tag{13.1.1}$$

Here E is Young's modulus, ρ is the density, both assumed constant; $A(x)$ is the cross-sectional area at section x, $I(x)$ is the second moment of this area about the axis through the centroid at right angles to the plane of vibration (the neutral axis). We put

$$x = Ls, \quad u(s) = u(x), \quad r(s) = \frac{I(x)}{I(x_c)}, \quad a(s) = \frac{A(x)}{A(x_c)}, \tag{13.1.2}$$

$$\lambda = A(x_e)\rho L^4\omega^2/(EI(x_c)), \tag{13.1.3}$$

where $x_c \in [0, L]$. Equation (13.1.1) then becomes

$$(r(s)u''(s))'' = \lambda a(s)u(s), \quad 0 \le s \le 1, \tag{13.1.4}$$

where $' = d/ds$. From now on we will use x rather than s for the dimensionless independent variable. Both $r(x)$ and $a(x)$ are positive, i.e.,

$$r(x) > 0, \quad a(x) > 0, \quad x \in [0, 1].$$

We shall assume throughout that $a(x), r(x) \in C^2[0, 1]$; they are twice continuously differentiable in $[0, 1]$.

For a beam, the most common end-conditions are

$$\begin{array}{rcll} \text{free} & : & u''=0=u''', & (13.1.5)\\ \text{pinned} & : & u=0=u'', & (13.1.6)\\ \text{sliding} & : & u'=0=(ru'')', & (13.1.7)\\ \text{clamped} & : & u=0=u'. & (13.1.8) \end{array}$$

There are certain combinations of these end conditions which allow movement of the beam as a rigid body:

$$\begin{array}{rcll} \text{free-free} & : & u=1,\ \text{and}\ u=x & (13.1.9)\\ \text{free-sliding} & : & u=1, & (13.1.10)\\ \text{sliding-sliding} & : & u=1, & (13.1.11) \end{array}$$

Note that the free-free beam has two *rigid-body modes*. The two which are given above are not orthogonal, but a combination $ax+c$ can be found that is orthogonal to the first, $u=1$. (Ex. 13.1.1).

The ends may be restrained by translational and rotational spring devices. In this case the end conditions are

$$(r(x)u''(x))'_0 + h_1u(0) = 0 = (r(x)u''(x))'_1 - h_2u(1), \quad (13.1.12)$$
$$r(0)u''(0) - k_1u'(0) = 0 = r(1)u''(1) + k_2u'(1). \quad (13.1.13)$$

Here h_1, h_2 are the translational and k_1, k_2 are the rotational stiffnesses. The conditions (13.1.5)-(13.1.8) correspond respectively to $h=0=k$; $h=\infty, k=0$; $h=0, k=\infty$; $h=\infty=k$. We shall say that the system groverned by equations (13.1.4), (13.1.12), (13.1.13) is *positive* if

$$h_1+h_2>0, \quad k_1+k_2>0. \quad (13.1.14)$$

Since $h_1, h_2, k_1, k_2 \geq 0$, this means that one of h_1, h_2 and one of k_1, k_2 must be strictly positive; this rules out rigid-body modes.

Papanicalaou (1995) [261] considers spectral theory for a *periodic* beam; we do not discuss this.

Theorem 13.1.1 *The Euler-Bernoulli operator*

$$\mathcal{B}u \equiv (r(x)u''(x))''$$

is self-adjoint, i.e.,

$$(\mathcal{B}u, v) = (u, \mathcal{B}v)$$

under the end conditions (13.1.12), (13.1.13).

Proof. $(\mathcal{B}u,v) - (u,\mathcal{B}v) = \int_0^1\{(ru'')''v - (rv'')''u\}dx = [(ru'')'v - ru''v' - (rv'')'u + rv''u']_0^1$.

Under any of the conditions (13.1.12), (13.1.13), the bracketed term is zero at each end. ■

Theorem 13.1.2 *Eigenvalues of an Euler-Bernoulli system are non-negative, and are positive iff the system is positive.*

Proof. Suppose $u(x)$ is an eigenfunction of equation (13.1.4) corresponding to λ, then

$$\mathcal{B}u = \lambda a u.$$

Thus $(\mathcal{B}u, \bar{u}) = \lambda(au, \bar{u})$. But $(\mathcal{B}u, \bar{u}) = (u, \mathcal{B}\bar{u}) = (\overline{\bar{u}, \mathcal{B}u}) = (\overline{\bar{u}, \lambda a u}) = \bar{\lambda}(u, a\bar{u}) = \bar{\lambda}(au, \bar{u})$. Thus $\lambda = \bar{\lambda}$ and λ is real. Now

$$(\mathcal{B}u, u) = [(ru'')'u - ru''u']_0^1 + \int_0^1 r(u'')^2 dx$$

so that

$$\begin{aligned} \lambda(au, u) &= h_1 u^2(0) + h_2 u^2(1) + k_1[u'(0)]^2 + k_2[u'(1)]^2 \\ &\quad + \int_0^1 r(u'')^2 dx. \end{aligned} \tag{13.1.15}$$

Since $u(x)$ is an eigenfunction, $(au, u) > 0$. There can be a zero eigenvalue only if the right hand side of (13.1.15) is identically zero. The integral is zero only if $u''(x) \equiv 0$, i.e., $u(x) = cx + d$. Each of the other terms must be separately zero, so that

$$h_1 d^2 = 0 = h_2(c + d)^2 = k_1 c^2 = k_2 c^2. \tag{13.1.16}$$

Suppose h_1, h_2, k_1, k_2 are *finite*. Equation (13.1.16) implies that *either* $k_1 = 0 = k_2$, in which case the system is not positive, *or* $c = 0$. If $c = 0$, then *either* $h_1 = 0 = h_2$, in which case the system is not positive; *or* $d = 0$, in which case $u(x) \equiv 0$, so that $u(x)$ is not an eigenfunction. We conclude that if h_1, h_2, k_1, k_2 are finite, then the eigenvalues are positive only if the system is positive. The cases when one or more of the h_i, k_i are infinite, may be considered similarly (Ex. 13.1.2).

Before introducing the *Green's function* in general, we consider the special case of a cantilever beam, i.e., a beam clamped at $x = 0$, free at $x = 1$. If a unit concentrated load (made dimensionless as in (13.1.2)) is applied to the beam at $x = s$ $(0 < s \le 1)$, then the deflection $u(x)$ and its first two derivatives will be continuous in $[0, 1]$, while its third derivative will have a jump discontinuity at $x = s$. Equilibrium demands

$$[(r(x)u''(x))']_{x=s-}^{x=s+} = 1.$$

The end conditions at $x = 1$, namely

$$u''(1) = 0 = u'''(1),$$

then yield

$$(r(x)u''(x))' = \begin{cases} -1 & , \quad 0 \le x < s, \\ 0 & , \quad s \le x \le 1, \end{cases}$$

and

$$r(x)u''(x) = \begin{cases} s-x & , \quad 0 \le x < s, \\ 0 & , \quad s \le x \le 1. \end{cases}$$

Thus

$$u'(x) = \int_0^{x_0} \frac{(s-t)}{r(t)} dt,$$

where $x_0 = \min(x,s)$, so that the displacement, i.e., the Green's function $G(x,s)$, is

$$G(x,s) = \int_0^{x_0} \frac{(x-t)(s-t)}{r(t)} dt. \tag{13.1.17}$$

Under general end-conditions of the form (13.1.12), (13.1.13), the Green's function has the following properties:

1. $G(x,s)$ is, for fixed s, a continuous function of x, and satisfies the end-conditions (13.1.12), (13.1.13).

2. Except at $x-s$, the first four derivatives of $G(x,s)$ w.r.t. x are continuous in $[0,1]$. At $x = s$, the third derivative has a jump discontinuity given by

$$\left[\frac{\partial}{\partial x}\left(r(x)\frac{\partial^2 G(x,s)}{\partial x^2}\right)\right]_{x=s^-}^{x=s^+} = 1. \tag{13.1.18}$$

3. $B_x G(x,s) = 0$ for $0 \le x < s$ and $s < x \le 1$. ∎

Theorem 13.1.3 *The Green's function is symmetric, i.e., $G(x,s) = G(s,x)$.*

Proof. See Ex. 13.1.3. ∎

Theorem 13.1.4 *If $f(x)$ is piecewise continuous then*

$$u(x) = \int_0^1 G(x,s)f(s)ds \tag{13.1.19}$$

is a solution of

$$\mathcal{B}u = f(x), \tag{13.1.20}$$

and satisfies the end conditions (13.1.12), (13.1.13). Conversely, if $u(x)$ satisfies (13.1.20) and the end conditions (13.1.12), (13.1.13), then it can be represented by (13.1.19).

This follows immediately from the properties (1)-(3).

The construction procedure used for the cantilever beam can be generalised. It can be shown (Ex. 13.1.5) that

$$G(x,s) = \begin{cases} \phi(x)\theta(s) + \psi(x)\chi(s), & 0 \le x \le s, \\ \phi(s)\theta(x) + \psi(s)\chi(x), & s \le x \le 1, \end{cases} \tag{13.1.21}$$

where $\phi(x), \psi(x)$ are linearly independent solutions of $\mathcal{B}u = 0$ satisfying the end-conditions at $x = 0$, while $\theta(x), \chi(x)$ are linearly independent solutions of

$\mathcal{B}u = 0$ satisfying the end-conditions at $x = 1$. Note that for (13.1.17) these functions are

$$\phi(x) = \int_0^x \frac{(s-t)dt}{r(t)}, \quad \psi(x) = \int_0^x \frac{t(x-t)dt}{r(t)}, \quad \theta(x) = x, \quad \chi(x) = -1. \tag{13.1.22}$$

Theorem 13.1.4 allows us to replace the differential equation (13.1.4) and end-conditions (13.1.12), (13.1.13) by the integral equation

$$u(x) = \lambda \int_0^1 a(s)G(x,s)u(s)ds. \tag{13.1.23}$$

Exercises 13.1

1. $\phi_{1,1} = 1$ and $\phi_{1,2} = cx + d$ will be orthogonal rigid-body modes of a free-free beam if

$$\int_0^1 a(x)\phi_{1,1}(x)\phi_{1,2}(x)dx = 0.$$

   Show that when $a(x)$ is symmetrical about $x = \frac{1}{2}$, i..e., $a(x) = a(1-x)$, then $\phi_{1,2}(x) = c(x - \frac{1}{2})$.

2. Show that eigenfunctions $\phi_i(x), \phi_j(x)$ of (13.1.4), (13.1.12), (13.1.13) corresponding to different eigenvalues λ_i, λ_j are orthogonal, i.e.,

$$(\phi_i, a\phi_j) = \int_0^1 a(x)\phi_i(x)\phi_j(x)dx = \delta_{ij}.$$

   Show also that

$$(\phi_i'', r\phi_j'') = \int_0^1 r(x)\phi_i''(x)\phi_j''(x)dx = \lambda_i\delta_{ij}.$$

3. Show that the Green's function $G(x,s)$ for (13.1.4) under the end-conditions (13.1.12), (13.1.13) is symmetric.

4. Show that the Green's function for a pinned-pinned beam is

$$G(x,s) = \left\{ s\int_s^1 \frac{(1-t)^2dt}{r(t)} \right\} + \left\{ (1-x)\int_0^x \frac{t^2dt}{r(t)} + x\int_x^1 \frac{t(1-t)dt}{r(t)} \right\} (1-s)$$

   when $x \le s$. Identify θ, ϕ, ψ, χ for this $G(x,s)$. Show that $G(x,s) > 0$ when $x, s \in (0,1)$, and that $y(x) = G(x,s)$ satisfies $y'(0) > 0, y'(1) < 0$.

5. Establish (13.1.21). Use the fact that $u(x) = G(x,s)$ has the forms

$$u(x) = \begin{cases} c(s)\phi(x) + d(s)\psi(x), & 0 \le x < s, \\ e(s)\theta(x) + f(s)\chi(x), & s < x \le 1. \end{cases}$$

   Now use the facts that u, u', u'' are continuous at s while the third derivative has the jump given by (13.1.18).

13.2 Oscillatory properties of the Green's function

First we prove some preliminary results.

Theorem 13.2.1 *Under the end conditions (13.1.12), (13.1.13) for finite positive h_1, h_2, k_1, k_2, the Green's function $u(x) = G(x, s)$ for $0 \le s \le 1$ satisfies*

$$M'(x) := (r(x)u''(x))' = \begin{cases} -c & 0 \le x < s \\ 1-c & s < x \le 1 \end{cases} \tag{13.2.1}$$

where $0 < c < 1$.

Proof. Properties (2) and (3) of the Green's function imply that (13.2.1) hold for some c; we prove that $0 < c < 1$.

Consider the values of $M(x) \equiv r(x)u''(x)$. The end conditions (13.1.13) preclude the case in which $M(0) \le 0$ and $M(1) \le 0$, for then $u'(0) \le 0$, $u'(1) \ge 0$ so that $u''(x_0) \ge 0$, i.e., $M(x_0) \ge 0$ for some $x_0 \in (0,1)$. But $M(x)$ is linear in each of $(0, s)$ and $(s, 1)$, so that $M(s) \ge 0$, $M'(s-) = (M(s) - M(0))/s \ge 0$, $M'(s+) = (M(1) - M(s))/(1-s) \le 0$, and therefore $[M'(s)]_-^+ \le 0$, contradicting (13.1.18).

Secondly, if $c \le 0$ or $c \ge 1$, i.e., if $M'(x)$ has the same weak sign throughout $[0,1]$, then the case $M(0) \ge 0$, $M(1) \ge 0$ is excluded. For then $M(x) \ge 0$ throughout $[0,1]$, while the end conditions yield $u'(0) \ge 0$, $u'(1) \le 0$ which contradicts $u''(x) \ge 0$ for all $x \in [0,1]$. Clearly, $u(x) \equiv 0$ is excluded.

Suppose $c \le 0$, so that $M'(x) \ge 0$, $x \in [0,1]$, then $M(0) \le 0$, $M(1) \ge 0$, since all other cases are excluded. Thus, the end condition (13.1.13) yields $u'(0) \le 0$, $u'(1) \le 0$ and, since $M(x)$ is piecewise linear, we may argue as before that $u'(x) \le 0$ for all $x \in [0,1]$. But $M'(0) \ge 0$, $M'(1) \ge 0$ and the end condition (13.1.12) imply $u(0) \le 0$, $u(1) \ge 0$ contradicting $u'(x) \le 0$ for all $x \in [0,1]$. The case $u'(x) \equiv 0$ is excluded by (13.1.18).

If $c \ge 1$ then $M'(x) \le 0$, $x \in [0,1]$ and $M(0) \ge 0$, $M(1) \le 0$ and the end conditions yield $u'(0) \ge 0$, $u'(1) \ge 0$ and, as before, $u'(x) \ge 0$ for $x \in [0,1]$. But now $u(0) \ge 0$, $u(1) \le 0$, which is again contradictory.

We conclude that $0 < c < 1$. ■

Corollary 13.2.1 *$M(x)$ cannot have the same sign throughout $[0,1]$.*

Proof. If $M(x) \le 0$, $x \in [0,1]$, then $M(0) \le 0$ and $M(1) \le 0$, which has been excluded. If $M(x) \ge 0$ then $M(0) \ge 0$, $M(1) \ge 0$ so that the end conditions (13.1.13) yield $u'(0) \ge 0$, $u'(1) \le 0$ which contradicts $M(x) \ge 0$. ■

In this Theorem and Corollary, we have assumed that h_1, h_2, k_1, k_2 are finite and positive, but the results still hold even if some or all of them are infinite, and $h_1 + h_2 > 0, k_1 + k_2 > 0$, i.e., provided the system is *positive*. See Ex. 13.2.1.

Theorem 13.2.2 *Under the end conditions (13.1.12), (13.1.13), the Green's function satisfies*

$$G(x,s) \geq 0, \quad x,s \in [0,1], \tag{13.2.2}$$
$$G(x,s) > 0, \quad x,s \in I. \tag{13.2.3}$$

Proof. Here I has the same meaning as in Chapter 10: it is $[0,1]$ if h_1, h_2 are finite, $(0,1]$ if $h_1 = \infty$, i.e., $u(0) = 0$, etc.

Theorem 13.2.1 and the Corollary show that $M(x)$ cannot have the same sign throughout $[0,1]$; there is one zero to the left of s and/or one zero to the right. If $u(x) = G(x,s)$, then the three possible forms of $M'(x), M(x), u'(x), u(x)$ are shown in Figure 13.2.1. ■

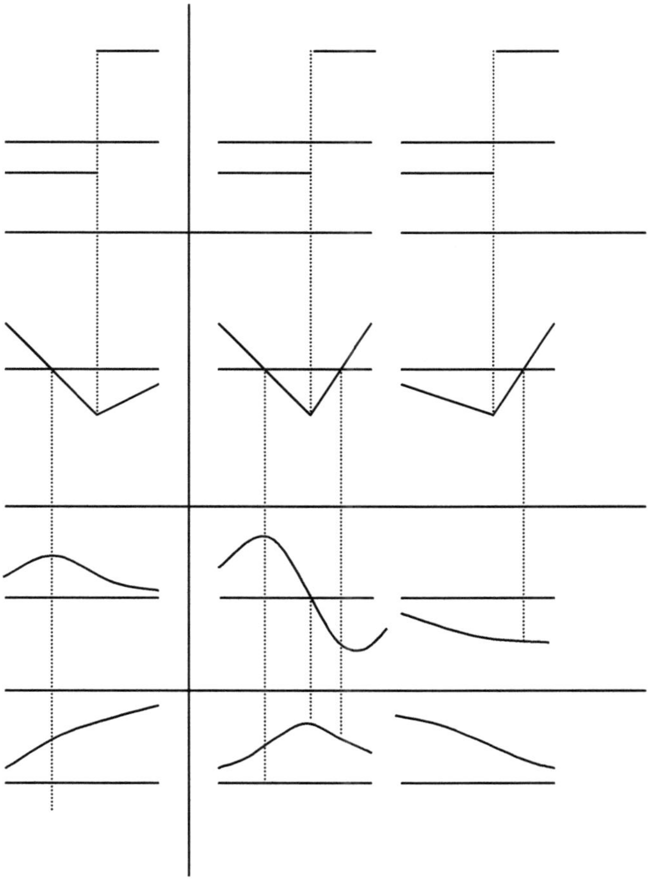

Figure 13.2.1 - The formation of the Green's function, showing $M'(x)$, $M(x)$, $u'(x)$, $u(x)$ in $[0,1]$.

In anticipation of the next result, we recall a classical theorem and prove a refinement.

Theorem 13.2.3 *(Rolle) Suppose $\phi(x)$ is continuous in $[a,b]$ and differentiable in (a,b). If $\phi(a) = 0 = \phi(b)$ then $\exists\ c \in (a,b)$ such that $\phi'(c) = 0$, i.e., $\phi'(x)$ has a zero place in (a,b). We need the following refinement.*

Theorem 13.2.4 *Suppose $\phi(x)$ is continuous in $[a,b]$ and differentiable in (a,b). If $\phi(a) = 0 = \phi(b)$ and $\phi(x)$ is not identically zero in $[a,b]$, then $\phi'(x)$ has a* ***nodal place*** *in (a,b).*

We recall that $f(x)$ is said to have a *node* at c if in any two-sided vicinity of c there are points ξ_1, ξ_2 such that $\xi_1 < c < \xi_2$ and $f(\xi_1)f(\xi_2) < 0$. Alternatively $f(x)$ can have a *nodal interval* $[c,d]$ such that in any two-sided vicinity of $[c,d]$ there are points ξ_1, ξ_2 such that $\xi_1 < c < d < \xi_2$ and $f(\xi_1)f(\xi_2) < 0$.

Proof. Since $\phi(x)$ is continuous in $[a,b]$, it assumes its maximum and minimum values in $[a,b]$. Since $\phi(x)$ is not identically zero, one of these must be non-zero. Without loss of generality we may suppose that it is the maximum; it will therefore be assumed at one or more points $\xi \in (a,b)$, or in an interval $[c,d] \in (a,b)$. In the former case ξ is a node of $\phi'(x)$, in the latter $[c,d]$ is a nodal place. ■

Theorem 13.2.5 *Suppose $\phi(x), \phi'(x)$ are continuous in $[a,b]$, and $\phi'(x)$ has n nodes $(\xi_i)_1^n$ such that $a = \xi_0 < \xi_1 < \xi_2 < \cdots < \xi_n < \xi_{n+1} = b$, then the function $\phi(x)$ has at most one zero place in each of the intervals $[a,\xi_1], [\xi_1,\xi_2], \ldots [\xi_n, b]$, $n+1$ in all. If $\phi(a)\phi'(a) > 0$ then $\phi(x)$ has no zero in $[a,\xi_1]$, while if $\phi(b)\phi'(b) <$ 0 it has no zero in $[\xi_n, b]$. The satisfaction of each of these inequalities thus reduces the number of zero places of $\phi(x)$ by 1.*

Proof. The first part follows from Theorem 13.2.4: if $\phi(x)$ had two zeros in $[\xi_i, \xi_{i+1}]$, then $\phi'(x)$ would have a nodal place in (ξ_i, ξ_{i+1}), contrary to hypothesis.

For the second we note that if $\phi(a)\phi'(a) > 0$ then $\phi(a)\phi'(\xi) > 0$ for $\xi \in [a, \xi_1)$. The mean value theorem states that for every $x \in [a,\xi_1]$ there is a $\xi \in (a,x)$ such that

$$\phi(a)\phi(x) = \phi(a)[\phi(a) + (x-a)\phi'(\xi)] > 0.$$

Thus $\phi(x)$ has no zero in $[a,\xi_1]$. Similarly, if $\phi(b)\phi'(b) < 0$ then $\phi(x)$ has no zero in $[\xi_n, b]$. ■

Corollary 13.2.2 *If instead of being continuous and having nodes at $(\xi_i)_1^n$, $\phi'(x)$ is continuous and of one sign in each of the intervals $[a,\xi_1), (\xi_1,\xi_2), \ldots, (\xi_n, b]$, and has jumps and may thus change sign only at $(\xi_i)_1^n$, then the results concerning $\phi(x)$ still hold.*

We are now ready for

Theorem 13.2.6 *Under the action of n forces $(F_i)_1^n$ acting at $(s_i)_1^n$, where $0 \le s_1 < s_2 < \cdots < s_n \le 1$, the beam can reverse its sign at most $n-1$ times.*

Proof. We assume that the beam is a positive system, as described in Section 13.1. First, we assume that h_1, h_2, k_1, k_2 are positive, $s_1 > 0$ and $s_n < 1$. The deflection of the beam is

$$u(x) = \sum_{i=1}^{n} F_i G(x, s_i),$$

and because of (13.1.18) it satisfies

$$M'(x) = \begin{cases} c_0, & x \in [0, s_1) \\ c_i & x \in (s_i, s_{i+1}), \quad i = 1, \ldots, n-1 \\ c_n & x \in (s_n, 1] \end{cases}$$

where

$$c_i = c_0 + \sum_{j=1}^{i} F_j, \quad i = 1, 2, \ldots, n.$$

Thus $M'(x)$ has the property stated in the Corollary to Theorem 13.2.5, so that $M(x)$ has at most $n+1$ zero places, at most one in each of $[0, s_1], [s_1, s_2] \ldots [s_n, 1]$. Thus $M(x)$, and therefore $u''(x)$, has at most $n+1$ nodes in $(0, 1)$, so that by Theorem 13.2.5, $u'(x)$ has at most $n+2$ nodes and $u(x)$ has at most $n+3$ nodes in $(0, 1)$.

Now consider the sequences

$$M'(0), M(0), u'(0), u(0) \text{ and } M'(1), M(1), u'(1), u(1)$$

i.e.,

$$-h_1 u(0), k_1 u'(0), u'(0), u(0) \text{ and } h_2 u(1), -k_2 u'(1), u'(1), u(1).$$

First, suppose that $u(0), u'(0), u(1), u'(1)$ are all non-zero, then equation (13.1.13) shows that

$$u'(0)u''(0) > 0 \text{ and } u'(1)u''(1) < 0.$$

But then Theorem 13.2.5 states that $u'(x)$ has at most n nodes, $u(x)$ has at most $n+1$ nodes. Now

$$\begin{aligned} M'(0)M(0) &= -h_1 k_1 u'(0)u(0) \\ M'(1)M(1) &= -h_2 k_2 u'(1)u(1) \end{aligned}$$

so that *either* $M'(0)M(0) > 0$ *or* $u'(0)u(0) > 0$ and either $M'(1)M(1) > 0$ or $u'(1)u(1) > 0$. If one of the left hand inequalities is satisfied the $M(x)$ has one less node than before, while if one of the right hand inequalities is satisfied then $u(x)$ has one less node. Thus in any case $u(x)$ has at most $n-1$ nodes.

A detailed consideration of special cases is left to the exercises, but in the typical case $h_1 = 0$, k_1 finite and non-zero, we may argue as follows. $M'(0) = 0$, so that $M'(x) \equiv 0$ in $[0, s_1]$, $M(x) = k_1 u'(0)$ in $[0, s_1]$, so that $M(x)$ has no node in $[0, s_1]$; $u''(0)u'(0) > 0$ so that the remainder of the argument holds. ■

Theorem 13.2.6 holds the key to the proof that the Green's function is an oscillatory kernel. However, in order to prove this, we must continue the investigation of oscillatory systems of functions started in Section 10.5. First, we introduce

Definition 13.2.1 *The function $\phi(x)$ is said to reverse its sign k times in the interval I, and this is denoted by $s\phi = k$, if there are $k+1$ points $(x_i)_1^{k+1}$ in I such that $x_1 < x_2 < \cdots < x_{k+1}$ and*

$$\phi(x_i)\phi(x_{i+1}) < 0 \quad i = 1, 2, \ldots, k+1$$

and there do not exist $k+2$ points with this property. Evidently, if $\phi(x)$ is continuous in $[0,1]$ and $s\phi = k$, then $\phi(x)$ has k nodal places in $(0,1)$.

Before going further we state a basic composition formula. Suppose $\{\phi_i(x)\}_1^n$ are continuous in $[0,1]$, $M(x,s)$ is continuous in $[0,1] \times [0,1]$,

$$\psi_i(x) = \int_0^1 M(x,s)\phi_i(s)ds$$

then

$$\Psi(\mathbf{x};\boldsymbol{\theta}) = \int\!\!\int_V\!\!\int \cdots \int M(\mathbf{x};\mathbf{s})\Phi(\mathbf{s};\boldsymbol{\theta})d\mathbf{s} \qquad (13.2.4)$$

where V is the simplex defined by $0 \le s_1 < s_2 < \cdots < s_n \le 1$,

$$\begin{aligned}
\Psi(\mathbf{x};\boldsymbol{\theta}) &= \det(\psi_i(x_j)) = \Psi(x_1, x_2, \ldots, x_n; 1, 2, \ldots, n) \\
M(\mathbf{x};\mathbf{s}) &= \det(M(x_i, s_j)) \\
\Phi(\mathbf{s};\boldsymbol{\theta}) &= \det(\phi_i(s_j)) = \Phi(s_1, s_2, \ldots, s_n; 1, 2, \ldots, n)
\end{aligned}$$

$d\mathbf{s} = ds_1 ds_2 \ldots ds_n$, and $\boldsymbol{\theta} = \{1, 2, \ldots, n\}$.

Theorem 13.2.7 *Suppose $\phi(x)$ is continuous in $[0,1]$ and $s\phi \le n-1$. If $M(x,s)$ is a continuous kernel with the property*

$$M(\mathbf{x};\mathbf{s}) > 0 \textit{ for } \mathbf{x};\mathbf{s} \in Q,$$

then

$$\psi(x) = \int_0^1 M(x,s)\phi(s)ds$$

does not vanish more than $n-1$ times in $[0,1]$.

Proof. Q is defined in Definition 10.5.1.

By assumption there are $n+1$ points $\{\xi_i\}_0^n$ such that $0 = \xi_0 < \xi_1 < \cdots < \xi_n = 1$ and $\phi(x)$ has one sign and is not identically zero on each interval (ξ_{i-1}, ξ_i), $1, 2, \ldots, n$. Put

$$\psi_i(x) = \int_{\xi_{i-1}}^{\xi_i} M(x,s)\phi(s)ds$$

then

$$\psi(x) = \sum_{i=1}^{n} \psi_i(x)$$

and for all $(x_i)_1^n$ such that $0 \le x_1 < x_2 < \cdots < x_n \le 1$ we have (Ex. 13.2.2)

$$\Psi(\mathbf{x};\boldsymbol{\theta}) = \int_{\xi_{n-1}}^{\xi_n} \cdots \int_{\xi_0}^{\xi_1} M(\mathbf{x};\mathbf{s})\phi(s_1)\phi(s_2)\ldots\phi(s_n)d\mathbf{s}. \tag{13.2.5}$$

The integrand is not identically zero and its non-zero values have one and the same sign, so that $\Psi(\mathbf{x};\boldsymbol{\theta})$ is strictly of one sign, and hence the $(\psi_i(x))_1^n$ form a Chebyshev sequence (Definition 10.6), and $\psi(x)$ does not vanish more than $n-1$ times in $[0,1]$. ■

An important kernel that satisfies the conditions of Theorem 13.2.7 is provided by

$$M_\varepsilon(x,y) = \frac{2}{\sqrt{\pi}\varepsilon}\exp\{-(x-y)^2/\varepsilon^2\}.$$

This kernel has a remarkable property. Suppose $\phi(x) \in C[0,1]$, and define

$$\psi(x,\varepsilon) = \int_0^1 M_\varepsilon(x,s)\phi(s)ds. \tag{13.2.6}$$

If we define $\phi(s) = 0$ for $x > 1$, then

$$\begin{aligned}\psi(x,\varepsilon) &= \tfrac{2}{\sqrt{\pi}\varepsilon}\textstyle\int_0^\infty \exp\{-(x-s)^2/\varepsilon^2\}\phi(s)ds,\\ &= \tfrac{2}{\sqrt{\pi}}\textstyle\int_0^\infty \exp(-\xi^2)\phi(x+\varepsilon\xi)d\xi,\end{aligned}$$

and

$$\lim_{\varepsilon\to 0}\psi(x,\varepsilon) = \frac{2}{\sqrt{\pi}}\int_0^\infty \exp(-\xi^2)d\xi.\phi(x) = \phi(x). \tag{13.2.7}$$

Using this kernel we may prove

Theorem 13.2.8 *Let $\{\phi_i(x)\}_1^n$ be linearly independent functions in $C[0,1]$, and define*

$$\phi(x) = \sum_{i=1}^{n} c_i\phi_i(x).$$

The necessary and sufficient conditions for $s\phi \le n-1$ in $[0,1]$ for all c_i not all zero, is that

$$\Phi(\mathbf{x};\boldsymbol{\theta}) \equiv \Phi(x_1,x_2,\ldots,x_n;1,2,\ldots,n), \quad \mathbf{x} \in Q$$

should have fixed sign, i.e., one and the same sign for those points for which it is not zero.

Proof. If for certain c_i not all zero, $s\phi \le n-1$ and

$$\psi_i(x,\varepsilon) = \int_0^1 M_\varepsilon(x,s)\phi_i(s)ds$$

then Theorem 13.2.7 shows that

$$\psi(x,\varepsilon) = \sum_{i=1}^{n} c_i\psi_i(x,\varepsilon)$$

vanishes in $[0,1]$ not more than $n-1$ times.

Conversely, equation (13.2.7) shows that if $\psi(x,\varepsilon)$ vanishes not more than $n-1$ times, then $s\phi \le n-1$. Thus $s\phi \le n-1$ for all c_i, not all zero, iff $\{\psi_i(x,\varepsilon)\}_1^n$ form a Chebyshev system in $[0,1]$, i.e., iff

$$\mathbf{\Psi}_\varepsilon(\mathbf{x};\boldsymbol{\theta}) := \det(\psi_i(x_j,\varepsilon))$$

has strictly fixed sign when $\mathbf{x} \in Q$. If $\Psi_\varepsilon(\mathbf{x};\alpha)$ has strictly fixed sign, then

$$\mathbf{\Phi} \equiv \mathbf{\Phi}(\mathbf{x};\boldsymbol{\theta}) = \lim_{\varepsilon\to 0} \mathbf{\Psi}_\varepsilon(\mathbf{x};\boldsymbol{\theta})$$

will have strictly fixed sign. On the other hand, since the $\{\phi_i(x)\}_1^n$ are linearly independent, $\mathbf{\Phi}$ will not be identically zero. Thus (13.2.4) with $M = M_\varepsilon$ shows that if $\mathbf{\Phi}$ has fixed sign, then $\mathbf{\Psi}_\varepsilon$ will have fixed sign. ■

We have now established all the results needed to prove

Theorem 13.2.9 *The Green's function of a positive Euler-Bernoulli system is oscillatory.*

Proof. There are three conditions to be fulfilled in the Definition 10.5.1. Theorem 13.2.2 yields i), the argument via strain energy yields iii). It remains to prove ii). Theorem 13.2.5 states that if

$$u(x) = \sum_{i=1}^{n} F_i G(x,s_i)$$

then $su \le n-1$. Put $\phi_i(x) = G(x,s_i)$, then, since the $\{\phi_i(x)\}_1^n$ are linearly independent, Theorem 13.2.8 states that

$$\Phi(x_1,x_2,\ldots,x_n;1,2,\ldots,n) = G(\mathbf{x};\mathbf{s}) \ge 0$$

for $\mathbf{x},\mathbf{s} \in Q$. This is ii) ■

The following theorem states which of the determinants in ii) are zero and which are non-zero; it is the analogue of Theorem 10.5.4 for the beam

Theorem 13.2.10 $G(\mathbf{x};\mathbf{s}) > 0$ *iff* $\mathbf{x},\mathbf{s} \in \mathcal{I}$ *and* $x_i < s_{i+2}$ *and* $s_i < x_{i+2}$ *for* $i = 1,2,\ldots,n-2$.

Proof. The first condition is necessary; for if one of x_1, x_n, s_1, s_n is not in $\mathcal{I}$, e.g., $x_1 = 0$, then $G(x_1, s_i) = 0$ for $i = 1, 2, \ldots, n$ and the determinant is zero.

Now suppose there is an index k such that $1 \le k \le n-2$ and $x_k \ge s_{k+2}$. Then $x_i \ge s_j$ for $i = k, k+1, \ldots, n$ and $j = 1, 2, \ldots, k+2$. Consider entries in the submatrix taken from rows $k, k+1, \ldots, n$ and columns $1, 2, \ldots, k+2$ of the matrix $(G(x_i, s_j))$. Since $x_o \ge s_j$ for each entry, equation (13.1.21) shows that

$$G(x_i, s_j) = \phi(s_j)\theta(x_i) + \psi(s_j)\chi(x_i)$$

so that the matrix will have rank ≤ 2. If $n = 3$, then $k = 1$ and the submatrix is the complete matrix which has rank 2, and therefore has zero determinant. If $n \ge 4$, then we evaluate the $n \times n$ determinant using Laplace's expansion with minors of order $k+2$ taken from the first $k+2$ columns; each such minor, having $k+2 \ge 3$ rows, will be zero, so that the determinant will be zero. Thus $x_i < s_{i+2}$ is necessary for $G(\mathbf{x}; \mathbf{s})$ to be positive, and so similarly is $s_i < x_{i+2}$.

Now we prove the sufficiency. Suppose $\mathbf{x}, \mathbf{s} \in \mathcal{I}$. $x_i < s_{i+2}$, and $s_i < x_{i+2}$ for $i = 1, 2, \ldots, n-2$. We will prove the determinant is positive by induction. When $n = 1$, the results holds, by Theorem 13.2.2. Suppose that, if possible, it holds for $n-1$, but not for n, i.e., there exist $(x_i^0)_1^n, (s_i^0)_1^n$ in I and satisfying $x_i^0 < s_{i+2}, s_i^0 < x_{i+2}^0$ for $i = 1, 2, \ldots, n-2$ such that $G(x_1^0, x_2^0, \ldots, x_n^0; s_1^0, s_2^0, \ldots, s_n^0) = 0$, but $G(x_1^0, x_2^0, \ldots, x_{n-1}^0; s_1^0, s_2^0, \ldots, s_{n-1}^0) > 0$ and $G(x_2^0, x_3^0, \ldots, x_n^0; s_2^0, s_3^0, \ldots, s_n^0) > 0$. Now choose arbitrary points $(x_i)_1^n, (s_i)_1^n$ such that

$$x_1^0 \le x_1 < x_2 < \cdots < x_n \le x_n^0, \quad s_1^0 \le s_1 < x_2 < \cdots < s_n \le s_n^0$$

and renumber $(x_i)_1^n, (x_i^0)_1^n$ increasingly as $(x_i')_1^{2n}$. The $2n \times 2n$ matrix $(G(x_i', s_i'))$ is TN and the minors corresponding to $(x_i^0)_1^n$ and $(s_i^0)_1^n$ fit the criteria of Theorem 6.6.6. Therefore, the matrix has rank $n-1$, so that

$$G(x_1, x_2, \ldots, x_n; s_1, s_2, \ldots, s_n) = 0. \qquad (13.2.8)$$

There are two cases: $n \ge 3$ and $n = 2$. In the first case $x_1^0 < s_3^0 \le s_n^0$ and $s_1^0 < x_3^0 \le x_n^0$ imply that the intervals (x_1^0, x_n^0) and (s_1^0, s_n^0) overlaps. We may therefore take $x_i = s_i$ for $i = 1, 2, \ldots, n$, so that (13.2.8) yields $G(x_1, \ldots, x_n; x_1, \ldots, x_n) = 0$ contradicting condition iii) of Definition 10.5.1.

If $n = 2$, equation (13.2.8) states that $G(x_1, x_2, ; s_1, s_2) = 0$ for all x_1, x_2, s_1, s_2 satisfying $x_1^0 \le x_1 < x_2 \le x_2^0, s_1^0 \le s_1 < s_2 \le s_2^0$. Without loss of generality we can take $x_2^0 \le s_1^0$, so that $x_1 < s_1 < s_2, x_2 \le s_1 < s_2$ and

$$\begin{aligned} G(x_1, x_2, ; s_1, s_2) &= \begin{vmatrix} \phi(x_1)\theta(s_1) + \psi(x_1)\chi(s_1), \phi(x_1)\theta(s_2) + \psi(x_1)\chi(s_2) \\ \phi(x_2)\theta(s_1) + \psi(x_2)\chi(s_1), \phi(x_2)\theta(s_2) + \psi(x_2)\chi(s_2) \end{vmatrix} \\ &= \begin{vmatrix} \phi(x_1) & \psi(x_1) \\ \phi(x_2) & \psi(x_2) \end{vmatrix} \bullet \begin{vmatrix} \theta(s_1) & \chi(s_1) \\ \theta(s_2) & \chi(s_2) \end{vmatrix} = 0. \end{aligned}$$

One or other of the factors in this equation must be zero. Suppose that for some s_1, s_2 the second factor is not zero, then the first must be zero for all

x_1, x_2 such that $x_1^0 \le x_1 < x_2 \le x_2^0$. But that means that $\phi(x), \psi(x)$ are proportional, contradicting the fact that there are linearly independent solutions of $(EIy'')'' = 0$ satisfying the end conditions at $x = 0$. Similarly, if the first factor is not zero for some x_1, x_2, then the second must be identically zero, which again is impossible. Hence, we have arrived at a contradiction. The stated conditions are sufficient to ensure that $G(x_1, x_2, \ldots, x_n; s_1, s_2, \ldots, s_n) > 0$. ■

Exercises 13.2

1. Establish Theorem 13.2.1 when some of the h_i, k_i are 0 or ∞, but the system is still positive.

2. Verify equation (13.2.4) in the case $n = 2$. Show that

$$\begin{aligned}\Psi(x_1, x_2; 1, 2) &= \int_0^1 \int_0^1 M(x_1, x_2; s_1, s_2)\phi(s_1)\phi(s_2)ds_2ds_1 \\ &= \frac{1}{2!}\int_0^1 \int_0^1 M(x_1, x_2; s_1, s_2)\Phi(s_1, s_2; 1, 2)ds_2ds_1 \\ &= \int_0^1 \int_0^{s_1} M(x_1, x_2; s_1, s_2)\Phi(s_1, s_2; 1, 2)ds_2ds_1.\end{aligned}$$

3. Establish equation (13.2.5) for $n = 2$.

4. Verify the Corollary of Theorem 13.2.5.

5. Establish Theorem 13.2.6 when some of the h_i, k_i are 0 or ∞, but the system is still positive.

13.3 Nodes and zeros for the cantilever beam

For the cantilever beam the governing equations are

$$(r(x)u''(x))'' = \lambda a(x)u(x), \tag{13.3.1}$$

$$u(0) = 0 = u'(0), \quad u''(1) = 0 = u'''(1). \tag{13.3.2}$$

The theory of Section 13.2 shows that the Green's function for the beam is an oscillatory kernel on $I = (0, 1]$, so that the eigenvalues $(\lambda_i)_1^\infty$ are distinct, and the eigenfunctions $(\phi_i(x))_1^\infty$ have properties (1)-(3) stated in Theorem 10.6.4.

We need to strengthen this classical result. To do so, we suppose, as in Section 13.1 that $a(x), r(x) \in C^2[0, 1]$, and put $M(x) = r(x)u''(x)$. Equation (13.3.1)-(13.3.2) show that $M(x)$ satisifies

$$(b(x)M''(x))'' = \lambda s(x)M(x), \tag{13.3.3}$$

$$M''(0) = M'''(0), \quad M(1) = 0 = M'(1), \tag{13.3.4}$$

where $b(x) = 1/a(x)$, $s(x) = 1/r(x)$. Thus $M(x)$ is an eigenfunction of a reversed cantilever on $(0, 1)$, and is thus an eigenfunction of an oscillatory kernel on $[0, 1)$.

We now state

Theorem 13.3.1 *If $\{\phi_i(x)\}_1^\infty$ are the eigenfunctions of a cantilever beam, then*

1. $\phi_1(x), \phi_1'(x)$ have no zeros in $(0, 1]$,

2. $\phi_i(x), \phi_i'(x)$ have $(i-1)$ nodes in $(0, 1)$ and no other zeros in $(0, 1]$,

3. If
$$\phi(x) = \sum_{i=j}^{k} c_i\phi_i(x), \quad 1 \le j \le k, \quad \sum_{i=j}^{k} c_i^2 > 0,$$
then $\phi(x)$ and $\phi'(x)$ have not less than $(j-1)$ nodes and not more that $(k-1)$ zeros in $(0, 1]$,

4. $M_i(x) := r(x)\phi_i''(x)$ and $M_i'(x)$ have the properties 2) and 3) on $[0, 1)$.

Proof. The stated properties of $\phi_i(x), M_i(x)$ follow from Theorem 10.6.4. We verify those for $\phi_i'(x), M_i'(x)$.

1) $\phi_1'(0) = 0$; if $\phi_1'(x_0) = 0$ for some $x_0 \in (0, 1]$, then Rolle's Theorem 13.2.3 states that there is a $\xi \in (0, x_0)$ such that $\phi_1'(\xi) = 0$, contradicting the fact that $M_1(x)$ has no zero in $[0, 1)$.

2) $\phi_i(x)$ has $i-1$ nodes $(x_j)_1^{i-1}$ in $(0, 1)$ and a zero at $x_0 = 0$. By Theorem 13.2.4, $\phi_1'(x)$ has $i-1$ nodes $(\xi_j)_1^{i-1}$ satisfying $x_{j-1} < \xi_{j-1} < x_j$, $j = 1, \ldots, i-1$; it also has a zero at $x = 0$. If $\phi_i'(x)$ has any other zero in $(0, 1]$ then $\phi_i''(x)$ would have more than $i-1$ zeros in $(0, 1]$, contradicting 4) for M_i.

3) The part relating to the $\phi_i'(x)$ may be proved in a similar way. See Ex. 13.3.1.

4) This follows because $M_i(x)$ is an eigenfunction of the reversed cantilever. ■

Theorem 13.3.2 *If $\phi_i(x)$ is an eigenfunction of a cantilever beam then $\phi_i(1)\phi_i'(1) > 0$.*

Proof. Theorem 13.3.1 shows that $\phi_i(1)\phi_i'(1) > 0 \neq 0$; we show that $\phi_i(1)$ and $\phi_i'(1)$ have the same sign. The Green's function for the cantilever is given in equation (13.1.17), and
$$\phi_i(x) = \lambda_i \int_0^1 G(x, s)a(x)\phi_i(s)ds,$$
so that
$$\phi_i(x) = \lambda_i \int_0^1 \frac{\partial G(x, s)}{\partial x} a(s)\, \phi_i(s)ds.$$
Since
$$\frac{\partial G}{\partial x}(x, s) = \int_0^{\min(x,s)} \frac{(s-t)dt}{r(t)},$$
we find that
$$[\phi_i'(1)]_x^1 = \lambda_i \int_x^1 a(s) \left\{ \int_x^s \frac{(s-t)\phi_i(t)dt}{r(t)} \right\} ds. \tag{13.3.5}$$

Suppose x^* is the largest zero of $\phi_i(x)$; it will be a node if $i \geq 2$, and 0 if $i = 1$. Since $\phi_i(1) > 0$, we have $\phi_i'(x^*) > 0$, and thus $\phi_i(x) > 0$ for $x \in (x^*, 1]$. Thus equation (13.3.5) yields

$$\phi_i'(1) - \phi_i'(x^*) > 0$$

so that $\phi_i'(1) > 0$. ■

Corollary 13.3.1 *If $\phi_i(x)$ is an eigenfunction of a cantilever beam then $M_i(0)M_i'(0) > 0$.*

Exercises 13.3

1. Establish the part 3) of Theorem 13.3.1 relating to $\phi_i'(x)$.

13.4 The fundamental conditions on the data

We are now in a position to prove the fundamental

Theorem 13.4.1 *Suppose $a(x), r(x)$ has derivatives of all orders, (This restriction can be relaxed but it is sufficient for our purposes.) then the infinite matrix*

$$P = \begin{bmatrix} u_1 & u_2 & u_3 & \cdots \\ \theta_1 & \theta_2 & \theta_3 & \cdots \\ \lambda_1 u_1 & \lambda_2 u_2 & \lambda_3 u_3 & \cdots \\ \lambda_1 \theta_1 & \lambda_2 \theta_2 & \lambda_3 \theta_3 & \cdots \\ \lambda_1^2 u_1 & \lambda_2^2 u_2 & \lambda_3^2 u_3 & \cdots \\ \vdots & & & \end{bmatrix}$$

is TP. Here $u_i := \phi_i(1)$, $\theta_i := \phi_i'(1)$, and the $\phi_i(x)$ have been chosen so that $\phi_i(1) > 0$.

Before starting the proof proper, we give the gist of the argument in a simple case.

Consider the determinant

$$\phi(x) = \begin{bmatrix} \phi_2(x) & \phi_3(x) & \phi_4(x) \\ u_2 & u_3 & u_4 \\ \theta_2 & \theta_3 & \theta_4 \end{bmatrix};$$

since this may be written

$$\phi(x) = \sum_{i=2}^{4} c_i \phi_i(x),$$

Theorem 13.3.1 states that it has at least one node in $(0, 1)$, and at most 3 zeros in $(0, 1]$. In fact, since $\phi(1) = 0 = \phi'(1) = \phi''(1) = \phi'''(1)$, it has a fourfold zero at $x = 1$. Since Theorem 13.3.1 does not state how to count such a multiple zero, we must consider the zeros of $\phi'(x)$ and $\phi''(x)$. We know that $\phi(x)$ has one node in $(0,1)$ and has zeros at 0 and 1. Suppose $\phi(x)$ had two zeros a_1, a_2 in $(0,1)$. By using Theorem 13.2.3 we deduce that

$$\begin{aligned} & \phi'(x) \text{ has zeros } 0, b_1, b_2, b_3, 1 \\ M(x) \quad : \quad & = r\phi''(x) \text{ has zeros } c_1, c_2, c_3, c_4, 1. \end{aligned}$$

But $M(x) = \sum_{i=2}^{4} c_i M_i(x)$ and, by part 4) of Theorem 13.3.1, $M(x)$ has at most 3 zeros in $[0,1)$. This is a contradiction; $\phi(x)$ has just one zero, a node, in $(0,1)$.

Now $\phi_i(x)$ has exactly $i-1$ changes of sign in $(0,1)$, so that $\phi_i(1) > 0$ implies $(-)^{i-1}\phi_i(0+) > 0$. We will eventually prove the Theorem by induction on the order of the minors. Suppose therefore that all the 2×2 minors of P are positive then, since

$$(-)\phi_2(0+) > 0, \quad (-)^2\phi_3(0+) > 0, \quad (-)^3\phi_4(0+) > 0,$$

we see by expanding the determinant $\phi(x)$ along its first row that $(-)\phi(0+) > 0$ and hence $\phi(1-) > 0$. Now expand $\phi(x)$ for small x in a Taylor series about $x = 1$:

$$\phi(1-x) = \frac{x^4}{4!}\phi^{\prime v}(1) + O(x^5)$$

so that

$$\phi^{1v}(1) = \begin{vmatrix} \lambda_2 u_2 & \lambda_3 u_3 & \lambda_4 u_4 \\ u_2 & u_3 & u_4 \\ \theta_2 & \theta_3 & \theta_4 \end{vmatrix} = \begin{vmatrix} u_2 & u_3 & u_4 \\ \theta_2 & \theta_3 & \theta_4 \\ \lambda_2 u_2 & \lambda_3 u_3 & \lambda_4 u_4 \end{vmatrix} > 0.$$

We may treat

$$\psi(x) = \begin{vmatrix} \phi_2'(x) & \phi_3'(x) & \phi_4'(x) \\ \theta_2 & \theta_3 & \theta_4 \\ \lambda_2 u_2 & \lambda_3 u_3 & \lambda_4 u_4 \end{vmatrix}$$

in exactly the same way: $\psi(x)$ has just one zero, a node, in $(0,1)$; $\phi_i'(1) > 0$ implies $(-)^{i-1}\phi_i'(0+) > 0$, $(-)\psi(0+) > 0$ and hence $\psi(1-) > 0$. But

$$\psi(1-x) = \frac{x^4}{4!}\psi^{\prime v}(1) + O(x^5),$$

and

$$\psi^{1v}(1) = \begin{vmatrix} \theta_2 & \theta_3 & \theta_4 \\ \lambda_2 u_2 & \lambda_3 u_3 & \lambda_4 u_4 \\ \lambda_2 \theta_2 & \lambda_3 \theta_3 & \lambda_4 \theta_4 \end{vmatrix} > 0.$$

We now generalise this analysis to provide a formal proof of the theorem.

Proof. We use the Corollary to Theorem 6.8.2 to prove the theorem by induction on the order of the minors. All minors of order 1 are positive; assume that all minors of order p are positive; we will prove that all minors of order $p+1$ involving consecutive rows and columns are positive. Because of the repetitive nature of the rows of P, it is sufficient to consider just two types of minors:

those beginning with $u_m, u_{m+1}, \ldots, u_n$, and those beginning with $\theta_m, \theta_{m+1}, \theta_n$. As we showed in the example, both may be treated in a similar way; we consider just the first.

Consider

$$\phi(x) = \begin{vmatrix} \phi_m(x) & \phi_{m+1}(x) & \cdots & \phi_n(x) \\ u_m & u_{m+1} & \cdots & u_n \\ \theta_m & \theta_{m+1} & \cdots & \theta_n \\ \vdots & \vdots & \cdots & \vdots \end{vmatrix} = \sum_{i=m}^{n} c_i\phi_i(x).$$

Take $n = m+p$. If p is even, i.e., $p = 2q$, then the last row is $\lambda_m^{q-1}\theta_m, \ldots, \lambda_n^{q-1}\theta_n$; if p is odd, i.e., $p = 2q-1$, then the last row is $\lambda_m^{q-1}u_m, \ldots, \lambda_n^{q-1}u_n$. Theorem 13.3.1 states that $\phi(x)$ has at least $(m-1)$ nodes in $(0,1)$, and at most $(n-1)$ zeros in $(0,1]$. Suppose p is even, then $\phi(x)$ has a zero of multiplicity $2p$ at 1. Suppose $\phi(x)$ had j zeros in $(0,1)$, where $j \geq m-1$. Thus $\phi(x)$ has zeros $0, a_1, a_2, \ldots, a_j, 1$; $\phi'(x)$ has zeros $0, b_1, \ldots, b_{j+1}, 1$; $M(x)$ has zeros $c_1, c_2, \ldots, c_{j+2}, 1$. Now introduce the notation

$$M_{,1} := M', \quad M_{,2} := a^{-1}(x)M'', \quad M_{,3} := (a^{-1}M'')', \quad M_{,4} := r(a^{-1}M'')''$$

then equation (13.3.3) states that $M_{i,4} = \lambda_i M_i$. Now extend this notation: if $k = 4s+t$ then $M_{,k} := (M_{,4s})_{,t}$, so that $M_{i,k} = (M_{i,4s})_{,t} = \lambda_i^s M_{i,t}$. Clearly we may deduce from Theorem 13.2.3, that there is a zero of $M_{,k+1}$ between any two zeros of $M_{,k}$. Now we can extend our study of zeros.

$$\begin{array}{rlr} M(x) & \text{has zeros} & c_1, c_2, \ldots, c_{j+2}, 1 \quad , \\ M_{,1}(x) & \text{has zeros} & d_1, d_2, \ldots, d_{j+2}, 1 \quad , \\ M_{,2}(x) & \text{has zeros} & 0, e_2, \ldots, e_{j+2}, 1 \quad , \\ M_{,3}(x) & \text{has zeros} & 0, f_2, \ldots, f_{j+3}, 1 \quad , \\ M_{,4}(x) & \text{has zeros} & g_1, \ldots, g_{j+4}, 1 \quad , \quad \text{etc.} \end{array}$$

In each 4-cycle, two zeros appear at $x = 0$, for $M_{,4s+2}$ and $M_{,4s+3}$. Thus $M_{,4q-4}$ has $j+2q$ zeros in $[0,1)$. But $M_{i,4q} = \lambda_i^q M_i$, so that

$$M_{,4q-4} = \sum_{i=m}^{n} \lambda_i^{q-1} c_i M_i$$

and hence, by part 4) of Theorem 13.3.1, $M_{,4q-4}$ can have at most $n-1 = m+2q-1$ zeros in $[0,1)$. Therefore, $j+2q \leq m+2q-1$, and $j \leq m-1$ so that $j = m-1 : \phi(x)$ has just $m-1$ zeros, all nodes, in $(0,1)$.

We continue the argument as in the example. Assume that all minors of P of order p are positive. Since $\phi_i(1) > 0$, we have $(-)^{i-1}\phi_i(0+) > 0$ and by expanding $\phi(x)$ along its first row we find $(-)^{m-1}\phi(0+) > 0$, and hence since $\phi(x)$ has just $m-1$ changes of sign in $(0,1)$, $\phi(1-) > 0$. We now expand $\phi(1-x)$ for small x:

$$\phi(1-x) = \frac{x^{2p}}{(2p)!}\phi^{2p}(1) + O(x^{2p+1})$$

so that

$$\phi^{2p}(1) = \begin{vmatrix} u_m & u_{m+1} & \dots & u_n \\ \theta_m & \theta_{m+1} & \dots & \theta_n \\ \cdot & \cdot & \dots & \cdot \\ \lambda_m^q u_m & \lambda_{m+1}^2 u_{m+1} & \dots & \lambda_n^q u_n \end{vmatrix} > 0.$$

This is a minor of order $(p+1)$ in P. Since all the other cases may be analysed in a similar way, we deduce that all the minors of P involving $(p+1)$ consecutive rows and columns are positive; the Corollary to Theorem 6.8.1 states that P is TP. ∎

Exercises 13.4

1. Establish the generalisation of the argument used with $\psi(x)$ is equation (13.4.1).

13.5 The spectra of the beam

Suppose that the beam of equation (13.1.1) specified by length L, cross-section $A(x)$, and second moment of area $I(x)$, is transformed into one of length L^*, cross-section $A^*(x^*)$, second moment $I^*(x^*)$, where

$$x^* = \gamma x, \quad I^*(x^*) = \alpha I(x), \quad A^*(x^*) = \beta A(x), \quad L^* = \gamma L, \tag{13.5.1}$$

then the spectra of the new beam under any combination of the end conditions (13.1.5)-(13.1.8) will be the same as those of the original beam provided that

$$\gamma^4 = \alpha/\beta. \tag{13.5.2}$$

With this relationship, equation (13.5.1) defines a two-parameter family os isospectral beams.

Now consider a beam, clamped at $x = 0$, and acted on by a concentrated static force F and bending moment M at its free end $x = L$. The deflection is given by

$$(I(x)u''(x))'' = 0 \tag{13.5.3}$$

subject to

$$u(0) = 0 = u'(0), \quad (I(x)u''(x))'_{x=L} = -F, \quad I(L)u''(L) = M, \tag{13.5.4}$$

so that

$$u(x) = F\int_0^x \frac{(x-s)(L-s)}{I(s)}ds + M\int_0^x \frac{(x-s)ds}{I(s)}, \tag{13.5.5}$$

and the end displacement and slope are given by

$$u(L) = G_2F + G_1M, \quad u'(L) = G_1F + G_0M,$$

where the *receptances* G_i are given by

$$G_i = \int_0^L \frac{(L-s)^i ds}{I(s)} ds, \quad i = 0, 1, 2.$$

For the transformed beam the receptances will be

$$G_i^* = \frac{\gamma^{i+1}}{\alpha} G_i, \quad i = 0, 1, 2.$$

We conclude that equation (13.5.2) and *any two* of the four equations

$$L^* = L, \quad G_i^* = G_i, \quad i = 0, 1, 2 \tag{13.5.6}$$

demand that $\alpha = 1 = \beta = \gamma$, so that the beams are identical.

We now use the results of Section 13.4 to order the eigenvalues for a beam clamped at $x = 0$ and subject to different end conditions at $x = 1$. (We shall work with the dimensionless equation (13.1.4) and use the numbering 1,2,3,...) Consider the variational problem of finding the stationary values of the functional

$$J(u) = \frac{1}{2}\int_0^1 r(x)(u''(x))^2 dx - \frac{\lambda}{2}\int_0^1 a(x)u^2(x)dx - Fu(1) - Mu'(1). \tag{13.5.7}$$

Replace u by $u + \delta u$ and find $\delta J := J(u + \delta u) - J(u)$; after two integrations by parts we find

$$\begin{aligned} \delta J &= \int_0^1 \{(ru'')'' - \lambda a u\}\delta u dx + [r(1)u''(1) - M]\delta u'(1) \\ &\quad -[(r(x)u''(x))'_{x=1} + F]\delta u(1) \end{aligned} \tag{13.5.8}$$

so that the displacement that makes J stationary satisifes equation (13.1.4) and the end conditions

$$r(1)u''(1) = M, \quad (r(x)u''(x))'_{x=1} = -F \tag{13.5.9}$$

i.e., it is the displacement of the cantilever due to the concentrated static force F and moment M applied at $x = 1$. We now use the eigenfunctions $(\phi_i(x))_1^\infty$ of the cantilever to find an alternative expression for this displacement. The eigenfunctions of the cantilever are complete in $L^2(0,1)$. Write

$$u(x) = \sum_{i=1}^{\infty} c_i \phi_i(x)$$

and use Ex. 13.1.3 to give

$$J(u) = \frac{1}{2}\sum_{i=1}^{\infty} \lambda_i c_i^2 - \frac{\lambda}{2}\sum_{i=1}^{\infty} c_i^2 - \sum_{i=1}^{\infty} c_i\{Fu_i + M\theta_i\}$$

where $u_i = \phi_i(1)$, $\theta_i = \phi_i'(1)$ and the eigenfunctions have been normalised so that

$$\int_0^1 a(x)\phi_i^2(x)dx = 1.$$

$J(u)$ will be stationary if

$$(\lambda_i - \lambda)c_i = Fu_i + M\theta_i$$

i.e.,

$$u(x) = \sum_{i=1}^{\infty} \frac{(Fu_i + M\theta_i)}{\lambda_i - \lambda}\phi_i(x).$$

This yields the *end receptances* $\alpha_{x1}, \alpha_{x1'}, \alpha_{x'1}, \alpha_{x'1'}$ of the beam with the properties

$$\begin{aligned} u(x) &= \alpha_{x1}F + \alpha_{x1'}M, \quad u'(x) = \alpha_{x'1}F + \alpha_{x'1'}M; \\ \alpha_{x1} &= \sum_{i=1}^{\infty} \frac{u_i\phi_i(x)}{\lambda_i - \lambda}, \quad \alpha_{x1'} = \sum_{i=1}^{\infty} \frac{\theta_i\phi_i(x)}{\lambda_i - \lambda}, \qquad (13.5.10) \\ \alpha_{x'1} &= \sum_{i=1}^{\infty} \frac{u_i\phi_i'(x)}{\lambda_i - \lambda}, \quad \alpha_{x'1'} = \sum_{i=1}^{\infty} \frac{\theta_i\phi_i'(x)}{\lambda_i - \lambda}. \qquad (13.5.11) \end{aligned}$$

We now use these expressions to obtain equations for the eigenvalues of the beam corresponding to various conditions at $x = 1$.

The eigenvalues of the clamped-pinned beam are the values of ω^2 for which the application of an end force F alone (i.e., $M = 0$) produces no end displacement, i.e., $u(0) = 0$. They are thus the roots of the equation $\alpha_{11} = 0$, i.e.,

$$\sum_{i=1}^{\infty} \frac{u_i^2}{\lambda_i - \lambda} = 0. \qquad (13.5.12)$$

We will denote them by $(\mu_i)_1^\infty$. Since $u_i > 0$, they satisfy

$$\lambda_i < \mu_i < \lambda_{i+1}, \quad i = 1, 2, \ldots$$

Similarly, the eigenvalues $(\sigma_i)_1^\infty$ of the clamped-sliding beam are the values of ω^2 for which M alone (i.e., $F = 0$) produces no end slope, i.e., $u'(1) = 0$. They are the roots of $\alpha_{1'1'} = 0$, i.e.,

$$\sum_{i=1}^{\infty} \frac{\theta_i^2}{\lambda_i - \lambda} = 0 \qquad (13.5.13)$$

and since $\theta_i > 0$, they satisfy

$$\lambda_i < \sigma_i < \lambda_{i+1}, \quad i = 1, 2, \ldots$$

The anti-resonant eigenvalues $(\nu_i)_1^\infty$ are those at which F alone produces no slope, or equivalently M alone produces no displacement; they are the roots of $\alpha_{11'} = 0$, i.e.,

$$\sum_{i=1}^{\infty} \frac{u_i\theta_i}{\lambda_i - \lambda} = 0. \tag{13.5.14}$$

Since $u_i\theta_i > 0$ (Theorem 13.3.2), they satisfy

$$\lambda_i < \nu_i < \lambda_{i+1}, \quad i = 1, 2, \ldots$$

We can order μ_i, σ_i, ν_i by using Theorem 8.4.2 and the total positivity of the matrix P in Theorem 13.4.1: $u_j\sigma_i - u_i\sigma_j > 0$ for $i > j$ gives the ordering

$$\sigma_i < \nu_i < \mu_i. \tag{13.5.15}$$

Since the clamped end condition may be obtained by adding another constraint $u'(1) = 0$ to the pinned condition, and alternatively by adding the constraint $u(1) = 0$ to the sliding condition, the clamped-clamped eigenvalues $(\gamma_i)_1^\infty$ will satisfy

$$\mu_i < \gamma_i < \mu_{i+1}, \quad \sigma_i < \gamma_i < \sigma_{i+1}.$$

Putting all these inequalities together we find

$$\lambda_1 < \sigma_1 < \nu_1 < \mu_1 < (\lambda_2, \gamma_1) < \sigma_2 < \nu_2 < \mu_2 < (\lambda_3, \gamma_2) \ldots \tag{13.5.16}$$

As with the discrete beam (see Ex. 8.5.3) the relative position of γ_i and λ_{i+1} is indeterminate. Tables 7.2(b), (c) of Bishop and Johnson (1960) [34] show that for the uniform beam $\gamma_1 > \lambda_2, \gamma_2 < \lambda_3, \gamma_3 > \lambda_4$, etc. and that thereafter γ_i and λ_{i+1} are vertically identical.

In order to find the asymptotic forms of the eigenvalues corresponding to different end conditions we use the WKB approach, see for example Carrier, Krook and Pearson (1966) [49], p. 291. First we make a change of independent variable:

$$s = \int_0^x \left(\frac{a(t)}{r(t)}\right)^{\frac{1}{4}} dt,$$

and write

$$b(s) = \left(\frac{a(x)}{r(x)}\right)^{\frac{1}{4}}, \quad c^2(s) = (r^3(x)a(x))^{\frac{1}{4}},$$

then

$$\frac{d}{dx} = \frac{d}{ds}\frac{ds}{dx} = \left(\frac{a(x)}{r(x)}\right)^{\frac{1}{4}} \frac{d}{ds} = b(s)\frac{d}{ds},$$

and

$$r(x)\frac{d^2}{dx^2} = r(x)b(s)\frac{d}{ds}\left(b(s)\frac{d}{ds}\right) = c^2(s)\frac{d}{ds}\left(b(s)\frac{d}{ds}\right).$$

Thus equation (13.3.1) becomes

$$b\frac{d}{ds}\left(b\frac{d}{ds}\left(c^2\frac{d}{ds}\left(b\frac{du}{ds}\right)\right)\right) = \lambda b^3 c^2 u,$$

since $a = b^3c^2$. Thus putting $' = d/ds$ we find

$$(b(c^2(bu')')')' = \lambda b^2c^2u, \quad 0 \leq s \leq L, \tag{13.5.17}$$

where

$$L = \int_0^1 \left(\frac{a(t)}{r(t)}\right)^{\frac{1}{4}} dt. \tag{13.5.18}$$

The new end conditions are

$$u(0) = 0 = u'(0), \tag{13.5.19}$$

$$(bu')'(L) = 0 = (c^2(bu')')'(L). \tag{13.5.20}$$

Now suppose that λ is large positive, put $\lambda = z^4$ and expand the left hand side of equation (13.5.17) to give

$$p^2(s)u'^v(s) + 2p(s)p'(s)u'''(s) + f_1(s)u''(s) + f_2(s)u'(s) = z^4p^2(s)u(s), \tag{13.5.21}$$

where

$$p = bc, \quad f_1 = p'' + bc^2b'' + 2bcb'c', \quad f_2 = (b(c^2b')')'.$$

For large z it will be the first two terms of (13.5.21) that will be dominant. We look for a solution having the form

$$U(s) = \exp\left(\int (z\psi_1(s) + \psi_2(s))\, ds\right).$$

After substituting this into (13.5.21) and retaining only the terms involving z^4 and z^3 we find

$$\psi_1^4 = 1, \quad p^2(6\psi_1^2\psi_1' + 4\psi_1^3\psi_2) + 2pp'\psi_1^3 = 0$$

so that $\psi_1 = \pm 1, \pm i$ and $2p\psi_2 + p' = 0$, i.e., $\exp(\psi_2(s)) = p^{-\frac{1}{2}}(s)$.

There are thus four solutions corresponding to the four values of ψ_1, and we may write

$$u(s) = p^{-\frac{1}{2}}(s)\{A\cos zs + B\sin zs + C\cosh zs + D\sinh zs\}. \tag{13.5.22}$$

Apart from the factor $p^{-\frac{1}{2}}(s)$, this has exactly the same form as that for a uniform beam, so that for large z the eigenvalue equation will be the same as for a uniform beam of equivalent length L. Thus, for the cantilever the four end conditions (13.5.19); (13.5.20) will yield the eigenvalue equation
Bishop and Johnson (1960) [34], p. 382)

$$\cos zL \cosh zL + 1 = 0, \tag{13.5.23}$$

so that

$$\cos zL = -\text{sech} zL \simeq -2\exp(-zL),$$

and

$$z_iL \simeq (2i-1)\frac{\pi}{2}, \quad i = 1, 2, \ldots,$$

or

$$\lambda_i \simeq (2i-1)^4\pi^4/(16L^4).$$

In a similar way we find

$$\begin{array}{rcl} \mu_i & \simeq & (4i+1)^4\pi^4/(256L^4), \\ \nu_i & \simeq & (i-1)^4\pi^4/L^4, \\ \sigma_i & \simeq & (4i-1)^4\pi^4/(256L^4), \\ \gamma_i & \simeq & (2i+1)^4\pi^4/(16L^4). \end{array}$$

We note that these do obey the interlacing conditions (13.5.15), and that $\gamma_i \simeq \lambda_{i+1}$. Note also that, taking account of the change of notation, the values of μ_i, ν_i, γ_i agree with those given by Barcilon (1982) [21]; his values of σ_r^2, ω_r^2 (our σ_i, λ_i) are incorrect.

Exercises 13.5

1. Verify the statement (13.5.2).
2. Carry out the integration from equations (13.5.3), (13.5.4) to (13.5.5).
3. Derive the expression for δJ in equation (13.5.7) by replacing u by $u + \delta u$, in (13.5.6), neglecting the second order terms and integrating by parts twice.
4. Show that the asymptotic form for the eigenvalue equation for the clamped-clamped beam is $\cos zL \cosh zL - 1 = 0$, and use this equation and (13.5.23) to show that for large i, γ_i is alternately greater and less than λ_{i+1}.

13.6 Statement of the inverse problem

Inverse problems for the vibrating Euler-Bernoulli beam seem to have been studied first by Niordson (1967) [250]. He was not concerned with the reconstruction of a unique beam from sufficient data, in the sense to be described below. Rather, he was concerned with constructing a beam in a class having n arbitrary parameters so that it would have n specified eigenvalues which would be perturbations on the eigenvalues of the uniform cantilever beam.

The proper study of the inverse problem for the vibrating Euler-Bernoulli beam began with the work of Barcilon. He realised that there are three questions to be answered. First, what spectral (and other) data are required to determine the properties (cross-sectional area $A(x)$, second moment of area $I(x)$) of the beam? In Barcilon (1974b) [15], Barcilon (1974c) [16] he showed that three spectra, corresponding to three different end conditions, are required. Secondly, what are the necessary and sufficient conditions on the data to ensure that the beam properties will be realistic, i.e., $A(x) > 0, I(x) > 0$? Barcilon battled with this question in Barcilon (1982) [21], but it was not fully answered until Gladwell (1986d) [110]. Thirdly, how can the beam be reconstructed? Barcilon (1976) [18] answered this question for the case in which the spectra were small perturbations on these for the uniform beam, but a proper reconstruction procedure was not available until McLaughlin (1984b) [227].

As a result of the analysis described in Section 13.5 we may state that there is only a two-parameter family of beams which have three given spectra $\{\lambda_i, \mu_i, \sigma_i\}_1^\infty$ (or $\{\lambda_i, \mu_i, \nu_i\}_1^\infty$ or $\{\lambda_i, \nu_i, \sigma_i\}_1^\infty$). The particular member in the family may be found as in (13.5.1). The spectra $\{\lambda_i, \mu_i, \sigma_i\}_1^\infty$ will have to satisfy certain conditions, amongst which will be some asymptotic ones. The argument of Section 13.5 shows that to be given $\{\lambda_i, \mu_i, \sigma_i\}_1^\infty$, and some appropriate asymptotic conditions in equivalent to being given $\{\lambda_i, u_i, \theta_i\}_1^\infty$ and some other asymptotic conditions. We can, and shall, circumvent the asymptotic conditions with the way in which the problem is posed in practice - that only $(\lambda_i, u_i, \theta_i)_1^n$ are given, while the remainder are chosen so that

$$(\lambda_i, u_i, \theta_i)_{n+1}^\infty = (\lambda_i^o, u_i^o, \theta_i^o)_{n+1}^\infty \tag{13.6.1}$$

where the ^{o} quantities relate to a known beam which, without loss of generality, may be taken to be a uniform beam.

In this case, equation (13.5.12), for example, may be written

$$\sum_{i=1}^{n} \frac{u_i^2}{\lambda_i - \lambda} - \sum_{i=1}^{n} \frac{{u_i^o}^2}{\lambda_i^o - \lambda} + \sum_{i=1}^{\infty} \frac{{u_i^o}^2}{\lambda_i^o - \lambda} = 0. \tag{13.6.2}$$

The infinite sum is an end receptance of the uniform beam and may be expressed in closed form, in fact Bishop and Johnson (1960) [34] if $u_i^o = 1$, then

$$\sum_{i=1}^{\infty} \frac{{u_i^o}^2}{\lambda_i^o - \lambda} = \frac{\cos\phi \sinh\phi - \sin\phi \cosh\phi}{4\phi^3(\cos\phi\cosh\phi + 1)}, \phi = \lambda^{\frac{1}{4}}.$$

Thus, the statement that (13.6.2) is satisfied by $\lambda = (\mu_i)_1^n$ yields n simultaneous linear equations for $(u_i^2)_1^n$. The first set of necessary conditions is therefore as follows:

1) $\mu_i \neq \lambda_j$, $\mu_i \neq \lambda_j^o$ for all $i, j = 1, 2, \ldots$. Those ensures that the matrix of coefficients in the equations for $(u_i^2)_1^n$ in non-singular, and the right hand sides are well-defined.

2) the solution $(u_i^2)_1^n$ must be positive. (See Ex. 13.6.1) Similarly, if the $(\theta_i^2)_1^n$ are to be determined from equation (13.5.13) then we need

3) $\sigma_i \neq \lambda_j$, $\sigma_i \neq \lambda_j^o$ for all $i, j = 1, 2, \ldots$

4) the solution $(\theta_i^2)_1^n$ must be positive.

Provided that these conditions are satisfied, i.e., $(u_i, \theta_i)_1^n$ may be found, then we shall show that the positivity of the minors of P of Theorem 13.4.1, which has been shown to be necessary, is also a sufficient condition for the construction of a unique realistic beam.

The analysis shows that three properly chosen spectra are required to reconstruct a beam uniquely. Gottlieb (1987b) [140] made an exhaustive study of beams that have one or two spectra in common, and/or in common with a uniform beam, for various combinations of end conditions. His study thus highlights the need for three (properly chosen) spectra. See also Gottlieb (1988) [142].

Exercises 13.6

1. By retaining (13.6.2) in the form

$$\sum_{i=1}^{\infty} \frac{u_i^2}{\lambda_i - \lambda} = 0$$

show that $u_i^2 > 0$ if the roots of (13.6.2) interlace the λ_i, i.e., $\lambda_i < \mu_i < \lambda_{i+1}, \quad i = 1, 2, \ldots$.

13.7 The reconstruction procedure

The procedure is essentially the same as that described in Chapter 11 for the vibrating rod, and is due to McLaughlin (1976) [223], McLaughlin (1978) [224], McLaughlin (1981) [225], McLaughlin (1984a) [226], McLaughlin (1984b) [227]. Papanicalaou and Kravvaritis (1997) [262] consider the special case, $a(x)r(x) = 1$; in this case the problem can effectively be reduced to a second order problem. See also Gladwell (1991d) [119]. We use a transformation operator as described in Section 11.3.

We suppose that we wish to construct a cantilever beam, i.e., functions $r(x)$ and $a(x)$, such that the equation

$$(r(x)u''(x))'' = \lambda a(x)u(x), \tag{13.7.1}$$

subject to the end conditions

$$u(0) = 0 = u'(0), \quad u''(1) = 0 = u'''(1), \tag{13.7.2}$$

has specified eigenvalues $(\lambda_i)_1^\infty$, and has eigenfunctions $(\phi_i(x))_1^\infty$, normalised w.r.t. $a(x)$, i.e., such that

$$\int_0^1 a(x)\phi_i(x)\phi_j(x)dx = \delta_{i,j}, \quad i, j = 1, 2, \ldots$$

which have specified values of $(\phi_i(1), \phi_i'(1))_1^\infty$.

First make a change in the independent variable similar to that used in Section 13.5:

$$s = \int_x^1 \left(\frac{a(t)}{r(t)}\right)^{\frac{1}{4}} dt \tag{13.7.3}$$

and write

$$b(s) = \left(\frac{a(x)}{r(x)}\right)^{\frac{1}{4}}, \quad c^2(s) = (r^3(x)a(x))^{\frac{1}{4}}, \tag{13.7.4}$$

$$p(s) = b(s)c(s), \tag{13.7.5}$$

then equation (13.7.1) becomes

$$(b(c^2(bu')')')' - \lambda b^2c^2u = 0, \quad 0 \le s \le L, \tag{13.7.6}$$

where $' \equiv d/ds$ and

$$L = \int_0^1 \left(\frac{a(t)}{r(t)}\right)^{\frac{1}{4}} dt, \tag{13.7.7}$$

while the end conditions become

$$(bu')'(0) = 0 = (c^2(bu')')'(0) \tag{13.7.8}$$

$$u(L) = 0 = u'(L). \tag{13.7.9}$$

Without loss of generality, we assume that $b(0) = 1 = c(0)$.

Just as with the Sturm-Liouville reconstruction, we introduce a base problem

$$(b_0(c_0^2(b_0v')')')' - \lambda b_0^2c_0^2v = 0 \tag{13.7.10}$$

$$(b_0v')'(0) = 0 = (c_0^2(b_0v')')'(0) \tag{13.7.11}$$

$$v(L) = 0 = v'(L) \tag{13.7.12}$$

where $b_0(s), c_0(s)$ are known (e.g., $b_0(s) = 1 = c_0(s)$) $b_0(1) = 1 = c_0(0)$, and $p_0(s) = b_0(s)c_0(s)$. This base problem has a certain set of eigenvalues $(\lambda_i^0)_1^\infty$ and its eigenfunctions $\phi_i^0(s)$, normalised so that

$$\int_0^L p_0^2(a)\phi_i^0(s)\phi_j^0(s)\,ds = \delta_{ij}, \quad i,j = 1,2,\ldots \tag{13.7.13}$$

will have end values $(\phi_i^0(0), \phi_j^{0\prime}(0))_1^\infty$.

For given values λ, ξ, η we may define a unique function

$$v(s;\lambda,\xi,\eta) \equiv v(s) \tag{13.7.14}$$

which is the solution of equation (13.7.10) satisfying

$$v(0) = \xi, \quad v'(0) = \eta, \quad (b_0v')'(0) = 0 = (c_0^2(bv')')'(0). \tag{13.7.15}$$

Clearly

$$v(s;\lambda_i^0, \phi_i^0(0), \phi_i^{0'}(0)) = \phi_i^0(s). \tag{13.7.16}$$

The eigenfunctions $\{\phi_i^0(s)\}_1^\infty$ are orthogonal with weight function $p_0^2(s)$, as shown by equation (13.7.13). The eigenfunctions $\{\phi_i(s)\}_1^\infty$ of equations (13.7.6)-(13.7.9) are to be orthogonal w.r.t. $p^2(s)$, i.e.,

$$\int_0^L p^2(s)\phi_i(s)\phi_j(s)ds = \delta_{ij}, \quad i,j = 1,2,\ldots \tag{13.7.17}$$

Therefore, following (11.3.5) we construct (13.7.6) so that the solution (13.7.16) of equation (13.7.10) is transformed into a solution of equation (13.7.6) satisfying

$$u(0) = \xi, \quad u'(0) = \eta, \quad (bu')'(0) = 0 = (c^2(bu')')'(0) \tag{13.7.18}$$

by means of the equation

$$p(s)u(s) = p_0(s)v(s) + \int_0^s K(s,t)p_0^2(t)v(t)dt. \tag{13.7.19}$$

The eigenfunctions of equations (13.7.6)-(13.7.9) will be

$$\phi_i(s) = u(s;\lambda, \phi_i(s)|_{s=0}, \phi_i'(s)|_{s=0})$$

and we note that

$$\phi_i(s)|_{s=0} = \phi_i(x)|_{x=1}, \quad \left.\frac{d\phi_i(s)}{ds}\right|_{s=0} = -\left.\frac{d\phi_i(x)}{dx}\right|_{x=1}$$

If the eigenvalues λ_i and end values $\phi_i(0)$, $\phi_i'(0)$ (with variables s) are chosen so that

$$\lambda_i = \lambda_i^0, \quad \phi_i(0) = \phi_i^0(0), \quad \phi_i'(0) = \phi_i^{0'}(0), \quad i = n+1,\ldots$$

then the system $\{\phi_i(s)\}_1^\infty$ will form a complete orthogonal set with weight $p^2(s)$ iff $K(x,s)$ satisfies the analogue of equation (11.5.20), i.e.,

$$K(r,s) + \int_0^r p_0^2(t)K(r,t)F(t,s)dt + p_0(r)F(r,s) = 0, \quad 0 \le s \le r, \tag{13.7.20}$$

where

$$F(r,s) = \sum_{i=1}^n \{v_i(r)v_i(s) - v_i^0(r)v_i^0(s)\} \tag{13.7.21}$$

and

$$\begin{aligned} v_i(s) &= v(s;\lambda_i, \phi_i(0), \phi_i'(0)) \\ v_i^0(s) &= v(s,\lambda_i^0, \phi_i^0(0), \phi_i^{0'}(0)) \equiv \phi_i^0(s). \end{aligned}$$

We note that

$$u(s;0,1,0) = 1, \quad v(s;0,1,0) = 1$$

so that equation (13.7.19) gives

$$p(s) = p_0(s) + \int_0^s K(s,t)p_0^2(t)dt. \tag{13.7.22}$$

On the other hand, if

$$q(s) = \int_0^s \frac{dt}{b(t)}, \quad q_0(s) = \int_0^s \frac{dt}{b_0(t)} \tag{13.7.23}$$

then

$$u(s;0,0,1) = q(s), \quad v(s;0,0,1) = q_0(s)$$

so that

$$p(s)q(s) = p_0(s)q_0(s) + \int_0^s K(s,t)p_0^2(t)q_0(t)dt. \tag{13.7.24}$$

The reconstruction procedure is thus as follows:

- solve equation (13.7.20) for $K(s,t)$
- find $p(s), q(s)$ from equations (13.7.22), (13.7.24)
- find $b(s), c(s)$ from equations (13.7.4), (13.7.5)
- find $x, a(x), r(x)$ from equations (13.7.3), (13.7.4).

To justify this procedure we need to verify that when $p(s), q(s)$ are given by (13.7.22), (13.7.24) then

1) $p(s), q(s)$ are well-defined and positive, and $q(s)$ is an increasing function.

2) $u(s)$ satisfies the end conditions at $s = 0$.

3) $u(s)$ satisfies the differential equation (13.7.6).

4) $u_i(s)$ satisfies the end conditions at $s = L$.

We shall consider these points in the order 2,3,4,1.

Equation (13.7.22) yields $p(0) = p_0(0) = 1$, while equation (13.7.19) yields $p(0)u(0) = u(0) = p_0(0)v(0) = v(0) = \xi$. On differentiating equation (13.7.22) we obtain

$$p'(0) = p_0'(0) + K(0,0),$$

while on differentiating equation (13.7.19) we find

$$p'(0)u(0) + p(0)u'(0) = p_0'(0)v(0) + p_0(0)v'(0) + K(0,0)p_0^2(0)v(0),$$

which yields

$$u'(0) = v'(0) = \eta.$$

By continuing this differentiation we may establish the remainder of 2) and 3).

As in Section 11.5, the solution of equation (13.7.20) has the form

$$K(r,s) = \sum_{i=1}^{n} \{F_i(r)v_i(s) - G_i(r)v_i^0(s)\} \tag{13.7.25}$$

where $F_i(r), G_i(r)$ satisfy

$$p_o(r)v_i(r) + F_i(r) + \sum_{i=1}^{n}\{b_{ij}F_j(r) - c_{ij}G_j(r)\} = 0 \qquad (13.7.26)$$

$$p_o(r)v_i^0(r) + G_i(r) + \sum_{i=1}^{n}\{c_{ji}F_j(r) - d_{ij}G_j(r)\} = 0 \qquad (13.7.27)$$

where

$$\begin{aligned} b_{ij}(r) &= \int_0^r p_0^2(t)v_i(t)v_j(t)dt, \quad c_{ij}(r) = \int_0^r p_0^2(t)v_i(t)v_j^0(t)dt \\ d_{ij}(r) &= \int_0^r p_0^2(t)v_i^0(t)v_j^0(t)dt. \end{aligned}$$

In considering point 4) we need to discuss two cases, $i \leq n$ and $i > n$. For the first we note than on subsituting (13.7.25) into (13.7.19) and using (13.7.27) we may deduce

$$p(s)u_i(s) = -F_i(s), \quad i = 1, 2, \ldots, n. \qquad (13.7.28)$$

But equation (13.7.27) with $r = L$ and the orthogonality conditions $d_{ij}(L) = \delta_{ij}$ ($v_i^0(t)$ are normalised eigenfunctions of the base problem) yield

$$\sum_{i=1}^{n} c_{ji}(L)F_j(L) = 0 \qquad (13.7.29)$$

since $v_i^0(L) = 0$. Thus if $\mathbf{C} = (c_{ij}(L))$ is non-singular, and $p(L) \neq 0$, then

$$F_j(L) = 0 = u_j(L).$$

On differentiating (13.7.27) we find, under the same proviso, that

$$F_j'(L) = 0 = u_j'(L), \quad j = 1, 2, \ldots, n.$$

We shall return to the proviso, $p(L) \neq 0$, later.

When $i > n$, then $v_i(L) = v_i^0(L) = \phi_i^0(L) = 0$, so that equation (13.7.19) yields

$$p(L)u_i(L) = \int_0^L K(L,t)p_0^2(t)v_i^0(t)dt,$$

so that on substituting for $K(L,t)$ from equation (13.7.25) we find

$$p(L)u_i(L) = \sum_{i=1}^{n}\{F_j(L)\int_0^L p_0^2(t)v_j(t)v_i^0(t)dt - G_j(L)\int_0^L p_0^2(t)v_j^0(t)v_i^0(t)dt\}.$$

But since $i > n$, $v_i^0(t)$ is orthogonal to all of $\{v_j^0(t)\}_1^n$. Therefore, again, *if* $F_j(L) = 0, j = 1, 2, \ldots, n$ *and* $p(L) \neq 0$, then $u_i(L) = 0$. The satisfaction of $u_i'(L) = 0$ may be verified similarly (Ex. 13.7.3).

We now discuss point 1). The first step is the determination of $F_i(s), G_i(s)$ from equations (13.7.26), (13.7.27). The argument used in Section 11.5 shows that the matrix of coefficients in these equations is non-singular unless $r = L$. Thus there remains only the case $r = L$. Then the matrix of coefficients in (13.7.26), (13.7.27) takes the form

$$\mathbf{A} = \begin{bmatrix} \mathbf{I}+\mathbf{B}, & -\mathbf{C} \\ \mathbf{C}^T & \mathbf{0} \end{bmatrix}$$

so that $\det \mathbf{A} = (\det \mathbf{C})^2$. Thus $\det \mathbf{C} \neq 0$ is a necessary and sufficient condition for the $F_i(r), G_i(r)$, and hence $p(s)$, to be well-defined. We now enquire as to when and whether $p(s) > 0$. Suppose $p(s) = 0$ for some $s \in [0, L]$, then (13.7.28) and the similar equation

$$p(s)u_i^0(s) = p(s)u(s; \lambda_i^0, \phi_i^0(0), \phi_i^{0'}(0)) = -G_i(s), \quad i = 1, 2, \ldots, n$$

show that $F_i(s) = 0 = G_i(s), \quad i = 1, 2, \ldots, n$ and hence, on account of (13.7.26), (13.7.27), $p_0(s)v_i(s) = 0 = p_0(s)v_i^0(s), \quad i = 1, 2, \ldots, n$. But $p_0(s)$, corresponding to an actual beam, is always positive, and the only common zero of the $v_i^0(s)$, is $s = L$. Thus the only possible zero of $p(s)$ is $s = L$. At $s = L$, equation (13.6.27) reduces to

$$\sum_{i=1}^{n} c_{ji}(L)F_j(L) = 0$$

so that if $\det \mathbf{C}(L) \neq 0$ then $F_j(L) = 0, \quad j = 1, 2, \ldots, n$. Then (13.6.26) reduces to

$$p_0(L)v_i(L) - \sum_{i=1}^{n} c_{ij}(L)G_j(L) = 0, \quad i = 1, 2, \ldots, n. \tag{13.7.30}$$

Equations (13.7.22), (13.7.25) yield

$$p(L) = p_0(L) - \sum_{i=1}^{n} G_j(L) \int_0^L p_0^2(t)v_j^0(t)dt. \tag{13.7.31}$$

Put

$$\mathbf{C} = (c_{ij}(L)), \quad \mathbf{g} = [G_1(L), \ldots, G_n(L)], \quad \mathbf{v} = [v_1(L), \ldots, v_n(L)]$$

then (13.7.30) becomes

$$p_0(L)\mathbf{v} = \mathbf{Cg}$$

so that on multiplying (13.7.31) by $\mathbf{v}$ we have

$$p(L)\mathbf{v} = p_0(L)\mathbf{v} - \mathbf{Hg} = p_0(L)(\mathbf{C} - \mathbf{H})\mathbf{g},$$

where

$$c_{ij} - h_{ij} = \int_0^L p_0^2(t)(v_i(t) - v_i(L))v_j^0(t)dt = e_{ij}.$$

Thus

$$p(L)\mathbf{v} = p_0(L)\mathbf{Eg}.$$

This means that if $\mathbf{E}$ is non-singular then $p(L) \neq 0$. If $\mathbf{C}$ and $\mathbf{E}$ are non-singular then all the stated operations may be carried out to obtain $p(s), q(s)$ and hence $b(s), c(s)$. Since $b(s), c(s)$ are never zero, and $b(0)c(0) = p(0) = 1$, $b(s), c(s)$ are always positive, and hence $a(x), r(x) > 0$.

Lowe (1993) [217] considers a special case of equation (13.7.1) in which $r(x) = [a(x)]^2$, and uses a construction based on a Fourier series for $a(x)$.

Exercises 13.7

1. Verify the transformation of equations (13.7.1), (13.7.2) to (13.7.6)-(13.7.9). Show that the conditions (13.7.8) are equivalent to

$$q'(0)u''(0) - q''(0)u'(0) = 0 = q'(0)u'''(0) - q'''(0)u'(0).$$

2. Show that if $v(s)$ satisfies

$$q_0'(0)v''(0) - q_0''(0)v'(0) = 0 = q_0'(0)v'''(0) - q_0'''(0)v'(0)$$

then u satisfies the conditions in Ex. 13.7.1.

3. We established $u_i(L) = 0 = u_i'(L)$ for $i \leq n$; establish them for $i > n$.

13.8 The total positivity of matrix P is sufficient

In Section 13.4 we showed that the eigenvalues $(\lambda_i)_1^\infty$ and end values u_i, θ_i of a cantilever beam make the infinite matrix $\mathbf{P}$ of Theorem 13.4.1 totally positive. In Section 13.7 we found some sufficient conditions for the reconstruction of an actual beam from such data. We now show that the total positivity of the matrix $\mathbf{P}$ is not only necessary but sufficient.

It was shown in Section 13.7 that the reconstruction will proceed provided that the matrices $\mathbf{C}$ and $\mathbf{E}$ are non-singular. Here

$$\begin{aligned} c_{ij} &= \int_0^L p_0^2(s)v_i(s)v_j^0(s)ds \\ e_{ij} &= \int_0^L p_0^2(s)v_i^0(s)[v_j(s) - v_j(L)]ds \end{aligned}$$

and $i, j = 1, 2, \ldots, n$. Suppose, if possible, that $\mathbf{C}$ were singular. Its rows will be linearly dependent, i.e., there are multipliers $(\alpha_i)_1^n$, not all zero, such that

$$\sum_{i=1}^n \alpha_i c_{ij} = 0, \quad j = 1, 2, \ldots, n.$$

Thus

$$\int_0^L p_0^2(s)\left\{\sum_{i=1}^{n}\alpha_i v_i(s)\right\} v_j^0(s)ds = 0, \quad j = 1, 2, \ldots, n. \tag{13.8.1}$$

But since the $\{v_j^0(s)\}_1^\infty$, being the eigenfunctions of the base problem, form a complete orthogonal set with weight function $p_0^2(s)$, equation (13.8.1) means that the sum in the integrand is a linear combination of the remaining $\{v_j^0(s)\}_{n+1}^\infty$. Thus

$$f(s) := \sum_{i=1}^{n}\alpha_i v_i(s) + \sum_{i=n+1}^{\infty}\alpha_i v_i^0(s) = 0.$$

In particular, $f(s)$ and all its derivatives must be zero at $s = 0$, the free end. Consider the case $b_0(s) \equiv 1 \equiv c_0(s)$. Now

$$\begin{aligned} v_i(s)|_{s=0} &= \phi_i(x)|_{x=1} = u_i, \\ b(s)v_i'(s)|_{s=0} &= v_i'(s)|_{s=0} = -\frac{d\phi_i(x)}{dx}|_{x=1} = -\theta_i, \end{aligned}$$

while equation (13.7.10) gives

$$v_i^{iv}(0) = \lambda_i v_i(0) = \lambda_i u_i, \quad v_i^{v}(0) = -\lambda_i\theta_i.$$

Thus

$$v_i^{(m)}(0) = 0 = v_i^{0(m)}(0) \text{ for } m = 2, 3; 6, 7; \ldots,$$

so that $f^{(m)}(0)$ is identically zero for these values of m. The equations obtained by setting $f^{(m)}(0)$ to zero for the remaining values $0, 1; 4, 5; \ldots$ are therefore

$$\sum_{i=1}^{\infty}\lambda_i^j u_i\alpha_i = 0 \qquad \sum_{i=1}^{\infty}\lambda_i^j\theta_i\alpha_i = 0, \quad j = 0, 1, 2, \ldots \tag{13.8.2}$$

and here we have used the fact that $u_i^0 = u_i$, $\theta_i^0 = \theta_i$ for $i = n+1, \ldots$ But the matrix of coefficients for equations (13.8.2) is just the matrix $\mathbf{P}$ of Theorem 13.4.1, and every minor of $\mathbf{P}$ is positive so that $\mathbf{P}$ has infinite rank. Thus all the α_i are zero, contradicting the assumed singularity of $\mathbf{C}$. Thus $\mathbf{C}$ is non-singular. When $b_0(s), c_0(s)$ are not identically unity, the rows of the matrix are linear combinations of the rows of $\mathbf{P}$, so that the conclusion still follows.

A similar argument shows that if $\mathbf{E}$ is singular then there are multipliers $(\beta_i)_1^\infty$, not all zero, such that

$$g(s) := \sum_{i=1}^{n}\beta_i\{v_i(s) - v_i(0)\} + \sum_{i=n+1}^{\infty}\beta_i v_i^0(s) = 0$$

i..e.,

$$g'(s) = \sum_{i=1}^{n}\beta_i v_i'(s) + \sum_{i=n+1}^{\infty}\beta_i v_i^{0'}(s) = 0.$$

When $b_0(s) = 1 = c_0(s)$, the matrix of coefficients for the equations obtained by setting $g^{(m)}(0)$ to zero for $m = 1; 4, 5; 8, 9; \ldots$ is just the matrix formed from rows $2, 3, \ldots$ of matrix $\mathbf{P}$. We conclude that the β_i are identically zero and that $\mathbf{E}$ is non-singular. We conclude that the total positivity of the matrix $\mathbf{P}$ is a necessary and sufficient condition for the reconstruction of a realistic beam.

Note that this conclusion is subject to the condition (13.6.1), and that the total positivity of the matrix $\mathbf{P}$ ensures only that $a(x), r(x)$ will be *positive*; they may still vary wildly along the beam, and in this case the vibration of the beam will not be governed by the Euler-Bernoulli equation, which applies only to slender beams, i.e., ones for which $a(x), r(x)$ do not differ much from the values of a uniform beam. If the beam is not uniformly slender, i.e., if

$$a^* = \frac{\max a(x)}{\min(a(x))}, \quad r^* = \frac{\max r(x)}{\min(r(x))}, \quad x \in [0, 1],$$

are not nearly unity, then the vibration of the beam is affected by thickness effects, and the simple Euler-Bernoulli model is inadequate. See Gladwell, England and Wang (1987) [112]. Note also that experimental studies of the natural frequencies of even a slender uniform beam show that the natural frequencies start to depart from the classical Euler-Bernoulli values after about the fourth or fifth frequency. This means that although the study of the inverse problem yields valuable insights into the behaviour of the Euler-Bernoulli beam, it must in many ways be considered as an academic exercise, and should be used only to find a beam in which the departures from the uniform beam is small and only a very few frequencies are to be changed, and that only by small amounts. For such problems, perturbation methods combined with least squares approaches form an alternative avenue; but such numerical methods are outside the purview of this book.

Chapter 14

Continuous Modes and Nodes

If there were no obscurity, man would not be sensible of his corruption; if there were no light, man would not hope for a remedy.
Pascal's *Pensées*, 585

14.1 Introduction

Throughout most of the preceding chapters, the emphasis has been placed on eigenvalues; in this chapter, we turn our attention to eigenmodes and, in particular, to the nodes of eigenmodes. We will find that, in contrast to inverse *eigenvalue* problems, there are no easily stated inverse *nodal* problems. There are some uniqueness results pertaining to nodes, most of which are due to McLaughlin and Hald; there are also some approximate solution of inverse nodal problems, again mostly due to McLaughlin and Hald; both of these topics are studied in Section 14.4. It should, however, be stated from the outset that it is impossible to do justice either to these uniqueness results or to the approximate solutions in the space available in this chapter; all we can do is to give an introduction to the published papers, discuss the methods used and some of the results obtained.

We take this opportunity to point out a fundamental difference between eigenvalues and nodes of a continuous system, and consequently between inverse eigenvalue and inverse nodal problems. Eigenvalues are *global* quantities; they are properties of the system as a whole. By contrast, a node, in particular the position of a node, is related to the properties of the system around that node; it is a *local* property.

We begin our discussion of modes and nodes by making reference to Sturm's Theorems relating to the nodes of a second order equation. These theorems have wide applicability, are easily proved, and yield valuable insight into the properties of the solutions of Sturm-Liouville systems. Sturm's original results appeared in 1836. The most complete account was given by Bôcher (1917) [37]. A detailed account also appears in Chapter X of Ince (1927) [185].

14.2 Sturm's Theorems

In Section 10.1 we introduced three equations, (10.1.1), (10.1.3) and (10.1.11) that appear in vibration problems. These equations all contain the frequency parameter λ, and they must all be complemented by end conditions in order to yield a well-posed eigenvalue problem. Sturm's Theorems may be formulated for a wider class of equations that includes (10.1.1), (10.1.3) and (10.1.11), and apply to the equation without regard for end conditions.

Consider the equation

$$(Ay')' + By = 0, \tag{14.2.1}$$

and suppose that $A(x), A'(x)$ and $B(x)$ are continuous and $A(x) > 0$ throughout an interval $[a, b]$. These conditions are unnecessarily restrictive; we could suppose, say, that they were piecewise continuous with a finite number of points of discontinuity, or even in a wider class. We leave such niceties to the interested reader.

For the starting point of our discussion, we note that if $y(x)$ is a continuous solution of (14.2.1) and $y(c) = 0 = y'(c)$ for some $c \in [a, b]$, then y is identically zero. If A, B have derivatives of all orders then $y(c) = 0 = y'(c)$ implies $y''(c) = 0 = y'''(c) \ldots$, so that the Taylor expansion of $y(x)$ is identically zero. If only A, A', B are continuous then we may reach the same conclusion by converting (14.2.1) into an integral equation. Alternatively, we may approximate A, B arbitrary closely by $\tilde{A}(x), \tilde{B}(x)$ that do have derivatives of all orders, and reach the same conclusion.

From this result, we may deduce that *every zero (node) of a solution of (14.2.1) is simple*: if $y(c) = 0$, then $y'(c) \neq 0$, and y crosses the axis at $x = c$.

We may deduce also that *no continuous solution of (14.2.1) can have an infinity of nodes in* $[a, b]$. For if there were an infinity of nodes then, by the Bolzano-Weierstrass Theorem, they would have at least one limit point $c \in [a, b]$ and we can show (Ex. 14.2.1) that, at c, not only $y(c) = 0$ but $y'(c) = 0 : y \equiv 0$.

Now suppose that u, v are two solutions of (14.2.1), so that

$$(Au')' + Bu = 0 = (Av')' + Bv.$$

Multiplying the first by v, the second by u, subtracting and rearranging, we find

$$(A(vu' - uv'))' = 0,$$

so that

$$A(vu' - uv') = \text{ constant } = C. \tag{14.2.2}$$

Since, by hypothesis, $A(x) > 0$, the constant is zero iff the Wronskian, $vu' - uv'$, is zero, i.e., iff the solutions are proportional, i.e., $u = kv$. Henceforward, we will say that two solutions u, v are *the same* if $u = kv$, *different* if there is no k such that $u = kv$. From this we may immediately deduce

Theorem 14.2.1 *Two different solutions of (14.2.1) cannot have a common zero.*

Proof. $u(c) = 0 = v(c)$ implies $C = 0$ in (14.2.2) ■

Theorem 14.2.2 *The nodes of two* ***real*** *different solutions of (14.2.1) separate each other.*

Proof. First note that it is necessary to include the word 'real' in the statement because

$$y(x) = \cos x + i \sin x,$$

a solution of $y'' + y = 0$, has *no* nodes on the real axis.

Now suppose one solution of (14.2.1), u, has two nodes $x_1, x_2 \in [a, b]$, and v is a second, different, solution. By Theorem 14.2.1, $v(x_1), v(x_2) \neq 0$. Suppose $v(x)$ has no node in (x_1, x_2), then it must have the same sign in $[x_1, x_2]$, say positive. That means that $z = u(x)/v(x)$ is continuous, has a continous derivative in $[x_1, x_2]$, and is zero at the ends x_1 and x_2. Therefore, by Rolle's Theorem, $z'(\xi) = 0$ for some $\xi \in (x_1, x_2)$. But

$$z' = \frac{vu' - uv'}{v^2}.$$

The numerator of this expression is the Wronskian, which is not zero because u, v are different, and the denominator is v^2, which is positive by hypothesis. Thus, $z' \neq 0$ throughout (x_1, x_2). This contradiction implies that v has a node in (x_1, x_2). ■

Corollary 14.2.1 *If u, v are two different solutions of (14.2.1), then the numbers of nodes of u, v in any interval $[\alpha, \beta] \subset [a, b]$ cannot differ by more than one.*

Theorems 14.2.1, 14.2.2 concern two different solutions of the same equation (14.2.1); the next results concern the solutions of two different equations.

Theorem 14.2.3 *Suppose $u(x)$ is a solution of*

$$(Ay')' + B_1 y = 0,$$

and $v(x)$ is a solution of

$$(Ay')' + B_2 y = 0,$$

where $B_1 \leq B_2$ in $[a, b]$ and $B_1(x) < B_2(x)$ for some $x \in [a, b]$, then $v(x)$ has a node between any two nodes of $u(x)$.

Proof. Suppose x_1, x_2 are consecutive nodes of u, and suppose, if possible, that v has no node in (x_1, x_2). With no loss in generality, we may assume that $u(x), v(x) > 0$ in (x_1, x_2). The equations $(Au')' + B_1 u = 0, (Av')' + B_2 v = 0$ yield, as before,

$$-u(Av')' + v(Au')' = (B_2 - B_1)uv$$

so that

$$-(A(uv' - vu'))' = (B_2 - B_1)uv$$

which, on integration, gives

$$-[A(uv' - vu')]_{x_1}^{x_2} = \int_{x_1}^{x_2} (B_2 - B_1)uvdx. \quad (14.2.3)$$

Since $u(x_1) = 0 = u(x_2)$ the L.H.S. is

$$P \equiv -A(x_1)v(x_1)u'(x_1) + A(x_2)v(x_2)u'(x_2).$$

Since $u(x) > 0$ in (x_1, x_2) we have $u'(x_1) > 0, u'(x_2) < 0$ so that $P \leq 0$.

P can be zero only if $v(x_1) = 0 = v(x_2)$ in which case the Wronskian of u and v is zero, u and v are the same solution in $[x_1, x_2]$, and

$$\int_{x_1}^{x_2} (B_2 - B_1)uvdx = 0.$$

Since $uv > 0$ in (x_1, x_2) and B_1, B_2 are continuous, this forces $B_1 = B_2$ in (x_1, x_2). *Otherwise*, $P < 0$, and $B_1 \leq B_2$ implies that the R.H.S. of (14.2.3) is non-negative (≥ 0). This contradiction implies that $v(x)$ has a node between x_1 and x_2. ∎

Picone extended Sturm's Theorem 14.2.3 to give

Theorem 14.2.4 *Suppose $u(x)$ is a solution of*

$$(A_1y')' + B_1y = 0$$

and $v(x)$ is a solution of

$$(A_2y')' + B_2y = 0$$

where $A_1 \geq A_2 > 0$, $B_1 \leq B_2$ in $[a, b]$ and $A_1(\xi) > A_2(\xi)$, $B_1(\eta) < B_2(\eta)$ for some $\xi, \eta \in [a, b]$. Then $v(x)$ has a node between any two nodes of $u(x)$.

Proof. Picone wrote

$$\left(\frac{u}{v}(A_1u'v - A_2uv')\right)' = (B_2 - B_1)u^2 + (A_1 - A_2)u'^2 + A_2\left(u' - \frac{uv'}{v}\right)^2. \quad (14.2.4)$$

Suppose, as before, that x_1, x_2 are two consecutive nodes of u, that $u(x) > 0$ in (x_1, x_2) so that $u'(x_1) > 0, u'(x_2) < 0$. Suppose v has no node in (x_1, x_2) and that $v(x_1), v(x_2) > 0$. On integrating (14.2.4) over (x_1, x_2) we find

$$\begin{aligned}\left[\frac{u}{v}(A_1u'v - A_2uv')\right]_{x_1}^{x_2} &= \int_{x_1}^{x_2} (B_2 - B_1)u^2dx + \int_{x_1}^{x_2} (A_1 - A_2)u'^2dx + \\ &\quad \int_{x_1}^{x_2} A_2\left(u' - \frac{uv'}{v}\right)^2 dx. \qquad (14.2.5)\end{aligned}$$

The L.H.S. is zero because $u(x_1) = 0 = u(x_2)$, while the R.H.S. is positive. This contradiction implies that v has a node in (x_1, x_2). The L.H.S. is still zero even if $v(x)$ is zero at one or both of x_1, x_2. For if $v(x)$ is zero, say, at x_1 then

$$\lim_{x \to x_1} \frac{u}{v} = \frac{u'(x_1)}{v'(x_1)}$$

so that

$$\lim_{x \to x_1} \left[\frac{u}{v}(A_1 u'v - A_2 uv')\right]_x = (A_1 - A_2)uu'|_{x_1} = 0.$$

Note that, in exceptional cases, discussed in Ex. 14.2.2, the R.H.S. may be zero, in which cases we must modify our conclusion slightly. ■

We may use Picone's formula (14.2.4) to prove two separation theorems. The first is

Theorem 14.2.5 *Suppose $u(x)$ is the solution of*

$$(A_1 u')' + B_1 u = 0 \tag{14.2.6}$$

subject to

$$u(a) = \alpha, \quad u'(a) = \alpha' \tag{14.2.7}$$

and $v(x)$ is the solution of

$$(A_2 v')' + B_2 v = 0$$

subject to

$$v(a) = \beta, \quad v'(a) = \beta'.$$

We make the following assumptions:

1) $A_1 \geq A_2 > 0, \quad B_1 \leq B_2$ *in* $[a, b]$.

2) α, α' *are not both zero, nor are* β, β'.

3) *If* $\alpha \neq 0$, *then*
$$\frac{A_1(a)\alpha'}{\alpha} \geq \frac{A_2(a)\beta'}{\beta}$$
which implies $\beta \neq 0$.

4) *The identity* $A_1 \equiv 0 = A_2$ *is not satisfied in any finite part of* $[a, b]$.

If $u(x)$ has n nodes in $(a, b]$, then $v(x)$ has at least n nodes in $(a, b]$, and the ith node of $v(x)$ is less than the ith node of $u(x)$.

Proof. Let $x_1, x_2, \ldots, x_n$ be the nodes of $u(x)$ in $(a, b]$, so that

$$a < x_1 < x_2 < \cdots < x_n \leq b.$$

Sturm's Theorem 14.2.4 states that $v(x)$ has a node between any two consecutive nodes x_i, x_{i+1}. The Theorem holds therefore if we can show that $v(x)$ has a node between a and x_1.

If $u(x)$ is zero at the left hand end point, then, by Theorem 14.2.4, $v(x)$ has a node between a and x_1. We therefore suppose that $\alpha \neq 0$, so that condition 3) implies $\beta \neq 0$. Integrate the Picone formula (14.2.4) between a and x_1, assuming that $v(x)$ has no node in (a, x_1); it is

$$\left[\frac{u}{v}(A_1 u'v - A_2 uv')\right]_a^{x_1} = -u^2(a)\left(\frac{A_1(a)\alpha'}{\alpha} - \frac{A_2(a)\beta'}{\beta}\right)$$

which, by condition 3), is negative. The integral of the R.H.S. of (14.2.4) is positive. This contradiction implies that $v(x)$ has a node between a and x_1. ∎

This theorem allows us to deduce what happens to the nodes of $u(x)$, the solution of equations (14.2.6), (14.2.7) when $A(x)$ decreases continuously and $B(x)$ increases continuously, while α and α' are kept invariant: each new node enters at $x = b$ and moves towards $x = a$.

Exercises 14.2

1. Suppose $y(x)$ has an infinity of nodes in $[a, b]$ with limit point c. Use the Mean Value Theorem to show that $y'(c) = 0$.
2. Explore how the R.H.S. of (14.2.5) can actually be zero. Show that one can ensure that it is not zero by imposing the condition that B_1 and B_2 are not identically zero in any finite part of (a, b).
3. See how the nodes of $y'' + \omega^2 y = 0, y(0) = \alpha, y'(0) = \alpha'$ in $[0, 1]$ travel from 1 towards 0 as ω increases.

14.3 Applications of Sturm's Theorems

Sturm's Theorems describe what happens to a node of a solution of equation (14.2.1) when $A(x)$ or $B(x)$ change. In this section we look at the inverse question: what can we deduce about changes in $A(x), B(x)$ from changes in nodal positions?

First, consider the taut string governed by equation (10.1.1), namely

$$u'' + \lambda\rho^2 u = 0. \tag{14.3.1}$$

Recall that $\lambda = \omega^2$, $\rho^2(x)$ is the mass per unit length, and that the end conditions are

$$u'(0) - hu(0) = 0 = u'(1) + Hu(1).$$

Equation (14.3.1) has the form (14.2.1) with

$$A(x) = 1, B(x) = \lambda\rho^2(x).$$

Consider what happens when a small mass is removed from the string at some interior point c. We can imagine that the mass is removed continuously over a small interval $(c-\varepsilon, c+\varepsilon)$. Removal of mass increases (or at least does not decrease) the natural frequency. Denote the new natural frequency by $\omega^*, \lambda^* = \omega^{*^2}$, the new mass distribution by $\rho^{*^2}(x)$, and let v be the solution of

$$v'' + \lambda^* \rho^{*^2} v = 0$$

subject to

$$v'(0) - hv(0) = 0 = v'(1) + Hv(1).$$

Suppose u has a node $\xi \in (0, c-\varepsilon]$, and apply Theorem 14.2.5 to $[0, c-\varepsilon]$. In that interval $A_1 = A_2$, $\rho^* = \rho$ and $B_1 = \lambda\rho^2 \le \lambda^*\rho^2 = B_2$. Thus v has a node $\eta \in (0, c-\varepsilon]$, and $\eta \le \xi$. If u has n nodes $(\xi_i)_1^n \in (0, c-\varepsilon]$, then v has at least n nodes $(\eta_i)_1^n \in (0, c-\varepsilon]$, and $\eta_i \le \xi_i$. Thus nodes to the left of $c-\varepsilon$ move to the left. By physically turning the string around, we see that nodes to the right of $c+\varepsilon$ will move to the right: the nodes *move away* from c. We note that the result holds if mass is removed over *any* interval, small or not. (But the theorem does not yield information about the movement of nodes *in* the interval.) Also, if mass is *added* rather than *removed* then the nodes will move *toward* the added mass.

We may draw a conclusion regarding the inverse question: if nodes move away from (toward) an interval, then mass has been removed (added) in that interval. This holds only for one interval; if there are two or more intervals in which mass is removed (added), then there will be interaction between the two effects.

Note that in Theorem 14.2.5, and in our analysis in this section, we predicted the movement of nodes to the left of $c-\varepsilon$ by considering only the solution of the differential equation and the left-hand end conditions

$$u(0) = \alpha, \quad u'(0) = h\alpha.$$

We can make a crude estimate of the amount of mass added or removed in a small interval if we can identify two neighbouring nodes x_1, x_2 of a mode with frequency ω, such that after the mass is added, the node x_1 moves to the right to x_1^*, and x_2 to the left, to x_2^*, the frequency decreases to ω^*. Suppose that the original mass per unit length between x_1 and x_2 was constant, ρ^2, and that after the mass is added it is $(\rho+\sigma)^2$. Since x_1, x_2 are consecutive nodes of the initial mode

$$\omega\rho(x_2 - x_1) = \pi$$

and similarly

$$\omega^*(\rho+\sigma)(x_2^* - x_1^*) = \pi.$$

This means that, knowing $x_2 - x_1, x_2^* - x_1^*, \omega$ and ω^* we may find $(\rho+\sigma)/\rho$:

$$\frac{\rho+\sigma}{\rho} = \frac{\omega}{\omega^*} \cdot \frac{x_2 - x_1}{x_2^* - x_1^*} > 1.$$

For a slightly more challenging problem, let us consider the effect of point damage to a rod in longitudinal vibration, following Gladwell and Morassi (1999)

[128]. Recall that for an undamaged rod with cross-sectional area $A(x)$, the governing equations are (10.1.3), (10.1.4):

$$(Au')' + \lambda Au = 0, \tag{14.3.2}$$

$$u'(0) - hu(0) = 0 = u'(1) + Hu(1). \tag{14.3.3}$$

We note that the 'stiffness' term, $(Au')'$, has the same distribution, A, as the 'inertia' term Au. If the rod is damaged by a small notch at $x = c$, then the stiffness will be seriously affected while the inertia term will be almost unaffected. For this reason, we model the notch as a spring so that, at c,

$$[u'(c)] = 0, \tag{14.3.4}$$

$$k[u(c)] = A(c)u'(c). \tag{14.3.5}$$

where $[f(c)] := f(c+) - f(c-)$. The undamaged rod corresponds to $k \to \infty$, i.e., $\varepsilon = 1/k \to 0$. We may show, as expected, that the natural frequencies are increasing functions of k, i.e., decreasing functions of ε. We may find the first order variation of the natural frequencies with ε by taking

$$u(x) = u_0(x) + \varepsilon v(x), \quad \lambda = \lambda_0 + \varepsilon\mu,$$

in (14.3.2)-(14.3.5). We find that

$$(Au_0')' + \lambda_0 Au_0 = 0, \tag{14.3.6}$$

$$(Av')' + \lambda_0 Av + \mu Au_0 = 0, \tag{14.3.7}$$

$$[v'(c)] = 0, \tag{14.3.8}$$

$$[v(c)] = A(c)u_0'(c). \tag{14.3.9}$$

Multiplying (14.3.6) by v, (14.3.7) by u_0 and subtracting, and then integrating from 0 to 1, using (14.3.3), we find

$$(A(c)u_0'(c))^2 + \mu \int_0^1 Au_0^2 dx = 0,$$

which, with the normalising condition

$$\int_0^1 Au_0^2 dx = 1$$

gives

$$\mu = -(A(c)u_0'(c))^2. \tag{14.3.10}$$

This equation shows how the natural frequencies change with ε. We now show how the modes, and particularly the nodes, change with ε. To do that, we use Theorem 14.2.4 again. We consider the portion of the rod to the left of c; there the displacement is given by the solution of (14.3.2) and the first of equations

(14.3.3) with $\lambda = \lambda_0 + \varepsilon\mu < \lambda_0$. We identify B_2 with the undamaged case ($B_2 = \lambda_0 A$) and B_1 with the damaged case ($B_1 = \lambda A$). According to Theorem 14.2.5, the nodes corresponding to B_2 lie to the left of those corresponding to B_1. That is, due to the damage, nodes move *toward* the damage.

We now determine the first order change in the positions of the nodes. To do this, we estimate the first order changes in the nodes of u to the left of c. This means that we are looking at the first order change in the solution of

$$\begin{aligned}(A\theta')' + \lambda A\theta &= 0,\\ \theta'(0) - h\theta(0) &= 0.\end{aligned}$$

Note that we write the dependent variable as θ to emphasize that we are not looking to an eigenmode, just at the solution satisfying the left-hand end condition. This solution is uniquely determined apart from an arbitrary multiplicative constant. Put

$$\theta = \theta_0 + \varepsilon\psi, \quad \lambda = \lambda_0 + \varepsilon\mu,$$

and find

$$(A\psi')' + \lambda_0 A\psi + \mu A\theta_0 = 0, \tag{14.3.11}$$

$$\psi'(0) - h\psi(0) = 0. \tag{14.3.12}$$

To solve these equations we use the method of *variation of parameters*: we write $\psi = \theta_0 f$. After some manipulation, we find that this will satisfy (14.3.11), (14.3.12) if

$$\begin{gathered}(A\theta_0^2 f')' + \mu A\theta_0^2 = 0,\\ f'(0) = 0.\end{gathered}$$

Thus

$$A\theta_0^2 f' + \mu \int_0^x A\theta_0^2 dx = 0. \tag{14.3.13}$$

If x_0 is a node of θ_0, then the corresponding node of θ is $x_0 + \varepsilon\xi$ where, to first order,

$$\begin{aligned}0 = \theta(x_0 + \varepsilon\xi) &= \theta_0(x_0 + \varepsilon\xi) + \varepsilon\psi(x_0)\\ &= \theta_0(x_0) + \varepsilon\xi\theta_0'(x_0) + \varepsilon\psi(x_0).\end{aligned}$$

Thus

$$\xi = -\psi(x_0)/\theta_0'(x_0). \tag{14.3.14}$$

Now $\psi(x) = \theta_0(x)f(x)$, and since $\theta_0(x) \to 0$ as $x \to x_0$, we must have $f(x) \to \infty$ as $x \to x_0$. We must therefore evaluate $\psi(x_0)$ by writing $f(x) = 1/g(x)$ and using l'Hôpital's rule:

$$\psi(x_0) = \lim_{x \to x_0} \frac{\theta_0(x)}{g(x)} = \frac{\theta_0'(x_0)}{g'(x_0)}.$$

Putting $f = 1/g$ in (14.3.13) we find

$$\frac{-A\theta_0^2(x)}{g^2(x)} g'(x) + \mu \int_0^x A\theta_0^2 dx = 0,$$

and on taking the limit $x \to x_0$, we find

$$-A(x_0)\left(\frac{\theta_0'(x_0)}{g'(x_0)}\right)^2 g'(x_0) + \mu \int_0^{x_0} A\theta_0^2 dx = 0. \qquad (14.3.15)$$

To find the change in a node of a mode, we put $\theta_0(x) = u_0(x)$ and combine (14.3.15) with (14.3.14) and (14.3.10) to give

$$\varepsilon\xi = \varepsilon[A(c)u_0'(c)]^2 \int_0^{x_0} Au_0^2 dx / (A(x_0)[u_0'(x_0)]^2) \qquad (14.3.16)$$

as the change in the position of the node, from x_0 to $x_0 + \varepsilon\xi$; as expected, $\xi > 0$.

In the particular case of a uniform free-free rod, for which $A = 1, u_0 = \sqrt{2}\cos[(n-1)\pi x], n = 1, 2, \ldots$ we find that the mth mode moves from $x_0 = (2m-1)/(2n-2), m = 1, \ldots, n-1$, to $x_0 + \varepsilon\xi$ where

$$\varepsilon\xi = \varepsilon x_0 \sin^2[(n-1)\pi c].$$

The corresponding changes for nodes to the right of c are

$$\varepsilon\xi = -\varepsilon(1 - x_0)\sin^2[(n-1)\pi c].$$

These results show that, for a given mode, the changes in node positions increase as the node, x_0, approaches the damage position. The proportional changes for those nodes to the left of $c, \varepsilon\xi/x_0$, and for those to the right, $\varepsilon\xi/(1 - x_0)$, are the same for each node; they depend only on the position of the notch. This means that to find the position of the damaged point we look for two nodes of a mode that have moved towards each other; the notch lies between these nodes.

An experimental study based on these results may be found in Gladwell and Morassi (1999) [128], which also gives references to the related literature.

14.4 The research of Hald and McLaughlin

Both Ole Hald and Joyce McLaughlin have been studying inverse problems, amongst other topics, for many years, and we have referred to their individual researches on numerous occasions already. In this section we make a brief report on their joint work on *inverse nodal problems.*

Inverse nodal problems differ from the inverse eigenvalue problems that form the subject matter of most of this book, in many subtle ways. We have already noted that while an eigenvalue, a natural frequency, reflects the properties of a system as a whole, a nodal position relates to the properties of the system *near* the node. But there are other differences, differences in the paths from data to system properties. When the data consist of eigenvalues (and maybe some norming constants) there is usually some algorithm that gives the exact values of a set of parameters defining the properties of a unique system which has this spectral data. In contrast, any researcher approaching an inverse nodal problem soon realises that nodal positions, the totality of nodal positions for all the principal modes, provide *too much data.* For example, for a string fixed at its ends, the first mode has no node, the next has one, and so on; the first n modes have a total of $n(n-1)/2$ nodes. Somehow we must make a choice: choose all the nodes of one mode, or choose one node from each mode, for example. Clearly, different choices will yield different models. The situation is made more complex because a continuous system, like a string, has an infinity of modes, and thus of nodes. From a mathematical point of view it would be reassuring to know that if one chose nodes of more and more modes in a particular way, then the resulting systems would *converge* in some sense to a unique system, and that one could give numerical estimates of the error one would make by using a finite number, n, of properly chosen nodes.

There are thus three distinct parts to the 'solution' of an inverse nodal problem: finding an approximate system which has given nodal positions for certain mode(s); establishing that if any infinity of nodes, chosen in a certain way, is given, there is no more than one vibrating system, of a specified type, that has these nodes for certain of its modes; constructing bounds for the error in truncating the infinite set of nodes at a certain number n. The first part is relatively simple; Hald and McLaughlin provide a number of algorithms for the various types of Sturm-Liouville system described in Section 10.1. The other two parts are difficult, and require a daunting array of analytical tools; we shall therefore content ourselves with giving the gist of the methods used and theorems proved; the interested reader may consult the original papers that are readily available.

Our starting point is a fundamental paper by McLaughlin alone, McLaughlin (1988) [231]. This deals with part 2 of the problem, uniqueness. McLaughlin considers the Sturm-Liouville equation (10.1.14) with Dirichlet end conditions:

$$y'' + (\lambda - q)y = 0, \qquad (14.4.1)$$

$$y(0) = 0 = y(1), \qquad (14.4.2)$$

where $q \in L^2(0,1)$.

First recall that if q_1, q_2 are two potentials with $q_2 = q_1 + c$, where c is a constant, then noting that

$$\lambda - q_2 = (\lambda - c) - q_1,$$

we see that the eigenvalues of the two problems differ by c while the eigenfunctions, and thus the nodes of the eigenfunctions, remain the same. This means that nodal information alone can yield q only to within an arbitrary additivie constant: any uniqueness theorem related to nodal information must contain the added information

$$\int_0^1 q_1(x)dx = \int_0^1 q_2(x)dx. \qquad (14.4.3)$$

McLaughlin proves that if two potentials q_1, q_2 satisfy (14.4.3), and if the eigenfunctions $y_n(q_1, x), y_n(q_2, x)$ have a common set of nodes that is *dense* in (0,1) (see Section 10.3 for the definition of dense), then $q_1 = q_2$ in $L^2(0,1)$. The gist of the proof is as follows.

First consider (14.4.1), (14.4.2) with $q \equiv 0$. The eigenvalues are $\lambda_n = (n\pi)^2, n = 1, 2, \ldots$; the eigenfunctions are $y_n(x) = y_n(0, x) = \sin n\pi x$; the nodes of $y_n(x)$ are $x_{n,j}(0) = j/n, j = 1, 2, \ldots, (n-1)$. Note that $y_1(x)$ has *no* node.

Now group the numbers 2,3,4,... as follows: 2;4,3;8,7,6,5;... This is equivalent to writing

$$n = 2^{k+1} - m;\ k = 0, 1, 2, \ldots;\ m = 0, 1, \ldots 2^k - 1. \tag{14.4.4}$$

The $(m+1)$th node of the nth node is

$$x_{n,m+1}(0) = (m+1)/(2^{k+1} - m) \tag{14.4.5}$$

and the set of numbers $x_{n,m+1}(0)$ for $k = 0, 1, 2, \ldots;\ m = 0, 1, \ldots, 2^k - 1$ is dense in (0,1); the numbers are $\frac{1}{2};\ \frac{1}{4}, \frac{2}{3};\ \frac{1}{8}, \frac{2}{7}, \frac{3}{6}, \frac{4}{5}; \ldots$ The uniqueness result is

Theorem 14.4.1 *Let $q_1, q_2 \in L^2(0,1)$, and consider the eigenvalue problems*

$$y'' + (\lambda - q_i)y = 0,$$

$$y(0) = 0 = y(1),$$

$i = 1, 2$. For each $n \geq 2$, suppose that the positions of the nodes, chosen according to (14.4.5) satisfy

$$x_{n,j}(q_1) = x_{n,j}(q_2),\ n = 2, 3, \ldots$$

and that

$$\int_0^1 q_1(x)dx = \int_0^1 q_2(x)dx$$

then $q_1 = q_2$ in $L^2(0,1)$.

McLaughlin contrasts this inverse nodal problem with inverse eigenvalue problems for the Sturm-Liouville equation, and recalls that *two* spectra, corresponding to two different end conditions at one end (or some equivalent data, e.g., norming constants) are needed to determine q. She comments '*what can be shown... is that the position of one node, albeit judiciously chosen, for each eigenfunction, $n \geq 2$, is more than enough data to determine q uniquely (apart from a constant). It seems then that the nodal positions in some sense contain "more" information about this potential q than either a set of eigenvalues or a set of norming constants.*'

While McLaughlin (1988) [231] was concerned only with part 2, uniqueness, Hald and McLaughlin (1989) [167] consider all three aspects, approximation, uniqueness and error bounds. They consider a generic equation with free end conditions:

$$(pv')' + \omega^2\rho^2 v = 0, \tag{14.4.6}$$

$$v'(0) = 0 = v'(1). \tag{14.4.7}$$

If $p \equiv 1$, this is the string equation (10.1.1) (with density ρ^2). If $p \equiv \rho^2$, this is the rod equation (10.1.3) with $A \equiv p \equiv \rho^2$.

One problem that they consider is the string ($p \equiv 1$) with free ends. They construct a string with piecewise constant density as follows. Suppose the nodes of the nth ($n \geq 2$) mode $v_n(x)$ are $(x_j)_1^{n-1}$, where $0 < x_1 < x_2 < \ldots x_{n-1} < 1$. Consider $v_n(x)$ in an interval (x_{j-1}, x_j). In that interval $v_n(x)$ is the fundamental mode of the string fixed at the ends x_{j-1} and x_j, and ω_n is the fundamental frequency; in $(0, x_1)((x_{n-1}, 1))$ it is the fundamental mode for a string free at 0, fixed at x_1 (fixed at x_{n-1}, free at 1). Suppose, therefore, that the non-uniform string is replaced by a string with uniform density ρ_j^2 in the interval $(x_{j-1}, x_j), j = 1, \ldots, n$ where $x_0 = 0, x_n = 1$. For the jth $(2 \leq j \leq n-1)$ part of the string, the governing equation is

$$v'' + \omega_n^2 \rho_j^2 v = 0,$$

$$v(x_{j-1}) = 0 = v(x_j),$$

so that

$$v(x) = \sin\{\pi(x - x_{j-1})/(x_j - x_{j-1})\},$$

and

$$\rho_j = \pi/(\omega_n(x_j - x_{j-1})), \; j = 2, \ldots, n-1. \tag{14.4.8}$$

For the first segment $(0, x_1)$ we have the end conditions $v'(0) = 0 = v(x_1)$ so that

$$v(x) = \cos(\pi x/(2x_1)),$$

and

$$\rho_1 = \pi/(2\omega_n x_1). \tag{14.4.9}$$

Similarly for the last segment $(x_{n-1}, 1)$,

$$\rho_n = \pi/(2\omega_n(1 - x_{n-1})). \tag{14.4.10}$$

This is the approximation in the specific case $p \equiv 1$, and the end conditions (14.4.7). If the end conditions are $v(0) = 0 = v(1)$, then equation (14.4.8) holds for $j = 1, n$ also.

Hald and McLaughlin present similar algorithms to compute approximations to p and ρ in other cases, and for the Sturm-Liouville potential q. For the rod equation, in which $p = \rho^2$, they first find a potential q, and then reverse the transformation leading to (10.1.14) to find $A(x)$. They also point out a fundamental difference between the nodes of the string equation (10.1.3) and the rod equation or the Sturm-Liouville equation (10.1.14): a perturbation of ρ in the string equation may cause (relatively) large changes in the nodal positions; by contrast a perturbation in $A(x)$ or q may cause only miniscule changes in the nodes of a high mode. They comment, "*... the information in the nodal positions which we use to approximate the... impedance function* ($A(x)$) *sits much deeper in the data than the information about the density (of the string).*"

These remarks concern the part of the solution, the approximate construction. The greater part of Hald and McLaughlin (1989) [167] concerns error bounds and uniqueness theorems. For the simple case of the string with free ends, for instance, they show that the procedure outlined above gives a second order approximation to the density at the mid points of the interior intervals, i.e., $\rho((x_j + x_{j-1})/2) = \rho_j$ but only a first order approximation for the two end intervals. They give precise error bounds which show how the rate of convergence with increasing n depends on the smoothness of the density. They present numerous case studies here and also in Hald and McLaughlin (1988) [166].

The uniqueness results that they obtain are generalisations of those found in McLaughlin (1988) [231], a typical one is

Theorem 14.4.2 *Let $p \equiv 1$, and suppose that the second derivative of ρ is integrable. Then ρ is uniquely determined (up to a multiplicative constant) by any dense set of nodes.*

In Hald and McLaughlin (1998) [169] they return to the inverse nodal problem and develop a theory governing approximation, uniqueness and error bounds for (14.4.6), subject to Dirichlet end conditions, when p and ρ are functions of bounded variation.

Hald and McLaughlin (1996) [168] deal with inverse nodal problems for non-uniform rectangular membranes. Space does not allow us to consider these problems. We simply note that they pose two difficulties.

Consider a uniform rectangular membrane with sides a, b vibrating with fixed edges. Its eigenvalues are

$$\lambda = \omega^2 = \left(\frac{m^2}{a^2} + \frac{n^2}{b^2}\right)\pi^2.$$

If $\alpha = a/b$ and α^2 is rational, then there will be multiple eigenvalues, and one can find eigenvalues with multiplicity exceeding any stated number. If α^2 is irrational, each eigenvalue will be distinct, but one can find two eigenvalues as close as one wishes. This closeness poses problems in the search for error bounds.

The second difficulty relates to the shape of nodal domains, regions bounded by nodal lines. For the uniform rectangular membrane the nodal domains are themselves rectangles; the eigenfunction corresponding to $\lambda_{m,n}$ divides the rectangle into mn equal rectangles, as shown in Figure 14.4.1a. However, if the membrane density is perturbed from its uniform value, then the nodal domains may change dramatically, as shown in Figure 14.4.1b. This complicates the search for an approximation to the density; one would like to have a situation which is a generalisation of that for the string; the perturbed nodal domain is roughly a rectangle. One could then assume that the density was constant over that rectangle, and use the fact that the eigenfunction is the fundamental eigenfunction on the region bounded by the nodal lines. The major contribution of Hald and McLaughlin (1996) [168] is that they show how both these difficulties may be overcome and how one can find good approximations to the density, and how to obtain uniqueness theoreoms. McLaughlin (2000) [232] reconsiders inverse problems for a rectangular membrane. She considers three different approaches to the problem: in the first, the data consists of mode shape level sets and frequencies; in the second, it consists of frequencies and boundary mode shape measurements; in the third, the data consists of frequencies for four different boundary value problems. Local existence, and uniqueness results are established together with numerical results for approximate solutions.

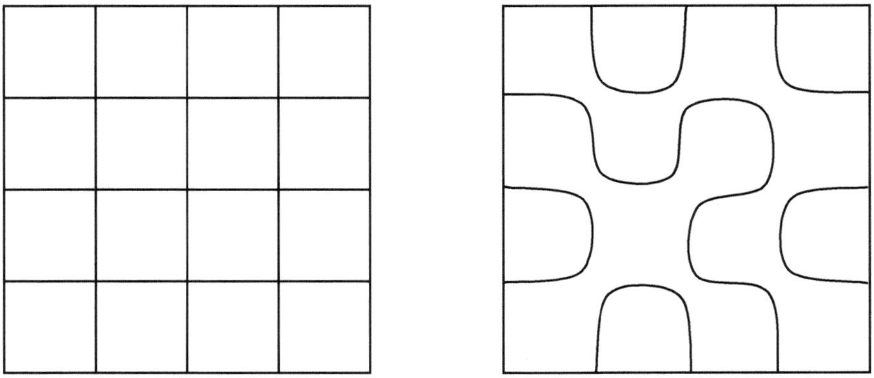

a) *b)*

Figure 14.4.1 - Nodal domains change from rectangles to irregular figures.

Chapter 15

Damage Identification

Chance gives rise to thoughts, and chance removes them; no art can keep or acquire them. A thought has escaped me. I wanted to write it down. I write instead, that it has escaped me.

Pascal's *Pensées*, 585

15.1 Introduction

As we mentioned in the Preface, the identification of damage in a vibrating structure from changes in vibratory behaviour is an inverse problem in a loose interpretation of the term. Since such damage identification has potentially important practical value, it is appropriate for it to be included in any treatment of inverse problems but, since it is essentially an application of inverse techniques and must be combined with numerical methods, it is only of marginal relevance in this book which, as we stated in the Preface, is concerned primarily with theoretical and qualitative matters. We will therefore confine our remarks in this Chapter to a survey of the literature, and an examination of the methods used, the assumptions that are made, and the conlusions that may be drawn regarding damage identification in certain simple cases.

We begin our discussion with some statements that may be grasped intuitively:

If a structure is damaged, its vibratory behaviour will change.

By *vibratory behaviour* we mean the response of a structure to time-varying forces. We will assume thta the structure is undamped so that we may speak about *frequency response*, the response of the structure to sinusoidal forces with a specific frequency ω. As usual, we focus our attention on the natural frequencies and corresponding principal mode shapes of the structure; these may be obtained (at least in theory) by applying standard modal analysis techniques to the frequency responses at various points of the structure. Thus, we make the following statement:

The vibratory behaviour of a structure may be characterised by its natural frequencies and corresponding principal mode shapes.

On the other hand

Structural damage may be characterised by its locations, intensities and types; we thus refer to a ***damage pattern****.*

Strictly speaking, a structure is said to be damaged when it undergoes a change that reduces its stiffness, or more generally reduces its strain energy. Under this definition, damage *reduces* the natural frequencies of the structure, or at least does not *increase* them. We shall loosen this definition and define damage as a (small) change in the structure; this would include (positive or negative) changes in stiffness or in mass.

Now consider the 'simple' forward problem: *Given a specified structure, find the changes in vibratory behaviour brought about by specified damage.*

The solution of this forward problem depends critically on there being *models* for the undamaged and damaged structures, from which the natural frequencies and mode shapes may be extracted using established methods. It is known that, under certain conditions, this problem may be well posed: specific damage will cause a unique set of changes to the natural frequencies and mode shapes; and these changes will be continuous functions of the damage parameters.

However, almost all the inverse problems, in which one tries to find the damage (i.e., its locations, intensities and types) which gave rise to specified behavioural changes, are ill-posed. Specifically, there may be *no* damage pattern (whether damage is interpreted strictly or loosely) that would give rise to a certain set of behavioural changes; or there may be *more than one* pattern that would produce the same set of behavioural changes; and there is no guarantee that the damage parameters will be continuous functions of the behavioural changes.

The fact that there may be more than one damage pattern giving rise to specific changes in natural frequencies is a consequence of the fact that natural frequencies are global constructs - they depend on the complete structure, its distribution of mass and stiffness, and the way in which it is supported. It is sometimes possible to identify a damage pattern because a specific damage pattern will affect different frequencies by differing amounts. We may make this statement precise. Suppose it is known that a structure is damaged just at one location, but the location, s, and magnitude, d, are unknown. Generally, for small d, the change in the ith frequency, $\delta\omega_i$, will have the form

$$\delta\omega_i = d.f_i(s) : \tag{15.1.1}$$

it depends linearly on d, and non-linearly on the position. Thus, if the position, s, is known, then the change, $\delta\omega_i$, in one frequency, may be enough to determine d (provided that $f_i(s)$ is known). If s is unknown, then we consider the ratio of the changes to two frequencies:

$$\frac{\delta\omega_i}{\delta\omega_j} = \frac{f_i(s)}{f_j(s)}. \tag{15.1.2}$$

Thus, if the form of $f_i(s)$ is known as a function of s, then it may be possible to find the value of s corresponding to a given value of $\delta\omega_i/\delta\omega_j$.

In any particular case, it will be necessary to determine whether there is no, one, or more than one value(s) of s satisfying (15.1.2) If (one or more values of) s is known, then d may be found from (15.1.1). Clearly, if the damage is not restricted to one location, then the identification procedure will be very complicated.

We divide our discussion into two parts: damage identification in rods, and in beams.

15.2 Damage identification in rods

For a rod in longitudinal vibration, we model damage as a crack that stays open; following Freund and Herrmann (1976) [91] or Cabib, Freddi, Morassi and Percivale (2001) [47] we model such a crack as a longitudinal spring of stiffness k, and write $1/k = d$. In one of the early papers, Adams, Cawley, Pye and Stone (1978) [2] (see also Cawley and Adams (1979) [50]; and Hearn and Testa (1991) [170] for references to engineering studies) considered a damaged one-dimensional system (a generalised rod) modelled as two parts B and C, linked by a spring of stiffness k. If $\beta_{ss} := \beta_{ss}(s,\omega)$ and $\gamma_{ss} := \gamma_{ss}(s,\omega)$ are direct receptances (Bishop and Johnson (1960) [34]) of B and C at $x = s$, then the usual receptance analysis gives the frequency equation of the damaged system as

$$\beta_{ss}(s,\omega) + \gamma_{ss}(s,\omega) + d = 0.$$

Thus, if $\omega_m = \omega_m^0 + \delta\omega_m, \quad \omega_n = \omega_n^0 + \delta\omega_n$, where ω_m^0, ω_n^0 are undamaged frequencies then

$$\beta_{ss}(s,\omega_m) + \gamma_{ss}(s,\omega_m) = \beta_{ss}(s,\omega_m^0) + \gamma_{ss}(s,\omega_m^0) + \delta\omega_m \frac{\partial}{\partial\omega}$$
$$\{\beta_{ss}(s,\omega) + \gamma_{ss}(s,\omega)\}|_{\omega=\omega_m^0}.$$

The first term is zero because ω_i^0 is a natural frequency of the undamaged system $(d = 0)$. Thus,

$$\delta\omega_m \frac{\partial}{\partial\omega}\{\beta_{ss}(s,\omega) + \gamma_{ss}(s,\omega)\}|_{\omega=\omega_m^0} + d = 0$$

which may be rearranged in the form (15.1.1). Narkis (1994) [247] used this approach for a uniform free-free rod. For a general rod, the perturbation analysis of Section 14.3 shows that if $\lambda = \omega^2$ then

$$\delta\lambda_m = \mu\varepsilon = \mu d = -(A(s)u_m'(s))^2 d. \tag{15.2.1}$$

Morassi (2001) [237] made extensive use of this result. He showed that the problem of determining the location s from changes in two natural frequencies is generally ill-posed: if the system is symmetrical, then damage at any one of a set of symmetrical points will produce identical changes in natural frequencies. Even if the system is not symmetrical, damage at different locations can still produce identical changes in two natural frequencies.

Morassi (2003) [238] obtains particular results for uniform rods under various end conditions and determines situations in which the knowledge of $\delta\lambda_m, \delta\lambda_n$ does, and does not, uniquely determine the location s. Thus, for example, for a rod under free end conditions, he defines

$$C_m^F = \frac{-\delta\lambda_m^F}{2m^2\pi^2}.$$

The mth ($m \geq 1$) mode shape is $u_m(x) = \sqrt{2}\cos(m\pi x)$ so that (15.2.1) gives

$$\delta\lambda_m = -2m^2\pi^2 \sin^2(m\pi s)d$$

so that

$$C_m^F = d\sin^2(m\pi s).$$

Now use the trigonometric identities to deduce that

$$\begin{aligned} \sin^2(2m\pi s) &= (2\sin n\pi s.\cos m\pi s)^2 \\ &= 4\sin^2 m\pi s - 4\sin^4 m\pi s \end{aligned}$$

and hence

$$d(4C_m^F - C_{2m}^F) = 4(C_m^F)^2$$

so that the amount and location of the damage are given by

$$d = \frac{C_m^F}{1 - C_{2m}^F/(4C_m^F)}, \qquad \sin^2 m\pi s = \frac{C_m^F}{d}.$$

Similarly, he shows that d and $\sin^2 m\pi s$ may be uniquely determined from C_{m+1}^F and C_m^S, defined as

$$C_{m+1}^F = \frac{-\delta\lambda_{m+1}^F}{2(m+1)^2\pi^2}, \qquad C_m^S = \frac{-\delta\lambda_m^S}{2m^2\pi^2} \tag{15.2.2}$$

where λ_m^S is the mth natural frequency of the rod when it is supported at both ends. (Ex. 15.2.1).

Morassi and Dilena (2002) [239] analyse the analogous problem of determining the magnitude and location of a *point mass* attached to a thin rod from its effect on the natural frequencies. Morassi (1997) [236] sets up the problem of crack detection in a rod as an inverse problem in the spirit of Chapter 11, following Hald (1984) [165]. He shows that the position of the crack is uniquely determined from the asymptotic form of the spectrum. Biscontin, Morassi and Wendel (1998) [32] study the asymptotic form of the spectrum for a uniform free-free rod of unit length with a spring of stiffness k at $x = c$. The eigenvalues $\lambda (= \omega^2)$, are the roots of

$$p(\lambda) = \lambda \sin\lambda c \sin\lambda(1-c) - k\sin\lambda. \tag{15.2.3}$$

We can study two kinds of asymptotics, for k large or k small. For $k \to \infty$ the two parts of the rod are firmly joined together: the rod is an undamped free-free rod with eigenvalues $\lambda_m = m\pi, m = 0, 1, 2, \ldots$ For large k, i.e., small $\varepsilon = 1/k$, the mth eigenvalue is $\lambda_m = m\pi + \varepsilon\mu_m$ where (Ex. 15.2.2)

$$\mu_m = (-)^m \sin m\pi c \sin m\pi(1-c).$$

This is the kind of small change that we have been observing in the analysis above. For small k, the asymptotic form is centred about $k = 0$; now the rod splits into two free-free rods, one of length c, the other $1 - c$; there are two branches

$$\lambda_{m_1} = m_1\pi/c, \quad \lambda_{m_2} = m_2\pi/(1-c).$$

We now perturb these branches and seek eigenvalues of the form $\lambda = \lambda_{m_1} + k\nu_{m_1}, \quad \lambda = \lambda_{m_2} + k\nu_{m_2}$ and find (Ex. 15.2.2), to first order, that

$$\nu_{m_1} = 1/(m_1\pi), \quad \nu_{m_2} = 1/(m_2\pi).$$

This gives the asymptotic form of the two branches as

$$\begin{aligned} \lambda_{m_1} &= \tfrac{m_1\pi}{c} + \tfrac{k}{m_1\pi} + o\left(\tfrac{1}{m_1}\right) \\ \lambda_{m_2} &= \tfrac{m_2\pi}{(1-c)} + \tfrac{k}{m_2\pi} + o\left(\tfrac{1}{m_2}\right). \end{aligned}$$

Biscontin et al. found experimental evidence of two such branches in some steel rods.

Our discussion so far has focused on the identification of damage from its effect on natural frequencies. We discussed the effect on nodal positions, for a rod, in Section 14.3. This is essentially a qualitative result which could be a useful adjunct in an experimental/numerical study, see Gladwell and Morassi (1999) [128].

Wu and Fricke (1989) [335], Wu and Fricke (1990) [336] and Wu and Fricke (1991) [337] discuss the problem of finding one or more small blockages in a duct.

Exercises 15.2

1. Consider a uniform rod of unit length under supported (S) and free (F) end conditions. Define C^F_{m+1}, C^S_m as in (15.2.2). Show that if the damage, d, is located at $x = s$, then

$$d = C^F_{m+1} + C^S_m, \quad \cos[2(m+1)\pi s] = -1 + \frac{2}{1 + C^F_{m+1}/C^S_m}.$$

2. Set up the eigenvalue equation (15.2.3) for the uniform free-free rod, governed by the equation

$$u'' + \lambda^2 u = 0, \quad u'(0) = 0 = u'(1),$$

when there is a spring of stiffness k connecting the parts to the left and right of $x = c$ (see equation (14.3.5)). Establish the asymptotic forms for the eigenvalues for small and large k in a uniform duct by using measured eigenfrequency shifts.

15.3 Damage identification in beams

A number of early papers, including Cawley and Adams (1979) [50], Hearn and Testa (1991) [170] used a sensitivity analysis based on the general discrete equation

$$(\mathbf{K} - \lambda\mathbf{M})\mathbf{u} = \mathbf{0}. \tag{15.3.1}$$

Suppose the stiffness and mass matrices are perturbed to $\mathbf{K} + \delta\mathbf{K}, \mathbf{M} + \delta\mathbf{M}$, respectively, and the solution is $\mathbf{u} + \delta\mathbf{u}, \lambda + \delta\lambda$. Then

$$(\mathbf{K} + \delta\mathbf{K} - (\lambda + \delta\lambda)(\mathbf{M} + \delta\mathbf{M}))(\mathbf{u} + \delta\mathbf{u}) = \mathbf{0}$$

and, to first order, this is

$$(\mathbf{K} - \boldsymbol{\lambda}\mathbf{M})\mathbf{u} + (\mathbf{K} - \lambda\mathbf{M})\delta\mathbf{u} + (\delta\mathbf{K} - \lambda\delta\mathbf{M})\mathbf{u} - \delta\lambda\mathbf{M}\mathbf{u} = \mathbf{0}$$

so that on premultiplying by $\mathbf{u}^T$ and using (15.3.1) and $\mathbf{u}^T\mathbf{M}\mathbf{u} = 1$, we find

$$\delta\lambda = \mathbf{u}^T\delta\mathbf{K}\mathbf{u} - \lambda\mathbf{u}^T\delta\mathbf{M}\mathbf{u}.$$

In particular, if there is only a change in the stiffness of the structure, then

$$\delta\lambda = \mathbf{u}^T\delta\mathbf{K}\mathbf{u}. \tag{15.3.2}$$

This equation shows that if the changes in $\mathbf{K}$ are known, then the changes in natural frequencies may be found. One way to solve the inverse problem is to compute the changes in the various frequencies produced by changes in each element of a finite element model of the structure, in turn, and then determine which element change yields a set of frequency changes closest (either by inspection or in some least squares sense) to that found or specified. Cawley and Adams (1979) [50], Yuen (1985) [341], Morassi and Rovere (1997) [241], Vestroni and Capecchi (1996) [327], Vestroni and Capecchi (2000) [328] follow this general approach. See Shen and Taylor (1991) [304] for a careful and detailed engineering study of the problem, treated in a least squares form. See also Liang, Hu and Choy (1992a) [213], Liang, Hu and Choy (1992b) [214], Davini, Gatti and Morassi (1993) [72], and Cerri and Vestroni (2000) [51], and Capecchi and Vestroni (1999) [48].

There are a number of papers that discuss, in many different ways, how a crack in a flexurally vibrating beam should be modelled, including Freund and Herrmann (1976) [91], Chondros and Dimarogonas (1980) [55], Gudmundson (1982) [157], Christides and Barr (1984) [54], Shen and Pierre (1990) [303], Rizos, Aspragathos and Dimarogonas (1990) [291], Ostachowicz and Krawczuk

(1991) [255], Chondros, Dimarogonas and Yao (1998) [56], and other papers cited in these. The simplest model of a crack is a rotational spring of stiffness k; see Chondros and Dimarogonas (1980) [55], Narkis (1994) [247], or Boltezar, Strancar and Kuhelj (1998) [38]. All these researchers approach the problem in their own ways; typically Boltezar et al. set up the frequency equation for a uniform beam with a crack, modelled as a rotational spring of stiffness k, at an interior location R, and find the value of R that will yield the same stiffness k deduced from the measured (actually numerically predicted) changes in the first six natural frequencies. See Wu (1994) [338] for a different approach, and Natke and Cempel (1991) [248] for a review of the subject.

Following Morassi (1993) [235] we set up a perturbation analysis for a beam with a rotational spring of stiffness k at location s, when $d := 1/k = \varepsilon$ is small. We follow the lines laid out for the rod in Section 13.1. The beam is governed by equation (13.1.4), the end conditions (13.1.12), (13.1.13), and the jump conditions at $x = s$:

$$[u] = 0 = [ru''] = [(ru'')'], \quad r(s)u''(s) = k[u']$$

where, as usual, $[f] := f(s+) - f(s-)$.

Writing

$$u(x) = u_0(x) + \varepsilon v(x), \quad \lambda = \lambda_0 + \varepsilon\mu,$$

in (13.1.4) we find

$$(ru_o'')'' = \lambda_0 a u_0, \tag{15.3.1}$$

$$(rv'')'' = \lambda_0 a v + \mu a u_0, \tag{15.3.2}$$

where both u_0 and v satisfy the end conditions (13.1.12), (13.1.13), and v satisfies the jump conditions

$$[v] = 0 = [rv''] = [(rv'')'], \quad [v'] = r(s)u_0''(s).$$

Multiplying (15.3.2) by u_0, (15.3.1) by v, subtracting and integrating over (0,1), using the end and jump conditions, we find (Ex. 15.3.1) that

$$\mu = -(r(s)u_0''(s))^2, \tag{15.3.3}$$

so that the change in the mth natural frequency is

$$\delta\lambda_m = -(r(s)u_m''(s))^2 d. \tag{15.3.4}$$

Morassi (1993) [235] noted that this shows that the change in $\lambda_m (= \omega_m^2)$ is proportional to the potential energy stored at location s in the undamaged beam; also, it is proportional to the square of the curvature of the undamaged beam at s. Morassi (2003) [238] uses (15.2.4) just as he used the corresponding equation (15.2.1) for the rod. He shows for instance that the severity and location of the damage in a uniform simply-supported beam is uniquely determined (except for symmetry) by the changes in the mth and $2m$th frequencies. An alternative identification is provided by the changes in the mth frequency of the beam under simply supported boundary conditions and the $(m+1)$th frequency of the beam under sliding-sliding end conditions. See Ex. 15.3.2. Clearly, he uses these conditions because they are the only ones for which the modes are simple sines or cosines; in the general case the modes involve both sinusoidal and hyperbolic terms. The procedure could easily be generalised to a consideration of the changes of frequencies under other end conditions. Morassi and Rollo (2001) [240] use (15.3.4) to estimate the positions of two cracks in a simply supported beam.

There have been a few papers devoted to damage identification from other effects, namely curvature, mode shape and nodal positions. Thus, Pandey, Biswas and Samman (1991) [260] noted that the curvature of a principal mode of a damaged beam increased in a region localised about the damaged zone; this was different from simply the change in a mode shape, which generally was not localised about the damaged zone, see Rizos, Aspragathos and Dimarogonas (1990) [291]. Pandey et al used this curvature effect to locate damage.

Dilena and Morassi (2002) [80] and Dilena (2003) [78] make a systematic study to see whether the conclusion of Gladwell and Morassi (1999) [128], for a rod, *that nodes move toward the damage location*, could be generalised to apply to a flexurally vibrating beam. The result for the rod follows from Sturm's theorems, as described in Section 14.3. The vibration modes of a beam are governed by the fourth order equation (13.1.4), and not by the simple second order equation (14.3.2) for the rod. As Leighton and Nehari (1958) [206] showed in their massive authoritative discussion of oscillatory properties of fourth order equations, there are no simple extensions of Sturm's results to such equations. There are points, called conjugate points, and it may be proved that conjugate points move toward damage, but conjugate points have no clear physical interpretation. To corroborate this conclusion, Dilena and Morassi (2002) [80] found counterexamples: nodes do *not* always move toward the damage location.

The simplest counterexample is shown in Figure 15.3.1, adapted from Dilena and Morassi (2002) [80]. Figure 15.3.1(a) shows the first (proper) bending mode (i.e., mode 3) of a free-free beam. It has two nodes ξ_1 and ξ_2. Figure 15.3.1(b) shows the sign of the change in position ξ_1 due to damage of magnitude d (measured in some way) and position s. We note that the sign depends almost entirely on s. The node ξ_1 is roughly $0\cdot 2$; the figure shows that if $s < 0\cdot 41$ then the node moves to the left (negative), while if $s > 0\cdot 41$ it moves to the right (positive). That means that when s is in $(0\cdot 2, 0\cdot 41)$ the node moves the 'wrong' way. Figure 15.3.1(c) shows that there is a similar interval near the second node in which damage causes the node to move the 'wrong' way. For a corresponding axially vibrating rod the sign change would occur *at* the node: if the damage is to the left of the (undamaged rod) node, the node will move left (negative); to the right it will move right (positive). For a beam it appears that one may state that *damage 'far away' from a node causes the node to move toward the damage.* Dilena and Morassi (2002) [79] extend their analysis to higher modes, and find that there is a difference in the effects of damage on the so-called *external* and *internal* nodes; an external node is an extreme node, one nearest an end of the rod. They complement their study with experimental tests.

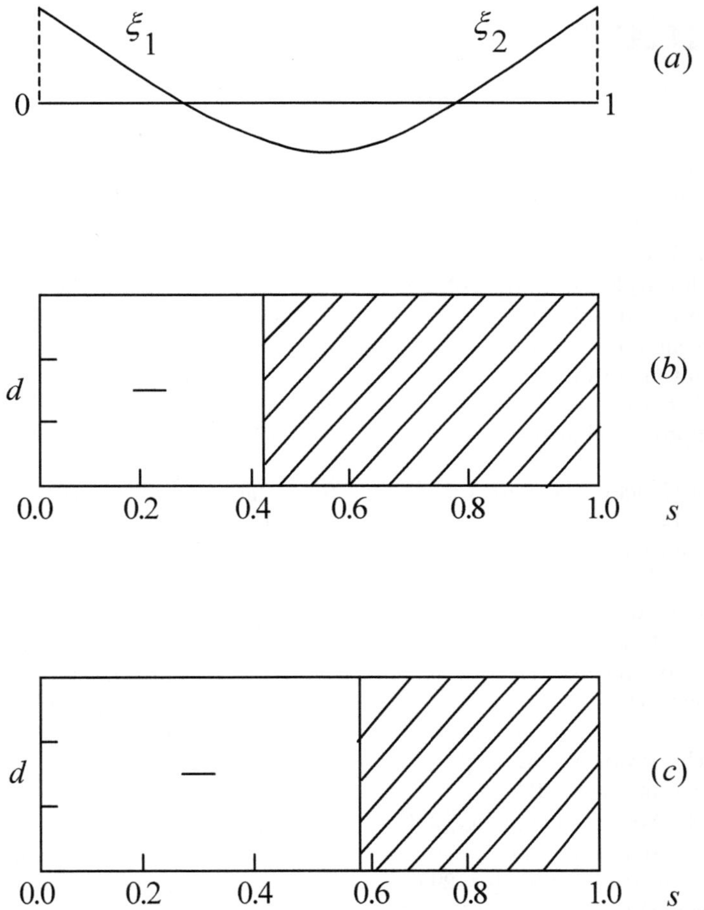

Figure 15.3.1 - When the damage is in the unshaded (shaded) region the node moves to the left (right).

Exercises 15.3

1. Derive equation (15.3.4) for the change in eigenvalue due to damage $d(= 1/k)$ at s.

2. Find the change in the mth eigenvalue of a simply-supported, and of a sliding-sliding uniform beam brought about by a rotational spring of stiffness k at $x = s$.

Index

Bibliography

[1] Abrate, S. (1995) Vibration of non-uniform rods and beams. [47], **185**, 703-716. *361.*

[2] Adams, R.D., Cawley, P., Pye, C.J. and Stone, B.J. (1978) A vibration technique for non-destructively assessing the integrity of structures. [44], **20**, 93-100. *419.*

[3] Ambarzumian, V. (1929) Űber eine Frager der Eigenwerttheorie. [92], **53**, 690-695. *290.*

[4] Ando, T. (1987) Totally positive matrices. [57], **90**, 165-219. *120, 133, 137, 143, 143, 144, 145, 145, 169.*

[5] Andersson, L.-E. (1970) On the effective determination of the wave operator in the case of a difference equation corresponding to a Sturm-Liouville differential equation. [41], **29**, 467-497. *293.*

[6] Andersson, L.-E. (1988a) Inverse eigenvalue problems with discontinuous coefficients. [29], **4**, 353-397. *305, 319.*

[7] Andersson, L.-E. (1988b) Inverse eigenvalue problems for a Sturm-Liouville equation in impedance form. [9], **4**, 929-971. *305, 319.*

[8] Andersson, L.-E. (1990) Algorithms for solving inverse eigenvalue problems for Sturm-Liouville equations in *Inverse Problems in Action*, ed. P.C. Sabatier, Berlin: Springer. *335.*

[9] Andrea, S.A. and Berry, T.G. (1992) Continued fractions and periodic Jacobi matrices. [57], **161**, 117-134. *105.*

[10] Andrew, A.L. and Paine, J.W. (1985) Correction of Numerov's eigenvalue estimates. [68], **47**, 289-300. *293.*

[11] Andrew, A.L. and Paine, J.W. (1986) Correction of finite element estimates for Sturm-Liouville eigenvalues. [68], **50**, 205-215. *293.*

[12] Arbenz, P. and Golub, G.H. (1995) Matrix shapes invariant under the symmetric QR algorithm. [67], **2**, 87-93. *170.*

[13] Ashlock, D.A., Driessel, K.R. and Hentzel, I.R. (1997) On matrix structures invariant under Toda-like isospectral flows. [57], **254**, 29-48. *180.*

[14] Barcilon, V. (1974a) Iterative solution of the inverse Sturm-Liouville equation. [42], **15**, 429-436. *293.*

[15] Barcilon, V. (1974b) On the uniqueness of inverse eigenvalue problems. [24], **38**, 287-298. *391.*

[16] Barcilon, V. (1974c) On the solution of inverse eigenvalue problems of high orders. [24], **39**, 143-154. *291, 391.*

[17] Barcilon, V. (1974d) A note on a formula of Gel'fand and Levitan. [41], **48**, 43-50. *362.*

[18] Barcilon, V. (1976) Inverse problems for a vibrating beam. [36], **27**, 346-358. *185, 392.*

[19] Barcilon, V. (1978) Discrete analog of an iterative method for inverse eigenvalue problems for Jacobi matrices. [42], **29**, 295-300. *71.*

[20] Barcilon, V. (1979) On the multiplicity of solutions of the inverse problem for a vibrating beam. [82], **37**, 605-613. *185.*

[21] Barcilon, V. (1982) Inverse problems for the vibrating beam in the free-clamped configuration. [69], **304**, 211-252. *185, 391, 392.*

[22] Barcilon, V. (1983) Explicit solution of the inverse problem for a vibrating string. [41], **93**, 222-234. *294, 362.*

[23] Barcilon, V. and Turchetti, G. (1980) Extremal solutions of inverse eigenvalue problems with finite spectral data. [90], **2**, 139-148. *65.*

[24] Barcilon, V. (1990) Two-dimensional inverse eigenvalue problem. [29], **6**, 11-20.

[25] Bellman, R. (1970) *Introduction to Matrix Analysis.* New York: McGraw-Hill. *131.*

[26] Benade, A.H. (1976) *Fundamentals of Musical Acoustics.* London: Oxford University Press. *345.*

[27] Berman, A. (1984) System identification of structural dynamic models - theoretical and practical bounds. [5], 84-0929, 123-129. *364.*

[28] Biegler-König, F.W. (1980) *Inverse Eigenwertprobleme.* Dissertation, Bielefeld. *108.*

[29] Biegler-König, F.W. (1981a) A Newton iteration process for inverse eigenvalue problems. [68], **37**, 349-354. *108.*

[30] Biegler-König, F.W. (1981b) Construction of band matrices from spectral data. [57], **40**, 79-84. *108.*

[31] Biegler-König, F.W. (1981c) Sufficient conditions for the solvability of inverse eigenvalue problems. [57], **40**, 89-100. *108.*

[32] Biscontin, G., Morassi, A. and Wendel, P. (1998) Asymptotic separation of the spectrum in notched rods. [53], **4**, 237-251. *420.*

[33] Bishop, R.E.D., Gladwell, G.M.L. and Michaelson, S. (1965) *The Matrix Analysis of Vibration.* Cambridge: Cambridge University Press. *12, 17, 19, 101.*

[34] Bishop, R.E.D. and Johnson, D.C. (1960) *The Mechanics of Vibration.* Cambridge: Cambridge University Press. *19, 40, 84, 190, 389, 390, 392, 419.*

[35] Boley, D. and Golub, G.H. (1984) A modified method for reconstructing periodic Jacobi matrices. [60], **42**, 143-150. *103, 105.*

[36] Boley, D. and Golub, G.H. (1987) A survey of matrix inverse eigenvalue problems. [29], **3**, 595-622. *103, 105, 106, 108, 108.*

[37] Bôcher, M. (1917) *Leçons sur les méthodes de Sturm dans la théorie des équations differentielles linéares et leurs dévelopements modernes.* Paris. *403.*

[38] Boltezar, M., Strancar, B. and Kuhelj, A. (1998) Identification of transverse crack locations in flexural vibrations of free-free beams. [47], **211**, 729-734. *423.*

[39] Borg (1946) Eine Umkehrung der Sturm-Liouvilleschen Eigenwertaufgabe. [1], **78**, 1-96. *290, 359.*

[40] Braun, S.G. and Ram. Y.M. (1991) Predicting the effect of structural modificiation: Upper and lower bounds due to modal truncation. [27], **6**, 199-211. *365.*

[41] Brown, B.M., Samko, V.S., Knowles, I.W. and Marletta, M. (2003) Inverse spectral problem for the Sturm-Liouville equation. [29], **19**, 235-252. *325.*

[42] Bruckstein, A.M. and Kailath, T. (1987) Inverse scattering for discrete transmission-line models. [87], **29**, 359-389. *334, 335, 343.*

[43] Bube, K.P. and Burridge, R. (1983) The one-dimensional inverse problem of reflection seismology. [86], **25**, 497-559. *334, 335.*

[44] Burak, S. and Ram, Y.M. (2001) The construction of physical parameters from spectral data. [63], **15**, 3-10. *367.*

[45] Burridge, R. (1980) The Gel'fand-Levitan, the Marchenko, and the Gopinath-Sondhi integral equations of inverse scattering theory, regarded in the context of inverse impulse-response problems. [90], **2**, 305-323. *293, 334.*

[46] Busacker, R.G. and Saaty, T.L. (1965) *Finite Graphs and Networks: an Introduction with Applications.* New York: McGraw Hill. *218.*

[47] Cabib, E., Freddi, L., Morassi, A. and Percivale, D. (2001) Thin notched beams. [39], **64**, 157-178. *419.*

[48] Capecchi, D. and Vestroni, F. (1999) Monitoring of structural systems by using frequency data. [23], **28**, 447-461. *422.*

[49] Carrier, G.F., Krook, M. and Pearson, C.E. (1966) *Functions of a Complex Variable.* New York: McGraw-Hill. *389.*

[50] Cawley, P. and Adams, R.D. (1979) The location of defects in structures from measurements of natural frequencies. [48], **14**, 49-57. *419, 422, 422.*

[51] Cerri, M.N. and Vestroni, F. (2000) Detection of damage in beams subjected to diffused cracking. [47], **234**, 259-276. *422.*

[52] Chadan, K. and Sabatier, P.C. (1989) *Inverse Problems in Quantum Scattering.* 2nd Ed. New York: Springer-Verlag. *334.*

[53] Cheng, S.Y. (1976) Eigenfunctions and nodal sets. [13], **51**, 43-55. *215.*

[54] Christides, S. and Barr, A.D.S. (1984) One-dimensional theory of cracked Euler-Bernoulli beams. [28], **26**, 639-648. *422.*

[55] Chondros, T.G. and Dimarogonas, A.D. (1980) Identification of cracks in welded joints of complex structures. [47], **69**, 531-538. *422, 423.*

[56] Chondros, T.G., Dimarogonas, A.D. and Yao, J. (1998) A continuous cracked beam vibration theory. [47], **215**, 17-34. *423.*

[57] Chu, M.T. (1984) The generalized Toda flow, the QR algorithm and the center manifold theory. [81], **5**, 187-201. *159.*

[58] Chu, M.T. (1998) Inverse eigenvalue problems. [86], **40**, 1-39. *108, 117.*

[59] Chu, M.T. and Golub, G.H. (2002) Structured inverse eigenvalue problems, [2], **11**, 1-71. *117.*

[60] Chu, M.T. and Norris, L.K. (1988) Isospectral flows and abstract matrix factorizations. [85], **25**, 1383-1391. *159.*

[61] Coleman, C.F. (1989) Inverse Spectral Problem with a Rough Coefficient. Ph.D. Thesis. Rensselaer Polytechnic Institute, Troy, N.Y. *319.*

[62] Coleman, C.F. and McLaughlin, J.R. (1993a) Solution of the inverse spectral problem for an impedance with integrable derivative I. [8], **46**, 145-184. *305, 319, 346.*

[63] Coleman, C.F. and McLaughlin, J.R. (1993b) Solution of the inverse spectral problem for an impedance with an integrable derivative II.' [8], **46**, 185-212. *305, 319, 346.*

[64] Courant, R. and Hilbert, D. (1953) *Methods of Mathematical Physics.* Vol. 1, New York: Interscience. *48, 214, 240.*

[65] Crum, M.M. (1955) Associated Sturm-Liouville systems. [76], **6**, 121-127. *350.*

[66] Cryer, C.W. (1973) The LU-factorization of totally positive matrices. [57], **7**, 83-92. *168, 176.*

[67] Cryer, C.W. (1976) Some properties of totally positive matrices. [57], **15**, 1-25. *168.*

[68] Dahlberg, B.E.J. and Trubowitz, E. (1984) The inverse Sturm-Liouville problem III. [16], **37**, 255-267. *346.*

[69] Darboux, G. (1882) Sur la représentation sphérique des surfaces. [17], **94**, 1343-1345. *347.*

[70] Darboux, G. (1915) *Leçons sur le Théorie Générale des Surfaces et les Applications Géométrique du Calcul Infinitesimal.* Vo. II. Paris: Gauthier Villars. *347.*

[71] Davies, E.B., Gladwell, G.M.L., Leydold, J. and Stadler, P.F. (2001) Discrete nodal domain theorems. [57], **336**, 51-60. *223, 224.*

[72] Davini, C., Gatti, F. and Morassi, A. (1993) A damage analysis of steel beams. [62], **28**, 27-37. *422.*

[73] Davini, C., Morassi, A. and Rovere, N. (1995) Modal analysis of notched bars: tests and comments on the sensitivity of an identification technique. [47], **179**, 513-527.

[74] Davini, C. (1996) Note on a parameter lumping in the vibrations of uniform beams. [79], **28**, 83-99. *37.*

[75] de Boor, C. and Golub, G.H. (1978) The numerically stable reconstruction of a Jacobi matrix from spectral data. [57], **21**, 245-260. *69.*

[76] de Boor, C. and Saff, E.B. (1986) Finite sequences of orthogonal polynomials connected by a Jacobi matrix. [57], **75**, 43-55. *68, 70.*

[77] Deift, P., Nanda, T., and Tomei, C. (1983) Ordinary differential equations and the symmetric eigenvalue problem. [85], **20**, 1-22. *159.*

[78] Dilena, M. (2003) On damage identification in vibrating beams from changes in node positions, in Davini, C. and Viola, E. (Eds) *Problems in Structural Identification and Diagnostics: General Aspects and Applications.* New York: Springer. *424.*

[79] Dilena, M. and Morassi, A. (2002) Identification of crack location in vibrating beams from changes in node positions. [47], **255**, 915-930. *424.*

[80] Dilena, M. and Morassi, A. (2002) The use of antiresonances for crack detection in beams. [47]. *424, 424, 424.*

[81] Duarte, A.L. (1989) Construction of acyclic matrices from spectral data. [57], **113**, 173-182. *110, 365.*

[82] Duval, A.M. and Reiner, V. (1999) Perron-Frobenius type results and discrete versions of nodal domain theorems. [57], **294**, 259-268. *223, 224.*

[83] Eisner, E. (1976) Complete solution of the 'Webster' horn equation. [92], **41**, 1126-1146. *345.*

[84] El-Badia, A. (1989) On the uniqueness of a bi-dimensional inverse spectral problem. [18], **308**, 273-276.

[85] Elhay, S., Gladwell, G.M.L., Golub, G.H. and Ram, Y.M. (1999) On some eigenvector-eigenvalue relations. [84], **20**, 563-574. *276.*

[86] Fekete, M. (1913) Über ein Problem von Laguerre. [78], **34**, 89-100, 110-120. *133, 143.*

[87] Ferguson, W.E. (1980) The construction of Jacobi and periodic Jacobi matrices with prescribed spectra. [60], **35**, 1203-1220. *103.*

[88] Fischer, E. (1905) Über quadratische Formen mit reelen Koeffizienten. [65], **16**, 234-249. *48.*

[89] Fix, G. (1967) Asymptotic eigenvalues of Sturm-Liouville systems. [41], **19**, 519-525. *283.*

[90] Forsythe, G.E. (1957) Generation and use of orthogonal polynomials for data fitting with a digital computer. *[54]*, **5**, 74-88. *54.*

[91] Freund, L.B. and Herrmann, G. (1976) Dynamic fracture of a beam or plate in plane bending. [37], **76**, 112-116. *419, 422.*

[92] Friedland, S. (1977) Inverse eigenvalue problems. [57], **17**, 15-51. *108.*

[93] Friedland, S. (1979) The reconstruction of a symmetric matrix from the spectral data. [41], **71**, 412-422. *108.*

[94] Friedland, S. and Melkman, A.A. (1979) Eigenvalues of non-negative Jacobi matrices. [57], **25**, 239-253. *68.*

[95] Friedland, S., Nocedal, J., and Overton, M. (1987) The formulation and analysis of numerical methods for inverse eigenvalue problems. [85], **24**, 634-667. *116.*

[96] Friedman, J. (1993) Some geometric aspects of graphs and their eigenfunctions. [22], **69**, 487-525. *224, 224.*

[97] Gantmacher, F.R. (1959) *The Theory of Matrices.* New York: Chelsea Publishing Co. *18, 118, 122, 123, 133.*

[98] Gantmakher, F.P. and Krein, M.G. (1950) *Oscillation Matrices and Kernels and Small Vibrations of Mechanical Systems.* 1961 Translation by U.S. Atomic Energy Commission, Washington, D.C. A revised edition was published in (2002) by AMS Chelsea Publishing, Providence, listing the first author as Gantmacher, not Gantmakher. *49, 63, 80, 118, 133, 133, 236.*

[99] Gasca, M. and Peña, J.M. (1992) Total positivity and Neville elimination. [57], **165**, 25-44. *138.*

[100] Gel'fand, I.M. and Levitan, B.M. (1951) On the determination of a differential equation from its spectral function. (In Russian). [31], **15**, 309-360. (In English). [54], **1**, 253-304. *293.*

[101] Gel'fand, I.M. and Levitan, B.M. (1953) On a simple identity for the characteristic values of a differential operator of the second order (in Russian). [21], **88**, 593-596. *362.*

[102] Gilbarg, D. and Trudinger, N.S. (1977) *Elliptic Partial Differential Equations of Second Order. Berlin, Springer. 337.*

[103] Gladwell, G.M.L. (1962) The approximation of uniform beams in transverse vibration by sets of masses elastically connected. *Proceedings of the 4th U.S. Congress of Applied Mechanics*, 169-176, New York: American Society of Mechanical Engineers. *38.*

[104] Gladwell, G.M.L. (1984) The inverse problem for the vibrating beam. [74], **393**, 277-295. *185.*

[105] Gladwell, G.M.L. (1985) Qualitative properties of vibrating systems. [74], **401**, 299-315. *192.*

[106] Gladwell, G.M.L. and Gbadeyan, J. (1985) On the inverse problem of the vibrating string and rod. [77], **38**, 169-174. *84.*

[107] Gladwell, G.M.L. (1986a) Inverse problems in vibration. [79], **39**, 1013-1018. *116.*

[108] Gladwell, G.M.L. (1986b) *Inverse Problems in Vibration.* Dordrecht: Martinus Nijhoff Publishers. *133, 133.*

[109] Gladwell, G.M.L. (1986c) The inverse mode problem for lumped-mass systems, [77], **39**, 297-307. *203, 209, 211.*

[110] Gladwell, G.M.L. (1986d) The inverse problem for the Euler-Bernoulli beam. [74], **407**, 199-218. *392.*

[111] Gladwell, G.M.L. and Dods, S.R.A. (1987) Examples of reconstruction of vibrating rods from spectral data. [47], **119**, 267-276. *319.*

[112] Gladwell, G.M.L., England, A.H. and Wang, D. (1987) Examples of reconstruction of an Euler-Bernoulli beam from spectral data. [47], **119**, 81-94. *401.*

[113] Gladwell, G.M.L. and Willms, N.B. (1988) The reconstruction of a tridiagonal system from its frequency response at an interior point. [29], **4**, 1018-1024. *87.*

[114] Gladwell, G.M.L. and Willms, N.B. (1989) A discrete Gel'fand-Levitan method for band-matrix inverse eigenvalue problems. [29], **5**, 165-179. *108.*

[115] Gladwell, G.M.L., Willms, N.B., He, B., and Wang, D. (1989) How can we recognise an acceptable mode shape for a vibrating beam? [77], **42**, 303-316. *211, 212, 213, 214, 365.*

[116] Gladwell, G.M.L. (1991a) Qualitative properties of finite element models I: Sturm-Liouville systems. [77], **44**, 249-265. *185.*

[117] Gladwell, G.M.L. (1991b) Qualitative properties of finite-element models II: the Euler Bernoulli beam. [77], **44**, 267-284. *185, 192.*

[118] Gladwell, G.M.L. (1991c) The application of Schur's algorithm to an inverse eigenvalue problem. [29], **7**, 557-565. *335.*

[119] Gladwell, G.M.L. (1991d) On the scattering of waves in a non-uniform Euler-Bernoulli beam. [72], **205**, 31-34. *61, 393.*

[120] Gladwell, G.M.L. (1993) *Inverse Problems in Scattering.* Dordrecht: Kluwer Academic Publishers. *334, 335.*

[121] Gladwell, G.M.L. (1995) On isospectral spring-mass systems. [29], **11**, 591-602. *160.*

[122] Gladwell, G.M.L. and Morassi, A. (1995) On isospectral rods, horns and strings. [29], **11**, 533-544. *347.*

[123] Gladwell, G.M.L. and Movahhedy, M. (1995) Reconstruction of a mass-spring system from spectral data I: Theory. [30], **1**, 179-189. *84.*

[124] Gladwell, G.M.L. (1996) Inverse problems in vibration-II. [9], **49**, 525-534. *116.*

[125] Gladwell, G.M.L. (1997) Inverse vibration problems for finite element models. [29], **13**, 311-322. *176.*

[126] Gladwell, G.M.L. (1998) Total positivity and the QR algorithm. [57], **271**, 257-272. *138, 167, 167, 175.*

[127] Gladwell, G.M.L. (1999) Inverse finite element vibration problems. [47], **211**, 309-324. *86, 87, 175.*

[128] Gladwell, G.M.L. and Morassi, A. (1999) Estimating damage in a rod from changes in node positions. [30], **7**, 215-233. *409, 411, 421, 424.*

[129] Gladwell, G.M.L. (2002a) Total positivity and Toda flow. [57], **350**, 279-284. *182.*

[130] Gladwell, G.M.L. (2002b) Isospectral vibrating beams. [74], **458**, 2691-2703. *175.*

[131] Gladwell, G.M.L. and Zhu, H.M. (2002) Courant's nodal line theorem and its discrete counterparts. [77], **55**, 1-15. *34, 224.*

[132] Golub, G.H. (1973) Some uses of the Lanczos algorithm in numerical linear algebra, in J.H.H. Miller (Ed) *Topics in Numerical Analysis*, New York: Academic Press. *67.*

[133] Golub, G.H. and Boley, D. (1977) Inverse eigenvalue problems for band matrices, in G.A. Watson (Ed.) *Numerical Analysis* Heidelberg, New York: Springer Verlag, 23-31. *70.*

[134] Golub, G.H. and Underwood, R.R. (1977) Block Lanczos method for computing eigenvalues, in Rice, J.R. (Ed.) *Mathematical Software* III. New York: Springer, 23-31. *108.*

[135] Golub, G.H. and Van Loan, C.F. (1983) *Matrix Computations.* Baltimore: The Johns Hopkins University Press. *12, 17, 67, 101, 156.*

[136] Gopinath, B. and Sondhi, M.M. (1970) Determination of the shape of the human vocal tract from acoustical measurements. [12], 1195-1214. *293, 331.*

[137] Gopinath, B. and Sondhi, M.M. (1971) Inversion of the telegraph equation and the synthesis of non-uniform lines. [25], **59**, 383-392. *293, 293, 331.*

[138] Gottlieb, H.P.W. (1986) Harmonic frequency spectra of vibrating stepped strings. [47], **108**, 63-72 and 345. *290, 291, 355, 355, 356, 359.*

[139] Gottlieb, H.P.W. (1987a) Multi-segment strings with exactly harmonic spectra. [47], **118**, 283-290. *356.*

[140] Gottlieb, H.P.W. (1987b) Isospectral Euler-Bernoulli beams with continuous density and rigidity functions. [74], **413**, 235-250. *359, 361, 393.*

[141] Gottlieb, H.P.W. (1988a) Isospectral operators: some model examples with discontinuous coefficients. [41], **132**, 123-137. *356.*

[142] Gottlieb, H.P.W. (1988b) Density distribution for isospectral circular membranes. [82], **48**, 948-951. *361, 393.*

[143] Gottlieb, H.P.W. (1989) On standard eigenvalues of variable-coefficient heat and rod equations. [37], **56**, 146-148.

[144] Gottlieb, H.P.W. (1991) Inhomogeneous clamped circular plates with standard vibration spectra. [37], **58**, 729-730. *361.*

[145] Gottlieb, H.P.W. (1992a) Examples and counterexamples for a string density formula in the case of a discontinuity. [41], **164**, 363-369. *363, 364.*

[146] Gottlieb, H.P.W. (1992b) Axisymmetric isospectral annular plates and membranes. [26], **50**, 107-112. *361.*

[147] Gottlieb, H.P.W. (1993) Inhomogeneous amular plates with exactly beam-like radial spectra. [26], **50**, 107-112. *361.*

[148] Gottlieb, H.P.W. (2000) Exact solutions for vibrations of some annular membranes with inhomogeneous radial densities. [47], **233**, 165-170. *361.*

[149] Gottlieb, H.P.W. (2002) Isospectral strings. [29], **18**, 971-978. *356.*

[150] Gottlieb, H.P.W. (2004a) Isospectral circular membranes. [29], **20**, 155-161. *361.*

[151] Gould, S.H. (1966) *Variational Methods for Eigenvalue Problems.* Toronto: University of Toronto Press. *48.*

[152] Gradshteyn, I.S. and Ryzhik, I.M. (1965) *Tables of Integrals, Series and Products*, 4th ed., Moscow 1963. English Translation, A. Jeffrey (Ed.) New York: Academic Press. *364.*

[153] Gragg, W.B. and Harrod, W.J. (1984) Numerically stable reconstruction of Jacobi matrices from spectral data. [68], **44**, 317-335. *108.*

[154] Gray, L.J. and Wilson, D.G. (1976) Construction of a Jacobi matrix from spectral data. [57], **14**, 131-134. *68.*

[155] Groetsch, C.W. (1993) *Inverse Problems in the Mathematical Sciences.* Braunschweig: Vieweg Verlag. *289.*

[156] Groetsch, C.W. (2000) *Inverse Problems: Activities for Undergraduates.* Washington, D.C.: Mathematical Association of America. *289.*

[157] Gudmundson, P. (1982) Eigenfrequency changes of structures due to cracks, notches or other geometrical changes. [51], **30**, 339-353. *422.*

[158] Halberg, C.J.A. and Kramer, V.A. (1960) A generalization of the trace concept. [22], **27**, 607-617. *362.*

[159] Hald, O.H. (1972) *On Discrete and Numerical Inverse Sturm-Liouville Problems.* Ph.D. Thesis, New York University, New York, NY. *293.*

[160] Hald, O.H. (1976) Inverse eigenvalue problems for Jacobi matrices. [57], **14**, 63-85. *68.*

[161] Hald, O.H. (1977) Discrete inverse Sturm-Liouville problems. [68], **27**, 249-256. *294.*

[162] Hald, O.H. (1978a) The inverse Sturm-Liouville problem with symmetric potentials. [1], **141**, 263-291. *291.*

[163] Hald, O.H. (1978b) The inverse Sturm-Liouville equation and the Rayleigh-Ritz method. [60], **32**, 687-705. *294.*

[164] Hald, O.H. (1983) Inverse eigenvalue problems for the mantle, II. [24], **72**, 139-164.

[165] Hald, O.H. (1984) Discontinuous inverse eigenvalue problems. [16], **37**, 539-577. *291, 293, 305, 420.*

[166] Hald, O.H. and McLaughlin, J.R. (1988) Inverse problems using nodal position data - uniqueness results, algorithms, and bounds. *Proceedings, Centre for Mathematical Analysis, Australian National University, Special Program in Inverse Problems*, ed. R.S. Anderssen and G.N. Newsam. **17**, 32-58. *415.*

[167] Hald, O.H. and McLaughlin, J.R. (1989) Solutions of inverse nodal problems. [29], **5**, 307-347. *413, 414.*

[168] Hald, O.H. and McLaughlin, J.R. (1996) Inverse nodal problems: finding the potential from nodal lines. [64], **119**, *415, 415.*

[169] Hald, O.H. and McLaughlin, J.R. (1998) Inverse problems: recovery of BV coefficients from nodes. [29], **14**, 245-273. *415.*

[170] Hearn, G. and Testa, R.B. (1991) Modal analysis for damage detection in structures. [49], **117**, 3042-3063. *419, 422.*

[171] Herrmann, H. (1935) Beziehungen zwischen den Eigenwerten und Eigenfunktionen verschiedener Eigenwertprobleme. [61], **40**, 221-241. *216.*

[172] Hochstadt, H. (1961) Asymptotic estimates of the Sturm-Liouville spectrum. [16], **14**, 749-764. *282.*

[173] Hochstadt, H. (1967) On some inverse problems in matrix theory. [10], **18**, 201-207. *68.*

[174] Hochstadt, H. and Kim, M. (1970) On a singular inverse eigenvalue problem. [11], **37**, 243-254. *290.*

[175] Hochstadt, H. (1973) The inverse Sturm-Liouville problem. [16], **26**, 715-729. *291, 291.*

[176] Hochstadt, H. (1974) On the reconstruction of a Jacobi matrix from spectral data. [57], **8**, 435-446. *68.*

[177] Hochstadt, H. (1975a) On inverse problems associated with Sturm-Liouville operators. [38], **17**, 220-235. *291, 345.*

[178] Hochstadt, H. (1975b) Well posed inverse spectral problems. [73], **72**, 2496-2497. *291.*

[179] Hochstadt, H. (1976) On the determination of the density of a vibrating string from spectral data. [41], **55**, 673-685. *291.*

[180] Hochstadt, H. (1977) On the well posedness of the inverse Sturm-Liouville problem. [38], **23**, 402-413. *291.*

[181] Hochstadt, H. and Lieberman, B. (1978) An inverse Sturm-Liouville problem with mixed given data. [82], **34**, 676-680. *291, 329.*

[182] Hochstadt, H. (1979) On the reconstruction of a Jacobi matrix from mixed given data. [57], **28**, 113-115. *74.*

[183] Horn, R.A. and Johnson, C.R. (1985) *Matrix Analysis.* Cambridge: Cambridge University Press. *1, 97, 130, 131.*

[184] Ikramov, Kh.D. and Chugunov, V.N. (2000) Inverse matrix eigenvalue problems. [43], **98**, 51-135. *117.*

[185] Ince, E.L. (1927) *Ordinary Differential Equations*, London: Longmans, Green. *236, 282, 403.*

[186] Isaacson, E.L. and Trubowitz, E. (1983) The inverse Sturm-Liouville problem I. [16], **36**, 767-783. *346.*

[187] Isaacson, E.L., McKean, H.P. and Trubowitz, E. (1984) The inverse Sturm-Liouville problem II. [16], **37**, 1-11. *346.*

[188] Jerison, D. and Kenig, C. (1985) Unique continuation and absence of positive eigenvalues for Schrödinger operators. [7], **121**, 463-494. *216.*

[189] Kailath, T. and Lev-Ari, H. (1985) On mappings between covariance matrices and physical systems. [20], **47**, 241-252. *342.*

[190] Karlin, S. (1968) *Total Positivity*, Vol. 1. Stanford: Stanford University Press. *133.*

[191] Kato, T. (1976) *Perturbation Theory for Linear Operators.* Springer Verlag, New York.

[192] Kautsky, J. and Golub, G.H. (1983) On the calculation of Jacobi matrices. [57], **52**, 439-455. *67, 68.*

[193] Kirsch, A. (1996) *An Introduction to the Mathematical Theory of Inverse Problems.* New York: Springer Verlag. *289, 291.*

[194] Knobel, R. and McLaughlin, J.R. (1992) A reconstruction method for the two spectra inverse Sturm-Liouville problem, preprint.

[195] Knobel, R. and Lowe, B.D. (1993) An inverse Sturm-Liouville problem for an impedance. [36], **44**, 433-450. *319.*

[196] Knobel, R. and McLaughlin, J.R. (1994) Reconstruction method for a two-dimensional inverse problem. [91], **45**, 794-826.

[197] Kobayashi, M. (1988) Discontinuous Inverse Sturm-Liouville Problems with Symmetric Potentials. Ph.D. Thesis. University of California at Berkeley. *305.*

[198] Krein, M.G. (1933) On the spectrum of a Jacobian matrix, in connection with the torsional oscillation of shafts. (in Russian) *[59]*, **40**, 455-466. *63.*

[199] Krein, M.G. (1934) On nodes of harmonic oscillations of mechanical systems of a certain special type. (in Russian) [59], **41**, 339-348. *63.*

[200] Krein, M.G. (1951a) Determination of the density of a non-homogeneous symmetric cord from its frequency spectrum. (In Russian). [21], **76**, 345-348. *293.*

[201] Krein, M.G. (1951b) On the inverse problem for a non-homogeneous cord. (In Russian). [21], **82**, 669-672. *293.*

[202] Krein, M.G. (1952) Some new problems in the theory of Sturm systems. (In Russian) [71], **16**, 555-563. *63, 293.*

[203] Lanczos, C. (1950) An iteration method for the solution of the eigenvalue problem of linear differential and integral operators. [46], **45**, 225-232. *67.*

[204] Landau, H.J. (1983) The inverse problem for the vocal tract and the moment problem. [83], **14**, 1019-1035. *293, 334.*

[205] Lebedev, L.P., Vorovich, I.I. and Gladwell, G.M.L. (1996) *Functional Analysis: Applications in Mechanics and Inverse Problems.* Dordrecht: Kluwer Academic Publishers. *240.*

[206] Leighton, W. and Nehari, Z. (1958) On the oscillation of solutions of self-adjoint linear differential equations of the fourth order. [88], **89**, 325-377. *424.*

[207] Levinson, N. (1949) The inverse Sturm-Liouville problem. [66], 25-30. *291.*

[208] Levinson, M. (1976) Vibrations of stepped strings and beams. [47], **49**, 287-291. *355.*

[209] Levitan, B.M. (1964a) *Generalized Translation Operators and Some of Their Applications.* Jerusalem: Israel Program for Scientific Translations. Chapters IV, V. *291.*

[210] Levitan, B.M. (1964b) On the determination of a Sturm-Liouville equation by spectra. (In Russian). [31], **28**, 63-68. (In English) [54], **68**, 1-20. *292.*

[211] Levitan, B.M. (1987) *Inverse Sturm-Liouville Problems.* Utrecht: VNU Science Press. *283, 291.*

[212] Levitan, B.M. and Sargsjan, I.S. (1991) *Sturm-Liouville and Dirac Operators.* Dordrecht: Kluwer Academic Publishers. *235, 282, 293.*

[213] Liang, R.Y., Hu, J. and Choy, F. (1992a) Theoretical study of crack-induced eigenfrequency changes on beam structures. [40], **118**, 384-396. *422.*

[214] Liang, R.Y., Hu, J. and Choy, F. (1992b) Quantitative NDE technique for assessing damages in beam structures. [40], **118**, 1469-1487. *422.*

[215] Lindberg, G.M. (1963) The vibration of non-uniform beams. [3], **14**, 387-395. *38.*

[216] Lowe, B.D., Pilant, M. and Rundell, W. (1992) The recovery of potentials from finite spectral data. [83], **23**, 482-504. *321.*

[217] Lowe, B.D. (1993) Construction of an Euler-Bernoulli beam from spectral data. [47], **163**, 165-171. *399.*

[218] Marchenko, V.A. (1950) On certain questions in the theory of differential operators of second order. (In Russian). [21], **72**, 457-460. *291, 293.*

[219] Marchenko, V.A. (1952) Some problems in the theory of one-dimensioned second order differential operators I (In Russian). [89], **1**, 327-420. *291.*

[220] Marchenko, V.A. (1953) Some problems in the theory of one-dimensioned second order differential operators II (In Russian). [89], **2**, 3-82. *291.*

[221] Markham, T. (1970) On oscillatory matrices. [57], **3**, 143-158. *138, 175, 184.*

[222] Mattis, M.P. and Hochstadt, H. (1981) On the construction of band matrices from spectral data. [57], **38**, 109-119. *108.*

[223] McLaughlin, J.R. (1976) An inverse problem of order four. [83], **7**, 646-661. *393.*

[224] McLaughlin, J.R. (1978) An inverse problem of order four - an infinite case. [83], **9**, 395-413. *393.*

[225] McLaughlin, J.R. (1981) Fourth order inverse eigenvalue problems, in Knowles, I.W. and Lewis, R.T. (Eds) *Spectral Theory of Differential Operators.* New York: North Holland, 327-335. Crum, M.M. (1995) [78], **6**, 121-127. *393.*

[226] McLaughlin, J.R. (1984a) Bounds for constructed solutions of second and fourth order inverse eigenvalue problems, in I.W. Knowles and T.R. Lewis (Eds) *Differential Equations.* New York: Elsevier/North Holland, 437-443. *393.*

[227] McLaughlin, J.R. (1984b) On constructing solutions to an inverse Euler-Bernoulli beam problem, in F. Santosa et al (Eds) *Inverse Problems of Acoustic and Elastic Waves.* Philadelphia: SIAM, 341-347. *392, 393.*

[228] McLaughlin, J.R. (1986) Analytical methods for recovering coefficients in differential equations from spectral data. [87], **28**, 53-72. *291, 305.*

[229] McLaughlin, J.R. (1986) Uniqueness theorem for second order inverse eigenvalue equations. [41], **118**, 38-41.

[230] McLaughlin, J.R. and Rundell, W. (1987) A uniqueness theorem for an inverse Sturm-Liouville problem. [42], **28**, 1471-1472. *305.*

[231] McLaughlin, J.R. (1988) Inverse spectral theory using nodal points as data - a uniqueness result. [41], **73**, 354-362. *412, 413, 415.*

[232] McLaughlin, J.R. (2000) Solving inverse problems with spectral data, in Colton, D., Engl, H.W., Louis, A.K., McLaughlin, J.R. and Rundell, W. (Eds) Surveys on Solution Methods for Inverse Problems. Vienna, Springer-Verlag. pp. 169-194. *415.*

[233] McNabb, A., Anderssen, R.S. and Lapwood, E.R. (1976) Asymptotic behaviour of the eigenvalues of a Sturm-Liouville system with discontinuous coefficients. [41], **54**, 741-751. *284.*

[234] Meirovitch, L. (1975) *Elements of Vibration Analysis.* New York: McGraw-Hill. *19.*

[235] Morassi, A. (1993) Crack-induced changes in eigenparameters on beam structures. [40], **119**, 1798-1803. *423, 423.*

[236] Morassi, A. (1997) An uniqueness result on crack localization in vibrating rods. [30], **4**, 231-254. *420.*

[237] Morassi, A. (2001) Identification of a crack in a rod based on changes in a pair of natural frequencies. [47], **242**, 577-596. *419.*

[238] Morassi, A. (2003) The crack detection problem in vibrating beams, in Davini, C. and Viola, E. (Eds) *Problems in Structural Identification and Diagnostics: General Aspects and Applications.* New York: Springer, 163-177. *420.*

[239] Morassi, A. and Dilena, M. (2002) On point mass identification in rods and beams from minimal frequency measurements. [30], **10**, 183-201. *420.*

[240] Morassi, A. and Rollo, M. (2001) Identification of two cracks in a simply supported beam from minimal frequency measurements. [53], **7**, 729-739. *424.*

[241] Morassi, A. and Rovere, N. (1997) Localizing a notch in a steel frame from frequency measurements. [40], **123**, 422-432. *422.*

[242] Movahhedy, M., Ismail, F. and Gladwell, G.M.L. (1995) Reconstruction of a mass-spring system from spectral data II: Experiment. [30], **1**, 315-327. *84.*

[243] Nabben, R. (2001) On Green's matrices for trees. [84], **22**, 1014-1026. *98.*

[244] Nachman, A., Sylvester, J. and Uhlmann, G. (1988) An n-dimensional Borg-Levinson theorem. [14], **115**, 595-605.

[245] Nanda, T. (1982) Ph.D. Thesis, New York University, New York. *159.*

[246] Nanda, T. (1985) Differential equations and the QR algorithm. [85], **22**, 310-321. *159.*

[247] Narkis, Y. (1994) Identification of crack location in vibrating simply-supported beams. [47], 172, 549-558. *419, 423.*

[248] Natke, H.G. and Cempel, C. (1991) Fault detection and localisation in structures: a discussion. [45], **5**, 345-356. *423.*

[249] Newton, R.G. (1983) The Marchenko and Gel'fand-Levitan methods in the inverse scattering problem in one and three dimensions, in J.G. Bednar, et al. (Eds.) *Conference on Inverse Scattering: Theory and Application.* Philadelphia: SIAM. 1-74. *289.*

[250] Niordson, F.I. (1967) A method of solving inverse eigenvalue problems, in B. Broberg, J. Hults and F.I. Niordson (Eds) *Recent Progress in Applied Mechanics: The Folke Odqvist* Volume. Stockholm: Almqvist and Wiksell, 373-382. *391.*

[251] Nocedal, J. and Overton, M.L. (1983) Numerical methods for solving inverse eigenvalue problems. [55], **1005**, 212-226. *116.*

[252] Nylen, P. and Uhlig, F. (1994) Realizations of interlacing by tree-patterned matrices. [58], **38**, 13-37. *116.*

[253] Nylen, P. and Uhlig, F. (1997a) Inverse eigenvalue problems associated with spring-mass systems. [57], **254**, 409-425. *79, 83, 92.*

[254] Nylen, P. and Uhlig, F. (1997b) Inverse eigenvalue problem: existence of special spring-mass systems. [29], **13**, 1071-1081. *83.*

[255] Ostachowicz, W.M. and Krawczuk, M. (1991) Analysis of the effect of cracks on the natural frequencies of a cantilever beam. [47], **150**, 191-201. *423.*

[256] Paine, J. (1982) Correction of Sturm-Liouville eigenvalue estimates. [60], **39**, 415-420. *293.*

[257] Paine, J. (1984) A numerical method for the inverse Sturm-Liouville problem. [81], **5**, 149-156. *293.*

[258] Paine, J.W. and de Hoog, F.R. (1980) Uniform estimation of the eigenvalues of Sturm-Liouville problems. [33], **21**, 365-383. *293.*

[259] Paine, J.W., de Hoog, F.R. and Anderssen, R.S. (1981) On the correction of finite difference eigenvalue approximations for Sturm-Liouville problems. [19], **26**, 123-139. *293.*

[260] Pandey, A.K., Biswas, M. and Samman, M.M. (1991) Damage detection from changes in curvature mode shapes. [47], **145**, 321-332. *424.*

[261] Papanicolaou, V.G. (1995) The spectral theory of the vibrating periodic beam. [14], **170**, 359-373. *369.*

[262] Papanicolaou, V.G. and Kravvaritis, D. (1997) An inverse spectral problem for the Euler-Bernoulli equation for the vibrating beam. [29], **13**, 1083-1092. *393.*

[263] Parker, R.L. (1977) Understanding inverse theory. [8], **5**, 35-64. *289.*

[264] Parlett, B.N. (1980) *The Symmetric Eigenvalue Problem.* Englewood Cliffs: Prentice Hall. *17, 365.*

[265] Parter, S. (1960) On the eigenvalues of a class of matrices. [54], **8**, 376-388. *113.*

[266] Pleijel, A. (1956) Remarks on Courant's nodal line theorem. [16], 543-550. *216.*

[267] Porter, B. (1970) Synthesis of lumped-parameter vibrating systems by an inverse Holzer technique. [44], **12**, 17-19. *208.*

[268] Porter, B. (1971) Synthesis of lumped-parameter vibrating systems using transfer matrices. [28], **13**, 29-34. *208.*

[269] Pőschel, J. and Trubowitz, E. (1987) *Inverse Spectral Theory.* Boston: Academic Press. *283, 354.*

[270] Pranger, W.A. (1989) A formula for the mass density of a vibrating string in terms of the trace. [41], **141**, 399-404. *363.*

[271] Protter, M.H. and Weinburger, H.F. (1984) *Maximum Principles in Differential Equations.* New York: Springer. *218.*

[272] Ram, Y.M., Braun, S. and Blech, J.J. (1988) Structural modification in truncated systems by the Rayleigh-Ritz method. [47], **125**, 203-209. *364, 365.*

[273] Ram, Y.M. and Braun, S.G. (1990a) Structural dynamic modification using truncated data: Bounds for the eigenvalues. [63], **4**, 39-52. *364.*

[274] Ram, Y.M. and Braun, S.G. (1990b) Upper and lower bounds for the natural frequencies of modified structures based on truncated modal testing results. [47], **137**, 69-81. *365.*

[275] Ram, Y.M., Blech, J.J. and Braun, S.G. (1990) Eigenproblem error bounds with application to the symmetric dynamic system modification. [84], **11**, 553-564. *365.*

[276] Ram, Y.M. (1993) Inverse eigenvalue problem for a modified vibrating system. [82], **53**, 1762-1775. *83, 365.*

[277] Ram, Y.M. and Blech, J.J. (1991) The dynamic behaviour of a vibrating system after modification. [47], **150**, 357-370. *83, 365, 365, 365.*

[278] Ram, Y.M. and Braun, S.G. (1991) An inverse problem associated with the dynamic modification of structures. [37], 58, 233-237. *365, 366.*

[279] Ram, Y.M. and Caldwell, J. (1992) Physical parameters reconstruction of a free-free mass-spring system from its spectra. [82], **52**, 140-152. *365, 366.*

[280] Ram, Y.M. and Braun, S.G. (1993) Eigenvector error bounds and their application to structural modification. [4], **31**, 759-764. *365.*

[281] Ram, Y.M. (1994a) Inverse mode problems for the discrete model of a vibrating beam. [47], **169**, 239-252. *365.*

[282] Ram, Y.M. (1994b) Enlarging a spectral gap by structural modification. [47], **176**, 225-234. *365.*

[283] Ram, Y.M. (1994c) An inverse mode problem for the continuous model of an axially vibrating rod. [37], **61**, 624-628. *366.*

[284] Ram, Y.M. and Elhay, S. (1996) The theory of a multi degree of freedom dynamic absorber. [47], **195**, 607-615. *366.*

[285] Ram, Y.M. and Elhay, S. (1995a) Dualities in vibrating rods and beams: continuous and discrete models. [47], **181**, 583-594. *162, 345, 366.*

[286] Ram, Y.M. and Elhay, S. (1995b) The construction of band symmetric models for vibrating systems from modal analysis data. [47], **184**, 759-766.

[287] Ram, Y.M. and Elhay, S. (1998) Constructing the shape of a rod from eigenvalues. [15], **14**, 597-608. *162, 366.*

[288] Ram, Y.M. and Elishakoff, I. (2004) Reconstructing the cross-sectional area of an axially-vibrating non-uniform rod from one of its mode shapes. [74]. *367.*

[289] Ram, Y.M. and Gladwell, G.M.L. (1994) Constructing a finite element model of a vibratory rod from eigendata. [47], **169**, 229-237. *87, 365, 367.*

[290] Rayleigh, Lord (1984) *The Theory of Sound.* London: Macmillan. *15.*

[291] Rizos, P.F., Aspragathos, N. and Dimarogonas, A.D. (1990) Identification of crack location and magnitude in a cantilever beam from the vibration modes. [47], **138**, 381-388. *422, 424.*

[292] Rundell, W. and Sacks, P.E. (1992a) Reconstruction techniques for classical inverse Sturm-Liouville problems. [60], **58**, 161-183. *321.*

[293] Rundell, W. and Sacks, P.E. (1992b) The reconstruction of Sturm-Liouville operators. [29], **8**, 457-482. *321.*

[294] Rundell, W. (1997) Inverse Sturm-Liouville problems, in Chadan, K., Colton, D., Pãivãrinta, L. and Rundell, W., (Eds.) *An Introduction to Inverse Scattering and Inverse Spectral Problems.* Philadelphia: SIAM. 67-131. *283, 321, 326.*

[295] Sabatier, P.C. (1978) Spectral and scattering inverse problems. [42], **19**, 2410-2425. *289.*

[296] Sabatier, P.C. (1979a) On some spectral problems and isospectral evolutions connected with the classical string problem. I. Constants of motion. [56], **26**, 477-482. *291.*

[297] Sabatier, P.C. (1979b) On some spectral problems and isospectral evolution connected with the classical string problem. II. Evolution equations. [56], **26**, 483-486. *291.*

[298] Sabatier, P.C. (1985) Inverse problems - an introduction. [29], **1**, i-iv. *289.*

[299] Sakata, T. and Sakata, Y (1980) Vibrations of a taut string with stepped mass density. [47], **71**, 315-317. *355.*

[300] Schur, J. (1917) Űber Potenzreihen, die im Innern des Einheitskreises beschrãnkt sind. [34], **147**, 205-232. *337.*

[301] Seidman, T. (1985) Converent approximation scheme for the inverse Sturm-Liouville problem. [29], **1**, 251-262. *291.*

[302] Seidman, T. (1988) An inverse eigenvalue problem with rotational symmetry. [29], **4**, 1093-1115.

[303] Shen, M.-H.H. and Pierre, C. (1990) Natural modes of Bernoulli-Euler beams with symmetric cracks. [47], **138**, 115-134. *422.*

[304] Shen, M.-H.H. and Taylor, J.E. (1991) An identification problem for vibrating cracked beams. [47], **150**, 457-484. *422.*

[305] Sinha, J.K., Friswell, M.I. and Edwards, S. (2002) Simplified models for the location of cracks in beam structures using measured vibration data. [47], **251**, 13-38.

[306] Sivan, D.D. and Ram, Y.M. (1997) Optimal construction of a mass-spring system from prescribed modal and spectral data. [47], **201** 323-334. *366.*

[307] Sivan, D.D. and Ram, Y.M. (1999) Physical modifications to vibratory systems with assigned eigendata. [37], **66**, 427-432. *366.*

[308] Sondhi, M.M. and Gopinath, B. (1971) Determination of vocal-tract shape from impulse response of the lips. [32], **49**, 1867-1873. *331.*

[309] Sondhi, M.M. (1984) A survey of the vocal tract inverse problem: theory, computation and experiments, in F. Santosa, Y.-H. Pao, W.W. Symes and C. Holland. (Eds.) *Inverse Problems of Acoustic and Elastic Waves.* Philadelphia: SIAM. *293, 331.*

[310] Stieltjes, T.J. (1918) *Oevres Completes. Vol. 2*, Groningen: Noordhoff. *63.*

[311] Strang, G. and Fix, G.J. (1973) *An Analysis of the Finite Element Method.* Prentice-Hall, Englewood Cliffs, NJ. *26.*

[312] Sussman-Fort, S.E. (1982) Reconstruction of bordered-diagonal and Jacobi matrices from spectral data. [50], **314**, 271-282. *103.*

[313] Sweet, R.A. (1969) Properties of a semi-discrete approximation to the beam equation with a second order term. [35], **5**, 329-339. *185.*

[314] Sweet, R.A. (1971) Oscillation properties of a semi-discrete approximation to the beam equation with a second order term. [35], **7**, 119-125. *185.*

[315] Symes, W.W. (1980) Hamiltonian group actions and integrable systems. [70], **1**, 339-374. *159.*

[316] Symes, W.W. (1982) The QR algorithm and scattering for the finite nonperiodic Toda lattice. [70], **4**, 275-280. *159.*

[317] Takewaki, I. and Nakamura, T. (1995) Hybrid inverse mode problems for FEM-shear models. [40], **121**, 873-880. *92.*

[318] Takewaki, I., Nakamura, T. and Arita, Y. (1996) A hybrid inverse mode problem for fixed-fixed mass-spring models. [52], **118**, 641-648. *92.*

[319] Takewaki, I. and Nakamura, T. (1997) Hybrid inverse mode problem for structure-foundation systems. [40], **123**, 312-321. *92.*

[320] Takewaki, I. (1999) Hybrid inverse eigenmode problem for top-linked twin shear building models. [28], **41**, 1133-1153. *92.*

[321] Takewaki, I. (2000) *Dynamic Structured Design: Inverse Problem Approach.* Southampton, UK: WIT Press. *92.*

[322] Temple, G. and Bickley, W.G. (1933) *Rayleigh's Principle and its Applications to Engineering.* London: Oxford University Press. *41.*

[323] Titchmarsh, E.C. (1962) *Eigenfunction Expansions.* Part I. Oxford: Oxford University Press. *276.*

[324] Toda, M. (1970) Waves in nonlinear lattices. [75], **45**, 174-200. *159.*

[325] Underwood, R.R. (1975) *An Iterative Block-Lanczos Method for the Solution of Large Sparse Symmetric Eigenproblems.* Ph.D. Thesis, Stanford University. *108.*

[326] van der Holst, H. (1996) *Topological and Spectral Graph characterizations.* Ph.D. Thesis. Universiteit van Amsterdam. *224.*

[327] Vestroni, F. and Capecchi, D. (1996) Damage evaluation in cracked vibrating beams using experimental frequencies and finite element models. [53], **2**, 69-86. *422.*

[328] Vestroni, F. and Capecchi, D. (2000) Damage detection in beam structures based on frequency measurements. [59], **126**, 761-768. *422.*

[329] Vijay, D.K. (1972) *Some Inverse Problems in Mechanics.* M.A.Sc. Thesis, University of Waterloo. *203.*

[330] Washizu, K. (1982) *Variational Methods in Elasticity and Plasticity.* 3rd Edition, Oxford: Pergamon Press. *41.*

[331] Watkins, D.S. (1984) Isospectral flows. [87], **26**, 379-391. *159.*

[332] Weinberger, H. (1974) Variational Methods for Eigenvalue Approximation. Regional Conf. Ser. in Appl. Math., **15**, SIAM.

[333] Willis, C. (1986) An inverse method using toroidal mode data. [29], **2**, 111-130.

[334] Willis, C. (1985) Inverse Sturm-Liouville problems with two discontinuities. [29], **1**, 263-289. *305.*

[335] Wu, Q.L. and Fricke, F. (1989) Estimation of blockage dimensions in a duct using measured eigenfrequency shifts. [47], **133**, 289-301. *421.*

[336] Wu, Q.L. and Fricke, F. (1990) Determination of blocking locations and corss-sectional area in a duct by eigenfrequency shifts. [32], **87**, 67-75. *421.*

[337] Wu, Q.L. and Fricke, F. (1991) Determination of the size of an object and its location in a rectangular cavity by eigenfrequency shifts - 1st order approximations. [47], **144**, 131-147. *421.*

[338] Wu, Q.L. (1994) Reconstruction of crack function of beams from eigenvalue shifts. [47], **173**, 279-282. *423.*

[339] Xu, S.F. (1998) *An Introduction to Inverse Algebraic Eigenvalue Problems.* Braunschweig: Vieweg. *105, 117.*

[340] Yen, A. (1978) *Numerical Solution of the Inverse Sturm-Liouville Problem.* Ph.D. Thesis, University of California at Berkeley.

[341] Yuen, M.M.F. (1985) A numerical study of the eigenparameters of a damaged cantilever. [47], **103**, 301-310. *422.*

[342] Zhu, H.M. (2000) *Courant's Nodal Line Theorem and its Discrete Counterparts.* Ph.D. Thesis, University of Waterloo. *34, 224.*

[343] Zienkiewicz, O.Z. (1971) *The Finite Element Method in Engineering Science*, London: McGraw-Hill. *26.*

List of Journals

	Full Journal Name	Abbreviated Journal Title	Call #
1	acta mathematica	acta math	QA1 .A185
2	acta numerica	acta numerica	QA297. A327
3	aeronautical quarterly	aeron q	TL501 .R7
4	AIAA journal.	aiaa j	TL501.A688 A2
5	american institute of aeronautics and astronautics paper	am inst aeronaut astronaut pap	TL512. A66
6	american mathematical society translations series 2	amer math soc trans ser 2	QA 1.A522
7	annals of mathematics	ann math	QA1 .A6
8	annual review of earth and planetary sciences	annu rev earth planet sci	QE1 .A674
9	applied mechanics reviews	appl mech rev	TA1 .A639
10	archiv der mathematik	arch math	QA 1.A66
11	archive for rational mechanics and analysis	arch rat mech anal	QA801 .A6
12	bell systems technical journal	bell sys tech j	TK 1.B425
13	commentarii mathematici helvetici	comment math helvetici	QA1 .C7
14	communications in mathematical physics	commun math phys	QC1 .C6
15	communications in numerical methods in engineering	comm numer methods engrg	TA335 .C65
16	communications on pure and applied mathematics	math comm pure appl	QA1 .C718
17	comptes rendus academie des sciences	cr acad sci paris	AS162.P315
18	comptes rendus des seances academie des sciences serie 1 mathematique	cr acad sci paris sec I math	Q46 .A14
19	computing	co	QA 76.C582
20	contemporary mathematicians	contemp mathematicians	monographic series
21	doklady akademii nauk sssr	dokl ak sssr	Q60.A3
22	duke mathematical journal	duke math j	QA1 .D8
23	earthquake engineering and structural dynamics	earthquake eng struct dyn	TH1095 .E27x
24	geophysical journal royal astronomical society	geophys j r astr soc	QD96.A643x
25	ieee transactions on sonics and ultrasonics	ieee trans sonics ultrason	QC244 .I53
26	ima journal of applied mathematics	ima j appl math	QA 1.I522
27	international journal of analytical and experimental modal analysis	intl j analyt exptl modal analysis	TA654.15.I56

	Full Journal Name	Abbreviated Journal Title	Call #
28	international journal of mechanical sciences	intl j mech sci	TJ1 .I59
29	inverse problems	inverse pr	QA370 .I52x
30	inverse problems in engineering	inverse probl eng	TA347.D45 I582
31	izvestiia akademii nauk sssr seriya matematicheskaya	izv akad nauk sssr ser mat	G 3271 .C55
32	journal acoustical society of america	j acoust soc am	QC 221.A4
33	journal australian mathematical society series b applied mathematics	j austral math soc series b	QA1 .J97645
34	journal fuer die reine und angewandte mathematik	j reine angew math	QA 1.J95
35	journal institute of mathematics and its applications	j inst math appl	QA1 .I552
36	journal of applied mathematics and physics	j appl math phys	QA1 .Z5
37	journal of applied mechanics	j app mech	TA1 .J6
38	journal of differential equations	j diff equa	QA371 .J73
39	journal of elasticity	j elast	QA931 .J6x
40	journal of engineering mathematics	j eng math	TA1 .A5233
41	journal of mathematical analysis and applications	j math anal appl	QA1 .J596
42	journal of mathematical physics	j math phys	QA1 .J598
43	journal of mathematical sciences	j math sci	QA1.J63x
44	journal of mechanical engineering science	j mech eng sci	TJ1.J6
45	journal of mechanical systems and systems processing	mech syst signal processing	TA654 .M38
46	journal of research, united states national bureau of standards, section b. mathematical sciences	j res nat bur standards sect b	QA1 .U571
47	journal of sound and vibration	j sound vib	QC221 .J6
48	journal of strain analysis	j strain anal	TG265 .J6
49	journal of structural engineering asce	j struct eng asce	TA1 .A5235
50	journal of the franklin institute b engineering and applied mathematics	j franklin inst b	T1 .F8
51	journal of the mechanics and physics of solids	j mech phys solids	TA350 .J68
52	journal of vibration and acoustics. Transactions of the asme	j vib accoust trans asme	TJ1 .J68x
53	journal of vibration and control	j vib control	TJ212 .J68x

	Full Journal Name	Abbreviated Journal Title	Call #
54	journal society of industrial and applied mathematics	jsiam	QA1 .S73
55	lecture notes in mathematics	lecture notes in math	QA3 .L28
56	lettere al nuovo cimento	lett nuovo c	QC 1.L4
57	linear alegbra and its applications	lin alg app	QA 251.L52
58	linear and multilinear algebra	linear multilin algebra	QA251 .L524x
59	matematicheskii sbornik	mat sb	QA1 .M4
60	mathematics of computation	math comp	QA 47.M29
61	mathematische zeitschrift	math z	QA1 .M799
62	meccanica journal of the italian association of theoretical and applied mechanics	meccanica j ital assoc theoret appl mech	QA801 .M4x
63	mechanical systems and signal processing	mech syst signal processing	TA654 .M38
64	memoirs of the american mathematical society	mono series	QA1 .A514
65	monatshefte fuer mathematik und physik	monatsh math phys	QA1 .M877
66	nordisk mathematisk tidsskrift b	nord mat tidsskr b	QA1 .N83
67	numerical linear algebra with applications	numer linear algebra appl	QA184. N88
68	numerische mathematik	numer math	QA1 .N8
69	philosophical transactions royal society of london series a mathematical and physical sciences	phil trans roy soc lond a	Q 41.L8
70	physica d. nonlinear phenomena	phys d	QC1 .P3834
71	prikladnaya matematika i mekhanika	prik mat mekh	TA350 . U34
72	proceedings of the institution of mechanical engineers	proc inst mech eng	TJ 1.I5
73	proceedings of the national academy of science	proc nat acad sci	Q11 .N26
74	proceedings royal society of london series a mathematical and physical science	proc roy soc lond a	Q41 .L72
75	progress of theoretical physics supplement	prog theor phys suppl	QC1 .P8852x
76	quarterly journal of mathematics oxford second series	q j math oxford ser 2	QA1 .Q22
77	quarterly journal of mechanics and applied mathematics	q j mech appl math	QA1 .Q23

	Full Journal Name	Abbreviated Journal Title	Call #
78	Rendiconti del Circolo matematico di Palermo	rend circ mat palermo	QA1 .C6
79	rendiconti dell'istituto di matematica dell'universita di trieste	rend istit mat univ trieste	QA1 .T82a
80	shock and vibration digest	shock vib dig	US1 DH 41 S37
81	siam journal on algebraic and discrete methods	siam j alg disc math	QA1 .S732x
82	siam journal on applied mathematics	siam j appl math	QA1 .S73
83	siam journal on mathematical analysis	siam j math anal	QA 1.S25
84	siam journal on matrix analysis and applications	siam j matrix anal appl	QA1 .S732x
85	siam journal on numerical analysis	siam j num anal	QA297.S52x
86	siam journal on scientific and statistical computing	siam j sci stat comput	QA264 .S53x
87	siam review	siamr	QA1 .S5
88	transactions of the american mathematical society	n y am mth s t	QA1 .A522
89	trudy moskovskogo matematicheskogo obshchestva	trudy mosk mat obsch	QA1 .M988
90	wave motion	wamod	QA927.W3x
91	zeitschrift fuer angewandte mathematik und physik	zamp	QA1 .Z5
92	zeitschrift fuer physik	z phys	QC1 .Z4

FSC
www.fsc.org
MIX
Papier aus verantwortungsvollen Quellen
Paper from responsible sources
FSC® C105338

If you have any concerns about our products,
you can contact us on
ProductSafety@springernature.com

In case Publisher is established outside the EU,
the EU authorized representative is:
Springer Nature Customer Service Center GmbH
Europaplatz 3, 69115 Heidelberg, Germany

Printed by Libri Plureos GmbH
in Hamburg, Germany